2017 IEEE Electron Devices Technology and Manufacturing Conference (EDTM 2017)

Toyama, Japan
28 February - 2 March 2017

IEEE Catalog Number:	CFP17J58-POD
ISBN:	978-1-5090-4661-4

Copyright © 2017 by the Institute of Electrical and Electronics Engineers, Inc
All Rights Reserved

Copyright and Reprint Permissions: Abstracting is permitted with credit to the source. Libraries are permitted to photocopy beyond the limit of U.S. copyright law for private use of patrons those articles in this volume that carry a code at the bottom of the first page, provided the per-copy fee indicated in the code is paid through Copyright Clearance Center, 222 Rosewood Drive, Danvers, MA 01923.

For other copying, reprint or republication permission, write to IEEE Copyrights Manager, IEEE Service Center, 445 Hoes Lane, Piscataway, NJ 08854. All rights reserved.

*** *This is a print representation of what appears in the IEEE Digital Library. Some format issues inherent in the e-media version may also appear in this print version.*

IEEE Catalog Number: CFP17J58-POD
ISBN (Print-On-Demand):   978-1-5090-4661-4
ISBN (Online):   978-1-5090-4660-7

Additional Copies of This Publication Are Available From:

Curran Associates, Inc
57 Morehouse Lane
Red Hook, NY 12571 USA
Phone: (845) 758-0400
Fax: (845) 758-2633
E-mail: curran@proceedings.com
Web: www.proceedings.com

2017 IEEE
Electron Devices Technology and Manufacturing Conference (EDTM)

Proceedings of Technical Papers

Toyama, Japan

February 28 - March 2, 2017

IEEE Electron Devices Society

Organized by

IEEE Electron Devices Society

Supported by

Toyama Prefecture

Toyama City

Publication Office

JTB Communication Design, Inc.

Celestine Shiba Mitsui Bldg.

3-23-1 Shiba, Minato-ku, Tokyo 105-8335, Japan

Copyright ©2017 by IEEE.

All rights reserved.

Committees

Steering Committee

EDS Vice President of Technical Committees & Meetings (Chair)	Ravi Todi	GLOBALFOUNDRIES
EDS President	Samar Saha	Prospicient Devices
EDS Junior Past President	Albert Wang	Univ. of California, Riverside
EDS Senior Past President	Paul Yu	Univ. of California, San Diego
EDS President-Elect	Fernando Guarrin	GLOBALFOUNDRIES
EDS Treasurer	Subramanian S. Iyer	Univ. of California, Los Angeles
General Chair	Shuji Ikeda	tei Solutions Inc.

Executive Committee

General Chair	Shuji Ikeda	tei Solutions Inc.
Technical Program Chair, Publication Chair	Hitoshi Wakabayashi	Tokyo Tech
J-EDS Special Issue Chair	Masaaki Niwa	Tohoku Univ.
J-EDS Special Issue Vice Chair (J-EDS Editor-in-Chief)	Mikael Ostling	KTH, Royal Inst. of Tech.
Education Chair	Akira Toriumi	The Univ. of Tokyo
Education Vice Chair	Mansun Chan	Hong Kong Univ. of Sci. and Tech.
Publicity Chair	Keiji Ikeda	Toshiba
Publicity Vice Chair	Tian Ling Ren	Tsinghua Univ.
Emerging Technologies Chair	Bin Zhao	ON Semiconductor
Emerging Technologies Vice Chair	Hiro Akinaga	AIST
Financial Chair	Iriya Muneta	Tokyo Tech
Finacial Vice Chair (EDS Treasurer)	Subramanian S. Iyer	Univ. of California, Los Angeles
International Advisory Chair (EDS VP SRC)	M. K. Radhakrishnan	NanoRel
International Advisory Vice Chair (EDS Secretary)	Simon Deleonibus	CAT-LETI
Exhibits / Sponsorship Chair	Kazunari Ishimaru	Toshiba
Exhibits / Sponsorship Vice Chair	Reza Arghavani	Lam Research
Support Relationship Chair	Seiichiro Kawamura	JST

Education Committee

Chair	Akira Toriumi	The Univ. of Tokyo
Vice Chair	Mansan Chan	Hong Kong Univ. of Sci. and Tech.
Members	Toshihide Nabatame	NIMS
	Jiro Ida	Kanazawa Inst. of Tech.
	Carmen Lilly	Univ. of Illinois
	Chandan Sarkar	Jadavpur Univ.
	Daniel Camacho	Intel

International Advisory Committee

Chair	M. K. Radhakrishnan	NanoRel
Vice Chair	Simon Deleonibus	CAT-LETI
Members	Angus Rockett	Univ. of Illinois
	Tim Anderson	Univ. of Masschusetts
	Illesanmi Adesida	Univ. of Illinois
	Patrick Fay	Univ. of Notre Dame
	Juin J. Liou	Univ. of Central Florida
	Rajendra Singh	Clemson Univ.
	Christofer Hierold	ETH Zürich

Cor Claeys	Imec / KU Leuven
Ming Liu	Chinese Academy of Sciences
T. Y. Chiu	SMIC
Steve Chung	National Chiao Tung Univ.
Kwyro Lee	KAIST
Chennupati Jagadish	Australian National Univ.
Firdaus Abdullah	Silterra
Juzer Vashi	IIT Bombay
Ramgopal Rao	IIT Delhi
Raj Jammy	Carl Zeiss
Jim Plummer	Stanford Univ.
Chenming Hu	Univ. of California, Berkeley
John Hu	NVIDIA
Sunit Tyagi	igrenEnergi, Inc
Jamal Deen	McMaster Univ.
Anisul Haque	East West Univ.
Albert Chin	National Chiao Tung Univ.
Kaustav Banerjee	Univ. of California, Santa Barbara
Sandeep Bahl	Texas Instruments
Monuko du Plessis	Univ. of Pretoria
John Suehle	NIST
Jeffrey J. Welser	IBM Almaden Research Center
Min Yang	IBM TJ Watson Research Center
D. Nirmal	Karunya Univ.
Serge Biesmans	TEL Europe
Eric Beyne	Imec
Jinho Ahn	Hanyang Univ.
Meyya Meyyappan	NASA's Ames Research Center
Leda Lunardi	North Carolina State Univ.
Xing Zhou	Nanyang Technological Univ., Singapore
Ru Huang	Peking Univ.
Hiroshi Iwai	Tokyo Tech

Exhibits / Sponsorship Committee

Chair	Kazunari Ishimaru	Toshiba
Vice Chair	Reza Arghavani	Lam Research
Chair Support	Masumi Saitoh	Toshiba
Members	Katsumi Ohmori	Tokyo Ohka Kogyo Co., LTD.
	Atanu Kundu	Heritage Inst. of Tech.
	Sachin Sonkusale	PDF Solutions
	Shima Sasaki	TEL
	Masaya Asai	SCREEN

Industrial Advisory Committee

Chair	Kazunari Ishimaru	Toshiba
Members	Masaki Momodomi	Toshiba
	Yukimasa Yoshida	Lam Research Japan
	Tadahiro Suhara	SCREEN
	Yoshinobu Mitano	TEL
	Harutoshi Sato	Tokyo Ohka Kogyo Co., LTD.
	Satoru Yamada	Samsung
	Seok-Hee Lee	SK Hynix
	Anne Cirkel	Mentor Graphics

2017 IEEE Electron Devices Technology and Manufacturing Conference (EDTM)

Wednesday, March 1
Main Hall, 3F

8:30-9:00 Opening

Opening Remarks, Shuji Ikeda, EDTM 2017 General Chair, tei SOLUTIONS Inc.

9:00-11:00   Plenary Session

Chair: H. Wakabayashi, Tokyo Institute of Technology

PL-1   9:00-9:40 (Invited)
Dimensions of Innovation to Enable the Next Era of Intelligent Systems, John G. Pellerin, GLOBALFOUNDRIES Inc. 1

PL-2   9:40-10:20 (Invited)
Flexible and Printed OTFT Devices for Emerging Electronic Applications, Shizuo Tokito, Yamagata University 2

PL-3   10:20-11:00 (Invited)
Advanced Heterogeneous Integration Technology Trend for Cloud and Edge, Douglas C. H. Yu, Taiwan Semiconductor Manufacturing Company..4

11:00-12:00   Exhibition Talks

Chairs: K. Ishimaru, Toshiba Corporation
        R. Arghavani, Lam Research Corporation

Ex-1   Hitachi Kokusai Electric Inc.
Ex-2   Toshiba Corporation, Storage & Electronic Devices Solutions Company
Ex-3   Atomera Inc.
Ex-4   National Institute of Advanced Industrial Science and Technology (AIST)
Ex-5   TowerJazz Panasonic Semiconductor Co., Ltd.
Ex-6   Yokogawa Solution Service Corporation

13:35-15:40   Session 3M - Emerging: Emerging Technologies

Chairs: B. Zhao, ON Semiconductor
        H. Akinaga, AIST

3M-1   13:35-14:00 (Invited)
Future Computing Devices – Excitation, Physarum, Fluerics, Actin, Andrew Adamatzky, University of the West of England, Bristol ....N/A

3M-2   14:00-14:25 (Invited)
Neuromorphic Technologies for Next-Generation Cognitive Computing, Robert M. Shelby[1], Pritish Narayanan[1], Stefano Ambrogio[1], Hsinyu Tsai[1], Kohji Hosokawa[2], Scott C. Lewis[3], and Geoffrey W. Burr[1], [1]IBM Research-Almaden, [2]IBM Tokyo Research Laboratory, [3]IBM T. J. Watson Research Center ···· 8

3M-3   14:25-14:50   (Invited)
Stateful Logic Circuit and Material Using Memristors, Nuo Xu[1,2], Xinglong Shao[1], Kyung Jean Yoon[1], Hae Jin Kim[1], Kyung Min Kim[3], and Cheol Seong Hwang[1], [1]Seoul National University, [2]National University of Defense Technology, [3]Hewlett Packard Enterprise ·· 10

3M-4   14:50-15:15   (Invited)
New-Paradigm CMOS Ising Computing for Combinatorial Optimization Problems, Masanao Yamaoka, Hitachi, Ltd.·· 13

3M-5   15:15-15:40   (Invited)
Understanding the Limit and Potential in Emerging Perovskite Solar Cells, Wolfgang Tress and Michael Graetzel, Swiss Federal Institute of Technology (EPFL) ··· 15

Wednesday, March 1
Room A (201+202), 2F

13:35-15:15   Session 3A - Process: Process Technology for Advanced Devices

Chairs: J. Yugami, Hitachi Kokusai Electric Inc.
         X. Guo, Shanghai Jiao Tong University

3A-1   13:35-14:00   (Invited)
The Impact of Fin Number on Device's Performance and Reliability in Tri-Gate FinFETs, Wen-Kuan Yeh[1,2], Po-Ying Chen[3], Chia-Hung Shih[4], Wenqi Zhang[1], and Yi-Lin Yang[5], [1]National University of Kaohsiung, [2]National Applied Research Laboratories, [3]National Chin-Yi University of Technology, [4]National Pingtung University, [5]National Kaohsiung Normal University·· 17

3A-2   14:00-14:25
Impact of e-SiGe S/D Processes on FinFET PFET TDDB Reliability, R. Ranjan, S. Uppal, H. Yu, B. Parameshwaran, T. Nigam, A. Kerber, C. LaRow, and M. I. Natarajan, GLOBALFOUNDRIES Inc. ············· 20

3A-3   14:25-14:50
In Content Dependence of Pre-Treatment Effects on Al_2O_3/In_xGa_{1-x} as MOS Interface Properties, C. Yokoyama, C.-Y. Chang, M. Takenaka, and S. Takagi, The University of Tokyo ·································· 23

3A-4   14:50-15:15
Deep Junction by Low Thermal Budget Process for Advanced Si Power Electronics, Inès Toqué-Trésonne, Toshiyuki Tabata, Sébastien Halty, Fulvio Mazzamuto, Karim Huet, and Yoshihiro Mori, SCREEN Semiconductor Solutions Co., Ltd ··· 26

Wednesday, March 1
Room B (203+204), 2F

13:35-15:40   Session 3B - Modeling: Reliability Analysis

Chairs: A. Oates, TSMC
         S. Koul, Indian Institute of Technology Delhi

3B-1   13:35-14:00   (Invited)
Assessing Device Reliability Margin in Scaled CMOS Technologies Using Ring Oscillator Circuits, A. Kerber, S. Cimino, F. Guarin, and T. Nigam, GLOBALFOUNDRIES Inc.································· 28

3B-2   14:00-14:25
The Impact of RTN-Induced Temporal Performance Fluctuation Against Static Performance Variation, Takashi Matsumoto[1], Kazutoshi Kobayashi[2], and Hidetoshi Onodera[3], [1]The University of Tokyo, [2]Kyoto Institute of Technology, [3]Kyoto University ·· 31

3B-3　14:25-14:50

Critical Discussion on Temperature Dependence of BTI in Planar and FinFET Devices, P. Srinivasan and Tanya Nigam, GLOBALFOUNDRIES Inc. ·······33

3B-4　14:50-15:15

Characterization of Critical Peak Current and Model of Cu/Low-k Interconnects under Short Pulse-Width Conditions, M. H. Lin, W. S. Chou, Y. T. Yang, Y. C. Peng, and A. S. Oates, Taiwan Semiconductor Manufacturing Company ·······36

3B-5　15:15-15:40

New Analytical Equations for Skin and Proximity Effects in Interconnects Operated at High Frequency, Haojun Zhang, Jian-Hsing Lee, Natarajan Mahadeva Iyer, and Linjun Cao, GLOBALFOUNDRIES Inc. ·······39

Wednesday, March 1
Main Hall, 3F

15:55-18:25　　Session 4M - Package: Subsystem Integration & Packaging

Chairs:　S. Yamamichi, IBM

Y. Kurita, Toshiba Corporation

4M-1　15:55-16:20 (Invited)
System Integration in a Package for Cloud and Edge, Tadahiro Kuroda, Keio University ·······42

4M-2　16:20-16:45 (Invited)
III-V/Si Low Temperature Direct Bonding Technology for Photonic Device Integration on SOI, Nobuhiko Nishiyama, Yusuke Hayashi, Junichi Suzuki, and Shigehisa Arai, Tokyo Institute of Technology ····44

4M-3　16:45-17:10 (Invited)
Focused Technologies in Near Future from OSAT View Point, Akio Katsumata, Akira Takashima, Kazuhiro Sawada, Hideki Sumihara, and Norio Ito, J-DEVICES Corporation ·······46

4M-4　17:10-17:35
BGA Packaging Process for a Device Made by Minimal Fab, Sommawan Khumpuang[1,2], Fumito Imura[1,2], and Shiro Hara[1,2], [1]National Institute of Advanced Industrial Science and Technology (AIST), [2]Minimal Fab·······49

4M-5　17:35-18:00
Electrodeposited Cobalt for Advanced Packaging Applications, Bryan Buckalew, Justin Oberst, and Thomas Ponnuswamy, Lam Research Corporation·······51

4M-6　18:00-18:25 (Invited)
Packaging Design Considerations for Mobile and Internet of Things (IOT), Piyush Gupta, Qualcomm Inc. ·······53

Wednesday, March 1
Room A (201+202), 2F

15:55-18:00   Session 4A - Material: Advanced FEOL Materials

Chairs: I. Muneta, Tokyo Institute of Technology
         P. Li, National Chiao Tung University

4A-1   15:55-16:20 (Invited)
Biocompatible ALD Barrier Coatings for Medical Devices, Mikko Matvejeff[1], Satu Ek[1], Riina Ritasalo[1], Jesse Kalliomäki[1], Päivi Järvinen[1], Oili Ylivaara[2], and Erik Östreng[1], [1]Picosun, [2]VTT ·········· 56

4A-2   16:20-16:45 (Invited)
Enablement of Cost Effective CVD/ALD Processing through Precursor Design, Jean-Marc Girard, Air Liquide Advanced Materials Inc. ··· 59

4A-3   16:45-17:10
Impact of Hydrogen Annealing Behavior of C_3H_5 Carbon Cluster Ion Implanted Projection Range Using Microwave Heat Treatment, Takeshi Kadono, Ryosuke Okuyama, Ayumi Masada, Ryou Hirose, Yoshihiro Koga, Hidehiko Okuda, and Kazunari Kurita, SUMCO Corporation ·················· 61

4A-4   17:10-17:35
New Opportunity of Ferroelectric Tunnel Junction Memory with Ultrathin HfO_2-Based Oxides, Xuan Tian and Akira Toriumi, The University of Tokyo ··· 63

Wednesday, March 1
Room B (203+204), 2F

15:55-18:25   Session 4B - Device: Advanced FET Technology

Chairs: J. Ida, Kanazawa Institute of Technology
         N. Horiguchi, imec

4B-1   15:55-16:20 (Invited)
Withdrawn

4B-2   16:20-16:45
Punch-Through Stop Doping Profile Control via Interstitial Trapping by Oxygen-Insertion Silicon Channel, Robert J. Mears[1], Hideki Takeuchi[1], Robert J. Stephenson[1], Marek Hytha[1], Richard Burton[1], Nyles W. Cody[1], Doran Weeks[1], Dmitri Choutov[1], Nidhi Agrawal[2*], and Suman Datta[2**] , [1]Atomera Inc., [2]Pennsylvania State University, *currently at Micron Technology Inc., **currently at Notre Dame University ····· 65

4B-3   16:45-17:10 (Invited)
FinFET/Nanowire Design for 5nm/3nm Technology Nodes: Channel Cladding and Introducing a "Bottleneck" Shape to Remove Performance Bottleneck, Victor Moroz, Joanne Huang, and Munkang Choi, Synopsys, Inc. ··· 67

4B-4   17:10-17:35
A Computational Study of Fundamentals and Design Considerations for Vertical Tunneling Field-Effect Transistor, Sheng Luo, Kain Lu Low, Xiaoyi Zhang, Qianyu Zhao, Hsin Lin, and Gengchiau Liang, National University of Singapore ··· 70

4B-5   17:35-18:00
Analysis of Break-Even Time for Nonvolatile SRAM with SOTB Technology, Daiki Kitagata, Yusuke Shuto, Shuu'ichirou Yamamoto, and Satoshi Sugahara, Tokyo Institute of Technology ·················· 72

4B-6   18:00-18:25

Role of Floating Body Effect on Super Steep Subthreshold Slope PN-Body Tied SOI FET,
Takahiro Yoshida, Jiro Ida, Syouta Inoue, Syougo Uchikura, Atsushi Hashimoto, Keisuke Hayashi,
Taichi Iwasaki, Masanori Kamako, and Takashi Horii, Kanazawa Institute of Technology ·····················75

18:25-19:05   Authors Interview in Main Hall, 3F
Poster Session in Foyer, 3F

*For Poster Session, please see the detailed information from page (17).

Thursday, March 2
Main Hall, 3F

8:30-10:10   Session 5M - Process: Innovative Process Tools

Chairs: K. Nojiri, Lam Research Corporation
Y. Kawasaki, SMIT

5M-1   8:30-8:55 (Invited)
EUV Lithography Insertion for High Volume Manufacturing: Status and Outlook, Alek Chen[1] and
Junji Miyazaki[2], [1]ASML US Inc., [2]ASML Japan Co., Ltd.·····················77

5M-2   8:55-9:20
Local, Isotropic, and Damageless Doping to Oxide Semiconductors by Using Electrochemistry,
Takeaki Yajima, Tomonori Nishimura, and Akira Toriumi, The University of Tokyo ·····················80

5M-3   9:20-9:45 (Invited)
Process Development for CMOS Fabrication Using Minimal Fab, Sommawan Khumpuang[1,2],
Kazuhiro Koga[1,2], Yongxun Liu[1], and Shiro Hara[1,2], [1]National Institute of Advanced Industrial Science
and Technology (AIST), [2]Minimal Fab·····················82

5M-4   9:45-10:10
New Compact ECR Plasma Source for Silicon Nitride Film Formation in Minimal Fab System,
Tetsuya Goto[1], Kei-ichiro Sato[2], Yuki Yabuta[3], Shigetoshi Sugawa[1], and Shiro Hara[4], [1]Tohoku University,
[2]Kotec Co., Ltd., [3]Seinan Industries Co., Ltd., [4]National Institute of Advanced Industrial Science and
Technology (AIST)·····················84

Thursday, March 2
Room A (201+202), 2F

8:30-10:10   Session 5A - Device: More-than-Moore Technologies

Chairs: R. Huang, Peking University
K. Uchida, Keio University

5A-1   8:30-8:55
A Scalable Si-Based Micro Thermoelectric Generator, Takanobu Watanabe[1], Shuhei Asada[1],
Taiyu Xu[1], Shuichiro Hashimoto[1], Shunsuke Ohba[1], Yuya Himeda[1], Ryo Yamato[1], Hui Zhang[1], Motohiro
Tomita[1], Takashi Matsukawa[2], Yoshinari Kamakura[3], and Hiroya Ikeda[4], [1]Waseda University, [2]National
Institute of Advanced Industrial Science and Technology (AIST), [3]Osaka University, [4]Shizuoka University·····86

5A-2   8:55-9:20 (Invited)
Health Monitoring of Houses and Communities by Recording Earthquake Response of Buildings, Minoru Yoshida[1], Yoichi Tanaka[1], Shoichi Ikeda[1], Soichiro Murata[2], Yoshiya Oda[3], and Masatake Ushiro[4], [1]Hakusan Corporation, [2]SAP Japan, [3]Tokyo Metropolitan University, [4]Business Break-Through University ···88

5A-3   9:20-9:45 (Invited)
Smart Biosensing Technologies to Detect Single Bacteria and Viruses, Masateru Taniguchi, Osaka University ···90

5A-4   9:45-10:10
Strain-Engineering in Germanium Membranes Towards Light Sources on Silicon, D. Burt[1], A. Z. Al-Attili[1], Z. Li[1], F. Liu[1], K. Oda[2], N. Higashitarumizu[3], Y. Ishikawa[3], O. M. Querin[4], F. Gardès [1], R. W. Kelsall[4], and S. Saito[1], [1]University of Southampton, [2]Hitachi, Ltd., [3]The University of Tokyo, [4]University of Leeds···92

Thursday, March 2
Room B (203+204), 2F

8:30-9:45 Session 5B - Modeling: Multi Physics Simulation

Chairs: A. Schenk, ETH Zurich
M. Natarajan, GLOBALFOUNDRIES Inc.

5B-1   8:30-8:55
Transient Characterization of Graphene NEMS Switch ESD Protection Structures, Qi Chen[1], Cheng Li[1], Jimmy Ng[2], Fei Lu[1], Chenkun Wang[1], Feilong Zhang[1], Rui Ma[3], Ya-Hong Xie[2], and Albert Wang[1], [1]University of California, Riverside, [2]University of California, Los Angeles, [3]Intel ·····························95

5B-2   8:55-9:20 (Invited)
Reliability Modeling of RF MEMS Switches and Phase Shifters for Microwave and Millimeter Wave Applications, Shiban K. Koul and Sukomal Dey, Indian Institute of Technology ······························N/A

5B-3   9:20-9:45 (Invited)
Multi-Scale and Multi-Domain Simulation of Electrical Power System, Shimeng Huang[1], Xiao Li[1], Tinghao Yeh[2], Shanghsun Mao[2], Takayuki Sekisue[3], Vel Ambalavanar[1], and Sameer Kher[1], [1]ANSYS Inc., [2]ANSYS Taiwan, [3]ANSYS Japan ··100

Thursday, March 2
Main Hall, 3F

10:25-12:05   Session 6M - Material: Advanced BEOL Materials

Chairs: S. Kim, Yeungnam University
P. R. Berger, Ohio State University

6M-1   10:25-10:50 (Invited)
Nano-Structure-Controlled Very Low Resistivity Cu Wires Formed by High Purity Electrolyte and Optimized Additives, Jin Onuki[1], Kunihiro Tamahashi[1], Takashi Inami[1], Takatoshi Nagano[1], Yasushi Sasajima[1], and Shuji Ikeda[2], [1]Ibaraki University, [2]tei SOLUTIONS··103

6M-2   10:50-11:15
Withdrawn

6M-3   11:15-11:40 (Invited)
Advanced Materials and Interconnect Technologies for Next Generation Smart Devices, Rozalia
Beica, Dow Electronic Materials

6M-4   11:40-12:05 (Invited)
**Proximity Gettering Technology for Advanced CMOS Image Sensors Using C_3H_5 Carbon Cluster
Ion Implantation Techniques**, Kazunari Kurita, Takeshi Kadono, Ryosuke Okuyama, Ayumi Masada,
Ryou Hirose, Yoshihiro Koga, and Hidehiko Okuda, SUMCO Corporation ··· 105

Thursday, March 2
Room A (201+202), 2F

10:25-12:05   Session 6A - Device: FET Reliability

Chairs: G. Xiao, National University of Singapore
        Y.-C. King, National Tsing Hua University

6A-1   10:25-10:50 (Invited)
New Visions for IC Yield Detractor Detection, Bill Nehrer, Kelvin Doong, and Dennis Ciplickas, PDF
Solutions ··· 107

6A-2   10:50-11:15
Comparative Study on RTN Amplitude in Planar and FinFET Devices, Zexuan Zhang[1], Zhe Zhang[1],
Shaofeng Guo[1], Runsheng Wang[1], Xingsheng Wang[2], Binjie Cheng[2], Asen Asenov[2,3], and Ru Huang[1],
[1]Peking University, [2]Synopsys, Inc., [3]University of Glasgow ·· 109

6A-3   11:15-11:40 (Invited)
A BTI Analysis Tool (BAT) to Simulate p-MOSFET Ageing under Diverse Experimental Conditions,
Souvik Mahapatra, Narendra Parihar, Subrat Mishra, Beryl Fernandez, and Ankush Chaudhary, Indian
Institute of Technology Bombay ·· 111

6A-4   11:40-12:05
Accurate Mapping of Oxide Traps in Highly-Stable Black Phosphorus FETs, Yu. Yu. Illarionov[1,2],
G. Rzepa[1], M. Waltl[1], T. Knobloch[1], J.-S. Kim[3], D. Akinwande[3], and T. Grasser[1], [1]Technische Universtät
Wien, [2]Ioffe Physical-Technical Institute, [3]The University of Texas at Austin ·· 114

Thursday, March 2
Room B (203+204), 2F

10:25-12:05   Session 6B - Modeling: Reliability & Modeling

Chairs: S. Huang, Peking University
        M. Miura-Mattausch, Hiroshima University

6B-1   10:25-10:50 (Invited)
ESD Performance Enhancement Methodologies for CMOS Power Transistors, Mahadeva Iyer
Natarajan and Jian-Hsing Lee, GLOBALFOUNDRIES Inc. ··· 116

6B-2   10:50-11:15
Aging Simulation of SiC-MOSFET in DC-AC Converter, Kenshiro Sato, Shinya Sekizaki, Dondee
Navarro, Yoshifumi Zoka, Naoto Yorino, Hiroshi Zenitani, Hans Jürgen Mattausch, and Mitiko Miura-
Mattausch, Hiroshima University··· 119

6B-3 11:15-11:40
Degradation Caused by Negative Bias Temperature Instability Depending on Body Bias on NMOS or PMOS in 65 nm Bulk and Thin-BOX FDSOI Processes, Ryo Kishida and Kazutoshi Kobayashi, Kyoto Institute of Technology ⋯⋯⋯⋯⋯⋯⋯⋯⋯⋯⋯⋯⋯⋯⋯⋯⋯⋯⋯⋯⋯⋯⋯⋯⋯⋯⋯⋯⋯⋯⋯ 122

6B-4 11:40-12:05
JFETIDG: A Compact Model for Independent Dual-Gate JFETs, Kejun Xia, Colin C. McAndrew, and Hanyu Sheng, NXP Semiconductors ⋯⋯⋯⋯⋯⋯⋯⋯⋯⋯⋯⋯⋯⋯⋯⋯⋯⋯⋯⋯⋯⋯⋯⋯⋯⋯⋯⋯⋯ 124

13:00-14:20   Poster Session in Foyer, 3F

*For Poster Session, please see the detailed information from page (17).

Thursday, March 2
Main Hall, 3F

14:20-16:00   Session 7M - Modeling: Device Characterization

Chairs: P. Su, National Chiao Tung University
    D. Navarro, Hiroshima University

7M-1 14:20-14:45 (Invited)
How Non-Ideality Effects Deteriorate the Performance of Tunnel FETs, Andreas Schenk[1], Saurabh Sant[1], Kirsten Moselund[2], and Heike Riel[2], [1]ETH Zurich, [2]IBM Research-Zurich ⋯⋯⋯⋯⋯⋯⋯⋯⋯⋯ 126

7M-2 14:45-15:10
Charge Splitting In-situ Recorder (CSIR) for Monitoring Plasma Damage in FinFET BEOL Processes, Ting-Huan Hsieh, Yi-Pei Tsai, Chrong Jung Lin, and Ya-Chin King, National Tsing Hua University ⋯⋯⋯ 128

7M-3 15:10-15:35
Geometric Variation: A Novel Approach to Examine the Surface Roughness and the Line Roughness Effects in Trigate FinFETs, E. R. Hsieh[1], Y. C. Fan[2], C. H. Liu[2], Steve S. Chung[1], R. M. Huang[3], C. T. Tsai[3], and T. R. Yew[3], [1]National Chiao Tung University, [2]National Taiwan Normal University, [3]UMC ⋯⋯⋯⋯⋯⋯⋯⋯⋯⋯⋯⋯⋯⋯⋯⋯⋯⋯⋯⋯⋯⋯⋯⋯⋯⋯⋯⋯⋯⋯⋯⋯⋯⋯⋯⋯⋯ 130

7M-4 15:35-16:00
A Novel Method to Characterize DRAM Process Variation by the Analyzing Stochastic Properties of Retention Time Distribution, Min Hee Cho, Namho Jeon, Moonyoung Jeong, Sungsam Lee, Satoru Yamada, and Hyeongsun Hong, Samsung Electronics Co., Ltd. ⋯⋯⋯⋯⋯⋯⋯⋯⋯⋯⋯⋯⋯⋯ 132

Thursday, March 2
Room A (201+202), 2F

14:20-16:00   Session 7A - Device: Non-conventional Material-based FET Technologies

Chairs: K. Xia, NXP Semiconductors
    H. Lv, Chinese Academy of Sciences

7A-1 14:20-14:45
Analysis of Subthreshold Swing and Internal Voltage Amplification for Hysteresis-Free Negative Capacitance FinFETs, Pin-Chieh Chiu and Vita Pi-Ho Hu, National Central University ⋯⋯⋯⋯ 134

7A-2   14:45-15:10
Design Space Exploration Considering Back-Gate Biasing Effects for Negative-Capacitance Transition-Metal-Dichalcogenide (TMD) Field-Effect Transistors, Wei-Xiang You and Pin Su, National Chiao Tung University ·· 136

7A-3   15:10-15:35
A Simple Way to Grow Large-Area Single-Layer MoS_2 Film by Chemical Vapor Deposition, Yan-Cong Qiao, Zhen Yang, Hai-Ming Zhao, Xue-Feng Wang, Lu-Qi Tao, Yi Yang, and Tian-Ling Ren, Tsinghua University ·· 138

7A-4   15:35-16:00
High-Mobility and H_2-Anneal Tolerant InGaSiO/InGaZnO/InGaSiO Double Hetero Channel Thin Film Transistor for Si-LSI Compatible Process, Nobuyoshi Saito, Kentaro Miura, Tomomasa Ueda, Tsutomu Tezuka, and Keiji Ikeda, Toshiba Corporation ··· 141

Thursday, March 2
Room B (203+204), 2F

14:20-16:00   Session 7B - Process: Process Innovation in MEMS and Sensors

Chairs: S. Tanaka, Tohoku University
          M. Miura, Hitachi High-Technologies Corp.

7B-1   14:20-14:45 (Invited)
Growing Market of MEMS and Technology Development in Process and Tools Specialized to MEMS, Hiroshi Yanazawa and Kohji Homma, MEMS CORE Co., Ltd. ··· 143

7B-2   14:45-15:10
Development of MEMS Vibrating Sensor with Phase-Shifted Optical Pulse Interferometry, Yusaku Ohe[1], Hitoshi Kimura[1], Norio Inou[1], Yoshiharu Hirayama[2], and Minoru Yoshida[2], [1]Tokyo Institute of Technology, [2]Hakusan Corporation ··· 145

7B-3   15:10-15:35
Microstructuring Polydimethylsiloxane Elastomer Film with 3D Printed Mold for Low Cost and High Sensitivity Flexible Capacitive Pressure Sensor, Bengang Zhuo, Sujie Chen, and Xiaojun Guo, Shanghai Jiao Tong University ·· 148

7B-4   15:35-16:00
Effective Performance of a Tiny-Chamber Plasma Etcher in Scallop Reduction, Sommawan Khumpuang[1,2], Hiroyuki Tanaka[1,2], and Shiro Hara[1,2], [1]National Institute of Advanced Industrial Science and Technology (AIST), [2]Minimal Fab ·· 150

Thursday, March 2
Main Hall, 3F

16:15-17:55   Session 8M - Device: Memory Technology

Chairs: M. Saitoh, Toshiba Corporation
          K. Tateiwa, TowerJazz Panasonic Semiconductor Co., Ltd.

8M-1   16:15-16:40
Exploiting NbO_x Metal-Insulator-Transition Device as Oscillation Neuron for Neuro-Inspired Computing, Ligang Gao, Pai-Yu Chen, and Shimeng Yu, Arizona State University ····················· 152

8M-2　16:40-17:05

Ohmic Contact Formation Between $Ge_2Sb_2Te_5$ Phase Change Material and Vertically Aligned Carbon Nanotubes, Panni Wang, Suwen Li, Yihan Chen, Lining Zhang, and Mansun Chan, Hong Kong University of Science and Technology ⋯⋯⋯⋯⋯⋯⋯⋯⋯⋯⋯⋯⋯⋯⋯⋯⋯⋯⋯⋯⋯⋯⋯ 154

8M-3　17:05-17:30

Impact of Current Distribution on RRAM Array with High and Low I_{ON}/I_{OFF} Devices, Mohammed Zackriya V.[1,2], Albert Chin[1], and Harish M. Kittur[2], [1]National Chiao Tung University, [2]VIT University ⋯⋯⋯ 156

8M-4　17:30-17:55

3D Time-Contingent Physical Unclonable Function Array on 16nm FinFET Dielectric RRAM, Yi-Hung Chang[1], Po Shao Yeh[1], Yue-Der Chih[2], Jonathan Chang[2], Ya-Chin King[1], and Chrong Jung Lin[1], [1]National Tsing Hua University, [2]Taiwan Semiconductor Manufacturing Company ⋯⋯⋯⋯⋯⋯⋯⋯⋯⋯⋯⋯ 158

Thursday, March 2
Room A (201+202), 2F

16:15-17:55　　Session 8A - Process: Process Innovation on Ge Surface and Interface

Chairs:　Y. Akasaka, TEL

　　　　O. Nakatsuka, Nagoya University

8A-1　16:15-16:40

Surface Preparation and Wet Cleaning for Germanium Surface, Masayuki Otsuji[1], Yukifumi Yoshida[1], Hiroaki Takahashi[1], Farid Sebaai[2], Kurt Wostyn[2], Frank Holsteyns[2], Masanobu Sato[1], and Hajime Shirakawa[1], [1]Screen Semiconductor Solutions Co., Ltd, [2]imec vzw ⋯⋯⋯⋯⋯⋯⋯⋯⋯⋯⋯⋯ 160

8A-2　16:40-17:05

Oxidation Mechanism and Surface Passivation of Germanium by Ozone, Xiaolei Wang[1], Jinjuan Xiang[1], Chao Zhao[1,2], Tianchun Ye[1,2], and Wenwu Wang[1,2], [1]Chinese Academy of Sciences, [2]University of Chinese Academy of Sciences ⋯⋯⋯⋯⋯⋯⋯⋯⋯⋯⋯⋯⋯⋯⋯⋯⋯⋯⋯⋯⋯⋯⋯⋯⋯⋯ 162

8A-3　17:05-17:30

The Impact of Atomic Layer Depositions on High Quality Ge/GeO_2 Interfaces Fabricated by Rapid Thermal Annealing in O_2 Ambient, Laura Žurauskaitė, Per-Erik Hellström, and Mikael Östling, KTH Royal Institute of Technology ⋯⋯⋯⋯⋯⋯⋯⋯⋯⋯⋯⋯⋯⋯⋯⋯⋯⋯⋯⋯⋯⋯⋯⋯⋯⋯⋯⋯⋯ 164

8A-4　17:30-17:55

Experimental Investigation on Growth Mechanism of GeO_x Layer Formed by Plasma Post Oxidation Based on Angle Resolved X-Ray Photoelectron Spectroscopy, Zhiqian Zhao[1], Xiaolei Wang[2], Jing Zhang[1], Chao Zhao[2,3], Tianchun Ye[2,3], and Wenwu Wang[2,3], [1]North China University of Technology, [2]Chinese Academy of Sciences, [3]University of Chinese Academy of Sciences ⋯⋯⋯⋯⋯⋯⋯ 167

17:55-18:25　　Authors Interview in Main Hall, 3F

Wednesday, March 1 18:25-19:05
Thursday, March 2 13:00-14:20
Foyer, 3F

Poster Session

P-1
UV-Annealing-Enhanced Stability in High-Performance Printed InO$_x$ Transistors, William J. Scheideler and Vivek Subramanian, University of California Berkeley 169

P-2
Random-Telegraph-Noise by Resonant Tunnelling at Low Temperatures, Z. Li[1], M. Sotto[1], F. Liu[1], M. K. Husain[1], I. Zeimpekis[1], H. Yoshimoto[2], K. Tani[2], Y. Sasago[2], D. Hisamoto[2], J. D. Fletcher[3], M. Kataoka[3], Y. Tsuchiya[1], and S. Saito[1], [1]University of Southampton, [2]Hitachi, Ltd., [3]National Physical Laboratory 172

P-3
Fabrication of E-Mode InGaN/AlGaN/GaN HEMT Using FIB Based Lithography, Shubhankar Majumdar[1,3], Chitrakant Sahu[2], and Dhrubes Biswas[3], [1]National Institute of Technology Raipur, [2]Malaviya National Institute of Technology Jaipur, [3]Indian Institute of Technology Kharagpur 175

P-4
Endurance Characterization of the Cu-Dope HfO$_2$ Based Selection Device with One Transistor-One Selector Structure, Qing Luo, Xiaoxin Xu, Hangbing Lv, Tiancheng Gong, Shibing Long, Qi Liu, Ling Li, and Ming Liu, Chinese Academy of Sciences 178

P-5
Investigation on Direct-Gap GeSn Alloys for High-Performance Tunneling Field-Effect Transistor Applications, Lei Liu[1], Renrong Liang[1], Guilei Wang[2], Henry H. Radamson[2], Jing Wang[1], and Jun Xu[1], [1]Tsinghua University, [2]University of Chinese Academy of Sciences 180

P-6
Online Training on RRAM Based Neuromorphic Network: Experimental Demonstration and Operation Scheme Optimization, Peng Yao[1], Huaqiang Wu[1*], Bin Gao[1], Ning Deng[1], Shimeng Yu[2], and He Qian[1], [1]Tsinghua University, [2]Arizona State University 182

P-7
Uniformity Improvements of Low Current 1T1R RRAM Arrays through Optimized Verification Strategy, Shan Wang, Xinyi Li, Huaqiang Wu, Bin Gao, Ning Deng, Dong Wu, and He Qian, Tsinghua University 184

P-8
Electroluminescence Characteristics of Rare Earth Doped Silicon Based Light Emitting Device, Fumihiro Hattori[1], Hideyuki Iwata[1], Toshihiro Matsuda[1], and Takashi Ohzone[2], [1]Toyama Prefectural University, [2]Dawn Enterprise Co., Ltd 187

P-9
High Photoresponsivity Germanium Nanodot PhotoMOSFETs for Monolithically- Integrated Si Optical Interconnects, Ming-Hao Kuo[1], Morris M. Lee[2], and Pei-Wen Li[1,2], [1]National Central University, [2]National Chiao Tung University 189

P-10

The Impact of Oxygen Insertion Technology on SRAM Yield Performance, A. Marshall[1], S. Nimmalapudi[1], W. K. Loh[2], J. Krick[3], L. Hutter[4], H. Takeuchi[5], R. Stephenson[5], R. J. Mears [5], and S. Ikeda[6], [1]The University of Texas at Dallas, [2]Consultant, [3]Logix Consulting, Inc., [4]Lou Hutter Consulting, [5]Atomera Inc., [6]tei SOLUTIONS Inc. ··· 191

P-11

Transport Properties in Silicon Nanowire Transistors with Atomically Flat Interfaces, F. Liu[1], M. K. Husain[1], Z. Li[1], M. S. H. Sotto[1], D. Burt[1], J. D. Fletcher[2], M. Kataoka[2], Y. Tsuchiya[1], and S. Saito[1], [1]University of Southampton, [2]National Physical Laboratory ··· 193

P-12

InAs MOS-HEMT Power Detector for 1.0 THz on Quartz Glass, Eiji Kume[1], Hiroyuki Ishii[2], Hiroyuki Hattori[2], Wen-Hsin Chang[2], Mutsuo Ogura[1], Haruichi Kanaya[3], Tanemasa Asano[3], and Tatsuro Maeda[2], [1]IRspec Corporation, [2]National Institute of Advanced Industrial Science and Technology (AIST), [3]Kyushu University ·· 196

P-13

Analysis and Modeling of Capacitances in Halo-Implanted MOSFETs, Chetan Gupta[1], Harshit Agarwal[3], Sagnik Dey[2], Chenming Hu[3], and Yogesh S. Chauhan[1], [1]Indian Institute of Technology Kanpur, [2]Texas Instruments Inc., [3]University of California Berkeley ······························· 198

P-14

Design and Performance of Thin-Film μTEG Modules for Wearable Device Applications, Tsuyoshi Kondo, Nana Chiwaki, and Satoshi Sugahara, Tokyo Institute of Technology ·································· 201

P-15

Analysis of Spin Accumulation in a Si Channel Using CoFe/MgO/Si Spin Injectors, Taiju Akushichi, Daiki Kitagata, Yusuke Shuto, and Satoshi Sugahara, Tokyo Institute of Technology ·····················204

P-16

SER Scaling and Trends in Planar Submicron Technology Nodes, Krishna Mohan Chavali, GLOBALFOUNDRIES Inc. ··206

P-17

Fully Packaged Compliant CMOS Electronic Systems for IoT and IoE Applications, G. A. Torres Sevilla, M. D. Cordero, J. M. Nassar, A. T. Kutbee, and M. M. Hussain, King Abdulla University of Science and Technology ··209

P-18

A Metal Micro-Casting Method for Through-Silicon Via (TSV) Fabrication, Jiebin Gu, Bingjie Liu, Heng Yang, and Xinxin Li, Shanghai Institute of Microsystem and Information Technology ·····················211

P-19

Acoustic Emission Wave Sensor with Thermally Controllable Force-Enhancement Mechanism for Acoustic Emission Source Detection and Biomedical Application, Guo-Hua Feng and Wei-Ming Chen, National Chung Cheng University ·· 213

P-20

Electroactive Polymer Actuated Gripper Enhanced with Iron Oxide Nanoparticles and Water Supply Mechanism for Millimeter-Sized Fish Roe Manipulation, Guo-Hua Feng and Shih-Chieh Yen, National Chung Cheng University ·· 216

P-21

Low Temperature Hermetic Sealing by Aluminum Thermocompression Bonding Using Tin Intermediate Layer, Shiro Satoh, Hideyuki Fukushi, Masayoshi Esashi, and Shuji Tanaka, Tohoku University 219

P-22

Low-Carrier Density Sputtered-MoS$_2$ Film by H$_2$S Annealing for Normally-Off Accumulation-Mode FET, Jun'ichi Shimizu[1], Takumi Ohashi[1], Kentaro Matsuura[1], Iriya Muneta[1], Kuniyuki Kakushima[1], Kazuo Tsutsui[1], Nobuyuki Ikarashi[2], and Hitoshi Wakabayashi[1], [1]Tokyo Institute of Technology, [2]Nagoya University 222

P-23

Impact of Ferroelectric Domain Switching in Nonvolatile Charge-Trapping Memory, Chia-Chi Fan[1], Yu-Chien Chiu[1], Chien Liu[1], Guan-Lin Liou[2], Wen-Wei Lai[1], Yi-Ru Chen[1], Chun-Hu Cheng[2], and Chun-Yen Chang[1,3], [1]National Chiao Tung University, [2]National Taiwan Normal University, [3]Academia Sinica 224

P-24

An In-Line MOSFET Process with Photomask Fabrication Process in a Minimal Fab, Norio Umeyama[1,2], Sommawan Khumpuang[1,2], and Shiro Hara[1,2], [1]National Institute of Advanced Industrial Science and Technology (AIST), [2]Minimal Fab 226

P-25

Development of a Half-Inch Wafer for Minimal Fab Process, Norio Umeyama[1,2], Atsushi Yamazaki[3], Takaaki Sakai[3], Sommawan Khumpuang[1,2], and Shiro Hara[1,2], [1]National Institute of Advanced Industrial Science and Technology (AIST), [2]Minimal Fab, [3]Fujikoshi Machinery 228

P-26

Helium Ion Microscopy (HIM) for Imaging Fine Line Features Patterned Organic Film with Less Damage, Shinichi Ogawa[1], Tomoya Ohashi[2], Shigeki Oyama[2], and Yuki Usui[2], [1]National Institute of Advanced Industrial Science and Technology (AIST), [2]Nissan Chemical Industries, Ltd. 230

P-27

Supercritical Fluid Deposition of Conformal Oxide Films: 3-Dimentionally-Stacked RuO$_2$/TiO$_2$/RuO$_2$ Structures for MIM Capacitors, Yu Zhao, Yusuke Shimoyama, Takeshi Momose, and Yukihiro Shimogaki, The University of Tokyo 232

P-28

Crystallinity Improvement Using Migration-Enhancement Methods for Sputtered-MoS$_2$ Films, Shin Hirano, Jun'ichi Shimizu, Kentaro Matsuura, Takumi Ohashi, Iriya Muneta, Kuniyuki Kakushima, Kazuo Tsutsui, and Hitoshi Wakabayashi, Tokyo Institute of Technology 234

P-29

Low Power Transparent Resistive Switching Device with Memristive Behavior, Amitesh Kumar, Mangal Das, Rohit Singh, Pankaj Sharma, Abhinav Kranti, and Shaibal Mukherjee, Indian Institute of Technology Indore .... N/A

P-30

Photoresist Development for Wafer-Level Packaging Process, Makiko Irie, Toshiaki Tachi, and Atushi Sawano, Tokyo Ohka Kogyo Co., Ltd. 238

P-31

Second-Harmonic Susceptibility Enhancement in Gallium Nitride Nanopillars, Kangwei Wang, Haoliang Qian, Zhaowei Liu, and Paul K. L. Yu, University of California, San Diego 240

P-32

Defect Formation in SiO$_2$ Formed by Thermal Oxidation of SiC, Kenta Chokawa, Masaaki Araidai, and Kenji Shiraishi, Nagoya University ⋯⋯⋯⋯⋯⋯⋯⋯⋯⋯⋯⋯⋯⋯⋯⋯⋯⋯⋯⋯⋯⋯⋯ 242

P-33

Current Enhanced Solid Phase Precipitation (CE-SPP) for Direct Deposition of Multilayer Graphene on SiO$_2$ from a Cu Capped Co-C Layer, Hiroyasu Ichikawa and Kazuyoshi Ueno, [1]Shibaura Institute of Technology ⋯⋯⋯⋯⋯⋯⋯⋯⋯⋯⋯⋯⋯⋯⋯⋯⋯⋯⋯⋯⋯⋯⋯⋯⋯⋯⋯ 244

P-34

Enhanced Photoresponse of InGaZnO TFT to Ultraviolet Illumination by Using a High-k Dielectric, Libin Liu, Renrong Liang, Jing Wang, and Jun Xu, Tsinghua University ⋯⋯⋯⋯⋯⋯ 247

P-35

Development of in-$situ$ Sb-Doped Ge$_{1-x}$Snx Epitaxial Layers for Source/Drain Stressor of Strained Ge Transistors, Jihee Jeon, Akihiro Suzuki, Kouta Takahashi, Osamu Nakatsuka, and Shigeaki Zaima, Nagoya University ⋯⋯⋯⋯⋯⋯⋯⋯⋯⋯⋯⋯⋯⋯⋯⋯⋯⋯⋯⋯⋯⋯⋯⋯⋯ 249

P-36

Novel in-situ Passivation of MoCl$_5$ Doped Multilayer Graphene with MoO$_x$ for Low-Resistance Interconnects, K. Kawamoto, Y. Saito, M. Kenmoku, and K. Ueno, Shibaura Institute of Technology ⋯⋯ 252

P-37

Thickness-Independent Behavior of Coercive Field in HfO$_2$-Based Ferroelectrics, Shinji Migita[1], Hiroyuki Ota[1], Hiroyuki Yamada[1], Akihito Sawa[1], and Akira Toriumi[2], [1]National Institute of Advanced Industrial Science and Technology (AIST), [2]The University of Tokyo ⋯⋯⋯⋯⋯⋯⋯⋯⋯⋯⋯ 255

P-38

Crystallinity Study of Si Single Crystal Stripe on Bended Glass Substrate Fabricated by Micro-Chevron Laser Beam Scanning Method, Wenchang Yeh and Seigo Moriyama, Shimane University ⋯⋯ 257

P-39

Physics Based System Simulation for Robot Electro-Mechanical Control Design, T. K. Maiti[1], L. Chen[1], M. Miura-Mattausch[1], S. K. Koul[2], and H. J. Mattausch[1], [1]Hiroshima University, [2]Indian Institute of Technology Delhi ⋯⋯⋯⋯⋯⋯⋯⋯⋯⋯⋯⋯⋯⋯⋯⋯⋯⋯⋯⋯⋯⋯⋯⋯⋯⋯⋯⋯ 259

P-40

Accurate Modeling of MOSFET Aging Based on Trap-Density Increase for Predicting Circuit-Performance Aging, Lei Chen, Hidenori Miyamoto, Tapas Kumar Maiti, Dondee Navarro, Takahiro Iizuka, Hans Jürgen Mattausch, and Mitiko Miura-Mattausch, Hiroshima University ⋯⋯⋯⋯⋯⋯ N/A

P-41

Modeling and Analysis of Depletion-Mode NMOS Transistor as Transmitter/Receiver RF Switch, Runtao Ning, Tianjiao Liu, and Z. John Shen, Illinois Institute of Technology ⋯⋯⋯⋯⋯⋯⋯⋯ N/A

P-42

An Analytical Charge-Sheet Drain Current Model for Monolayer Transition Metal Dichalcogenide Negative Capacitance Field-Effect Transistors, Renrong Liang, Chunsheng Jiang, Jing Wang, Jun Xu, and Tian-Ling Ren, Tsinghua University ⋯⋯⋯⋯⋯⋯⋯⋯⋯⋯⋯⋯⋯⋯⋯⋯⋯⋯⋯⋯⋯⋯⋯ N/A

P-43

A Simple Test Method for Electromigration Reliability of Solder/Cu Pillar Bumps Using Flat Cables, Naoki Azuma[1], Misaki Owada[1], Takumi Abe[1], Tsutomu Nakada[2], Makoto Kubota[2], and Kazuyoshi Ueno[1,3], [1]Shibaura Institute of Technology, [2]EBARA Corporation, [3]Research Center for Green Innovation ⋯⋯⋯⋯⋯⋯⋯⋯⋯⋯⋯⋯⋯⋯⋯⋯⋯⋯⋯⋯⋯⋯⋯⋯⋯⋯⋯⋯⋯⋯ N/A

Teruo Hirayama	Sony
Atsuyoshi Koike	SanDisk Japan
Steven Johnston	Intel
Min Cao	TSMC
Hitoshi Nakao	Applied Materials Japan, Inc
Ken Sugimoto	JSR
William Chen	ASE
Sachin Sonkusale	PDF Solutions

Technical Program Committee

| Chair | Hitoshi Wakabayashi | Tokyo Tech. |

Sub-Committee: Devices and Manufacturing for "Cloud and Edge"

Chair	Ken Uchida	Keio Univ.
Members	Kenji Tateiwa	TowerJazz Panasonic Semiconductor
	Jiro Ida	Kanazawa Inst. of Tech.
	Hidetoshi Ohnuma	Sony
	Kazuaki Sawada	Toyohashi Univ. of Tech.
	Masumi Saitoh	Toshiba
	Takeshi Yanagida	Kyushu Univ.
	Adrian Ionescu	EPFL
	Ru Huang	Peking Univ.
	Mei-Kei Ieong	ASTRI
	Gong Xiao	National Univ. of Singapore
	Kejun Xia	NXP Semiconductors
	Ya-Chin King	National Tsing Hua Univ.
	Osbert Cheng	UMC
	Hangbing Lv	Chinese Academy of Sciences
	Naoto Horiguchi	Imec
	Jong-Ho Lee	Seoul National Univ.
	Jae-kyu Lee	Samsung
	Wilman Tsai	TSMC

Sub-Committee: Packaging and Manufacturing for "Cloud and Edge"

Chair	Shintaro Yamamichi	IBM
Members	Takayuki Ohba	Tokyo Tech
	Makoto Murai	Sony
	Hisashi Kaneko	JSR
	Yoichiro Kurita	Toshiba
	Katsumi Kikuchi	NEC
	Michihiro Inoue	AIST
	Jean Trewhella	GLOBALFOUNDRIES
	Kyeong Sool Seong	J-Devices
	Naoki Oka	Applied Materials Japan, Inc

Sub-Committee: Process, Tools and Manufacturing

Chair	Jiro Yugami	Hitachi Kokusai Electric Inc.
Members	Nobuyoshi Kobayashi	ASM
	Makoto Miura	Hitachi High-Technologies Corp.
	Yoji Kawasaki	SMIT
	Shuji Tanaka	Tohoku Univ.
	Kazuo Nojiri	Lam Research
	Yasushi Akasaka	TEL
	Sean Cunnigham	Intel

Xiaojun Guo	Shanghai Jiao Tong Univ.
Masanobu Sato	SCREEN
Wenwu Wang	Chinese Academy of Sciences
Kyoichi Suguro	Toshiba
Osamu Nakatsuka	Nagoya Univ.
Yuichi Setsuhara	Osaka Univ.
Byoung Hun Lee	Gwangju Inst. of Sci. and Tech.
Yasuhiro Morikawa	ULVAC
Hyosan Lee	Samsung
Gill Lee	Applied Materials,Inc

Sub-Committee: Materials

Chair

Soo-Hyun Kim	Yeungnam Univ.
Members Takao Enomoto	TKK
Toshihide Nabatame	NIMS
Iriya Muneta	Tokyo Tech
Kazuhito Matsukawa	SUMCO
Toshiya Sakata	The Univ. of Tokyo
Atsuhiro Tsukune	TAIYO NIPPON SANSO Corp.
Satoshi Fujimura	Tokyo Ohka Kogyo CO., LTD
Hyungjun Kim	Yonsei Univ.
Muhammad Mustafa Hussain	King Abdullah Univ. of Sci. and Tech. (KAUST)
Paul Berger	Ohio State Univ.
Srabanti Chowdhury	Univ. of California, Davis
Satoshi Kamo	JSR
Pei-Wen Li	National Chiao Tung Univ.

Sub-Committee: Modeling & Reliability

Chair

Mitiko Miura-Mattausch	Hiroshima Univ.
Members Dondee Navarro	Hiroshima Univ.
Colin MacAndrew	NXP
Durga Misra	New Jersey Inst. of Tech.
Wladyslaw Grabinski	MOS-AK
Mansun Chan	Hong Kong Univ. of Sci. and Tech.
Xing Zhou	NTU Singapore
Zhiping Yu	Tsinghua Univ.
Tian-Ling Ren	Tsinghua Univ.
Natrain Arora	Siprosys Inc.
Tibor Grasser	TU Wien
Perumal Ratnam	SanDisk / Western Digital
Sadayuki Yoshitomi	Toshiba
Lining Zhang	Hong Kong Univ. of Sci. and Tech.
Kenji Okada	TowerJazz Panasonic Semiconductor
Mahadeva Iyer Natarajan	GLOBALFOUNDRIES
Pin Su	National Chiao Tung Univ.
Kin Leong Pey	Singapore Univ. of Tech. and Design
Anthony S. Oates	TSMC
Inkook Jang	Samsung
Hisayo Sasaki Momose	Yokohama National Univ.

2017 IEEE Electron Devices Technology and Manufacturing Conference (EDTM)
Short Course 1
"Memory-Based Technology"

Tuesday, February 28, 2017

Moderator: A. Toriumi, *The University of Tokyo*

13:00 **Fundamentals for Resistance Switching RAM**

C. S. Hwang, *Seoul National University*

14:00 **ReRAM and PCM Device Structure and Technology**

D. Ielmini, *DEIB – Politecnico di Milano and IU.NET*

15:00 Break

15:15 **Nano-Scale Analog Synaptic Devices for Neuromorphic System**

H. Hwang, *POSTECH*

16:15 **Emerging Non-Volatile Memory (NVM) based Computing System**

S. Yu, *Arizona State University*

2017 IEEE Electron Devices Technology and Manufacturing Conference (EDTM)
Short Course 2
"Advanced FET Technology"

Tuesday, February 28, 2017

Moderator: J. Ida, *Kanazawa Institute of Technology*

13:00   **Device Challenges in Advanced CMOS Technologies**

       F. Bœuf, *STMicroelectronics, The University of Tokyo*

14:00   **Gate Stack Reliability**

       T. Grasser, *Technische Universität Wien*

15:00   Break

Moderator: T. Nabatame, *NIMS*

15:15   **Contact and Doping Technology for Advanced CMOS**

       K.-W. Ang, *National University of Singapore*

16:15   **Characterization-Based Yield Engineering**

       K. Nemoto, *Hitachi High-Technologies Corporation*

Program at a glance

Day 1 - Tuesday, February 28

Time	Room A (201+202), 2F	Room B (203+204), 2F
8:30-8:45	Introduction	Introduction
8:45-10:15	Tutorial 1 (in Japanese) Nano Imprint	Tutorial 2 (in English) 3D Design and Process
10:15-10:30	Break	
10:30-12:00	Tutorial 1 (in Japanese) Adhesion and Bonding	Tutorial 2 (in English) 3D Design and Process
12:00-13:00	Break	
13:00-15:00	Short Course 1 Memory-Based Technology	Short Course 2 Advanced FET Technology
15:00-15:15	Break	
15:15-17:15	Short Course 1 Memory-Based Technology	Short Course 2 Advanced FET Technology
17:15-18:00	Lecturers Interview	
18:00-19:00	Reception @ Foyer, 3F / Exhibition	

Day 2 - Wednesday, March 1

Time	Main Hall, 3F	Room A (201+202), 2F	Room B (203+204), 2F
8:30-9:00	Opening Remarks		
9:00-11:00	Plenary Session		
11:00-12:00	Exhibition Talks		
12:00-13:35	Break (Poster display starts)	Luncheon Seminar	Break
13:35-15:40	Session 3M - Emerging: Emerging Technologies	Session 3A - Process: Process Technology for Advanced Devices	Session 3B - Modeling Reliability Analysis
15:40-15:55	Break		
15:55-18:25	Session 4M - Package: Subsystem Integration & Packaging	Session 4A - Material: Advanced FEOL Materials	Session 4B - Device: Advanced FET Technology
18:25-19:05	Authors Interview @ Main Hall / Poster Session @ Foyer, 3F		
19:15-21:15	Banquet @ Banquet Room "OHTORI", 3F, ANA Crowne Plaza Toyama		

Day 3 - Thursday, March 2

Time	Main Hall, 3F	Room A (201+202), 2F	Room B (203+204), 2F
08:30-10:10	Session 5M - Process: Innovative Process Tools	Session 5A - Device: More-than-Moore Technologies	Session 5B - Modeling: Multi Physics Simulation
10:10-10:25	Break		
10:25-12:05	Session 6M - Material: Advanced BEOL Materials	Session 6A - Device: FET Reliability	Session 6B - Modeling: Reliability & Modeling
12:05-13:00	Break		
13:00-14:20	Poster Session / Exhibition @ Foyer, 3F		
14:20-16:00	Session 7M - Modeling: Device Characterization	Session 7A - Device: Non-conventional Material-based FET Technologies	Session 7B - Process: Process Innovation in MEMS and Sensors
16:00-16:15	Break		
16:15-17:55	Session 8M - Device: Memory Technology	Session 8A - Process: Process Innovation on Ge Surface and Interface	
17:55-18:25	Authors Interview @ Main Hall		
18:25-19:25	Farewell Session / Best Poster Award Ceremony @ Foyer, 3F		

Dimensions of Innovation to Enable the Next Era of Intelligent Systems

Dr. John G. Pellerin
Deputy Chief Technology Officer and Vice President of
Worldwide Research and Development
GLOBALFOUNDRIES, Malta, New York

ABSTRACT — The pace of technological change has again reached an inflection point setting the stage for a new era of dramatic innovation in the electronics industry. This pace is driven by the growth of wireless communications from sensors throughout our everyday environment connecting and transmitting data across ever-expanding wireless networks. As a result, the demands upon computing for artificial intelligence and signal processing are also being aggressively pushed in order to act upon this tremendous growth of information. For the semiconductor industry, connectivity, the Internet of Things and high bandwidth data processing now promise to be the enablers for the next phase of growth. However, the industry has reached a new level of maturity that demands new innovations in computing, connectivity, system integration and ultra-low power applications. One technology doesn't fit all and many differentiated technologies are needed to meet all of these innovations. Although traditional Moore's Law scaling remains important, the complexity of this next inflection point requires a new perspective on silicon scaling. Changing market demands are also driving innovation in differentiated technologies and system-level integration approaches that will be combined with silicon technology scaling.

This presentation will examine the economic trends reshaping the industry and explore the technology directions that can successfully produce the needed innovations. Collaborative innovation will be necessary across many industries and market segments and this presentation will also provide insight into GLOBALFOUNDRIES' vision to enable technology for a new digital era.

Flexible and Printed OTFT Devices for Emerging Electronic Applications

Shizuo Tokito

Research Center for Organic Electronics, Yamagata University, Yonezawa, Japan

tokito@yz.yamagata-u.ac.jp

Abstract

The latest development results for flexible and printed electronics technology based on organic thin-film transistor (OTFT) devices as well as printable semiconductors and metal nanoparticles are briefly reported in this paper. The successful fabrication and operation of printed OTFT devices and potential integrated circuit applications such as flip-flop logic gates and operational amplifiers will be presented. The application of OTFT devices to biosensor applications will be also described.

(Keywords: Organic electronics, Printed electronics, Flexible electronics, Organic thin-film transistor)

Introduction

Smart labels and wearable biosensors are garnering a significant amount of attention in research and development because of immense need to deploy the Internet of Things (IoT), Internet of Everything (IoE) and Trillion Senor Universe frameworks. These applications are envisaged as thin-film transistor (TFT) devices fabricated on thin plastic film substrates with various printing processes. In particular, TFT devices based on organic semiconductors (OSC) can be fabricated at low temperatures and are more compatible with printing methods and low-cost plastic substrates than inorganic semiconductors. Therefore, the printed OTFT devices are expected to have great potential in these applications.

Here, we report briefly on the printable materials and OTFT devices used in integrated circuit and biosensor applications [please see references].

Printable Materials

Silver (Ag) nanoparticle inks have become important materials for the fabrication of electrodes and interconnect layers in printed electronics. Accordingly, we developed Ag nanoparticle ink that was optimized for organic TFT applications. Finely patterned Ag lines with line widths of 25 μm were fabricated by inkjet printing and sintered at temperatures lower than those reported for other Ag nanoparticle inks. A low resistivity of less than 10 μΩ-cm could be obtained in the printed lines by thermal sintering at 120° C or with photonic sintering for 1msec.

We adopted a newly developed p-type OSC material based on a dithienobenzodithiophene derivative (DTBDT), which is soluble in common organic solvents and highly crystalline. In the OTFT device fabrication, the surfaces of the source and drain were treated with self-assembled monolayer (SAM) materials to reduce the contact resistance. A high carrier mobility of 3 cm^2/Vs and a high on/off current ratio of 10^7 were obtained in a typical bottom-contact top-gate OTFT device. A new "TU series" N-type OSC material, based on a benzobisthiadiazole moiety was also developed. A mobility of 2 cm^2/Vs was demonstrated in a top-contact bottom-gate OTFT device.

Printed OTFT Devices

Inkjet printing, nozzle dispensing and reverse-offset printing methods were employed in forming the electrode, bank and OSC layers, resulting in fully-printed OTFT devices. Printed OTFT arrays (30x30) were successfully fabricated employing DTBDT on plastic film substrates. By optimizing the OSC ink formulation, semiconducting layer crystal growth, excellent p-type electrical performance with a high mobility of 1.9 cm^2/Vs and a high on/off current ratio of over 10^7 were achieved. Exceptionally uniform device characteristics were also observed within in the array. Ultra-thin OTFT devices can also be fabricated using ultra-thin Parylene films, resulting in inverter circuits that are extremely lightweight, flexible and compressible.

Integrated Circuits

The important applications of OTFT devices are integrated circuits for RFID devices and in microprocessors. We successfully fabricated pseudo-CMOS inverters using the p-type OTFT devices, as well as NAND logic gates, which exhibited ideal characteristics at low operating voltages, and very high gains over 250 were obtained. True CMOS inverters using both p-type and n-type OSC materials are important for low power operation and in circuit design. Using our n-type OSC material (TU-3) and commonly used p-type OSC material (diF-TES-ADT) we fabricated a CMOS inverter that employed a stacked TFT device construction. Good switching characteristics were observed and a high gain was obtained at an operating voltage of 10 V. Based on this CMOS inverter design, three–stage ring oscillator and D-flip flop circuits were also fabricated on an ultra-thin Parylene film substrates.

In order to realize very short-channel OTFT devices and their integrated circuits we employed the reverse offset printing method. The typical channel length was below 10 μm. The good characteristics were obtained in the printed CMOS inverters.

Biosensors

We are also challenging to a smart biosensor device, a combination of printed integrated circuits and a detection electrode, which can be connected wirelessly to the Internet. We recently succeeded in detecting several biomarkers related to various diseases and types of infections. The basic construction of the biosensor is an organic transistor with an extended gate electrode, upon which receptors are immobilized.

Immunoglobulin A was successfully detected in an aqueous solution by the organic transistor-based biosensor (FET-based biosensor). It is known that nitrate oxide ion (NO^{3-}) in human saliva and is related to mental stress. We tried to detect the NO^{3-} in an aqueous solution and succeeded in the detection with a limit of detection (LOD) = 150ppb. Glucose is a well-known biomarker for diabetes, which was detected in the biosensor employing phenyl borate as the receptor.

More recently, we have succeeded in fabricating a flexible printed biosensor on a plastic film substrate, whereby the organic transistor and receptor as well as a reference electrode were integrated onto a plastic film. Biotylated immunoglobulin G was successfully detected in this single-chip biosensor.

Conclusion

Integrated circuits based on printed OTFT devices were successfully fabricated by optimizing printing conditions and selecting the appropriate printing materials, as well as by improving interface conditions. The performance of the integrated circuits could be further improved by addressing issues such as contact resistance and OSC layer uniformity. Ongoing materials development is also required to improve OTFT device electrical characteristics as well as process compatibility. The practical application of these devices to flexible displays, integrated circuits and wearable biosensors are anticipated in the near future.

Acknowledgments

The author gratefully acknowledges the contributions of D. Kumaki, H. Matsui, Y. Takeda, T. Minami (Tokyo University), T. Minamiki (Tokyo University) and K. Fukuda (RIKEN) to this project, and financial support from the Japan Science and Technology Agency (JST).

References

[1] K. Fukuda, et al., Organic Electronics 13, 1660 (2012).

[2] Y. Takeda, et al., Organic Electronics 14, 3362 (2013).

[3] K. Fukuda, et al., Nature Commun. 5, 243703 (2014).

[4] Y. Yoshimura, et al., Organic Electronics 15, 2696 (2014).

[5] Y. Takeda, et al., Jpn. J. Appl. Phys. 54, 04DK03 (2015).

[6] K. Fukuda, et al., Adv. Electro. Mater. 1, 1400052 (2015).

[7] K. Fukuda, et al., Adv. Electro. Mater. 1, 1500145 (2015).

[8] M. Mamada, et al., Chem. Mater. 27, 141 (2015).

[9] M. Mizukami, et al., IEEE Electron. Dev. Lett. 36, 841 (2015).

[10] Y. Takeda, et al., Scientific Reports 6, 25714 (2016).

[11] R. Shiwaku, et al., Scientific Reports 6, 34723 (2016).

[12] J. Kwon, et al., ACS Nano, DOI :10.1021 /acsnano.6b06041 (2016).

[13] T. Minamiki, et al., Materials 7, 6843 (2014).

[14] T. Minami, et al., Chemical Communications 50, 15613 (2014).

[15] T. Minamiki, et al., Analytical Sciences 31, 725 (2015).

[16] T. Minami, et al., Biosens. Bioelectron. 81, 87 (2016).

PL-3

Advanced Heterogeneous Integration Technology Trend for Cloud and Edge

Douglas C.H. Yu

Taiwan Semiconductor Manufacturing Company R&D, 168 Park Ave. 2, Science Based Industrial Park, Hsinchu, Taiwan, R.O.C.

Abstract

Advanced heterogeneous integration (HI) technology is much needed for applications from edge to cloud to meet the stringent system-level requirements on performance, power, profile, cycle-time and cost (P3C2). In addition to 3DIC with TSV, innovative packaging technologies such as silicon interposer (2.5D) and fan-out wafer-level-packaging (2D/3D) become new paradigm for the semiconductor industry to realize the system integration. In this paper, we will discuss the new trend of advanced packaging technology- a strong need for application-specific integration solutions. Many of those are proposed. The solution with higher performance at lower cost will prevail. Furthermore, the solutions that readily integrate multi-chip to enable chip-partition to extend Moore's Law effectively have long-term advantages.

Application-specific integration technologies

Ubiquitous computing from ultrahigh performance clouds such as data center all the way down to high performance power efficient edge devices such as hand-helds, mobiles, and wearable, are transforming the way we create, access, and live-by information. Besides the well advocated Moore's Law, which drives for transistor shrinkage with performance/power leaps, an un-sung hero is system integration and packaging that gradually taking the stage as driving Moore's Law alone becoming costly. 3DIC had emerged as a potential solution for high performance computing (HPC) but suffered from heat dissipation bottleneck. Silicon interposer with heterogeneous integration and better thermal dissipation is now becoming the favorable HPC integration platform. On the edge computing side, hand-held devices are driving for thin form-factor, low power, and low cost system solutions. When flip-chip package starts to show bandwidth and density limitations, multi-chip 2D/3D fan-out platforms, such as InFO, is now taking center stage for edge device heterogeneous system integration. The needs for application specific solutions are trending toward using different technology platforms, which are shown in Fig 1 for cloud and edge. Interposer and fan-out technologies have been introduced [1] and delivered to production for cloud and edge applications, respectively, with optimized P3C2 (Fig 2.)

CoWoS and Ultra-Large Si Interposer

CoWoS has been the first silicon interposer delivered for cloud and HPC system integration. This technology is moving toward higher performance, reaching an interposer size that is 1.75 times of scanner reticle size. It enables split-dies and chip-partition to increase function and reduce cost, as shown in Fig 3. Industry also proposed several other interposer-like technologies. EMIB uses miniaturized silicon bridges, which are embedded in organic substrates, as shown in Fig 4, for high density interconnection [3]. TSV was eliminated to help reduce its cost. Other proposals targeting low cost include integrated- Thin film High density Organic Package (iTHOP) [4], which is also known as 2.1D organic interposer as shown in Fig 5, and Silicon-Less Interconnect Technology (SLIT and SLIM) which using silicon based RDLs but with silicon removed [5, 6], Fig 6. InFO on substrate (InFO_oS) for with high density RDL enables heterogeneous integration of chip partition/split-dies without silicon interposer [7], Fig 7. These developments showed the trend for requirements of higher performance and lower cost for HPC system integration. In general, the yield and reliability are the main challenges associated with those newly proposed HI technologies.

Integrated Fan-Out (InFO) and Package-on-Package (PoP)

InFO_PoP is the first proposed and the first delivered 2D/3D high density, high performance FOWLP heterogeneous integration technology for mobile applications [1,2]. It is shown to have better P3C2 optimization than flip-chip and 3DIC/TSV. One of the InFO's cost advantages is using die face-up RDL-last process flow, Fig 8. It is not only lower in process complexity and cost than RDL-first approach, but also with significant performance advantage [8]. Fan-out technology is proposed more recently to be processed in larger size panel form, Fig 9, for further cost reduction. However, larger panel size represents more challenging substrate warpage and uniformity issues. Its feasibility for high density, high performance applications remain to be proven. The feasibility includes process tool, integration flow and the risks involved in relatively higher investment as compared with wafer form technology which has been well proven [8].

Multi-chips Integration

SoC (system on chip) scaling is facing stronger challenge every day especially on production cost. Chip-splitting or partitioning can reduce chip size and, therefore, its cost. The integration solution that readily integrates multiple chips to reduce total cost without compromising performance can provide critical supports to extend Moore's Law, Fig 10. This remains as an important heterogeneous integration trend.

Summary

Advanced system packaging technology has developed towards application-specific solutions for cloud and edge computing. Industry continues it attempt to drive for lower cost and higher performance HI technology. This SiP driven solution will enable chip-partition to work hand-in-hand with Moore's Law effectively.

References

[1] D. Yu, 2011 Semicon Taiwan, 3D-IC Technology Forum. [2] D. Yu, 2012 iMAP and Device Package Conference, Scottsdale, Az [3] R. Mahajan, ECTC 2016. [4] Shinko Watanabe, i-NEMI 2014. Cisco Li, ECTC 2016 [5] Kwon, IMAPS 2014. [6] R. Huemoeller, Amkor, 2015 [7]D. Yu, Semicon Taiwan 2014 & Tseng, ECTC 2016 [8] D. Yu, IEDM 2014, VLSI Symposium 2015 and iMAPS Pasadena 2016.

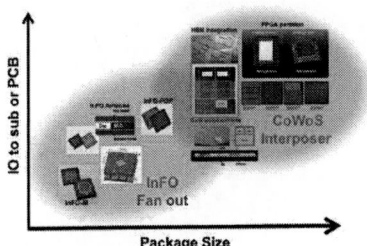

Fig. 1 Application Specific Solutions for Heterogeneous Integration. CoWoS for HPC and InFO for edges and mobiles.

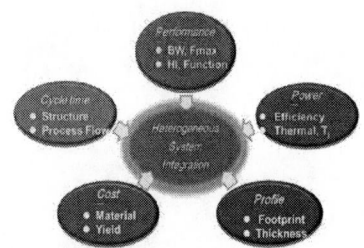

Fig. 2 System key performance indexes: electrical performance, consumption and dissipation of power, package profile, design and manufacturing cycle time, and cost.

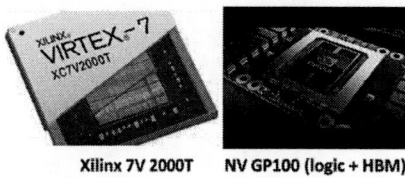

Fig. 3 CoWoS silicon interposer enables logic die splitting, chip partitioning and heterogeneous integration. [e-element.com and techtimes.com]

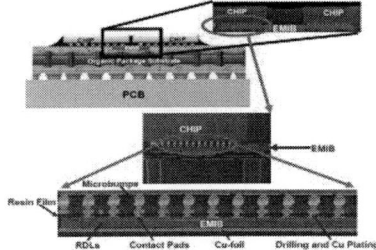

Fig. 4 EMIB using miniature silicon bridges to achieve high density interconnection without TSV [Ravi Mahajan et al., IEEE/ECTC2016].

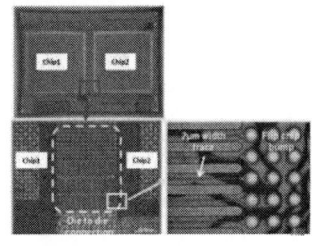

Fig. 5 iTHOP enables ASIC and the HBM assembled to the organic interposer with the u-pillar inter-connect and Chip-2-Chip joining [Shinko, IEEE/ECTC2014].

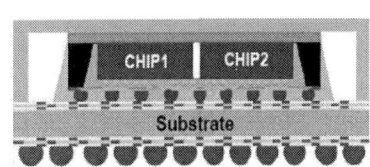

Fig 6.. Non TSV interposer (NTI), with interposer Si removed and only RDLs remained [Fan-Yu Liang et al., IEEE/ECTC2016].

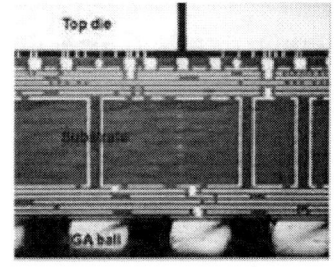

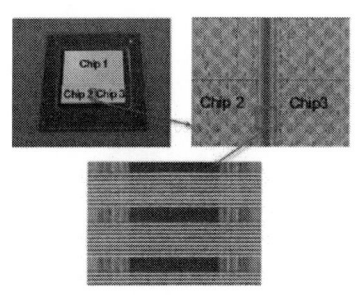

Fig. 7. InFO_oS multi-chips integration with RDL min. L/S = 2/2 um enables chip partition and/or splitting dies heterogeneous integration [7].

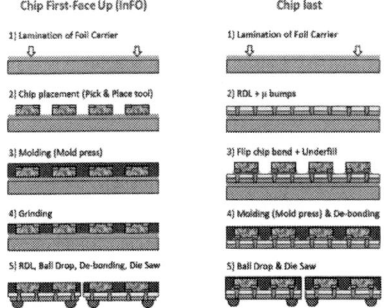

Fig. 8 InFO chip-first flow removes flip-chip bonding, underfill and achieve better P3C2 optimization.

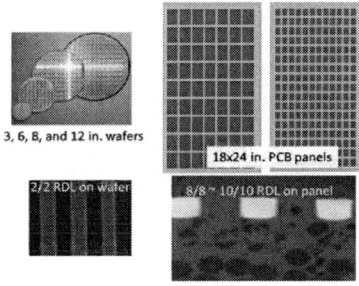

Fig. 9 Fan-out Panel Level Packaging aims for lower cost with larger panel substrate. It is more suitable for lower pin-count and smaller size packages. Its feasibility for higher performance, higher density, and multi-layer RDL remains to be shown.

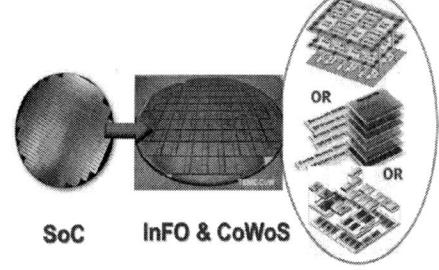

Fig.10 2D/3D heterogeneous integration technologies that integrate multi-chips cost-effectively with less performance degradation can provide more values. Those technologies can increase system function, reduce system cost, and support Moore's Law extension.

978-1-5090-4661-4/17 $31.00 © 2017 IEEE

Gap in pagination due to unavailable paper.

Pages 6-7

This page intentionally left blank.

3M-2 (Invited)

Neuromorphic Technologies for Next-Generation Cognitive Computing

Robert M. Shelby[1], Pritish Narayanan[1], Stefano Ambrogio[1], Hsinyu Tsai[1],

Kohji Hosokawa[2], Scott C. Lewis[3], and Geoffrey W. Burr[1]

[1]IBM Research–Almaden, San Jose, CA USA, *gwburr@us.ibm.com*

[2]IBM Tokyo Research Laboratory, Japan   [3]IBM T. J. Watson Research Center, Yorktown Heights, NY USA

Abstract

We describe IBM's roadmap for Neuromorphic Technologies to drive next-generation cognitive computing, ranging from nanodevice-based hardware for accelerating well-known supervised-learning algorithms (which happen to rely on static, labeled data), to emerging, biologically-inspired algorithms capable of learning from temporal, unlabeled data. The various hardware-centric neuromorphic projects currently underway at IBM Research will be surveyed, with a focus on the use of Non-Volatile Memory (NVM) for on-chip acceleration of the training of Deep Neural Networks (DNNs). (Keywords: Cognitive Computing, Non-von Neumann Computing)

Introduction

The extreme flexibility of digital circuits has allowed modern processors based on the von Neumann architecture to not only efficiently implement algorithms for a wide variety of problems, but to consistently improve system performance at an exponential rate. However, with continued device scaling constrained by power- and voltage-considerations, the time and energy spent transporting data between memory and processor (across the so-called "von-Neumann bottleneck") has become problematic for data-centric applications such as real-time image recognition and natural language processing.

Non-von Neumann Computing

One example of Non-von Neumann computing is the human brain, which can outperform modern processors on many tasks involving unstructured data classification and pattern recognition. Deep neural networks accelerated by VN-based GPUs perform computations in a similarly parallel fashion. IBM's TrueNorth chip is a flexible and modular non-VN tool for implementing forward inference of such DNNs at ultra-low power [1]. Synaptic weights are trained off-line and transferred into SRAM.

One path for extending such non-VN systems towards full on-chip learning – and thus to potentially accelerate DNN training at lower power than GPUs – is to replace the reliable but binary SRAM memory cells used in TrueNorth with dense and analog (but less reliable) NVM. By performing computation at the location of data, such on-chip training of largescale DNN using NVM-based synapses [2-8] could potentially provide significant power and speed benefits on this specific yet very important task. Such an implementation can realize the multiply-accumulate (MAC) operations at the heart of most neural network algorithms extremely efficiently, using physics – Ohm's law followed by current summation (Kirchoff's current law) – for locally analog computation at the location of the weight data.

Energy-Efficient Computing through Sparsity

Spiking Neural Networks, in which synapses are modified based on the timing of sparse upstream and downstream neuronal spikes [9-12], offer the potential for ultra-low energy computing through spatially- and temporally-sparse computing. However, the local learning rules observed in biology will need to be harnessed by a global learning algorithm that can exhibit scalability (bigger networks with more data offer more performance) and robust convergence.

Finally, brain-inspired learning algorithms such as Hierarchical Temporal Memory (HTM) [13] or Context-Aware Learning, based on the Sparse Distributed Representations observed in the human cortex, offer a potential path to Machine Intelligence through continuous learning of temporal sequences.

References

[1] P. A. Merolla, J. V. Arthur, et al., *Science*, **345**(6197), 668-673 (2014).

[2] G. W. Burr, R. M. Shelby, et al, *IEDM Technical Digest*, 2014, T29.5; G. W. Burr, R. M. Shelby, et al., *IEEE Trans. Electr. Dev.*, **62**(11), 3498–3507, 2015.

[3] G. W. Burr, P. Narayanan, et al., *IEDM Technical Digest*, 2015, T4.4.

[4] J.-W. Jang, S. Park, et al., *IEEE Electron Device Letters*, **36**(5), 457–459, 2015.

[5] S. Sidler, I. Boybat, et al., *ESSDERC*, 2016.

[6] A. Fumarola, S. Sidler, et al., *ICRC*, 2016.

[7] P. Narayanan, L. L. Sanches, et al., submitted to *ISCAS 2017*.

[8] P. Narayanan, A. Fumarola, et al., *IBM J. Res. Dev.*, to appear, 2017.

[9] G. Indiveri, B. Linares-Barranco, et al., *Frontiers in Neuroscience*, **5**(73), 2011.

[10] S. Kim, M. Ishii, et al., *IEDM Technical Digest*,

978-1-5090-4661-4/17 $31.00 © 2017 IEEE

T17.1, 2015.

[11] R. A. Nawrocki, R. M. Voyles, et al., *IEEE Trans. Electr. Dev.*, **63**(10), 3819-3829, 2016.

[12] G. W. Burr, R. M. Shelby et al., *Adv. Phys. X*, to appear, 2017.

[13] J. Hawkins and S. Blakeslee. *On intelligence.* Macmillan, 2007.

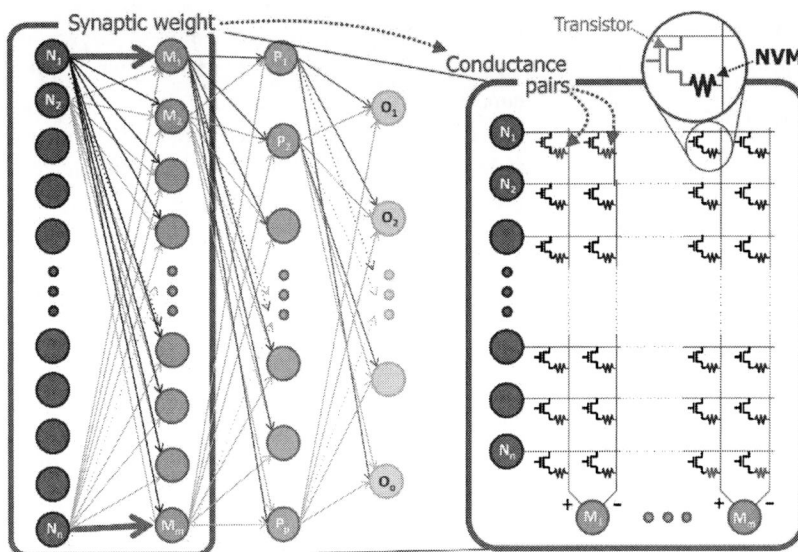

Fig. 1– Neuro-inspired non-von Neumann computing [1-9], in which neurons activate each other through dense networks of programmable synaptic weights, can be implemented using dense crossbar arrays of nonvolatile memory (NVM) and transistor device-pairs.

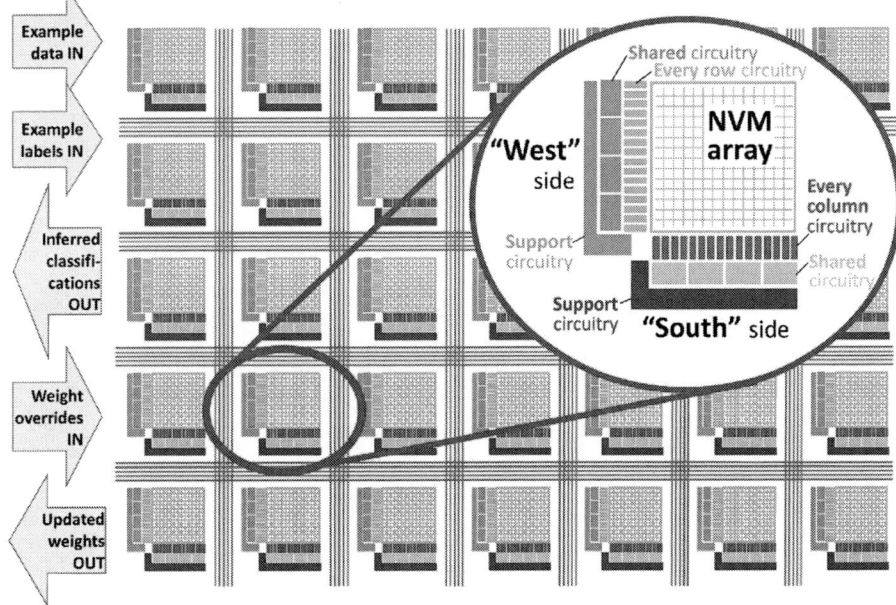

Fig. 2– Proposed chip architecture for the acceleration of neural network training with analog non-volatile memory. A flexible routing network has three tasks: 1) to convey chip inputs (such as example data, example labels and weight overrides) from the edge of the chip to the device arrays, 2) to carry chip outputs (such as inferred classifications and updated weights) from the arrays to the edge of the chip, and 3) to interconnect the various arrays in order to implement multi-layer neural networks. Each NVM array has input neurons (here shown on the "West" side of each array) and output neurons ("South" side), connected with a dense grid of synaptic connections. Peripheral circuitry is divided into circuitry assigned to individual rows and columns, circuitry shared between a number of neighboring rows and columns, and support circuitry [8].

978-1-5090-4661-4/17 $31.00 © 2017 IEEE

3M-3 (Invited)

Stateful logic circuit and material using memristors

Nuo Xu[1,2], Xinglong Shao[1], Kyung Jean Yoon,[1] Hae Jin Kim[1], Kyung Min Kim[3], and Cheol Seong Hwang[1,*]

[1]Department of Materials Science and Engineering, and Inter-University Semiconductor Research Center, Seoul National University, Seoul 139-743, South Korea, cheolsh@snu.ac.kr;

[2]State Key Laboratory of High Performance Computing, National University of Defense Technology, Changsha, Hunan 410073, China;

[3]Hewlett Packard Labs, Hewlett Packard Enterprise, Palo Alto, California 94304, USA

Abstract

By combining the functionalities of Boolean gates and non-volatile memory, stateful logic may enable significant savings in time and energy for computational processes where available power source is limited. In this talk, fundamental principles of stateful logic will be described first, and circuit level implementation of it using recently explored bi-functional memristor (coexistence of unipolar and bipolar switching), and conventional bipolar and complementary resistance switching devices are presented.

(Keywords: Stateful logic, memristor, bi-functional)

Introduction

In conventional von Neumann-based computer systems, information is stored in memory units and processed in central processing unit (CPU). This configuration requires huge data transition between memory and CPU, which limits the computation speed and energy efficiency (von Neumann bottleneck) [1]. To breakthrough this bottleneck, it is necessary to find new materials, new devices and new computing paradigms to increase the capability and capacity of future information technology [2]. Stateful logic is an attractive solution to solve this problem because of its inherent format of logic in memory. Borghetti et al. demonstrates a simple practice of this concept in 2010 via a material implication logic gate using two parallel memristors and a conditional write operation [3]. In this work, new sub-circuit structures based on various resistance switching devices is demonstrated to achieve a full functionality of stateful logic.

Results and Discussions

This work introduces three different types of stateful logic circuits which are based on (1) recently found bi-functional memeristor [4], where the unipolar and bipolar resistance switching (URS and BRS) functionalities are embedded in one device (TiN/ TiO$_2$/Al); (2) a complementary resistive switching device (CRS) and a BRS device; and (3) anti-parallel connected two BRS cells. Figure 1 shows the example structure of the circuit composed of CRS and BRS. In this logic concept, logic inputs and outputs are all represented by the resistance states of the two devices. Through combining a series resistor, the resistance states of the two devices can be conditionally written based on the specially designed voltage pulses, which enabled the full functionality of stateful logic in either CRS device or BRS device for the case of (2) and any of the two BRS cells for the case of (3). The sixteen Boolean logic functions can be reconfigured in a same sub-circuit in time dimension simply by combining corresponding series resistor and specially voltage pulses in the case of (2) (Fig.2). Furthermore, logic cascading is flexible to achieve in this suggested devices owing to the spatial reconfiguration characteristic of sub-circuit (Fig. 2d). With this spatial reconfiguration characteristic, the sixteen Boolean logic functions can also be achieved in the case of (3). Circuit configuration with (1) is simpler because only one device can be used to implement all the logic functionalities, but it also encompasses complications in writing and reading processes, which can be accomplished by a sequence of voltage pulses and different sense amplifiers. The suggested sub-circuit concept could contribute to the development of new generation of stateful logic computing.

Conclusion

Three circuit configurations were suggested to accomplish the stateful logic. Compared with the original suggestion by Borghetti et al. [3], all the sixteen logic operations were probable, and the logic cascading could also be demonstrated.

Acknowledgments

This work was supported by National Research Foundation of Korea through the Global Research Laboratory Program (2012K1A1A2040157) and Doyak program (2014R1A2A1A10052979).

References

[1] S. Borkar and A. A. Chien, "The future of microprocessors," Commun. ACM, vol. 54, no. 5, p. 67, May 2011.

[2] D. S. Jeong, K. M. Kim, S. Kim, B. J. Choi, and C. S. Hwang, "Memristors for Energy-Efficient New Computing Paradigms," Adv. Electron. Mater., pp. 1–27, 2016.

[3] J. Borghetti, G. S. Snider, P. J. Kuekes, J. J. Yang, D. R. Stewart, and R. S. Williams, "'Memristive'

978-1-5090-4661-4/17 $31.00 © 2017 IEEE

switches enable 'stateful' logic operations via material implication.," Nature, vol. 464, no. 7290, pp. 873–876, 2010.

[4] X. L. Shao, K. M. Kim, K. J. Yoon, S. J. Song, and J. H. Yoon, "A study of the transition between the non-polar and bipolar resistance switching mechanisms in the TiN/TiO2/Al memory," Nanoscale, p. DOI: 10.1039/c6nr02800d, 2016.

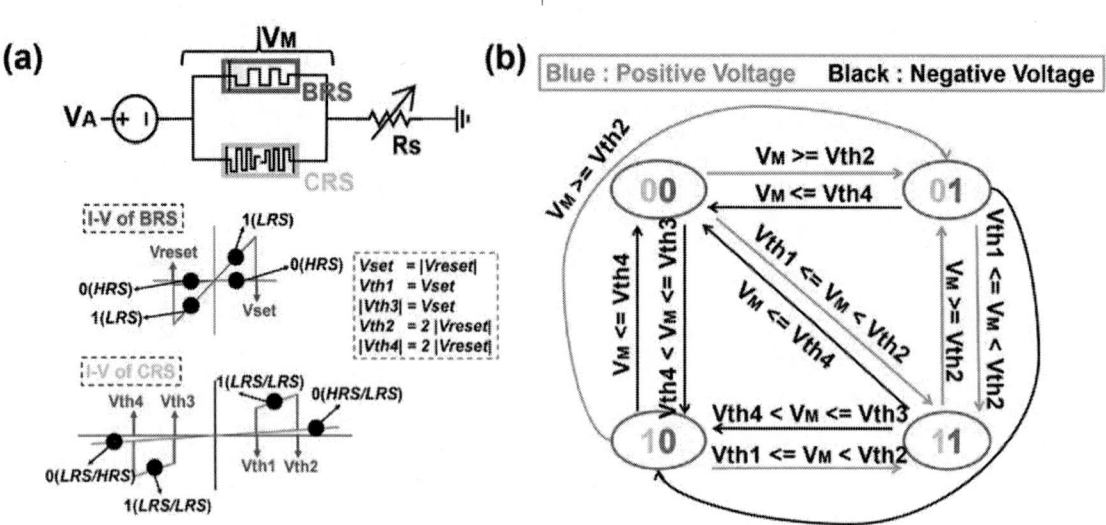

Fig 1. Basic principle of the logic concept using CRS and BRS cells. (a) The suggested sub-circuit structure and the schematic current – voltage (I-V) curves of BRS. (b) State transfer diagram of the two devices based on the divided voltage V_M.

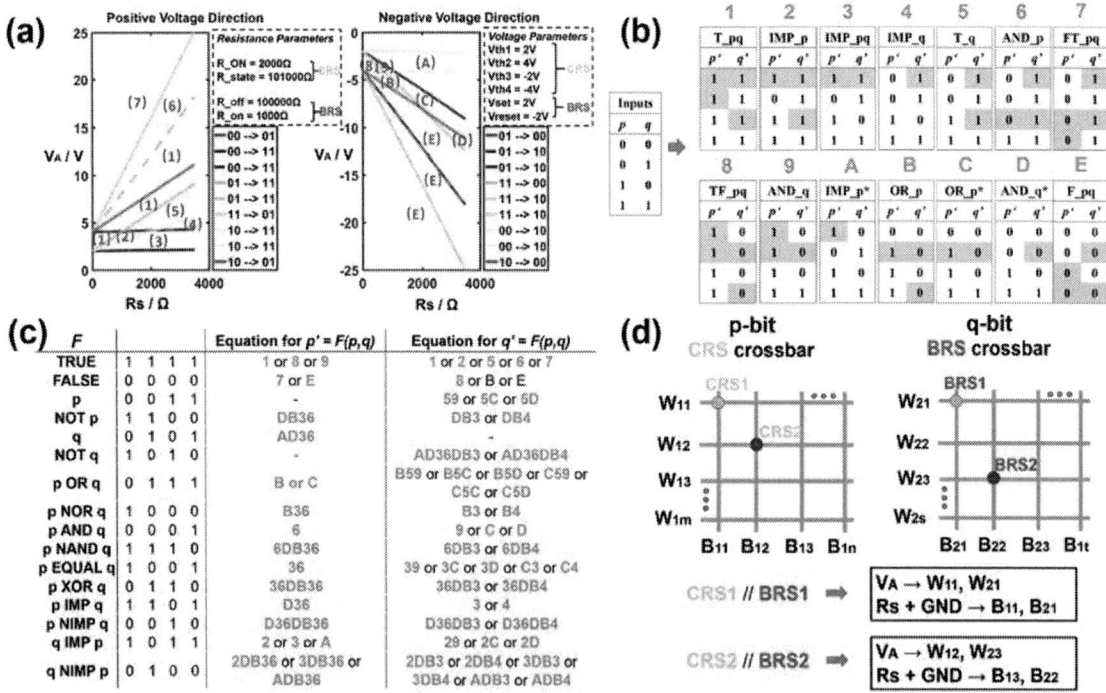

Fig 2. Method of achieving the possible unit logic operations and complete Boolean logic in the suggested CRS+BRS sub-circuit. (a) Switching voltage boundaries for twelve single-step transitions as a function of series resistance (R_S). Each section corresponds to a different unit logic operation. (b) The truth tables of the fourteen unit operations. (c) Sixteen Boolean logic equations for p (CRS bit) and q (BRS bit) by combining the fourteen unit operations. (d) A schematic of configuring sub-circuit in two crossbar arrays.

978-1-5090-4661-4/17 $31.00 © 2017 IEEE   12

3M-4 (Invited)

New-paradigm CMOS Ising Computing for Combinatorial Optimization Problems

Masanao Yamaoka

Hitachi, Ltd., Tokyo, Japan, masanao.yamaoka.ns@hitachi.com

Abstract

A new computing using Ising model that effectively solves combinatorial optimization problems is proposed. The computing maps problems to an Ising model, a model to express the behavior of magnetic spins, and solves the problems by its own convergence property. We fabricated a prototype computing chip and confirmed the power efficiency of the chip is 1800-times higher than that of the conventional von-Neumann computers.
(Keywords: Ising model, combinatorial optimization problem, SRAM, non von-Neumann computing)

Introduction

Today, the performance growth of von-Neumann computing will slow down due to the end of semiconductor scaling. Natural computing, which maps problems to natural phenomena and solves the problems by its own convergence property, is proposed (Fig. 1). The natural computing does not solve problems step by step, and therefore, it does not expect the operation speed acquired by the semiconductor scaling. The quantum computer [1] using superconductor is one of those computers and the neuron chip [2] is also one of them. We proposed a CMOS Ising computing [3][4] to solve combinatorial optimization problems.

CMOS Ising computing

The Ising computing maps the combinatorial optimization problems to Ising model, a model to express the behavior of magnetic spins (Fig. 2), and solves the problems by ground-state search operations with its convergence property. The energy of the system is expressed as the formula besides the diagram. Here, σ_i is spin status, J_{ij} is interaction coefficient, and h_i is external magnetic field coefficient. In the CMOS Ising computing, the operation of the ground-state search of Ising model is mimicked by CMOS circuits as indicated in Fig. 3. Here, the spin values of +1 or -1 are stored in SRAM cells as "0" or "1". The interactions between spins are realized by memory cells and digital circuits, and are performed in parallel, and the necessary steps for the ground-state search are smaller than that using the conventional sequential computing, von-Neumann architecture.

CMOS Annealing

For the ground-state search, we implement a CMOS annealing technique in CMOS Ising computer. In the CMOS annealing, the energy of Ising model is reduced along the line of energy landscape of the Ising model (solid arrows in Fig. 4).

It is achieved by the digital circuits in the CMOS Ising computer. Only using the digital circuits, the state is trapped in the local minimum point of the energy landscape. To avoid trapping to local minimum points, we added random change of the state (broken arrows in Fig. 4). To change the state, we proposed two methods, random number injection and using the variation of SRAM cells. In SRAM circuits, application of lower supply voltage to the memory cells causes error bits. These error bits are used to change the energy states randomly. Fig. 5 shows the measurement results of SRAM array. At the 0.7-V VDDM, about 30% cells have errors.

Prototype of Ising Computing

We fabricated prototype Ising chips with 65-nm process as shown in Fig. 6. In the Ising chip, 20 1k-spin sub-arrays are placed and 20k spins are embedded. Fig. 7 shows the energy transition when the CMOS annealing with random number injection is operated. As time goes, the energy is lowered. Fig. 8 shows a transition of spin status. Black and white points show spin value "+1" and "-1". In this problem, at the global minimum, three characters, "ABC", is clearly appeared. After 10 ms, 1,000,000 interaction steps, "ABC" is clearly appeared. Fig. 9 shows the transition of spin status when using SRAM variation to change spin states. The picture (a) is local minimum and the "ABC" pattern is appeared with noises. Then SRAM variation is used to randomly destroy the spin status and "ABC" pattern is almost diminished as shown in the picture (b). Some interaction and destruction operations are activated, and a better solution is acquired as shown in picture (e). The results show the SRAM variation contributes to avoid local minimum sticking. Fig. 10 shows the power efficiency compared to a conventional method, an approximation algorithm "SG3" run on a CPU when solving randomly generated MAX-cut problems. When the problem size is 20k spins, the energy efficiency is 1800 times higher than that of the conventional method.

Conclusion

It is confirmed by prototype chip that the CMOS Ising computing can solve combinatorial optimization problems. The power efficiency is 1800-times higher than the conventional computing architectures.

References

[1]   W. Johnson et al., Nature 473, pp. 194–198, May 2011.
[2]   R. F. Service, Science, Vol. 345, Issue 6197, August 2014.
[3]   C. Yoshimura et al., 21st ECCT D, September 2013.
[4]   M. Yamaoka et al., ISSCC 2015, pp. 432-433, Feb., 2015.

978-1-5090-4661-4/17 $31.00 © 2017 IEEE

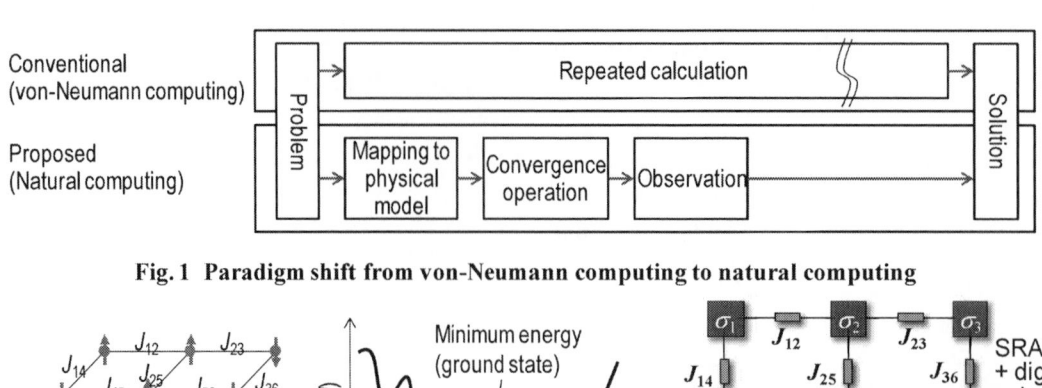

Fig. 1 Paradigm shift from von-Neumann computing to natural computing

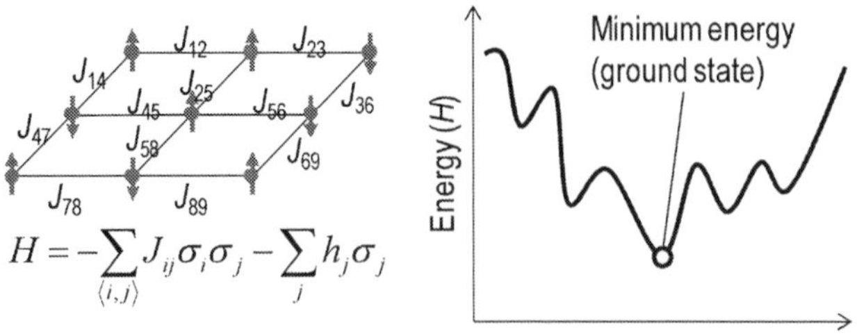

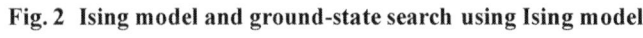

$$H = -\sum_{\langle i,j \rangle} J_{ij}\sigma_i\sigma_j - \sum_j h_j\sigma_j$$

Fig. 2 Ising model and ground-state search using Ising model

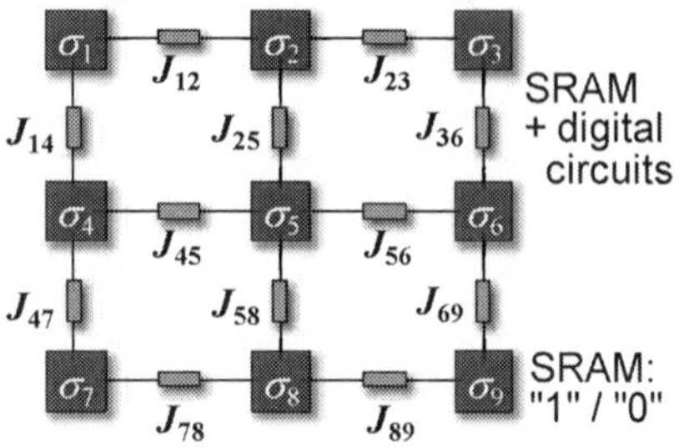

Fig. 3 CMOS Ising computer

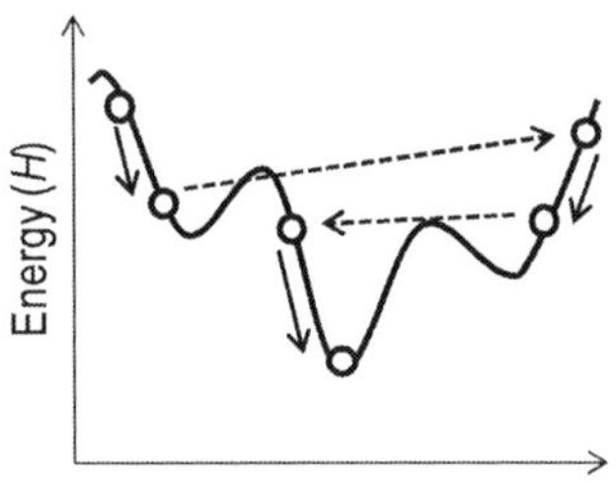

Fig. 4 CMOS annealing

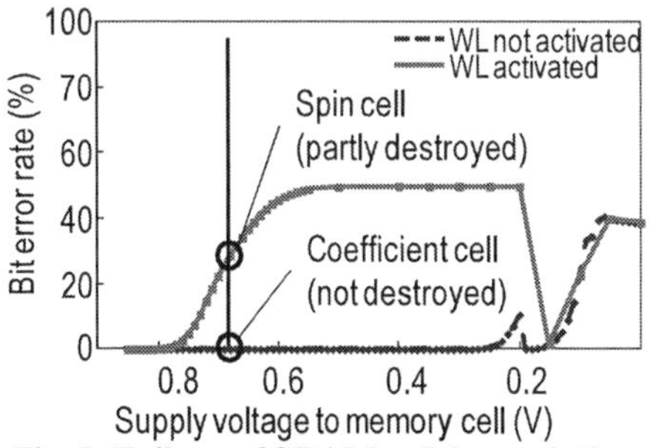

Fig. 5 Failure of SRAM cell by variation

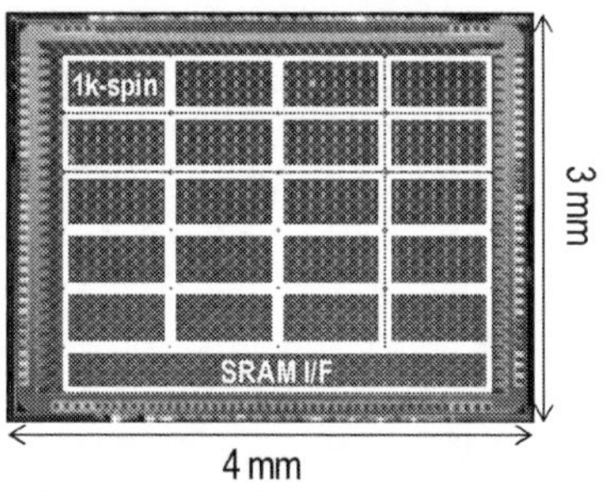

Fig. 6 Ising chip photograph

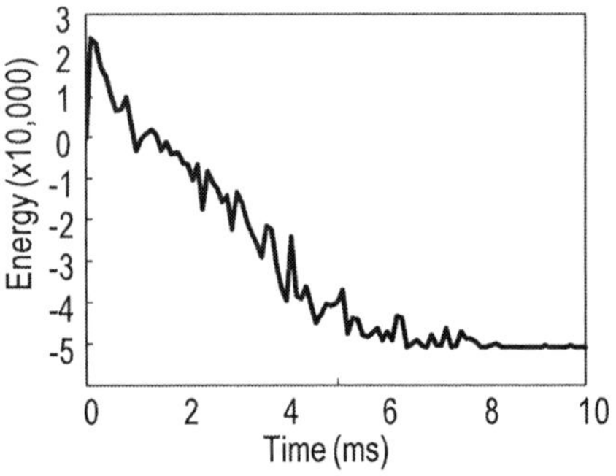

Fig. 7 Energy transition

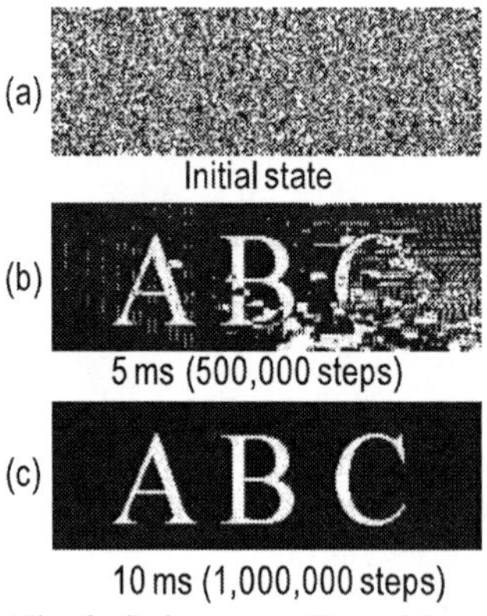

(a) Initial state

(b) 5 ms (500,000 steps)

(c) 10 ms (1,000,000 steps)

Fig. 8 Spin status Transition

(a)

(b)

(c)

(d)

(e)

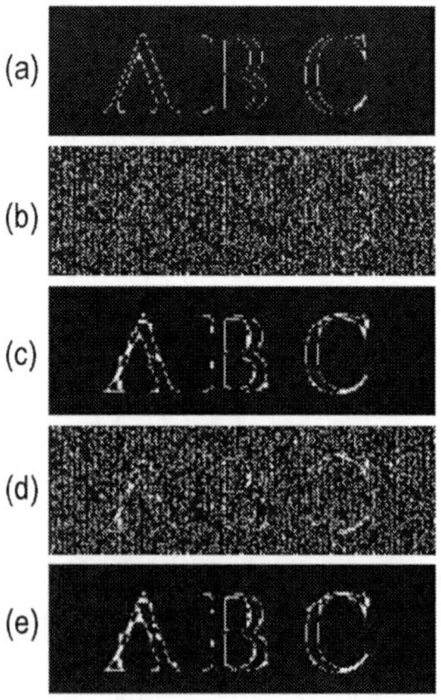

Fig. 9 Spin status transition with SRAM variation

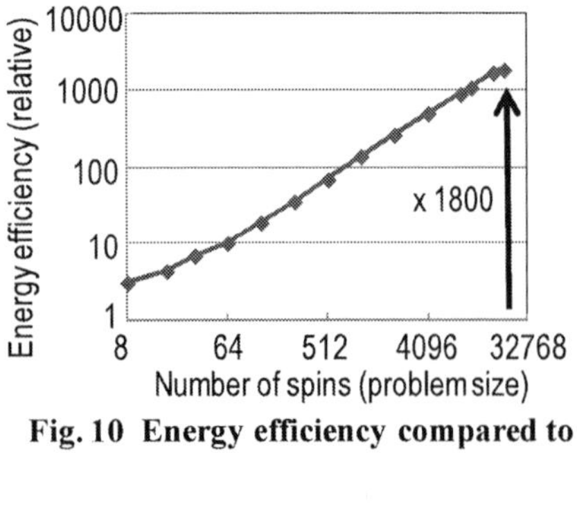

x 1800

Fig. 10 Energy efficiency compared to an approximation algorithm on CPU

978-1-5090-4661-4/17 $31.00 © 2017 IEEE

3M-5 (Invited)

Understanding the Limit and Potential in Emerging Perovskite Solar Cells

Wolfgang Tress and Michael Graetzel

Laboratory of Photonics and Interfaces, Swiss Federal Institute of Technology (EPFL), Lausanne, CH-1015, Switzerland, wolfgang.tress@epfl.ch, michael.graetzel@epfl.ch

Abstract

In this contribution we give an overview of the recent development of metal-halide perovskite solar cells. We focus on the mixed electronic-ionic conductivity of this material, which causes hysteresis in the current-voltage curve. We discuss latest results on recombination obtained by electroluminescence measurements. We report on high luminescence yields (1%) and open-circuit voltages larger than 1.2 V, which are exceptional for a material with a bandgap of 1.6 eV and yield to efficiencies >21%. We identify the efficiency limiting processes and propose measures to overcome those.

(Keywords: photovoltaics, recombination)

Introduction

Perovskite photovoltaics is an emerging technology with the potential of combining low fabrication costs with high performance. The first $CH_3NH_3PbI_3$ perovskite solar cell reported in 2009 showed a power-conversion efficiency of almost 4% with an open-circuit voltage of 0.61 V [1]. Now, seven years later, efficiencies exceed 22% [2] and open-circuit voltages (V_{oc}) reach 1.24 V [3] for a material with a bandgap of approximately 1.6 eV. This V_{oc} value is remarkable in particular considering that the perovskite films are fabricated by solution processing, which is a method that comes along with various challenges such as inhomogeneous films, remnants of solvents, impurities in precursors, and contaminations introduced during fabrication under non-cleanroom conditions. A key factor for the progress was a better control of the film morphology by engineering deposition techniques and precursor compositions [4]–[8]. In particular, mixed solvents and various precursor components allowed for a preparation of novel multiple-cation mixed-halide perovskite films with excellent optoelectronic properties.

On the other hand, perovskite solar cells show hysteresis in their current-voltage curve measured at a certain voltage sweep rate [9]. Coinciding with a slow transient current response, the hysteresis is attributed to a slow voltage-driven (ionic) charge redistribution in the perovskite solar cell. Thus, the electric field profile and in turn the electron/hole collection efficiency become dependent on the biasing history.

In this contribution underlying processes for both, charge carrier recombination and the hysteresis are discussed.

Results and Discussion

A. Recombination

The results of electroluminescence measurements are shown in Fig. 1. The current-voltage curve (blue) shows an ideality factor of 2 indicative of Shockley-Read-Hall recombination, whereas the emitted photon flux (red) with an ideality factor of 1 results from band-to-band recombination of electrons and holes. The external luminescence quantum efficiency (green) increases with driving voltage reaching 0.5% for that device. Recently, we achieved values larger than 1% [3].

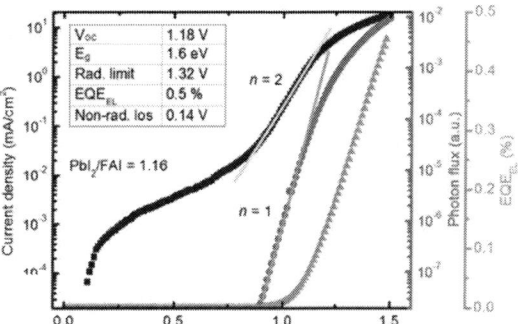

Figure 1. Luminescence of perovskite solar cell. Reprinted with permission from AAAS. From [7].

B. Hysteresis

Hysteresis in the current-voltage (JV) curve means that the measured results depend on the direction and scan rate [9] (Fig. 2). Therefore, extracting the efficiency from the JV curve is complicated. We find that hysteresis results from slowly moving mobile ionic charge in the perovskite that redistributes upon applied voltage. Therefore, the ratio between charge extraction and recombination becomes a function of the biasing history [10]. For the normal hysteresis (Fig. 2, left), fill factor and open-circuit voltage are larger for a backward scan because the positive prebias removes the driving force for charge to pile up at the electrodes, which screen the electric field.

Hysteresis can also be inverted (Fig. 2 right) indicating that the piled-up charge can also be beneficial. It increases the probability for electron extraction in the case of extraction barriers due to an enhanced electric field allowing for tunneling or dipole formation at the perovskite/electrode interface. In that case, we observe an inverted hysteresis, resulting in higher performance metrics

978-1-5090-4661-4/17 $31.00 © 2017 IEEE

for a voltage sweep starting at low prebias [11]. Transient measurements of V_{oc} under varied prebias indicate that mobile ionic charges might act as electronically active traps as well.

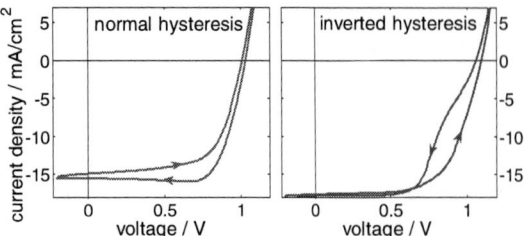

Figure 2. Normal and inverted hysteresis in the *JV* curve of perovskite solar cell (10 mV/s). From [11]

Conclusion

We have shown that *JV*-hysteresis in perovskite solar cells results from a complex interplay between charge transport in the perovskite layer, charge extraction at the contacts, and a slow process changing the electric field in the perovskite and at interfaces to the electrodes. The slow process, most likely ion migration, changing dipole charge at interfaces, is the real origin of the hysteresis. Whether and in which way (normal or inverted) the hysteresis becomes apparent is related to the relevance of the electric field on charge collection. The lower charge carrier mobilities and extraction rates are, the more prone is the device to exhibiting hysteresis.

Recombination in perovskite solar cells is slow due to low defect densities and cross sections allowing for high V_{oc} and luminescence yields. Further research is required to identify the nature of the remaining defects and the role of surface recombination.

References

[1] A. Kojima, K. Teshima, Y. Shirai, and T. Miyasaka, "Organometal Halide Perovskites as Visible-Light Sensitizers for Photovoltaic Cells," *J. Am. Chem. Soc.*, vol. 131, no. 17, pp. 6050–6051, May 2009.

[2] "NREL Chart." [Online]. Available: http://www.nrel.gov/ncpv/images/efficiency_chart.jpg. [Accessed: 23-Dec-2016].

[3] M. Saliba *et al.*, "Incorporation of rubidium cations into perovskite solar cells improves photovoltaic performance," *Science*, vol. 354, no. 6309, pp. 206–209, Oct. 2016.

[4] J. Burschka *et al.*, "Sequential deposition as a route to high-performance perovskite-sensitized solar cells," *Nature*, vol. 499, no. 7458, pp. 316–319, Jul. 2013.

[5] N. J. Jeon, J. H. Noh, Y. C. Kim, W. S. Yang, S. Ryu, and S. I. Seok, "Solvent engineering for high-performance inorganic–organic hybrid perovskite solar cells," *Nat. Mater.*, vol. 13, no. 9, pp. 897–903, Sep. 2014.

[6] N. J. Jeon *et al.*, "Compositional engineering of perovskite materials for high-performance solar cells," *Nature*, vol. 517, no. 7535, pp. 476–480, Jan. 2015.

[7] D. Bi *et al.*, "Efficient luminescent solar cells based on tailored mixed-cation perovskites," *Sci. Adv.*, vol. 2, no. 1, p. e1501170, Jan. 2016.

[8] M. Saliba *et al.*, "Cesium-containing triple cation perovskite solar cells: improved stability, reproducibility and high efficiency," *Energy Environ. Sci.*, vol. 9, no. 6, pp. 1989–1997, Jun. 2016.

[9] H. J. Snaith *et al.*, "Anomalous Hysteresis in Perovskite Solar Cells," *J. Phys. Chem. Lett.*, vol. 5, no. 9, pp. 1511–1515, May 2014.

[10] W. Tress, N. Marinova, T. Moehl, S. M. Zakeeruddin, M. K. Nazeeruddin, and M. Grätzel, "Understanding the rate-dependent J–V hysteresis, slow time component, and aging in CH3NH3PbI3 perovskite solar cells: the role of a compensated electric field," *Energy Environ. Sci.*, vol. 8, no. 3, pp. 995–1004, Mar. 2015.

[11] W. Tress, J. P. Correa Baena, M. Saliba, A. Abate, and M. Graetzel, "Inverted Current–Voltage Hysteresis in Mixed Perovskite Solar Cells: Polarization, Energy Barriers, and Defect Recombination," *Adv. Energy Mater.*, vol. 6, no. 19, p. 1600396, Oct. 2016.

3A-1 (Invited)

The Impact of Fin Number on Device's Performance and Reliability in Tri-gate FinFETs

Wen-Kuan Yeh[1,2], Po-Ying Chen[3], Chia-Hung Shih[4], Wenqi Zhang[1], and Yi-Lin Yang[5]

1. Department of Electrical Engineering, National University of Kaohsiung, Kaohsiung, Taiwan
2. National Nano Device Laboratories (NDL), National Applied Research Laboratories, Taiwan
3. Department of Electronic Engineering, National Chin-Yi University of Technology, Taichung, Taiwan
4. Bachelor Program in Robotics, National Pingtung University, Taiwan
5. Department of Electronic Engineering, National Kaohsiung Normal University, Kaohsiung, Taiwan

Abstract

In this paper, the impact of fin number on device performance and hot carrier induced device degradation was investigated for n-channel tri-gate multi-fin FinFET with different fin numbers. The threshold voltage (V_{TH}) shift, transconductance, and subthreshold swing degradation were extracted to determine the degradation of device. It was found that the device with fewer fins shows better device performance, but suffer from more serious hot carrier induced device degradation. It is suggested that the existed coupling effect between the fins reduces the equivalent electric field in the multi-fins devices, thus shows better reliability than the single-fin device does after hot carrier stress. (Keywords: FinFET, multi-fins, hot carrier Injection)

Introduction

FinFET is the most promising candidates for device scaling because its stronger electrostatic control of the channel allows effective suppression of the short channel effect (SCE) [1]–[2]. In order to enhance FinFETs driving capability, multiple parallel fins between the source and drain are proposed to introduce a large total channel width [3]. However, the multi-fins structure FinFET will not only increase the total drive current but also introduce coupling effect between the nearby Si fins. Hot carrier injection (HCI) is a critical issue for submicron n-type MOSFET due to the high lateral electric fields, and this issue is even more severe in high-k/metal gate MOSFETs [4]. The lifetime of hot carrier is commonly predicted by stressing devices at high drain voltages, setting the gate voltage to correspond to the maximum substrate current (I_{SUB}), and measuring the degradation of the device over stress time [5]. Hot carrier degradation is especially important for nFinFETs because of the higher probability of carrier capture in tri-gate structure and higher density of available Si-H bonds at (110) fin sidewalls [6]. However, fundamental understanding in reliability characteristics to FinFET architecture is still lacking [7], and there is no study about the effects of fin numbers on FinFET reliability. In this work, the impact of fin numbers on hot carrier injection stress induced device degradation for n-type FinFET device

was investigated for the first time

Experimental

N-type tri-gate FinFET was fabricated on a bulk-Si wafer. The surface orientation of the fin sidewall is (110) and the channel direction is <110>. Figure 1(a) shows the structures of tri-gate FinFET. Related TEM picture of this fabricated high-k/metal tri-gate FinFET with multi-fins is shown in Fig. 1(b). The fin height and fin pitch of the device are 40nm and 60nm, respectively. The current-voltage (I-V) curves were measured at certain intervals of stress time by using a semiconductor parameter analyzer (Agilent-B1500A) at room temperature. As the HCI stress was applied, the gate and drain voltages were fixed at 2V and 1.9 V, respectively, with the source and body grounded.

Results and Discussion

For multi-fins nFinFET, after characteristic being normalized, Figure 2 shows device with fewer fins possess (a) higher driving current, (b) higher transconductance, and (c) lower subthreshold swing. Various fin number FinFETs were inspected to find out the fin number dependence of HCI induced instability. Figure 3(a) and (b) show the typical variations of threshold voltage (V_{TH}) shift and subthreshold swing (SS) versus stress time under a stress conditions of V_G=2V and V_D=1.9V. It could be observed that the single-fin device shows the most severe reliability degradation than the other devices in V_{TH} shift and SS degradation. The largest positive V_{TH} shift in FinFET with single-fin reveals that the injected electrons were trapped in the gate dielectric layer during HCI stress. The most revere SS degradation observed in single-fin device also confirmed that the strongest impact ionization, which causes the interface state generation, occurred on single-fin FinFET during HCI stress. On the other hand, the forty-fin device shows the smaller V_{TH} shift and SS degradation than the twenty-fin device does, which indicates that the reliability can be improved with fin number increasing. Figure 4 shows the degradation of transconductance (Gm) for FinFET with different fin numbers under HCI stress for 5000 seconds. It could be observed that the degradation in transconductance becomes more severe with the

978-1-5090-4661-4/17 $31.00 © 2017 IEEE

decreasing of fin numbers. It is well known that interface state generation, which is corresponding to the degradation of SS, results in the reduction of G_m. The single-fin device shows the most serious G_m degradation also indicates the strongest impact ionization occurs during HCI stress.

A charge density model is proposed in Fig. 5 to explain the mechanism about reliability improvement for the multi-fin FinFETs. If the fins are close to each other, it is suggested that the inversion charges are reduced in center fin because of that the two objects with the same charge repel each other. Compared with single fin FinFET, multi-fins FinFET shows fewer inversion charges in the channel will suppress the capability of channel inversion, degrading device performance with lower driving current, lower tranconductance, and higher subthreshold swing. On the other hand, for multi-fin FinFET with fewer inversion charges, lower the impact ionization during HCI stress thus lessen the stress induced damage and improve the reliability. With the fin number increasing, there are more fins which are affected by their nearby fins. As the results, device reliability improved with the increasing of fin numbers. In order to verify the proposed model, the distribution of electron density and electric field of multi-fin FinFET were simulated as shown in Fig. 6(a) and (b), respectively. As the gate voltage was applied, it could be observed that the lower electron density and weaker electric field appeared in the center fin. The simulation results are corresponding to our suggestion model. Therefore, the multi-fins devices show the better reliability than the single-fin device after stress.

Conclusion

The device characteristic and hot carrier induced FinFET degradation with different fin numbers was studied. For the FinFET with fewer fins shows better device performance, but suffer from more severe device degradation under HCI stressing. For the multi-fins FinFET, every fins are very close to each other, causing charge repel each other and further reducing the inversion charges in center fin, thus suppress impact ionization, resulting in lower hot carrier induced degradation.

Acknowledgment

This work was supported by the Ministry of Science and Technology under Contract MOST 104-2221-E-390-063-MY2.

Reference:

[1]. Paul, *et.al, IEEE IEDM*, p. 361, 2013

[2]. P. Kerber *et.al, IEEE Electron Dev. Lett.* vol. 34, p. 876, 2013

[3]. H.W. Chen *et.al, IEEE ISNE.*, 32 (2010).

[4]. M. Wang, *et.al, IEEE Electron Device Lett.* 34, 837 (2013).

[5]. J. Y. Tsai *et.al, Appl. Physics. Letters.* 103, 022106 (2013).

[6]. Y.L. Yang, *et.al, Appl. Physics. Letters.* 104, 083505 (2014).

[7]. M. Wang *et.al,* IRPS, pp.4A. 5.1-4A.5.7, IEEE, (2015

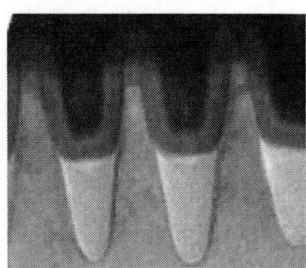

Fig. 1(a) The schematic illustration of tri-gate bulk FinFET structure, and (b) the TEM image for multi-fins FinFET.

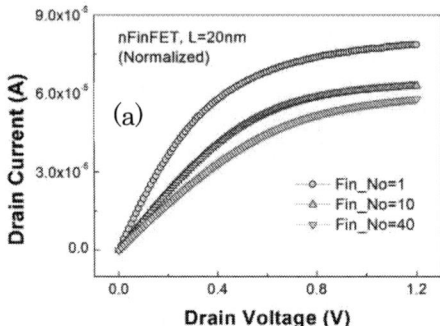

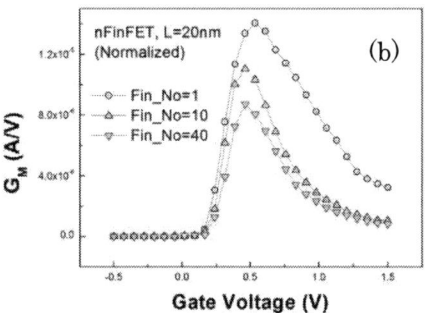

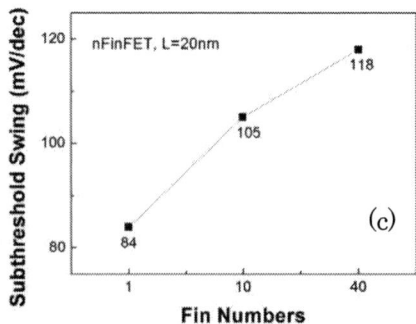

Fig. 2 (a) I_D-V_D characteristic, (b) G_m-V_G characteristic, and (b) subthreshold swing for single-fin and multi-fins nFinFET.

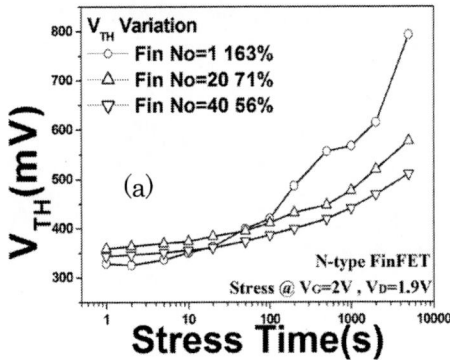

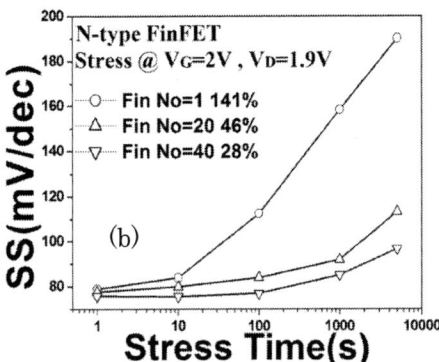

Fig. 3: The variation of (a) V_{TH} and (b) subthreshold swing versus hot carrier stress time for multi-fin nFinFET.

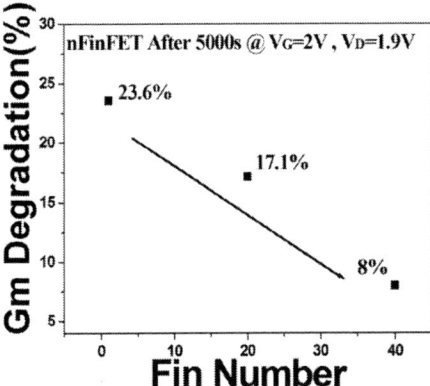

Fig. 4 The degradation of transconductance with various fin numbers nFinFETs.

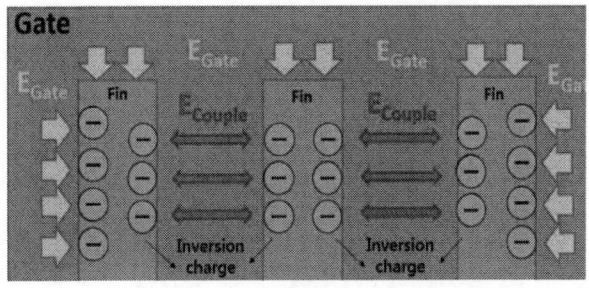

Fig. 5: Models of charge density distribution for the nFinFET with multi-fins.

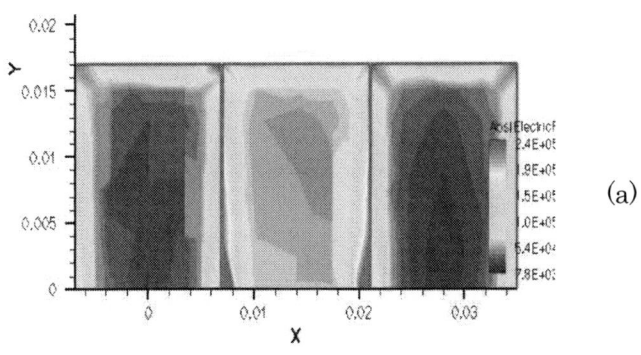

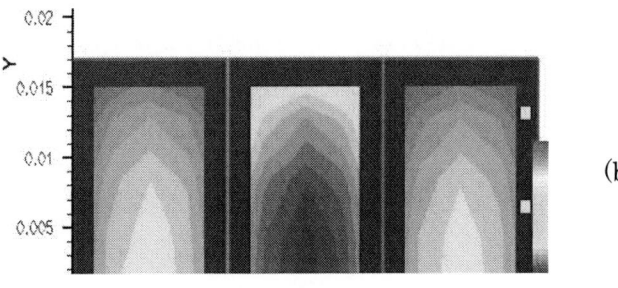

Fig. 6 The simulation results of (a) electron density and (b) electric field distribution of multi-fin nFinFETs.

978-1-5090-4661-4/17 $31.00 © 2017 IEEE   19

3A-2

Impact of e-SiGe S/D processes on FinFET PFET TDDB Reliability

R. Ranjan, S. Uppal, H. Yu, B. Parameshwaran, T. Nigam, A. Kerber, and C. LaRow, and M.I. Natarajan

GLOBALFOUNDRIES, US, Inc., 400 Stonebreak Rd. Ext., Malta, NY 12020

Phone: (+1)518-305-7408, e-mail address: rakesh.ranjan@globalfoundries.com

Abstract

The impact of source/drain e-SiGe process engineering on time dependent dielectric breakdown (TDDB) on core PFETs fabricated with bulk FinFET technology is evaluated. It is observed that thicker e-SiGe buffer layer improves the PFETs TDDB. Electrical and physical analysis revealed that with thinner buffer layer, Ge atoms migrate to gate dielectric and accelerate the breakdown mechanisms due to poor surface roughness and stoichiometry. In addition, the process optimization of pre-baking of e-SiGe trench can also improve the TDDB even for relatively thinner buffer layer.

(Keywords: e-SiGe process, TDDB and PFETs)

Introduction

With continued CMOS scaling, source/drain (S/D) epitaxial-SiGe (e-SiGe) process optimization for strain engineering is widely studied and reported by researchers to improve the performance of PFET devices for both planar and FinFET technologies [1-5]. In contrast, reliability study is mainly focused on high-k (HK)/interfacial layer (IL), metal gate (MG) process/composition, HK post deposition anneal (PDA) temperature and cleaning [6-11]. In this report, the impact of S/D e-SiGe buffer layer thickness and pre-baking of e-SiGe process on PFET TDDB is presented. The results clearly show that e-SiGe process optimization is required for achieving enhanced FinFET PFET TDDB reliability.

Experiments

HK/MG Core-PFETs (Fig. 1) are fabricated using bulk FinFET technology with two S/D e-SiGe processes (e-SiGe1 and e-SiGe2), having different buffer layer thickness (T1-T4) and various HK- post deposition annealing (PDA) processes. Pre-baking conditions of e-SiGe trench are different for e-SiGe1 and e-SiGe2 processes. Devices are stressed with constant voltage stress (CVS) and constant current stress (CCS) with pre and post-stress leakage characterization. X-ray Energy Dispersive Spectroscopy (XEDS) elemental mapping analysis is performed to understand the microstructural defects on unstressed and stressed devices post breakdown.

Results and Discussions

(A) Impact of Buffer layer thickness on TDDB - Fig. 2 shows the typical area scaled Weibull distribution plots of time to breakdown (t_{bd}) of core-PFETs

(shorter channel length L1) having different buffer layer thicknesses (T1-T4) with e-SiGe1 process stressed under CVS. It can be seen that thicker buffer layer increases the $t_{63\%}$ of time to breakdown, area scaled Weibull slope (β) and predicted Vmax [11] (Figs. 3-5). It is also observed that equivalent oxide thickness (T_{oxgl}) based on pre-stress gate leakage [7] is not significantly impacted by buffer layer thickness of e-SiGe1 (Fig. 6). It implies that improvement in TDDB model parameters, $t_{63\%}$ and Weibull β, with thicker buffer layer is not caused by a change in the equivalent gate oxide thickness but rather by the buffer layer thickness playing a role during the stressing of the devices.

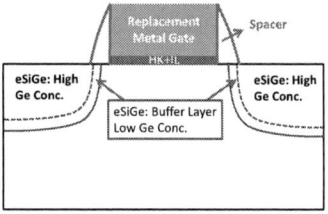

Fig.1. 2D schematics of core PFETs devices fabricated with

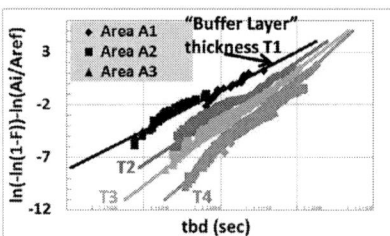

FinFET technology.

Fig. 2. Typical area scaled Weibull distribution with 3 areas (A1, A2= 10xA1 & A3 = 100xA1) of t_{bd} of PFETs (shorter channel length L1) having various e-SiGe1 buffer layer thicknesses (T1<T2<T3<T4) stressed under CVS.

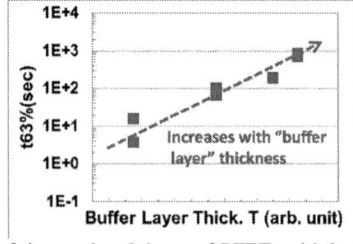

Fig. 3. $t_{63\%}$ of time to breakdown of PFETs with buffer layer thickness (T1-T4) of e-SiGe1 on several wafers. $t_{63\%}$ increases with thicker buffer layer.

Figs.7 & 8 show that TDDB model parameters ($t_{63\%}$ and Weibull slope) of short channel length p-FETs are mainly impacted with thinner buffer layer thickness of e-SiGe1. It indicates that the impact of e-SiGe process on TDDB model parameters is more

978-1-5090-4661-4/17 $31.00 © 2017 IEEE   20

prominent only on short channel length devices.

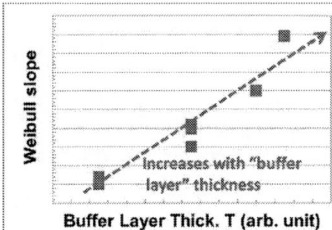

Fig. 4. Weibull slope (β) of PFETs with e-SiGe1 buffer layer thickness (T1-T4) on several wafers. Weibull β increases with thicker buffer layer.

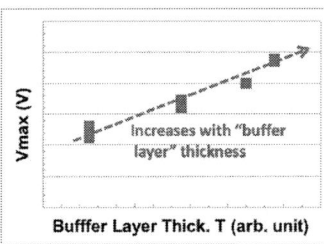

Fig. 5. Vmax [11] of PFETs, predicted with TDDB model parameters $t_{63\%}$ (Fig. 3) and β (Fig. 4), increases with thicker e-SiGe1 buffer layer.

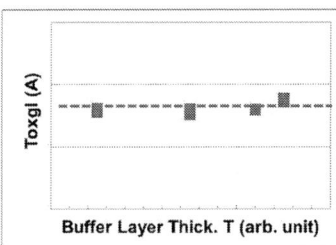

Fig. 6. No significant impact of e-SiGe1 buffer layer thickness on T_{oxgl} estimated by pre-stress gate leakage [7].

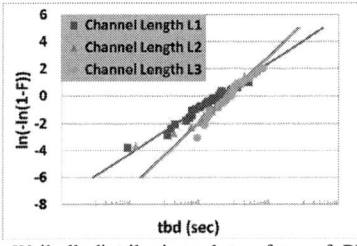

Fig. 7. Weibull distribution plots of t_{bd} of PFETs having different channel lengths (L1, L2 = 7x L1, L3 = 14x L1) with thinner buffer layer T2 of e-SiGe1.

To obtain a clear understanding on the relationship of buffer layer thickness of e-SiGe1 on PFETs TDDB model parameters, the microstructural analysis of the fresh and post breakdown devices with thinner buffer layer thickness of T1 was carried out with XEDS elemental mapping analysis (Figs. 9 (a) & (b)). The analyzed devices were stressed under constant current stress using a voltage threshold to determinate the stress at breakdown (voltage

threshold in a CCS provides a better control of the severity of breakdown as compared to devices stressed under constant voltage stress with compliance on sense current [12]). Fig. 9 (a) shows the presence of Ge particles under the gate, which is not observed in the unstressed devices (Fig. 9 (b)). It implies that Ge atoms migrate to gate dielectric during stress and accelerate the TDDB breakdown mechanisms for PFETs. Based on electrical and physical observations, it can be postulated that thinner buffer layer devices have poor surface roughness and stoichiometry, compared to thicker buffer layer devices, causing undesirable migration of Ge atoms to gate dielectric and results in lower $t_{63\%}$, Weibull slope and V_{max} (Figs. 2-5).

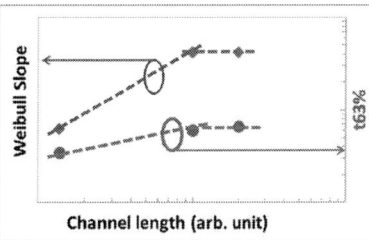

Fig. 8. Weibull slope and $t_{63\%}$ extracted from Fig. 7. Smaller channel length has lower $t_{63\%}$ and Weibull slope.

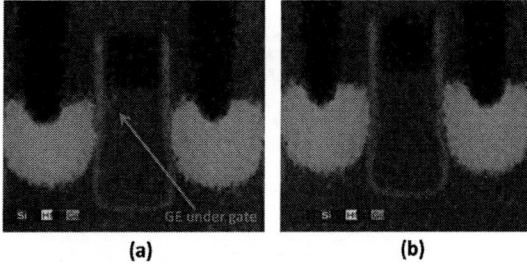

Fig.9. XEDS elemental mapping of (a) Stressed sample: Ge particles found under gate of PFET with thinner buffer layer thickness of T1 and e-SiGe1; (b) reference unstressed PFET.

(B) Impact of pre-baking process optimization of e-SiGe trench on TDDB - Fig. 10 shows the typical Weibull distribution plot of t_{bd} of PFETs processed with e-SiGe2 process having optimized pre-baking condition of e-SiGe trench with relatively thinner buffer layer thickness of T2 compared to thicker buffer layer thickness T4 of e-SiGe1 devices (T2<T4). It can be seen that the Weibull slope of e-SiGe2 process improved compared to e-SiGe1 process at same buffer layer thickness of T2 and it is comparable to thicker buffer layer T4 of e-SiGe1 devices (Fig. 11). This confirms that better surface

978-1-5090-4661-4/17 $31.00 © 2017 IEEE

roughness and stoichiometry of e-SiGe buffer layer improves the PFETs TDDB.

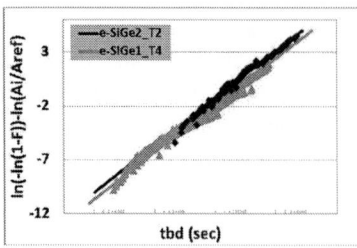

Fig. 10. Weibull plots of PFET t_{bd} having thinner buffer layer T2 of e-SiGe2 with optimized pre-baking process and compared with thicker buffer layer T4 with e-SiGe1.

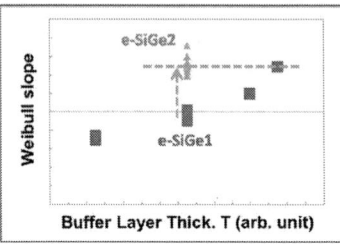

Fig. 11. Pre-baking process optimization of e-SiGe improves the Weibull slope even with thinner buffer layer.

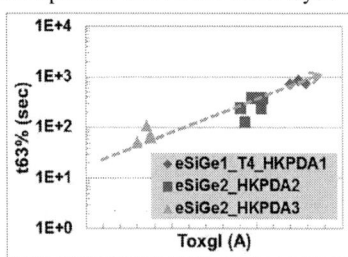

Fig. 12. $t_{63\%}$ plotted vs T_{oxgl} of PFETs having e-SiGe2 with thinner buffer layer T2 and e-SiGe1 with thicker buffer layer T4, processed with different HK thickness. $t_{63\%}$ follows T_{oxgl}.

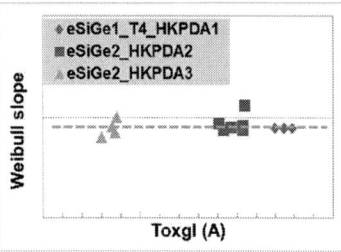

Fig. 13. Weibull β plotted vs T_{oxgl} of PFETs as shown in Fig. 12. No change of β with T_{oxgl}.

From Fig. 12, it can be seen that $t_{63\%}$ scales well with T_{oxgl} with better e-SiGe surface roughness and stoichiometry, as the dielectric thickness is controlled by high-k PDA temperatures (HKPDA1>HKPDA2>HKPDA3). Further, no impact of T_{oxgl} scaling on Weibull slope with

optimized e-SiGe process is observed (Fig. 13). This is also consistent to fundamental physics of gate dielectric breakdown in PFET TDDB with optimized e-SiGe process.

Conclusion

This work presents for the first time impact of S/D e-SiGe process on TDDB of core PFETs. e-SiGe with thicker buffer layer and optimized pre-baking process of e-SiGe trench improve the TDDB model parameters, highlighting the need for e-SiGe process optimization as apart of robust bulk FinFET technology development.

References

[1] T. Ghani *et al.*, "A 90-nm high volume manufacturing logic technology featuring 45-nm gate length silicon CMOS transistors", *IEDM Tech. Dig.* 2003, pg. 978.

[2] K. Mistry *et al.*, "A 45 nm logic technology with high-k/metal gate transistors", *IEDM Tech. Dig.* 2007, pg. 247.

[3] C. Auth *et al.*, "A 22 nm high performance and low-power CMOS technology featuring fully-depleted tri-gate transistors, self-aligned contacts and high density MIM capacitors", *VLSI Tech. Dig.* 2012, pg. 131.

[4] A. Nainani *et al.*, "Is strain engineering scalable in FinFET era?: Teaching the old dog some new tricks", *IEDM Tech. Dig.* 2012, pg. 427.

[5] H. Okamoto *et al.*, "A study on aggressive proximity of embedded SiGe with SiGe with comprehensive source drain extension engineering for 32 nm node high-performance pMOSFET technology", *Solid State Electronics* 2009, pg. 712.

[6] T. Nigam *et al.*, "Accurate model for time-dependent dielectric breakdown of high-k metal gate stacks", *IRPS Reliability Symp.* 2009, pg. 523.

[7] A. Kerber *et al.*, "TDDB failure distribution of metal gate/high-k CMOS devices on SOI substrates", *IRPS Reliability Symp.* 2009, pg. 505.

[8] S. Pae *et al.*, "Gate dielectric TDDB characterizations of advanced high-k and metal-gate CMOS logic transistor technology", *IRPS Reliability Symp.* 2012, pg. 5C.1.1.

[9] S. Pae *et al.*, "Reliability characterization of 32nm high-K and metal-gate logic transistor technology", *IRPS Reliability Symp.* 2010, pg. 3D.2.1.

[10] S. Pae *et al.*, "Considering physical mechanisms and geometry dependencies in 14nm FinFET circuit aging and product validations", *IEDM Tech. Dig.* 2015, pg. 557.

[11] R. Ranjan *et al.*, "Reliability-performance trade-off for work-function optimization in advanced node replacement metal gate technology", *IRPS Reliability Symp.* 2016, pg. DI-8-1.

[12] A. Ghetti, "Gate oxide reliability: Physical and computational models", *Springer Series in Materials Science, J. Dabrowski and E. R. Weber, Eds. Springer* 2004, pg. 201.

978-1-5090-4661-4/17 $31.00 © 2017 IEEE

3A-3

In content dependence of pre-treatment effects on Al$_2$O$_3$/In$_x$Ga$_{1-x}$As MOS interface properties

C. Yokoyama, C.-Y. Chang, M. Takenaka and S. Takagi

The Univ. of Tokyo, 7-3-1 Hongo, Bunkyo-ku, Tokyo, 113-0032, Japan

E-mail: cyokoyama@mosfet.t.u-tokyo.ac.jp

Abstract

We examine the electrical and physical properties of ALD Al$_2$O$_3$/In$_x$Ga$_{1-x}$As (x = 0.53, 0.7 and 1) MOS interfaces with (NH$_4$)S$_y$, BHF and HF pretreatment. It is found that, in higher In content (x), In$_x$Ga$_{1-x}$As MOS interfaces with BHF and HF cleaning exhibit better *C-V* characteristics and lower interface state density (D_{it}) than with (NH$_4$)S$_y$ pretreatment. Also, amounts of arsenic oxides, evaluated from x-ray photoelectron spectroscopy (XPS), are found to increase in the higher In content In$_x$Ga$_{1-x}$As MOS interfaces with BHF and HF cleaning, suggesting that arsenic oxides can contribute to passivation of Al$_2$O$_3$/InGaAs MOS interface defects.

Introduction

As the scaling rule of Si CMOS technology is approaching its limitation, other materials are studied for better performance. Among them, In$_x$Ga$_{1-x}$As with x = 0.53 to 1 is expected as channel materials for future n-channel MOSFET, because of the high electron mobility. However, there are still several issues for realizing InGaAs MOSFETs. One of the most critical issues is high D_{it} at InGaAs MOS interfaces [1], which causes the degradation of the drive current and the sub-threshold swing. Thus, a variety of pretreatments before forming gate insulators have been studied for appropriate InGaAs MOS interface passivation to realize low D_{it} [2,3]. While S cleaning ((NH$_4$)S$_y$ pretreatment) is commonly used for Al$_2$O$_3$/In$_{0.53}$Ga$_{0.47}$As MOS interfaces [4-6], it has recently been reported that HfO$_2$/InAs [7] and Al$_2$O$_3$/ultra-thin InAs/GaSb [8] MOS capacitors with BHF cleaning show lower D_{it} than with S cleaning. These results suggest that the optimum pretreatment could depend on the In content of InGaAs. However, there is no systematic work on the impact of the pretreatment and resulting interface structures on D_{it} for InGaAs with different In contents, which are quite informative to better understanding of physical origins of D_{it} at InGaAs MOS interfaces. In this study, thus, we study the MOS interfaces properties of ALD Al$_2$O$_3$/ In$_x$Ga$_{1-x}$As with different In contents (x = 0.53, 0.7, 1) by changing the pretreatment among (NH$_4$)S$_y$, BHF and HF. Here, we also perform HF cleaning in order to discriminate the effects of NH$_4$F and HF in BHF cleaning, because BHF is consist of NH$_4$F and HF.

Experiments

Al$_2$O$_3$/In$_x$Ga$_{1-x}$As MOS capacitors (x = 0.53, 0.7 and 1) were fabricated with S, BHF or HF cleaning. The each soaking time of the wafers in the S, BHF and HF cleaning was 5 min, 15sec and 15sec after acetone cleaning for 1 sec. The concentration of (NH$_4$)S$_y$, BHF, HF were 0.6-1%, 22% and 50%, and the temperatures of solutions were room temperature. The fabricated gate stack structures are shown in Fig. 1. 90-cycle-Al$_2$O$_3$ (~10nm, the growth rate of Al$_2$O$_3$ was estimated to be 0.11nm/cycle) was deposited by ALD at 200 °C on Si-doped *n*-In$_{0.53}$Ga$_{0.47}$As (N_D ~ 5×10^{15} cm^{-3}) and Zn-doped *n*-In$_{0.7}$Ga$_{0.3}$As (N_D ~ 3×10^{16} cm^{-3}) grown on (001) InP and un-doped n-InAs (N_D ~ 2×10^{16} cm^{-3}) after each cleaning. The precursors for Al$_2$O$_3$ were Al(CH$_3$)$_3$ (TMA) and H$_2$O as the liquid sources. Next, the Au gate electrode was formed by thermal evaporation and Al was deposited as the back contact. Post metallization annealing (PMA) in N$_2$ ambient at 350 °C for 1 min was performed for all the MOS capacitors to improve the interface properties.

In order to examine the electrical properties, we performed *C-V* measurements of the MOS capacitors and estimated D_{it} values by the Terman method. Also, XPS analyses were performed in order to study the interface structures.

Results and discussions

The *C-V* curves of the ALD Al$_2$O$_3$/In$_x$Ga$_{1-x}$As MOS capacitors (x = 0.53 at room temperature, x = 0.7 at 180 K and x = 1 at 180 K for 1 k, 10 k, 100k and 1 MHz) with each cleaning are shown in Fig. 2. The maximum capacitance of ~0.6μF for each sample showed almost no variation. The hysteresis in the *C-V* characteristics becomes larger with an increase in the In content, indicating that the amount of slow traps increases with increasing the In content. While there seems little difference among S, BHF and HF cleaning at x = 0.53, smaller frequency dispersion in inversion and depletion regions are observed at x = 0.7 with BHF/HF cleaning than with S cleaning. The smaller humps with BHF/HF cleaning than with S cleaning at x = 0.7 also indicate smaller D_{it}. Furthermore, the capacitance of InAs (x = 1) with S cleaning does not change significantly, meaning that the surface potential is almost pinned because of a large amount of D_{it}. In contrast, the capacitance change with BHF/HF cleaning is much larger and, thus, the interface properties are much better than those with S cleaning. It is also found that BHF and HF cleaning do not provide much difference in the *C-V* characteristics, irrespective of the In content.

The energy distributions of D_{it} of Au/Al$_2$O$_3$/ In$_x$Ga$_{1-x}$As (x = 0.53 at room temperature, x = 0.7 and x = 1 at 180 K), evaluated by the Terman method, are shown in Fig. 3. The energy of the conduction band edge (E_c) of

978-1-5090-4661-4/17 $31.00 © 2017 IEEE

In$_{0.53}$Ga$_{0.47}$As, In$_{0.7}$Ga$_{0.3}$As and InAs locates at 0.76eV, 0.59eV, and 0.36eV, respectively, as shown in Fig.3. In the analysis using the Terman method, we adopted the conventional semi-classical model, where the Poisson's equation and the Boltzmann distribution under parabolic band model considering only Γ valley in the conduction band are used for calculating carrier concentrations and potentials. Also, the quantum effect in MOS accumulation layers has not been included. As a result, the accuracy of D_{it} inside the conduction band is suspicious in the present calculation.

It is found in Fig. 3 that, when the In content is larger, D_{it} with BHF/HF cleaning becomes lower than with S cleaning. An interesting point for the results of x = 0.53 is that D_{it} near the conduction band edge, typically evaluated by the conductance method, has no difference among the three cleaning solutions, while D_{it} around the midgap and in the valence band side is lower in the MOS interfaces with BHF/HF cleaning than in those with S cleaning. In x = 0.7 and x = 1, on the other hand, the reduction in D_{it} with BHF/HF cleaning is quite obvious. In contrast, there is no significant difference in D_{it} between BHF and HF cleaning over all of the In contents. These results indicate that HF-based species play a more essential role in the improvement in the interface properties than NH$_4$F (NH$_3$) –based species.

The physical origin of the difference in D_{it} was studied by XPS at a take-off angle of 90° for the MOS interfaces (Fig. 4). Here, the Al$_2$O$_3$ thickness in the samples for XPS was taken to be 1 nm. The In 3d core level can be resolved into three components, In $3d_{3/2}$ (451.7eV), In$3d_{5/2}$ (444.3eV) and In$_2$O$_3$ (445.3eV), the Ga 2p core level into Ga (1118.0eV) and Ga$_2$O$_3$ (1119.0eV), and the As 2p core level into As (1323.5eV) and As$_2$O$_3$ (1327.0eV). The In 3d and the Ga 2p peak have asymmetric shapes towards higher binding energy, indicating the existence of oxides. However, there is almost no difference in the In $3d$ and Ga $2p$ spectra among S, BHF and HF cleaning, irrespective of the In content. It is found, on the other hand, in the As $2p$ spectra that, when the In content become higher, the amounts of As$_2$O$_3$ increase at the Al$_2$O$_3$/In$_x$Ga$_{1-x}$As interfaces especially with BHF and HF cleaning, which is correlated with the results of C-V and D_{it}. Much smaller amount of As$_2$O$_3$ with S cleaning is reasonable by considering of the strong surface passivation nature of S for InGaAs [9]. As a result, a possible interpretation for reduced D_{it} with BHF/HF cleaning is As$_2$O$_3$ passivation for InGaAs MOS interface defects.

Based on the above results, we propose a physical model of relationship between pretreatment and possible defects responsible for D_{it}, shown in Fig. 5. Here, the defect energy levels of possible origins of interface states such as dangling bonds and dimer bonds have

been taken from [10], while the energy distribution could originate in broading of descrete defect levels with variations of bond lengths and bond angles [11]. As a result, we can assume a continuous U-shape energy distribution of D_{it} originating from the discrete defect states. The downward-convex D_{it} curve near the conduction band edge corresponds to that obtained in the previous study [12]. We suppose that the change in the energy distributions with the In content can be associated with the change in the conduction band edge under an assumption that the valence band edges of In$_x$Ga$_{1-x}$As (x = 0.53, 0.7 and 1) lie at the same energy. Here, the interface states near the conduction band of In$_{0.53}$Ga$_{0.47}$As are attributable to As-As defect bonds, while those in the midgap and valence band side attributed to As dangling bonds. With increasing the In content, thus, the dominant origin of the interface states can move from As-As bonds to As dangling bonds. As a result, the present In content and cleaning dependencies of D_{it} can be explained by this transition of the physical origin of dominant defects to As dangling bonds and effective passivation of As dangling bonds with oxygen atoms, evident in the observation of As$_2$O$_3$, by BHF/HF cleaning.

Conclusion

We investigated the electrical and physical properties of the Al$_2$O$_3$/In$_x$Ga$_{1-x}$As (x = 0.53, 0.7 and 1) MOS interfaces with (NH$_4$)S$_y$, BHF and HF pretreatment. We have found that, as the In content becomes higher, BHF/HF cleaning is more effective in reducing D_{it} at Al$_2$O$_3$/In$_x$Ga$_{1-x}$As MOS interfaces than S cleaning. It was observed through XPS that the amount of As$_2$O$_3$ with BHF/HF cleaning is higher than with S cleaning and that it increases with increasing the In content. As a result, the reduction in D_{it} is attributable to the dominance of As dangling bonds on D_{it} with higher In content and the effective passivation of oxygen atoms with the dangling bonds, evident in the formation of As$_2$O$_3$.

Acknowledgements

This work was partly supported by JST-CREST. The authors would like to thank Drs. M. Yokoyama, O. Ichikawa, H. Yamada and T. Yamamoto in Sumitomo Chemical Corporation for their collaborations.

References

[1] P. D. Ye, J. Vac. Sci. Technol. A 26, 697 (2008). [2] H.-C. Chin et al., TED 57, 973 (2010). [3] R. Engel-Herbert et al., JAP 108, 124101 (2010). [4] Y. Xuam et al., IEDM 637 (2007). [5] Y. Urabe et al., IEDM 142 (2010). [6] R. Suzuki et al., SSDM 941 (2011). [7] D. Wheeler et al., Microelectron. Eng. 86, 1561 (2009). [8] K. Nishi et al., APEX 8, 061203 (2015). [9] M. Yokoyama et al., APL 109, 182111 (2016). [10] J. Robertson et al., JAP 117, 112806 (2015). [11] T. Sakurai et al., JAP 52, 2889 (1981). [12] N. Taoka et al., APEX 9, 111202 (2016).

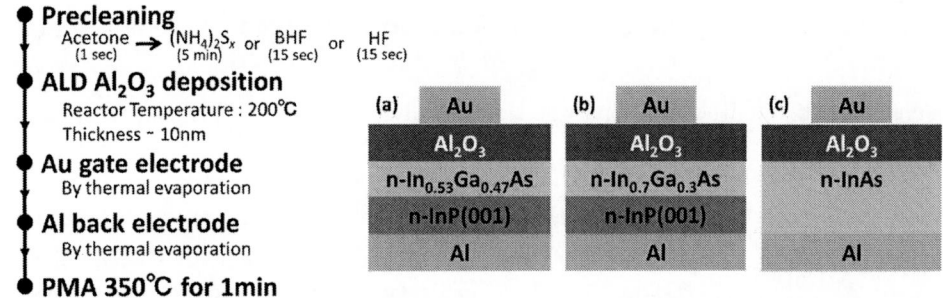

Precleaning
Acetone → (NH₄)₂Sₓ or BHF or HF
(1 sec) (5 min) (15 sec) (15 sec)

ALD Al₂O₃ deposition
Reactor Temperature : 200°C
Thickness ~ 10nm

Au gate electrode
By thermal evaporation

Al back electrode
By thermal evaporation

PMA 350°C for 1min

(a) Au / Al₂O₃ / n-In₀.₅₃Ga₀.₄₇As / n-InP(001) / Al
(b) Au / Al₂O₃ / n-In₀.₇Ga₀.₃As / n-InP(001) / Al
(c) Au / Al₂O₃ / n-InAs / Al

Fig. 1 Process flow of Au/Al₂O₃/In$_x$Ga$_{1-x}$As ((a) x=0.53, (b) x=0.7 and (c) x=1) MOSCAPs.

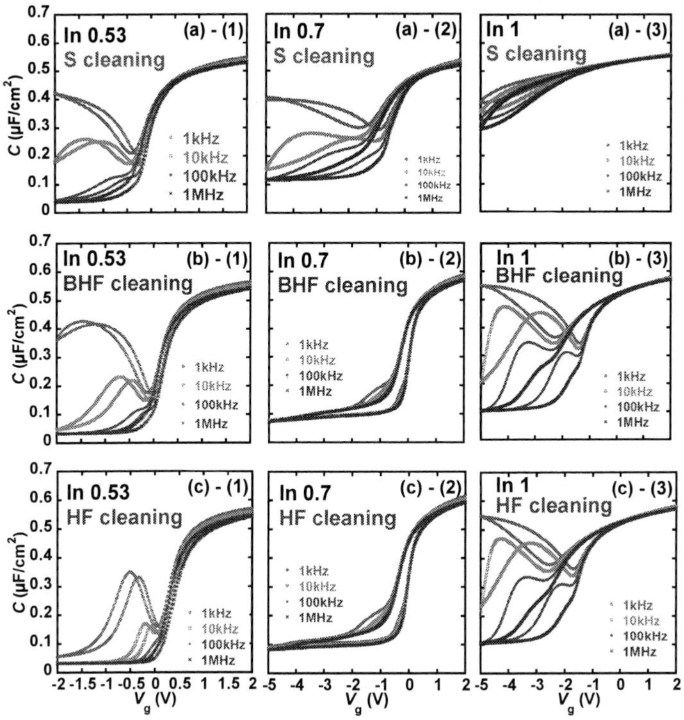

Fig. 2 C-V curves of Au/Al₂O₃/In$_x$Ga$_{1-x}$As ((1) x=0.53 at R.T., (2) x=0.7 at 180K and (3) x=1 at 180K) MOSCAPs with (a) S cleaning, (b) BHF cleaning and (c) HF cleaning.

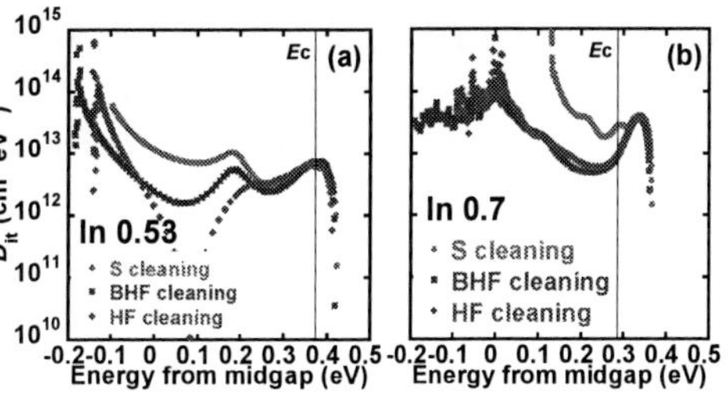

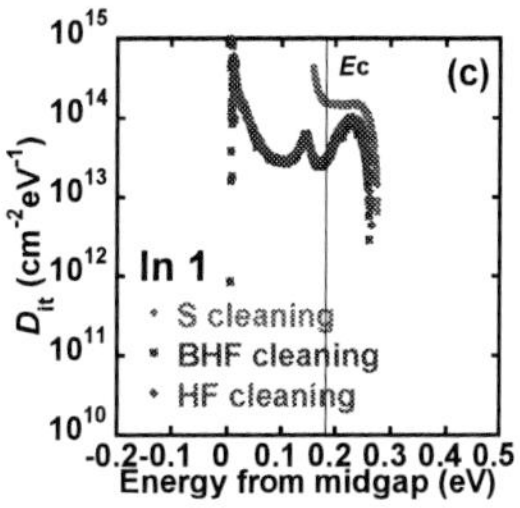

Fig. 3 D_{it} of Au/Al₂O₃/In$_x$Ga$_{1-x}$As ((a) x=0.53 at R.T., (b) x=0.7 at 180K and (c) x=1 at 180K) MOS capacitors with S cleaning, BHF cleaning and HF cleaning.

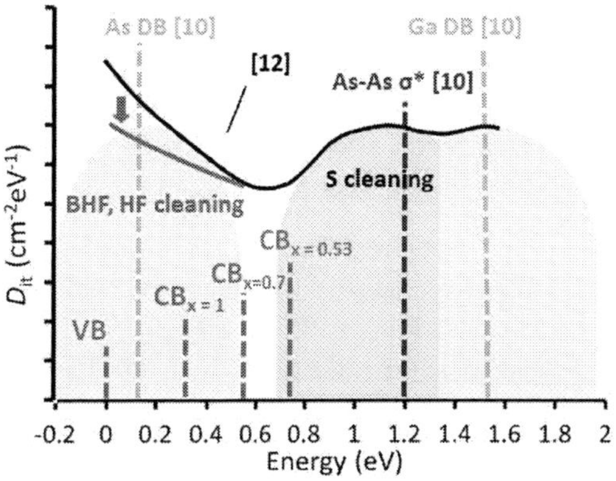

Fig. 5 Chemical trends of defect energy levels for dangling bonds and dimer bonds for In$_x$Ga$_{1-x}$As.

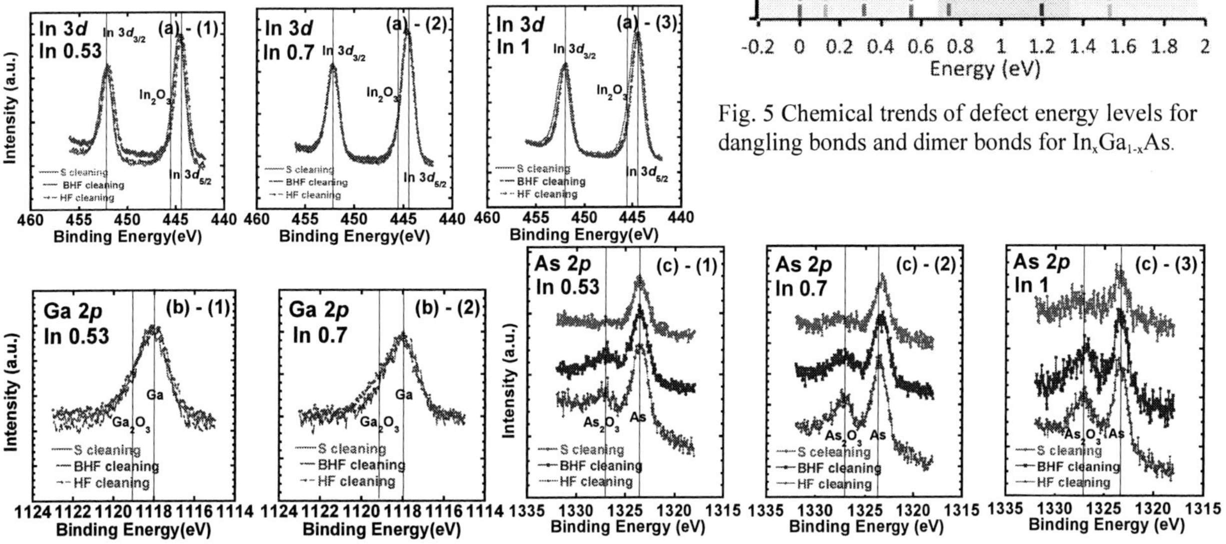

Fig. 4 (a) In 3d, (b) Ga 2p and (c) As 2p spectra in XPS analysis at the 90° take of angle on Al₂O₃/In$_x$Ga$_{1-x}$As ((1) x=0.53, (2) x=0.7 and (3) x=1).

978-1-5090-4661-4/17 $31.00 © 2017 IEEE

Deep junction by low thermal budget process for advanced Si power electronics

Inès Toqué-Trésonne[1], Toshiyuki Tabata[1], Sébastien Halty[1], Fulvio Mazzamuto[1], Karim Huet[1], and Yoshihiro Mori[1]

[1]Laser Systems and Solutions of Europe (LASSE)

SCREEN Semiconductor Solutions Co. Ltd, GENNEVILLIERS, France, ines.toque-tresonne@screen-lasse.com

Abstract

Two new processes for deep junction formation have been demonstrated with low thermal budget UV excimer laser annealing using the melting regime: (i) in-depth controllable activation after high energy implantation and (ii) diffusion and recrystallization after heavily-doped Si deposition.

(Keywords: Power Electronics, Excimer Laser Annealing, Deep Junction)

Introduction

In Si power electronics, the wafer's backside process presents many challenges due to the temperature constrain to not affect the already-formed front side area. Particularly, the formation of deep junction without standard annealing techniques is very challenging. UV excimer laser annealing can selectively anneal the backside only. [1-5] The two explored solutions consist of (i) deep Phosphorous implantation followed by UV laser annealing with substrate temperature control to extend the activation depth or (ii) heavily-doped poly-Si deposition on the substrate followed by melting laser to diffuse dopants and regrow the deposited layer in a single crystal from the silicon substrate seed.

Activation after Deep Implantation

Fig. 1(a) and (b) show the assist heating temperature dependence on SIMS and SRA profiles for the samples annealed once with 10 J/cm^2 at RT and 350°C. In the deep region of interest, the dopant profile is slightly different from the as-implanted case but the effect is small enough to be attributed essentially to dopant diffusion. Moreover, it can be clearly seen that an extra thermal budget can extend up to 2x the activation depth, resulting in near-perfect dopant activation over 4μm depth. The effect of assist heating is significantly stronger with higher laser energy density (data not shown). Typically, a very high activation rate is achievable up to the liquid/solid interface in a molten Si. Then, part of the energy diffuses into the underlying Si solid phase for additional dopant activation. Fig. 2(a) shows that the assist heating can increase both melting depth and annealing temperature in the solid phase. In addition, as shown in Fig. 2(b), the melting time has been significantly lengthened.

Junction Formation by Solid-State Doping

An alternative solution to avoid high energy implantation has been explored. 1μm of poly-silicon has been deposited with high boron concentration (above 1e19 atm/cm^3) on a standard silicon substrate. The sample has been exposed to high energy laser annealing with the purpose of transforming the polysilicon into a single crystal from the substrate seed and control the diffusion of the dopants from the deposited layer to the substrate. Fig. 3 shows the X-TEM images of the deposited heavily-doped poly-Si/Si structure before and after melting UV laser annealing. After the annealing, the deposited layer is fully recrystallized. However, in the as-deposited structure (Fig. 3(a)), some residue of native oxide at the interface between the substrate and deposited poly-Si layer can be observed. This trace remains after the annealing ((Fig. 3(a)) and could be the root cause of the defects remaining after the recrystallization. This is confirmed by the XRD measurement (Fig. 4). FWHM of the main silicon peak is slightly reduced after laser annealing confirming the crystallinity improvement, however a secondary tail appears and may indicate the presence of residual defects. Finally, Fig. 5 shows how the laser annealing diffuses the box-like profile of active dopants up to 1.9μm depth. Further investigation is ongoing to minimize residual defects and maximize the diffusion.

Conclusion

Melting UV laser annealing is very powerful for forming the deep junctions required in advanced Si power electronics. Using deep implantation, we have demonstrated a near-perfect activation over 4 μm in P-doped Si. Using heavily-doped poly-Si deposition, we have demonstrated a box-like B profile that is successfully extended with full recrystallization of the deposited layer.

References

[1] K. Ukawa et al., Jpn. J. Appl. Phys 49, 076503 (2010).

[2] S. Gupta et al., J. Mater. Sci., 46, pp. 196-206 (2011).

[3] M. Rahimo et al., Electron Dev. Lett., 33(11), pp. 1601-1603 (2012).

[4] K. Huet et al., Proceedings of RTP'09, p. 1, (2009)

[5] T. Gutt et al., Proceedings of The 22th International Symposium on Power Semiconductor Devices & ICs, Hiroshima, pp. 29-32 (2010).

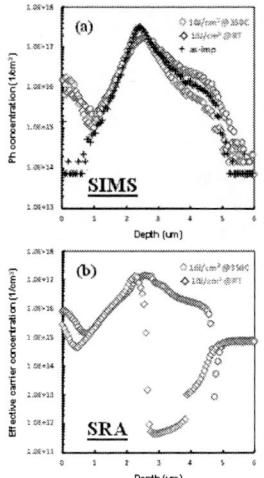 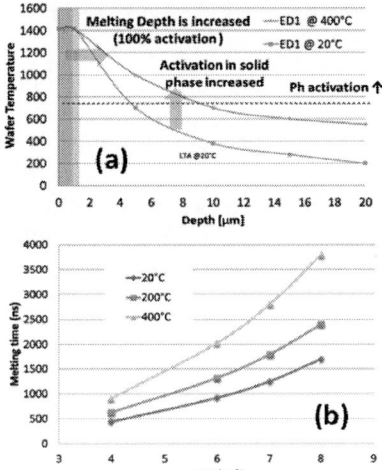

Fig. 1: Impact of the assist heating for a 10 J/cm² single shot on (a) SIMS and (b) SRA profiles.

Fig. 2: Impact of assist heating on simulation results of (a) temperature profile during laser annealing and (b) melting time at different laser energy densities.

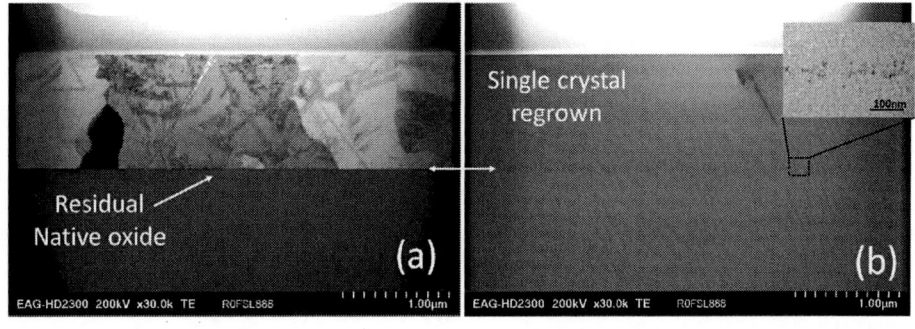

Fig. 3: TEM images of (a) heavily-doped poly-Si as deposited (before anneal) and (b) the same sample after single crystal regrowth using melting UV laser.

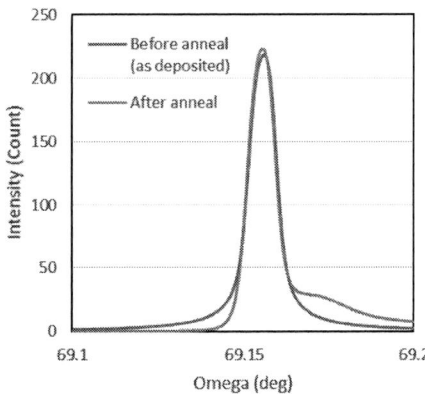 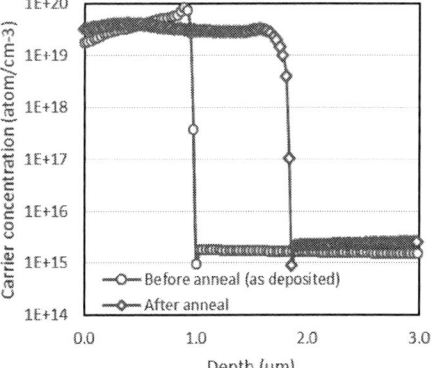

Fig. 4: XRD signal from the samples as deposited and after UV laser annealing

Fig. 5: Active carrier concentration profile of the samples as deposited and after UV laser annealing.

978-1-5090-4661-4/17 $31.00 © 2017 IEEE

3B-1 (Invited)

Assessing device reliability margin in scaled CMOS technologies using ring oscillator circuits

A. Kerber, S. Cimino, F. Guarin, T. Nigam

Reliability Engineering, GLOBALFOUNDRIES Inc.
400 Stone Break Road extension, Malta, NY 12020, USA
email: Andreas.Kerber@globalfoundries.com

Abstract — Device performance enhancement elements are frequently reducing device reliability margin in scaled CMOS technologies. To assess the impact of HCI degradation on digital CMOS logic we study the frequency degradation ($\Delta f/f$) of ring oscillator circuits using core and IO devices in 14nm FinFET technology and correlate the results with discrete device degradation using the conventional DC and a novel AC HCI stress methodology. While for IO devices the traditional scaling factor of 50x used to define DC equivalent HCI targets is applicable, additional HCI margin is demonstrated for core devices bringing relief for device optimization in scaled technology nodes.

Keywords- high-k dielectrics, metal gate, BTI, HCI, self-heating

I. INTRODUCTION

Bias Temperature Instability (BTI) is widely considered the most important contributor to logic CMOS aging in scaled technology nodes [1]. In addition to BTI degradation, which occurs when the device is biased in the on-state, hot carrier injection (HCI) during switching in logic gates also occurs contributing to the circuit degradation. HCI was the dominant aging mechanism in earlier CMOS technology nodes with high core supply voltages (5V, 3.3V, 2.5V and 1.8V) and remains a formidable challenge for input-output (IO) devices in scaled CMOS technologies like FinFET and fully-depleted SOI technologies. In this paper we examine the device reliability margin with focus on HCI contribution for core and IO devices utilizing 14nm bulk FinFET and 28nm planar bulk MG/HK CMOS technology. The discrete device degradation data is correlated to ring-oscillator (RO) degradation for core and IO devices to demonstrate HCI reliability margin in scaled technology nodes.

II. EXPERIMENTAL

Discrete core and IO transistors based on 14nm replacement metal gate bulk-FinFET technology and 28nm gate-first bulk MG/HK CMOS technology were subjected to conventional DC HCI and a newly introduced AC HCI stress methodology [2] at T=125°C or T=30°C. The conventional DC HCI stress methodology applied to ultra-scaled CMOS devices uses either Vg=Vd=Vstress or Vg=mid-Vd stress while the AC HCI stress methodology mimics switching of digital CMOS circuits employing time resolved digital waveforms wherein gate and drain voltage signals are offset by a time Δ using synchronized remote sense amplifier units. The offset in the voltage signal determines the point (Vx) where the gate and drain voltage cross over. It is found that DC HCI stress using Vg=mid-Vd and AC HCI stress with Vx of 0.7 to

0.8 are in good agreement for core and IO devices and represent relevant switching conditions in digital CMOS circuits.

To correlate the discrete device degradation to digital CMOS circuits, a time-resolved RO stress-and-sense methodology was applied [3] which captures the BTI component accurately at short stress times and is capable of tracking the change in time evolution arising from HCI contributions.

III. RESULTS AND DISCUSSION

The substrate current is considered a key factor when determining the appropriate HCI stress condition for CMOS devices at different supply voltages and gate length [4, 5]. IO devices typically exhibit a peak in the substrate current, Isub, around mid-Vg indicating maximum carrier generation at reduced gate bias while core devices for scaled CMOS technologies show the maximum carrier generation at Vg=Vd condition [6]. The data shown in Fig. 1 for 28nm and 14nm IO devices and 14nm FinFET devices are in agreement with these observations. When subjecting these devices to conventional DC HCI stress (see Fig. 2) it is noticed that 3.3V nFET IO devices show an enhanced degradation when the gate is biased at Vg=mid-Vd while for 1.8V nFET IO devices comparable degradation is observed for Vg=mid-Vd and Vg=Vd stress condition. In 14nm core devices the highest degradation is obtained when HCI stress is carried out at Vg=Vd, however, it does not represent a relevant bias condition for HCI degradation in digital CMOS circuits.

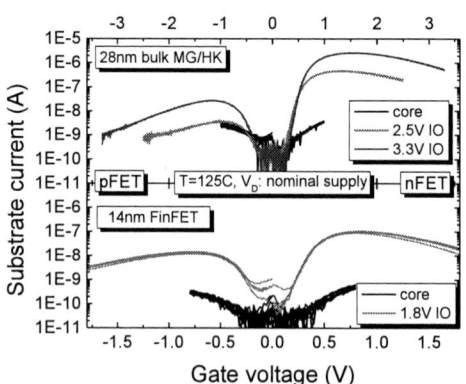

Fig. 1. Substrate currents versus gate voltage in saturation mode for core and IO devices in 28nm bulk MG/HK and 14nm FinFET technology. Note that the substrate current shows a peak for IO devices while for core devices a progressive increase is observed with increasing gate bias.

978-1-5090-4661-4/17 $31.00 © 2017 IEEE

To assess the impact of the stress temperature on the HCI degradation, 14nm core and IO devices were stressed at T=30°C and T=125°C using DC HCI stress at Vg=Vd and AC HCI stress with Vx=0.8. Core nFET, pFET and 1.8V IO pFET devices show higher degradation at T=125°C while 1.8V nFET devices show similar degradation at both temperatures. AC HCI degradation with Vx=0.8 is substantially lower for both core devices and also reduced for IO pFETs devices compared to DC HCI stress at Vg=Vd. In addition, IO pFET devices exhibit electron trapping leading to a deviation from power law at short stress times.

To identify HCI contributions to frequency degradation in 14nm core FinFET devices, various inverter, NAND, and NOR based RO with a fan-out of 1 to 4 and 31 or 501 stages were stressed at T=125°C. Circuit type, number of stages and fan-out have negligible impact on Δf/f suggesting that BTI is the dominant degradation mechanism at elevated stress temperature (see Fig. 4). When inverter based core RO with fan-out of 1 are stressed at T=30°C an increase in time evolution is observed for circuits with lower number of stages confirming the HCI contributions (Fig. 5). The change in time slope can be model by combining power law contribution from BTI and HCI.

stress voltage (V1 to V5 in Fig. 6) the time evolution of Δf/f follows a power law n~0.55 and the degradation saturates to Δf/f~20% consistent with the HCI mechanism. While at stress voltages close to use condition (V6 & V7) a shallower time slope is observed consistent with NBTI becoming the major contributing factor. In Fig. 7, the lifetime is plotted versus the stress voltage for the criteria of 2% and projected to 10% following the power law model. A 10 year lifetime target is met for HCI contributions to RO degradation with no additional margin, confirming HCI as a critical mechanism for CMOS devices operated at high voltage.

To determine the scale factor from operation time of circuits to DC equivalent HCI stress time in discrete devices, the RO degradation is compared to AC HCI transition times in core FinFET devices (Fig. 8) and IO FinFET devices (Fig. 9). The scale factor is 10^5 for core device considering a combined nFET and pFET HCI contribution while for IO devices a factor of ~30x is determined which is comparable to the historical 50x scaling factor used to define DC equivalent HCI targets. The additional HCI margin for core devices brings relief for device optimization in scaled CMOS technologies while for IO devices HCI is confirmed as a critical mechanism.

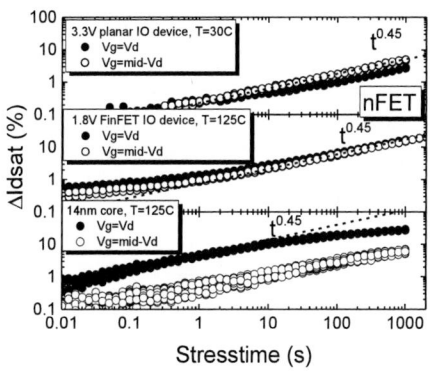

Fig. 2. DC conducting HCI degradation for planar 3.3V IO nFET devices, 1.8V FinFET IO devices and 14nm core FinFET devices stressed at the more stringent test temperature. Note the transition from mid-Vg to Vg=Vd stress as test condition resulting in higher degradation for 3.3V IO compared to 14nm core FinFET devices.

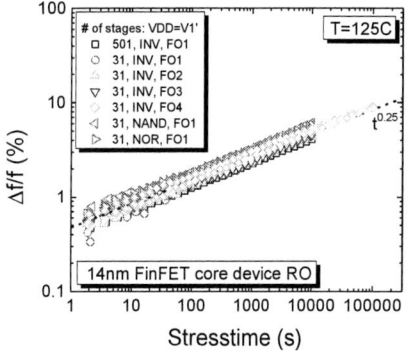

Fig. 4. Frequency degradation for various 14nm FinFET core device RO (inverter, NAND and NOR) tested at T=125°C. Negligible impact of inverter load (fan-out of 1 to 4) and circuit type on frequency degradation is observed and no increase in time-slope evident suggesting xBTI being the dominant degradation mechanism at high stress temperature and low stress voltage (V1').

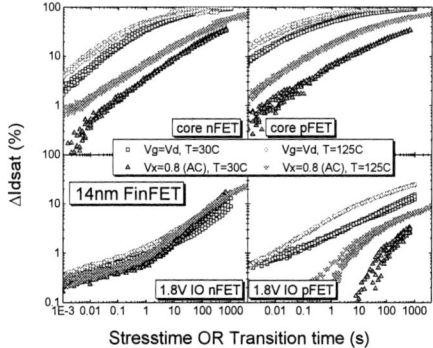

Fig. 3. Impact of test temperature on the DC HCI degradation at Vg=Vd stress condition and AC HCI degradation with a Vx cross-point of 0.8. 14nm core FinFET and 1.8V IO pFET devices show higher degradation at T=125°C while for 1.8V IO nFET devices the degradation shows minor temperature sensitivity.

HCI contributions to Δf/f in 1.8V IO devices can be readily seen at T=125°C when covering a wide stress voltage range. At high

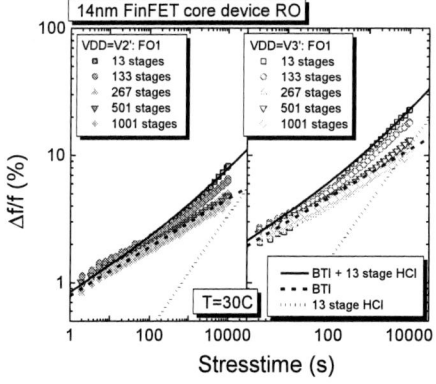

Fig. 5. Frequency degradation for 14nm FinFET inverter based core RO with number of stages varying from 13 to 1001 stressed at T=30°C. Note the systematic increase in time evolution for RO with lower number of stages confirming contributions from HCI to frequency degradation at medium (left) and high (right) stress voltages.

978-1-5090-4661-4/17 $31.00 © 2017 IEEE

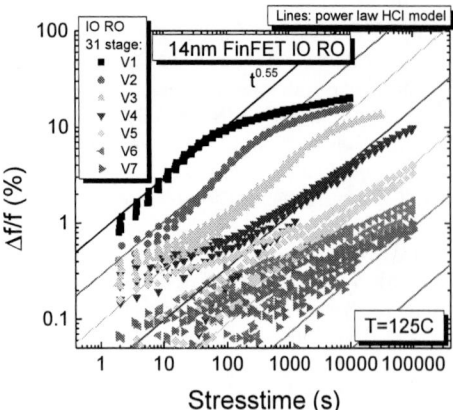

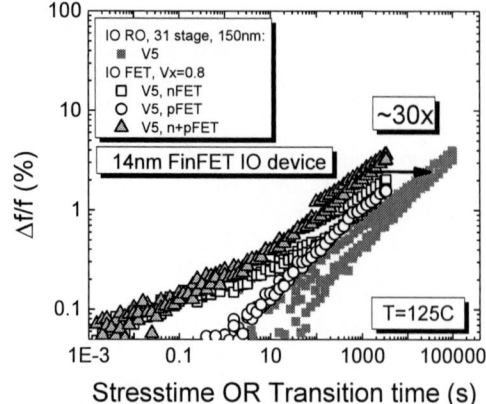

Fig. 6. Frequency degradation for 14nm FinFET inverter based IO RO (31 stage) stressed over a wide voltage range at T=125C. Note the transition from HCI dominated RO degradation (V1 to V5) to NBTI dominated degradation (V6, V7) close to the operation condition.

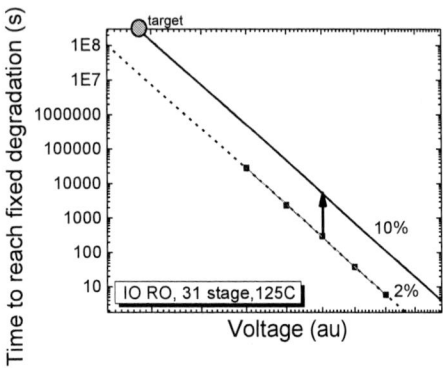

Fig. 7. Stress voltage versus time at constant frequency degradation for 14nm FinFET inverter based IO RO stress which is derived from data shown in Fig. 6. Note that 10% RO degradation projects to meet 10 year target with no additional reliability margin for HCI mechanism.

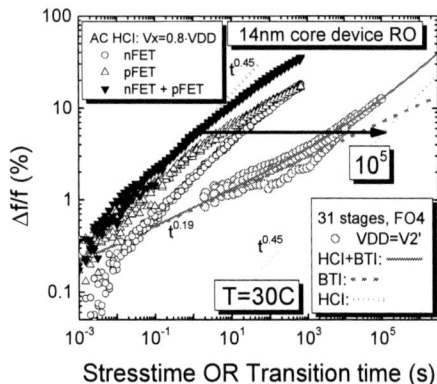

Fig. 8. Comparison of core RO frequency degradation with projected core nFET and pFET AC HCI (f=5kHz) contribution for a cross-point Vx=0.8. The gap between stress and transition time is ~10^5x considering combined contribution for nFET and pFET devices.

Fig. 9. Comparison of IO RO frequency degradation with projected IO nFET and pFET AC HCI (f=5kHz) contribution for a cross-point Vx=0.8. The gap between stress and transition time is ~30x which is comparable to the historical 50x scaling factor used to define DC equivalent HCI targets.

IV. CONCLUSION

The impact of HCI degradation on digital CMOS logic was assessed based on RO circuits using core and IO device in 14nm FinFET technology. HCI contributions are clearly visible in IO devices stressed at T=125ºC while for core device it is only seen at T=30ºC and high stress voltage. The correlation of RO degradation data with AC HCI in discrete devices reveals no additional reliability margin for IO devices while for core devices the DC equivalent HCI targets can be substantially reduced.

ACKNOWLEDGMENT

The authors would like to acknowledge the discussions with the Reliability Engineering team at GLOBALFOUNDRIES.

REFERENCES

[1] S. Mahapatra, V. Huard, A. Kerber, V. Reddy, S. Kalpat and A. Haggag, "Universality of NBTI – From Devices to Circuits and Products", IRPS pp. 3B.1.1-3B.1.8, 2014.

[2] A. Kerber and T. Nigam, "Determination of DC equivalent hot carrier stress times in scaled CMOS devices using novel AC stress methodology", submitted to IEEE EDL.

[3] A. Kerber, X. Wan, Y. Liu, T. Nigam, "Fast wafer-level stress-and-sense methodology for characterization of Ring-Oscillator degradation in advanced CMOS technologies", IEEE Trans. Electron Devices, vol. 62, no. 5, pp. 1427 - 1432, 2015.

[4] P. Heremans, R. Bellens, G. Groeseneken, H.-E. Maes, "Consistent Model for the Hot-Carrier Degradation in n-Channel and p-Channel MOSFET's", IEEE Trans. Electron Devices, vol. 35, no. 12, pp. 2194 - 2209, 1988.

[5] P. Fang, J. Tao, J.-F. Chen, C. Hu, "Design In Hot-Carrier Reliability for Higher Performance Logic Applications", IEEE Custom Integrated Circuits Conference, pp. 525-531, 1998.

[6] E. Li, E. Rosenbaum, J. Tao, G. C-F Yeap, M-R. Lin, and P. Fang, "Hot Carrier Effects in nMOSFETs in 0.1mm CMOS Technology", IRPS, pp. 253 - 258, 1998.

The Impact of RTN-induced Temporal Performance Fluctuation against Static Performance Variation

Takashi Matsumoto[1], Kazutoshi Kobayashi[2] and Hidetoshi Onodera[3]

[1] VLSI Design and Education Center (VDEC), The University of Tokyo, Tokyo, Japan
[2] Department of Electronics, Kyoto Institute of Technology, Kyoto, Japan
[3] Department of Communications and Computer Engineering, Kyoto University, Kyoto, Japan
Phone: +81-3-5841-6764, Email: takashi.matsumoto@cad.t.u-tokyo.ac.jp

Abstract

Random telegraph noise (RTN) is one of major recent transistor reliability concerns in designing reliable systems. In a circuit that contains a large number of small transiostors, the impact of RTN-induced fluctuation is considered to increase when it is compared with the static frequency variation caused by manufacturing process. The impact of RTN on process variation is described based on our measurement results from 40 nm test chip.

1. Introduction

Physical feature size of transistors has been minimized continually over time. One of the dominant issues on realizing reliable systems is transistor performance variations. In this paper, we discuss the impact of Random Telegraph Noise (RTN) affecting transistor performance variations. It was predicted in the past that the impact of RTN-induced drain current fluctuation might exceed manufacturing process variation in 22 nm technology[1]. The horizontal axis in Fig. 1 is the drain current fluctuation in linear scale. The vertical axis is the normal quantile. The dotted line is a large MOSFET case and the solid line is a small MOSFET case. Process variation usually follows a normal distribution. RTN does not follow normal distribution and its distribution has a long-tail part. The impact of RTN dominates process variation at the cross point depicted as red circle in Fig. 1.

2. Test Structure for RTN measurement

Figure 2 shows the test structure for RTN measurement [2][3]. Combinational logic circuit delay is measured by ring oscillator (RO) oscillation frequency. The power supply for RO (VDD_{RO}) and DFF (VDD_{DFF}) can be independently supplied. All logic gates except NAND2 with EN input are homogeneous. RTN-induced delay fluctuation is measured by the RO frequency fluctuation. There are 840 same ROs on 2 mm^2 area and the statistical nature of RTN can be evaluated by the RO array. This chip is fabricated in a commercial 40 nm CMOS technology. All measurements are done at room temperature. Figure 2 also shows measurement result example of 7-stage RO oscillation frequency for about 80 s at $VDD_{RO} = 0.65$V. The transistor width of the inverter (INV) is smallest in this technology. Measurement results show the large step-like frequency fluctuation caused by RTN. Here, F_{max} is defined as the maximum oscillation frequency and ΔF is defined as the maximum frequency fluctuation. $\Delta F/F_{max}$ is a good reference for the impact of RTN-induced frequency fluctuation for logic delay.

3. Impact of RTN on Process Variation

It is confirmed in [3] that the distribution of $\Delta F/F_{max}$ follows a log-normal distribution above 50% level in cumulative probability when data are collected over 15 chips (12,600 ROs) under 0.65 V operation. The distribution of RO frequencies (F_{max}) for the same ensemble is confirmed to follow normal distribution as expectedt[3]. In a circuit with a large number of small transiostors, the impact of RTN-induced temporal fluctuation is considered to increase when it is compared with the static frequency variation caused by manufacturing process. Figure 3 shows ΔF versus F_{max} plot over 12,600 ROs. The vertical axis is plotted with log scale. Figure 4 shows $\Delta F/F_{max}$ versus F_{max} plot over 12,600 ROs. It represents how the distribution of the impact of RTN correlates with process variation distribution. The triangle shape distribution suggests that there is no or weak correlation between RTN and process variations. Figure 5 can be obtained by plotting the vertical axis of Fig. 4 with log scale. The circle shape distribution suggests that there is no or weak correlation between the variation of the impact of RTN and process variation. Figure 6 shows the impact of RTN when it is compared with that of process variation. The impact of RTN on process variation is defined as

$$\frac{(\Delta F/F_{max})_{n\sigma}}{(n\sigma/\mu)}. \tag{1}$$

The plot for the minimum size 7-stage RO at 0.65 V ($\times$) can be obtained as follows. $\Delta F/F_{max}$ follows log-normal distribution. $(\Delta F/F_{max})_{n\sigma}$ can be obtained for each σ using the log-normal distribution. F_{max} follows normal distribution. $(n\sigma/\mu)$ can be obtained for each σ using the normal distribution. The dotted line is estimated from measured distributions of both RTN and process variation when log-normal distribution for RTN is assumed up to 7σ value. For the minimum size 7-stage RO at 0.65 V ($\times$), the impact grows exponentially when σ is increased. It is found that RTN becomes comparable to process variation around 7σ value. When the operating voltage is slightly increased to 0.75 V ($\triangle, \circ$), the RTN impact decreases rapidly. Finally, when the transistor size is increased from the minimum to the standard size at 0.75 V, RTN has small (and almost constant) impact on process variation ($\circ$).

4. Conclusions

The impact of RTN on process variation with respect to CMOS combinational circuit is estimated by experimental data. Measurement data suggests that there is no or weak correlation between RTN variation and process variation. It is found that the impact of RTN can be drastically increased when supply voltage and gate area become low and small.

978-1-5090-4661-4/17 $31.00 © 2017 IEEE

Acknowledgement

This work was supported in part by CREST, JST.

References

[1] N. Tega, *et al.*, IRPS2011, p.630.

[2] T. Matsumoto, *et al.*, IEDM2012, p.581.

[3] T. Matsumoto, *et al.*, CICC2014, Session 14-4.

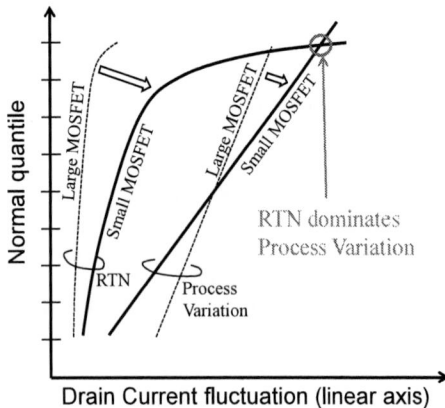

Figure 1: Conceptual figure of RTN vs Process Variation as for statistical distribution.

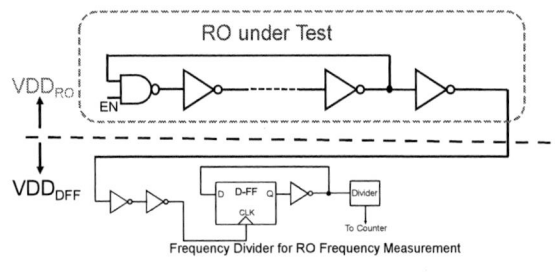

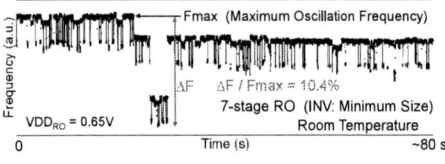

Figure 2: Test structure for RTN measurement. RO frequency fluctuation by RTN in one RO is also shown.

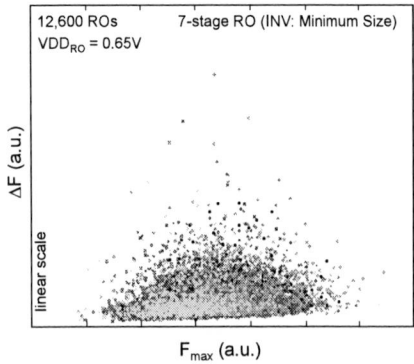

Figure 3: ΔF versus $F_{\max}$ plot over 12,600 ROs (vertical axis: log-scale).

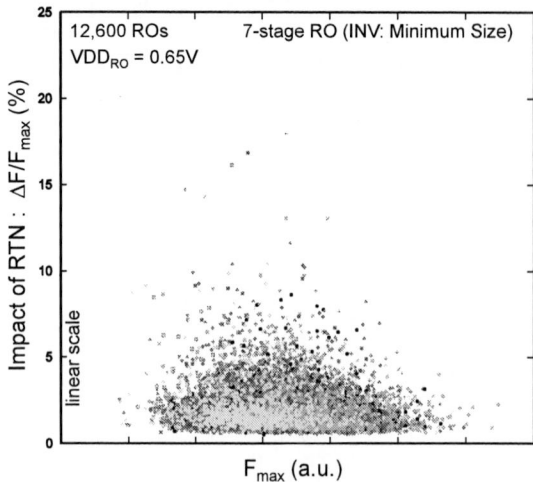

Figure 4: $\Delta F/F_{\max}$ versus $F_{\max}$ plot over 12,600 ROs.

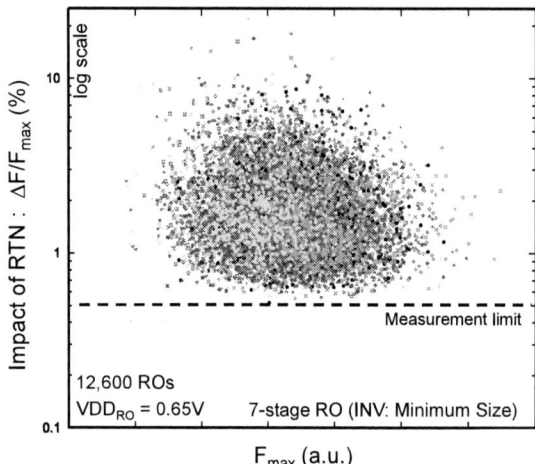

Figure 5: $\Delta F/F_{\max}$ versus $F_{\max}$ plot over 12,600 ROs (vertical axis: log-scale).

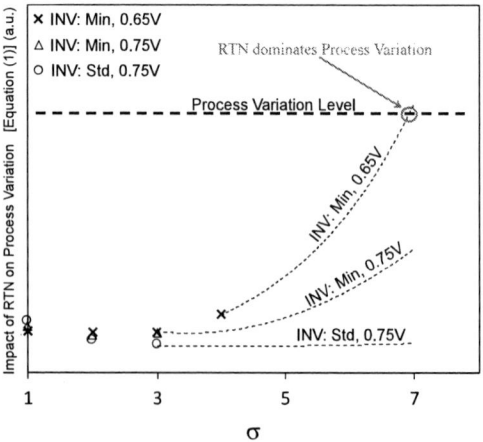

Figure 6: The impact of RTN on process variation.

3B-3

Critical Discussion on Temperature Dependence of BTI in Planar and FinFET devices

P .Srinivasan, Tanya Nigam

GLOBALFOUNDRIES Inc., Stonebreak Road Extension, Malta, NY, 12020, USA

Phone: +1 (518) 305 1325, email: Purushothaman.Srinivasan@globalfoundries.com

Abstract

The impact of temperature (T) on BTI time slope (n) and voltage acceleration (VAE) in MGHK planar and FinFETs are discussed here. Steeper time slopes and higher VAE are observed at lower 1/kT for thin gate oxide devices across different stress voltages. Dual time slope behavior in planar thick oxide PFETs is observed. Time slopes increase as sense delay is increased for lower 1/kT while a reverse trend is observed at higher 1/kT, mainly due to trapping behavior. A higher recovery ratio close to 100% is observed at higher 1/kT and is due to negligible degradation. The implications on the activation energy extraction are discussed. (Keywords: Planar, FinFET, High-k and BTI)

Introduction

Bias Temperature Instability (BTI) is a critical reliability issue [1] and meeting BTI technology requirement have been challenging both in planar and FinFET devices. In addition, to process and integration impact, BTI modeling is also a critical aspect. In this work, the impact of T on BTI is studied in thin and thick gate oxide RMG planar and FinFET MG/HK stacks. D. Varghese *et al.* [2], performed similar work in SiON based devices, but not comprehensively. The temperature effect is extensively studied here from -40 C to 250 C focusing on BTI time slopes and voltage acceleration. Ultra-fast sensing was performed to study sense delay impact which is used to understand the trapping behavior. The impact due to T on recovery is also studied. Trap distribution and its energy level are correlated to time slopes and the implications towards activation energy extraction are discussed.

Experimental

Short channel MG/HK planar and FinFET MOS devices with Hf-based gate dielectric were fabricated utilizing a gate-last process flow on (100)Si substrates. Stress-Measure-Stress (S-M-S) technique [4] was used to measure BTI for three stress voltages V_{G1}, V_{G2} and V_{G3} (>= 1ks) using ultrafast intermittent drain current sensing [3,4] for threshold voltage degradation (ΔV_T) extraction. The chuck T was varied from -40 C to 250 C and sample size/per condition of each individual data set was N ~ 10. A simple V_G-based model where

$$\Delta V_T = A * V_G{}^{VAE} * t^n * e^{-Ea/kT}$$

is used to extract n, VAE and E_a.

Results and Discussion

Time slope 'n' behavior: The impact of temperature in planar and FinFET devices is summarized in Fig 1 and Fig 2 where ΔV_T Vs stress time is plotted for planar and FinFET thin gate oxide devices. Both show consistent trend, where lower degradation and shallower time slopes are observed at lower T. In addition, the FinFET devices show higher dispersion at low T.

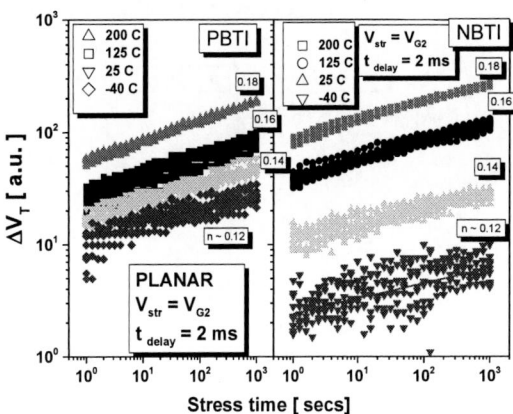

Fig 1. ΔV_T vs Stress time of (Left) nFET and (Right) pFET thin gate oxide PLANAR device for different temperatures. Lower temperature in these devices show lower degradation and have shallower time slopes. A sense delay of 2ms is used measured using S-M-S technique.

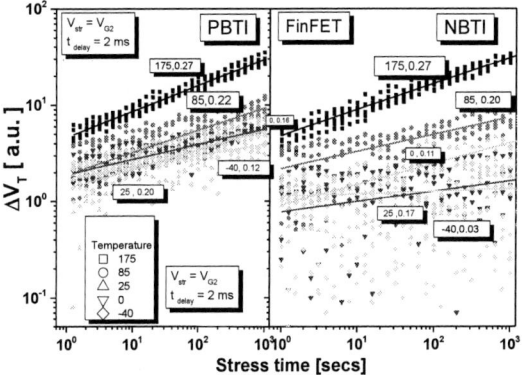

Fig.2. ΔV_T vs Stress time of (Left) nFET and (Right) pFET thin gate oxide FinFET device for different temperatures. Similar to planar devices, lower temperature in these devices have lower degradation and shallower time slopes. In addition, a higher dispersion is also noticed at lower temperature. S-M-S technique with sense delay of 2ms is used.

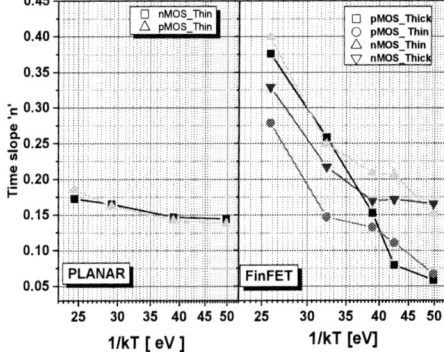

Fig 3. Time slope Vs 1/kT of (Left) PLANAR FET and (Right) FinFET thin and thick gate oxide at different temperatures. Consistent trend between planar and FinFET devices noticed where steeper time slopes are observed for higher temperature (lower 1/kT).

The time slopes 'n' were extracted using the model $\Delta V_T = A*t^n$ and plotted in Fig. 3 against 1/kT. Both planar and FinFET devices, show steeper time slopes at higher T where n~0.27 is observed at 175 C and reduces to n < 0.1 at -40 C. The impact of stress voltage conditions were studied using three different stress

978-1-5090-4661-4/17 $31.00 © 2017 IEEE   33

voltages and summarized in Fig 4 for both thin and thick gate oxide nFET and pFET. For every stress voltage, similar dependency of 'n' to T was observed. This indicates that there is no strong 'n' dependence due to stress voltage condition.

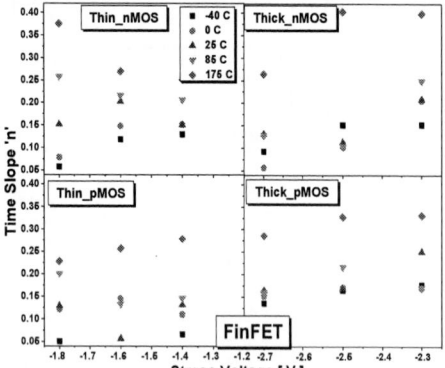

Fig 4. Similar time slope behavior with temperature observed across various stress voltages. Three stress voltages were evaluated for (a) NMOS and (b) PMOS. Both show steeper time slopes at higher temperature and shallower time slopes at lower temperature.

The impact of sense delay was also studied and Fig 5 shows ΔV_T Vs stress time for two sense delay conditions (2µs and 2ms) at three different T for planar devices. Shorter sense delays show shallower time slopes at all temperatures.

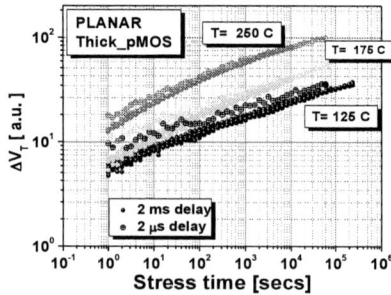

Fig. 5. Impact of sense delay on time slope in thick gate oxide planar devices. Time slopes increase as the sense delay is increased for lower 1/kT (higher temperature) values.

For FinFET devices, the impact of sense delay on thin and thick pMOS devices is plotted in Fig. 6. For both these device types, time slopes increase with increasing sense delay for T > 0 C. However a reverse trend for T < 0 C is observed, where the time slopes decrease.

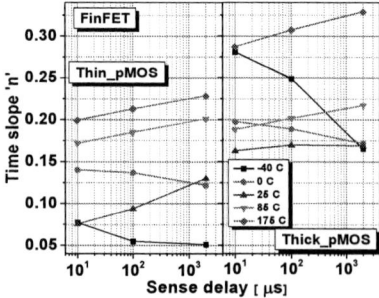

Fig 6. Impact of sense delay on time slopes in thin and thick gate pMOS FinFET devices. Time slopes increase as the sense delay is increased for lower 1/kT (higher temperature) values. However, a reverse trend is observed at higher 1/kT values, where longer delay show shallower time slopes.

Isolation of trapping component from permanent component:

The sense delay behavior is further understood by isolating the trapping (recoverable) component from the permanent degradation component of ΔV_T. The trapping component N_{OT} is isolated using $\Delta N_{OT} \sim (\Delta V_T)_{10\mu s} - (\Delta V_T)_{2ms}$ and is plotted in Fig 7. Clearly, the trapping behavior is dependent on T where higher T shows higher trapping behavior for both oxide thicknesses. In addition, negative (hole) trapping is observed for T > 0 C while positive (electron) trapping is observed for T < 0 C, causing negative ΔV_T shift. This reverse trapping behavior during shorter sense delay measurements leads to higher time slopes at lower T.

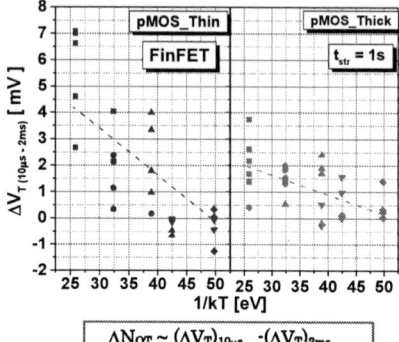

Fig. 7. Trapping behavior captured as ΔN_{OT} and is dependent on T. Higher temperature shows higher trapping behavior. Positive (electron) trapping occurs for T < 0 C and negative (hole) trapping for T > 0 C.

Dual slope behavior in thick gate oxide planar devices is observed at high T and the results are summarized in Fig. 8a-8d. At lower stress times (t_{str} < 100s), as seen in Fig 8b, the time slopes are steeper and increase from n ~0.16 to ~0.23 as T increases. In Fig.8c, shallower time slopes are observed for t_{str} > 100s, where n varies from ~0.12 to ~0.17 and saturates at ~125 C. These two sets of values are also plotted as a function of T as shown in Fig 8d.

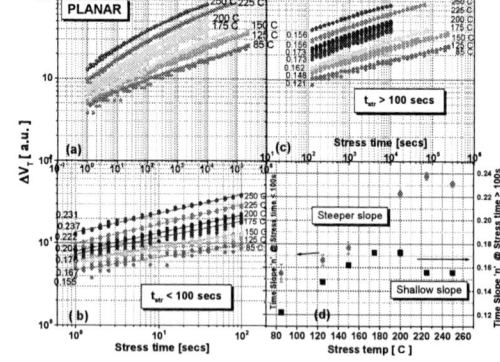

Fig 8. Dual time slope behavior in planar thick gate oxide PFETs at high T and lower stress times. However, time slopes become comparable at higher stress times across temperature for these devices.

Discussion on time slope behavior:
Overall, two kinds of behavior are observed as shown in Fig 9. If R-D model is assumed [1,5], a varying time slope with T would be observed for devices having uniform trap distribution across energy level,

as seen in most planar and FinFET devices. However, a finite trap distribution centered at a distance from the interface could cause dual slope behavior as seen in thick gate PFET planar devices.

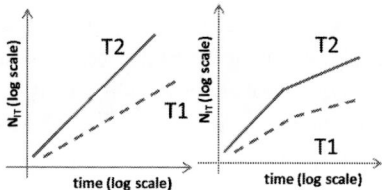

Fig 9. Trap distribution impact on time slopes.

Voltage Acceleration behavior: Impact of Voltage Acceleration (VAE) is studied in planar and FinFET devices using $\Delta V_T = A \cdot V_G^{VAE}$ and the results are plotted in Fig 10 for high (175 C) and low T (-40C), extracted at t_{str} = 1ks. Clearly, higher VAE is observed at lower T for both gate devices and vice versa. The overall summary is shown in Fig 11, where VAE increases with increasing 1/kT for both planar and FinFETs.

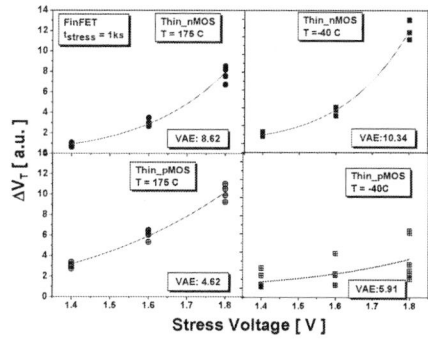

Fig 10. Impact of VAE due to temperature in thin gate oxide NFET and PFET FinFET devices. VAE is evaluated at t_{stress} of 1ks. VAE increases as temperature decreases (or 1/kT increases) for these device types.

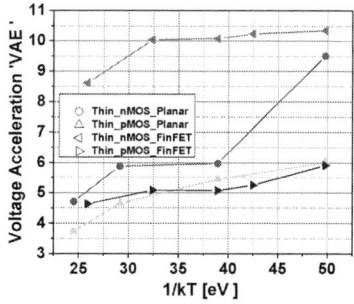

Fig 11. Impact of VAE due to T in thin gate oxide PLANAR and FinFET devices. VAE increases as 1/kT increases for both (a) NMOS and (b) PMOS in PLANAR and FinFET devices.

Recovery impact is also studied using two symmetric stress cycles – Cycle 1 and 2: a 1ks at $V_{g,str}$ is applied followed by an idle time of 1ks and the results are plotted in Fig 12. As indicated earlier, ΔV_T during cycle 1 is dependent on T. In addition, ΔV_T is not fully recovered at higher T during cycle 2 while it is close to fully recovery at low T. This is due to higher permanent component of ΔV_T at higher T coupled with low ΔV_T at T<0 C.

Recovery ratio is defined as percentage of ΔV_T that gets recovered after cycle 2 and is plotted in Fig 12. Both nFET and pFET devices show ~95% at low T while highest T show 40~50% recovery for pFET and 30~40% recovery for nFET.

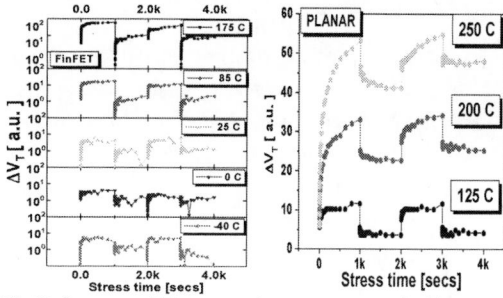

Fig 12. Impact on BTI recovery due to temperature in thin gate oxide FinFET and PLANAR devices. Two symmetric stress cycles – Cycle 1 and 2: a $V_{g,stress}$ of 1ks is applied followed by an idle time of 1ks.

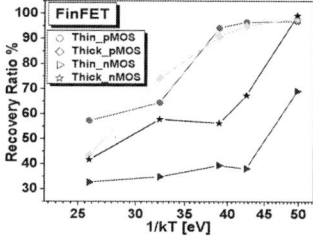

Fig 13. Recovery ratio in thick and thin NMOS and PMOS devices. A higher recovery ratio is observed at higher 1/kT and vice versa. Recovery ratio is close to 100% for all devices at higher 1/kT, primarily due to negligible degradation.

Implications on activation (E_A) energy estimation: For an operating T range (85 C to 125C), a constant time slope and VAE can be assumed during E_A estimation, as Arrhenius behavior is expected. However, for a wider T range (175 C to -40C), the behavior is non-Arrhenius and the effects of time slope and VAE should not be ignored. Adequate care should be given during extraction of E_A at these T ranges.

Acknowledgement

The authors would like to thank Peter Paliwoda for measurements, L. Pantisano, S. Uppal and Z. Chibli for comments and suggestions.

Summary

In summary, time slopes and voltage acceleration vary with kT in thin and thick oxide planar and FinFETs. Time slopes are sensitive to sense delay conditions. The trapping behavior associated with delay show dependence to T. Recovery study shows higher recovery at higher 1/kT. Adequate care should be given for E_A estimation due to non-Arrhenius behavior at such T.

References

[1] M. A. Alam *et al*, Micro. Rel. , vol. 45, no. 1, pp. 71-81, 2005. [2] D. Varghese *et al*, Proc. IEDM, 2005. [3] A. Kerber, *et al.,* IEEE TED, vol. 55, no. 11, pp. 3175–3183, 2008 [4] H. Reisinger, *et al.,* IRPS 2006, pp. 448-453. [5] S. Mahapatra *et al.,* IEEE TED, vol. 53, no. 7, pp. 1583-92, 2006.

3B-4

Characterization of critical peak current and model of Cu/low-k interconnects under short pulse-width conditions

M. H. Lin, W. S. Chou, Y. T. Yang, Y. C. Peng, and A. S. Oates

Taiwan Semiconductor Manufacturing Company. Ltd., email:mhlinza@tsmc.com

Abstract

We characterize the critical peak current which causes melting of Cu/low-k interconnects under short-pulse conditions. High-current with 100ps pulse width is achieved using an on-die pulse generator. A model incorporating the heating of the metal layer and heat diffusion through the insulator layer is supported by the experimental results. The model accurately describes the relationship between peak current and pulse width (20μs - 100ps), and can be used to generate reliable guidelines for high-current applications.

Introduction

High-current robustness of interconnects is an important issue for high-performance circuits today since aggressive interconnect scaling has resulted in increasing current densities and associated thermal effects. Metal interconnects in protection circuits frequently withstand a high current, short duration electrostatic discharge (ESD, <200ns event) and electrical overstress (EOS, >1us event). However, in high-performance circuits, interconnects need to sustain a transient current with pulse width smaller than 100ps which occurs during logic transition Apart from normal circuit conditions, instantaneous high current spikes occur during wrong power up/down sequences. These high current pulses can cause abrupt temperature increases in metal interconnects, and may result in instantaneous failure. The critical peak current is the current at which a metal line undergoing excessive Joule heating begins to melt. Extensive experimental investigations and modeling efforts have been made to characterize the critical peak current of Al metallization systems [1-4]. However, studies performed with Cu / low-k interconnects are very limited [3]. No peak current characterization beyond 1ns pulse width is reported to our knowledge. It is important to clarify physical mechanisms involved in high-current failure since this finding is central to the accurate characterization of critical peak current in high-speed conditions. Therefore, in this work, we investigate the effect of pulse width in the Cu / low-k interconnect system and demonstrate that the Wunsch-Bell model accurately describes the relationship between peak current and pulse width (20μs - 100ps) [1].

Experimental Details

Samples were fabricated using a Cu/low-k dual-damascene process. The interconnects consisted of a TaN based liner and Cu seed with ECP Cu trench layers. All test structures were fabricated with SiN based capping layers. A transmission line pulse (TLP) system was used to characterize peak current dependence in the range of 1ns to 20μs. The critical peak current is usually measured by increasing the pulse current magnitude until the metal line melts (fig. 1a). However, there is a limitation of pulse width with this technique due to a distortion of the waveform below 1ns TLP, shown in fig. 1b. Since it is nearly impossible to inject very high frequency signals through a bound pad to the test structure, an on-die voltage-controlled oscillator (VCO) fabricated in a 16nm FinFET process was designed to explore the extra-low pulse width region [5]. A small pulse width and a duty cycle can be generated with a tunable pulse then a minimum pulse width of 100ps can be achieved (fig. 2).

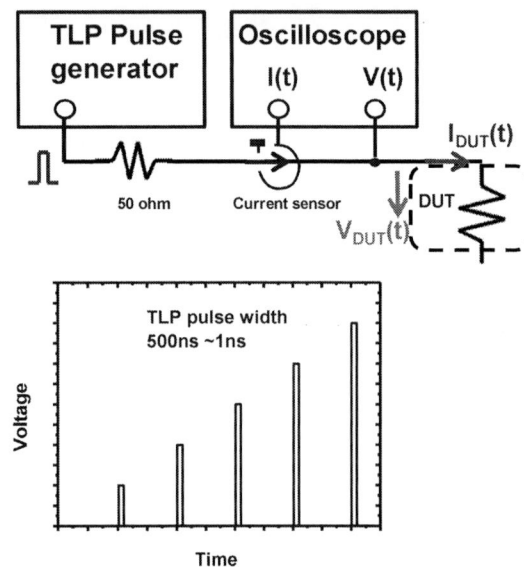

Fig. 1a: Schematic diagram of TLP and voltage ramp stress for critical peak current measurement.

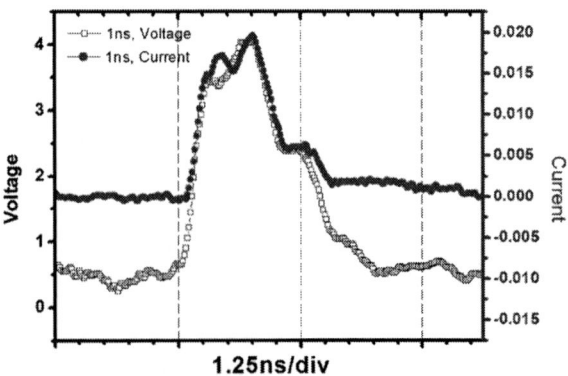

Fig. 1b: The waveform of TLP 1ns current/voltage shown using an external oscilloscope.

978-1-5090-4661-4/17 $31.00 © 2017 IEEE   36

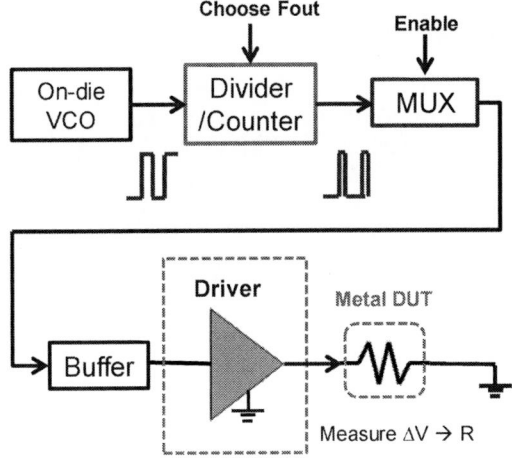

Fig. 2: On-die pulse generator system for short-pulse current measurement.

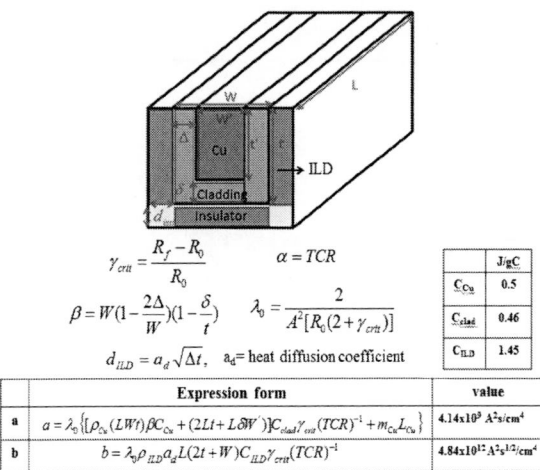

$$\gamma_{crit} = \frac{R_f - R_0}{R_0} \qquad \alpha = TCR$$

$$\beta = W(1 - \frac{2\Delta}{W})(1 - \frac{\delta}{t}) \qquad \lambda_0 = \frac{2}{A^2[R_0(2 + \gamma_{crit})]}$$

$$d_{ILD} = a_d \sqrt{\Delta t}, \quad a_d = \text{heat diffusion coefficient}$$

		J/gC
	C_{Cu}	0.5
	C_{clad}	0.46
	C_{ILD}	1.45

	Expression form	value
a	$a = \lambda_0\{[\rho_{Cu}(LWt)\beta C_{Cu} + (2Lt + L\delta W)]C_{clad}\gamma_{crit}(TCR)^{-1} + m_{Cu}L_{Cu}\}$	4.14×10^3 A^{2}s/cm^4
b	$b = \lambda_0 \rho_{ILD} a_d 2(t + W)C_{ILD}\gamma_{crit}(TCR)^{-1}$	4.84×10^{12} A^{2}s$^{1/2}$/cm^4
c	$c = \lambda_0 \rho_{ILD} 2(a_d^2 \Delta t)LC_{ILD}\gamma_{crit}(TCR)^{-1}$	3.00×10^{16} A^2/cm^4

Table.1: The expression forms of parameters in eq. (2) and values used in model fitting.

Model and results

The total energy required to melt the interconnects is comprised of the energy to reach the melting temperature and additional heat for fusion. The formula is given by: $E=C\Delta T+mCuL_f$, where E is total energy, C is the capacity, mCu is Cu mass and L_f is fusion heat. The input pulse energy is related to the I^2R joule-heating integrated over the pulse width shown below.

$$E = \int I^2 R dt \approx \frac{1}{2} I^2 \Delta t (R_0 + R_f) \qquad (1)$$

Where, Δt is pulse width, R_0 is initial resistance, and R_f is final resistance to failure. A physical model is presented that includes Cu, barrier metals, and insulator film heating. The relationship between the critical peak current and the pulse width can be obtained based on total energy and eq.(1). Three heat dissipation mechanisms describe the relationship of critical peak current and pulse width [4]. When pulse width is much shorter than the dielectric film thermal diffusion time, τd, the generated thermal energy is kept inside the metal line (adiabatic response). When pulse width is comparable to τd, heat will diffuse into the dielectric film. The thickness of the insulator sheath is assumed to be proportional to the square root of the pulse width based on heat diffusion theory. Critical peak current reaches steady-state when pulse width >> τd, and is then pulse width independent. So the maximum power in the pulse is limited by the rate at which the heat sink can remove energy from the interconnects. The critical peak current density is the summation of three pulse width regions.

$$J_{peak}^2 = \frac{a}{\Delta t} + \frac{b}{\sqrt{\Delta t}} + c \qquad (2)$$

Parameters for Cu/low-k interconnect are given in table 1.

Fig. 3 shows a typical TLP stress induced resistance increase until the metal line is open. Metal melting and dielectric film cracking are observed at the failure location. Fig. 4 shows the TLP I-V characteristic of nano Cu wires at different pulse widths. Higher critical peak currents can be sustained while pulse width decreases, as expected. The critical peak current as a function of pulse width is shown in fig. 5 and the values are consistent with the pulse width predicted by eq. (2). The adiabatic term and thermal diffusion term are included. It is clear that heat dissipation into the low-k film is significant at pulse width 1μs. There is some deviation between the experimental data and the model prediction at very short pulse widths. The deviation may be caused by the inaccuracy of heat capacity values for the thin Cu film and barrier layers used in the calculation of the adiabatic term. An overshooting transient current during logic transition can cause metal line melting, even though pulse width is smaller than 100ps. The on-die pulse generator provides an extra-low pulse width signal to characterize the peak current behavior in the GHz range. Fig. 6 shows critical peak current versus pulse width from 1μs to 100ps. The adiabatic term dominates the failure mechanism of critical peak current at very low pulse widths. The time-dependent critical peak current is demonstrated in comparison with the limited adiabatic regime, thermal diffusion, and steady-state conditions. The adiabatic approximation is found to be valid only for an extremely short pulse duration (<1ns). The results and model provide a guideline to prevent catastrophic failures of the metal line caused by high current pulses through a chip.

978-1-5090-4661-4/17 $31.00 © 2017 IEEE

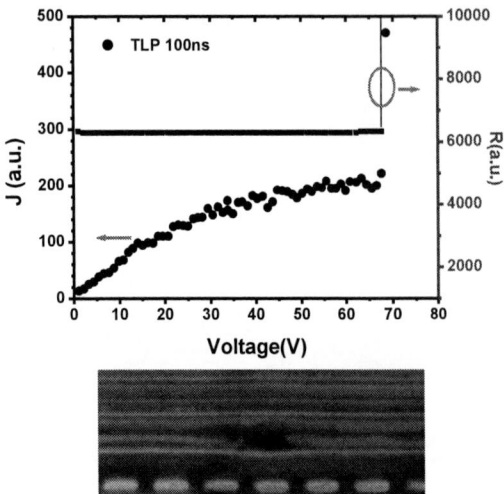

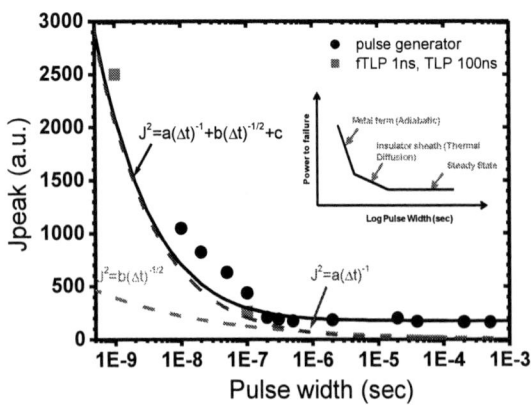

Fig. 3: Typical critical peak current and resistance as a function of TLP voltage. SEM plane view of post TLP stress failure mode after resistance jump.

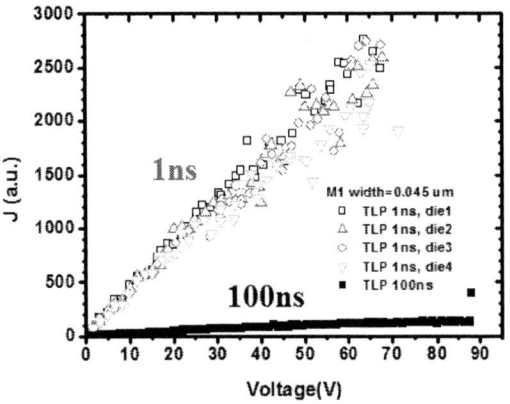

Fig. 4: The current density as a function of TLP voltage for 100ns and 1ns pulse widths.

Fig. 5: The critical current density as a function of pulse width. Model fits results well from steady-state to adiabatic region.

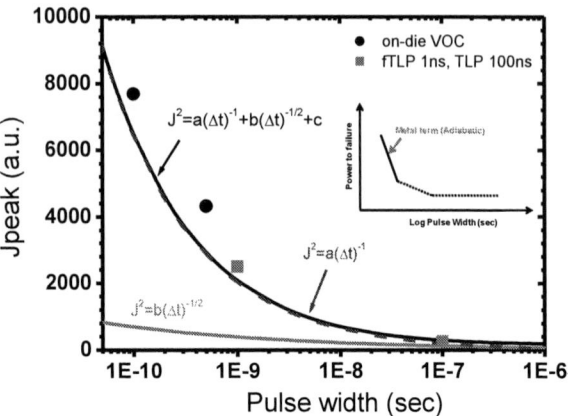

Fig. 6: The critical current density as a function of short-pulse width. Model fits results well in adiabatic region.

Conclusion

High-current induced metal melting has been characterized in Cu/low-k interconnects under short pulse width conditions. The observed critical peak current as a function of pulse width can be accurately described by Wunsch-Bell model down to 100ps. When pulse width is less than 1μs, thermal dissipation diminishes compared to the adiabatic response

References

[1] D. C. Wunsch, and R. R. Bell, IEEE Transcations on Nuclear Science, Vol. 15, p.244, 1968.

[2] D. M. Tasca, IEEE Transcations on Nuclear Science, Vol. 17, p.364, 1970.

[3] S. Voldman, et al., in Proc. Int. Reliability Physics Symp. (IRPS), p.144, 1999.

[4] K. Banerjee et al., IEEE Electron Device lett., Vol. 18, p.405, 1997.

[5] Y. T. Yang et al., in preparing CICC conference.

978-1-5090-4661-4/17 $31.00 © 2017 IEEE   38

3B-5

New Analytical Equations for Skin and Proximity Effects in Interconnects Operated at High Frequency

Haojun Zhang, Jian-Hsing Lee, Natarajan Mahadeva Iyer, and Linjun Cao

Reliability Engineering, GLOBALFOUNDRIES Inc., Malta, New York USA

Haojun.zhang@globalfoundries.com

Abstract

For the first time, analytical equations for skin and proximity effects are derived to successfully describe current distributions in advanced CMOS technology interconnects subject to high-frequency signals. The analytical solution matches simulations evaluating skin depth as a function of interconnect geometry and operating frequency.

(Keywords- Electromagnetic Induction, Skin effect, Skin depth, Proximity effect, AC Electromigration)

I. Introduction

In back-end interconnects of chips operated at GHz frequencies or higher, electromagnetic induction causes significant skin and proximity effects. These effects are not accounted for by traditional methods used to predict reliability. In particular, skin depth, which is the most important parameter to describe electrical skin effects, should be considered for AC electromigration [1] and inductor modeling [2], [3] since ~68% of current is confined in this region. The formula used to estimate skin depth in early CMOS technology literature [1]-[3] is only appropriate for large scale components ($\delta = 1 / \sqrt{\pi f \mu \sigma}$) [4].

As dimensions shrink down to micrometer scale or below, this classic formula is no longer valid. Moreover, there is no analytical equation to describe proximity effects for two parallel interconnectors interacting with each other at high frequencies.

In this work, we derive new analytical equations for skin and proximity effects based on the original Maxwell equations. We validate them with simulations. We show how these equations can be used to calculate current distributions and skin depth in interconnects as a function of dimensions (thickness, width) and proximity to other interconnects.

II. Analytical Equations

A. Single interconnect

Based on Maxwell equations $\nabla \times E = -\frac{\partial B}{\partial t}$ and $\nabla \times B = \mu_0 J$ and $J = \sigma E$, current density for an interconnect as shown in Fig. 1 is given by:

$$\nabla^2 J_z = -\mu_0 \sigma \frac{\partial J_z}{\partial t} \tag{1}$$

Let $J_z = e^{i(\omega t - \vec{k}\vec{r})}$, $\vec{k} = k_z \vec{z} - i(k_x \vec{x} + k_y \vec{y})$, we obtain

$$k^2 J_z = -i\omega \mu_0 \sigma J_z \tag{2}$$

Therefore,

$$k = \sqrt{\frac{\omega \mu_0 \sigma}{2}} - i\sqrt{\frac{\omega \mu_0 \sigma}{2}} = k_z - i\sqrt{k_x^2 + k_y^2} \tag{3}$$

Based on boundary condition $J_z(a, b, 0) = J_z(-a, -b, 0)$, solution is

$$J_z = J_0 e^{-i(wt - k_z Z)} \cosh(k_x x) \cosh(k_y y) \tag{4}$$

Based on continuity theory ($B_x = B_y$ at $x=a$, $y=b$) and $\nabla \times J_z = -\sigma \frac{\partial B}{\partial t}$, we get

$$\frac{k_x}{k_y} = \frac{\sinh(k_y b)}{\sinh(k_x a)} \frac{\cosh(k_x a)}{\cosh(k_y b)} \approx \frac{k_y b}{k_x a} \tag{5}$$

Substituting k_x and k_y into Eq.(3), we obtain

$$k_z = \sqrt{\pi f \mu_0 \sigma}, k_x = k_z \sqrt{\frac{b}{a+b}}, k_y = k_z \sqrt{\frac{a}{a+b}} \tag{6}$$

From Eq. (4) and the definition for skin depth δ ($J_z = e^{-1}J_{max}$), where δ is given by

$$\delta = a - \cosh^{-1}(e^{-1}\cosh(k_x a))/k_x \tag{7}$$

B. Two parallel interconnects

In two parallel interconnects separated by distance $2d$, currents flowing in the same direction (Fig. 2a), the current for the right hand side interconnect is:

$$J_z = J_0 e^{-i(wt - k_z z)} \left(\cosh\left(k_x\left(x - d - \frac{a}{2}\right)\right) - m\cosh\left(k_x\frac{a}{2}\right)e^{-k_x(x-d)} \right)\cosh(k_y) \tag{8}$$

Where $m = (a2+2b)/(a2+4da+2b(a+4d))$.

For currents flowing in opposite directions (Fig. 2b), it is:

$$J_z = J_0 e^{-i(wt - k_z z)} \left(\cosh\left(k_x\left(x - d - \frac{a}{2}\right)\right) + m\cosh\left(k_x\frac{a}{2}\right)e^{-k_x(x-d)} \right)\cosh(k_y) \tag{9}$$

Where $m = (a2+2b)/(a2+4da+2b(a+4d))$.

Based on continuity theory ($B_x = B_y$ at $x=a+d$, $y=b$) and $\times J_z = -\sigma \frac{\partial B}{\partial t}$, for currents flowing in the same direction (Fig. 2a), we get

$$\frac{k_x}{k_y} = \frac{\sinh(k_y b)}{\left(\sinh\left(k_x\frac{a}{2}\right) + m\cosh\left(k_x\frac{a}{2}\right)e^{-k_x(a+2d)}\right)} \frac{\left(\cosh\left(k_x\frac{a}{2}\right) - m\cosh\left(k_x\frac{a}{2}\right)e^{-k_x(a+2d)}\right)}{\cosh(k_y b)} \approx \frac{2k_y b}{k_x a} \tag{10}$$

And for currents flowing in opposite directions (Fig. 2b), we get

$$\frac{k_x}{k_y} = \frac{\sinh(k_y b)}{\left(\sinh\left(k_x\frac{a}{2}\right) - m\cosh\left(k_x\frac{a}{2}\right)e^{-k_x(a+2d)}\right)} \frac{\left(\cosh\left(k_x\frac{a}{2}\right) + m\cosh\left(k_x\frac{a}{2}\right)e^{-k_x(a+2d)}\right)}{\cosh(k_y b)} \approx \frac{2k_y b}{k_x a} \tag{11}$$

Substituting k_x and k_y into Eq.(3), we obtain

$$k_z = \sqrt{\pi f \mu_0 \sigma}, k_x = k_z \sqrt{\frac{2b}{a+2b}}, k_y = k_z \sqrt{\frac{a}{a+2b}} \tag{12}$$

III. Results

A. Single interconnect

Fig.3(a) shows simulated current distributions in a single copper metal line with 0.5μm thickness and 6μm width subject to a 10GHz pulse by 3-D electromagnetic simulation (Ansys). Fig.3(b) shows calculated current distributions in the metal line in Fig.3(a) at z=0μm, based

978-1-5090-4661-4/17 $31.00 © 2017 IEEE

on Eq.(4) and Eq.(6). Both show that current densities at the two edges are higher than in the center. By comparing normalized current densities in the center (y=0μm), the calculated result matches the simulation except at the two edges (Fig.4). This deviation might be caused by the fact that the analytical model does not consider electromagnetic field losses at boundaries.

Fig.5 shows skin depth versus frequency evaluated from simulation, classic model, and calculations based on the 36% (e^{-1}) maximum current density from Eq.(4) and Eq.(7). It shows that calculations based on the 36% maximum current density and Eq.(7) are very close to simulation results, while the classic model cannot fit the simulation. This is direct evidence that the classic model is invalid to estimate skin depth for advanced CMOS technology interconnects.

B. Two parallel interconnects

Fig.6(a) shows simulated current distributions in two $6\times0.5\mu m^2$ parallel copper metal lines, 3μm apart, under same-direction 10GHz pulses. Fig.6(b) shows calculated current distributions in the same two metal lines at z=0μm, based on Eq.(8) and Eq.(12). Because of the proximity effect, the minimum current density for each metal line is at its center. It also causes current to crowd at two outer edges, where current densities reach 2.7x and 1.5x that of the centers and the edges facing each other. By comparing normalized current densities at the center (y=0μm), calculated and simulation results match except at the two edges, due to electromagnetic field losses at boundaries (Fig.7).

Fig.8(a) shows simulated current distributions in two $6\times0.5\mu m^2$ parallel copper metal lines, 3μm apart; and under opposite direction 10GHz pulses. Fig.8(b) shows calculated current distributions in the same two metal lines at z=0μm, based on Eq.(9) and Eq.(12). Because of the proximity effect, the minimum current density for each metal line is also at its center. However, it causes current to crowd at edges facing each other, where current densities reach 2.0x and 1.3x that of centers and outer edges. By comparing normalized current densities at the center (y=0μm), calculated and simulation results match well, as shown in Fig. 9.

VI. Discussion and conclusion

For interconnects in advanced CMOS technologies, the classic formula used to estimate electrical skin depth in large scale components is no longer valid. We found that a more appropriate description of skin depth should account for both signal frequency and interconnect layout, such as conductor thickness, width and proximity to other conductors. Even when the width is narrower than two times the skin depth, there is a non-uniform current distribution to account for. For a single metal line operated at 10GHz, the current density at the edge nearly reaches

1.5x that at the center. If there is any neighboring interconnect, edge current densities become either 2.0x higher (same-direction pulses), or 2.7x higher (opposite-direction pulses).

Using the superposition principle, the analytical skin effect model can be extended to account for proximity effects. This model can describe current distributions in two parallel interconnects under same- or opposite-direction signals. This same methodology can further be used to extend the analytical model to describe proximity effects in multi-interconnects or multi-layer interconnects.

References

[1] W. Wu, J. S. Yuan, Solid-state electrons, 2002, pp. 2269-2272.

[2] C. P. Yue, S. S. Wong, IEEE Trans. ED., 2000, pp. 560-568.

[3] R. J. Chan, J. C. Guo, IEEE Proc. 9th European Microwave Integrated Circuit conference, 2014.

[4] Harold A Wheeler, Proceeding of the I.R.E., 1942, pp. 412-424.

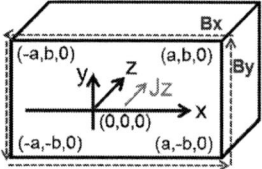

Fig.1: high frequency (HF) current flow through a single interconnect

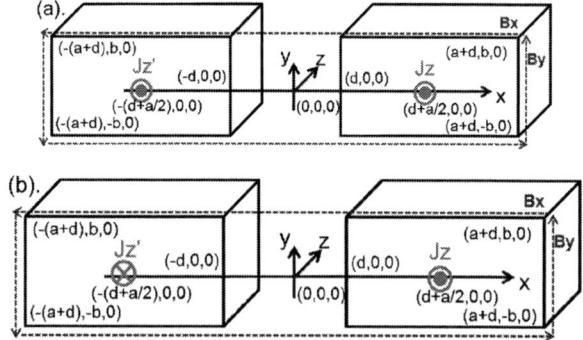

Fig.2(a): same-direction, and (b): opposite-direction HF current flow through two parallel interconnects.

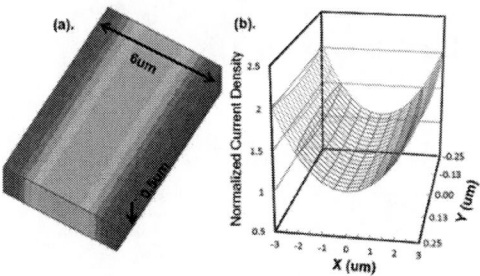

Fig.3(a): simulated, and (b): calculated current distributions in a single metal line under a 10GHz pulse.

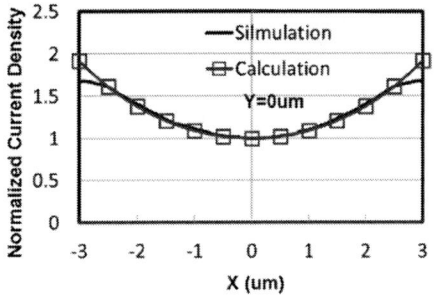

Fig.4: simulated and calculated current distributions in the metal line center (y=0um) in Fig. 3a.

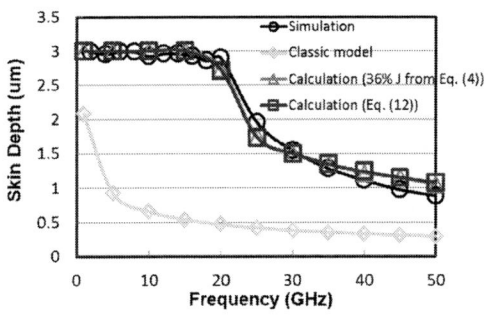

Fig. 5: skin depth vs. frequency comparisons for the simulation-based, classic model-based, and new equations-based results.

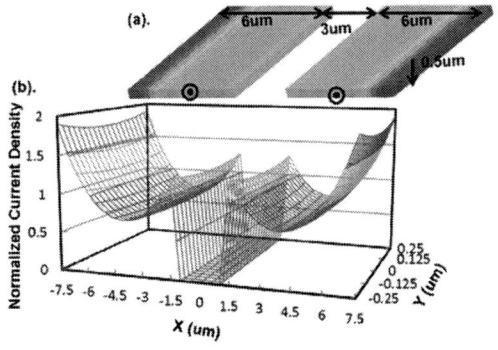

Fig.6(a): simulated, and (b): calculated current distributions in two parallel metal lines under same-direction 10GHz pulses.

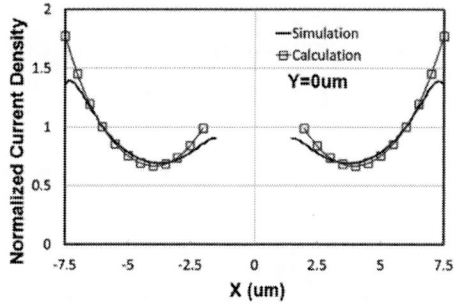

Fig.7: simulated and calculated current distributions in two parallel metal lines at the center region (y=0um) of Fig.6a.

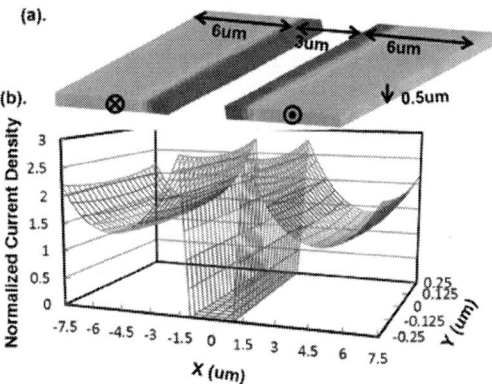

Fig.8(a): simulated, and (b): calculated current distributions in two parallel metal lines under opposite-direction 10GHz pulses.

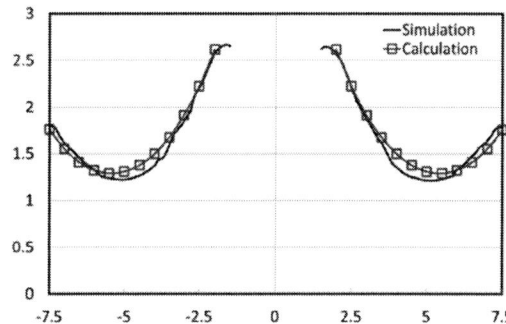

Fig.9: simulated and calculated current distributions in two parallel metal lines in the center region (y=0um) of Fig.8a.

978-1-5090-4661-4/17 $31.00 © 2017 IEEE

4M-1 (Invited)

System Integration in a Package for Cloud and Edge

Tadahiro Kuroda

Electronics and Electrical Engineering, Keio University

3-14-1, Hiyoshi, Kohoku-ku, Yokohama 223-8522, Japan

Introduction

Integrated Circuit (IC) was invented in 1958 in the process of challenges to "Tyranny of Numbers". We face the same challenge again with the end of Moore's Law and rise of IoT. A near-field coupling integration technology (Fig.1) is proposed as a new solution in very large system to replace mechanical connections by electrical ones. This paper presents two technologies. ThruChip Interface (TCI) replaces TSVs and enables 3D integrations to improve power efficiency in cloud computing. Transmission Line Couper (TLC) replaces mechanical connectors and enables LEGO-type module assembly for edge sensors.

Power Efficient Cloud Computer

A. ThruChip Interface (TCI)

TCI uses inductive coupling for data link of stacked chips (Fig.2)[1-25]. While TSV is a mechanical solution in package process, TCI is an electrical solution in wafer process (Fig.3). Power can be delivered by Highly Doped Silicon Via (HDSV).

B. 3D Integration

TCI and HDSV can make memory stacking very thin and lower IO energy significantly (Fig.4) [26].

C. 100 GFLOPS/W Computer

An ACCEL project [27] has been launched, a goal of which is to present a 512GB/s stacked DRAM by 2017 and Proof of Concept of 100GFLOPS/W computer (Fig.5) by 2019.

Design Platform for Edge Sensor

A. Transmission Line Coupler (TLC)

TLC is a distributed coupler using near field (Fig.6) [28-43]. It enables contactless connector that makes module assembly easy and modular design possible. Data rate is over 10Gb/s, and energy dissipation is less than 10pJ/b. (Fig.7).

B. Modular Design

A modular design (Fig.8) is made possible by TLC with inductive-coupling power transfer. TLC is placed close enough to the power coil as well as antennas of LTE and GPS with negligibly small interference.

C. Platform

A CREST project [44] has been launched, where platform design with modularization is investigated to improve development efficiency of various small sensors.

Future Challenge

Together with recent advances of Artificial Intelligence driven by Deep Neural Network with Deep Learning, the new IC technology may contribute to create silicon brains [45].

Acknowledgement

This work was supported by JST.

References

[01] *ISSCC 2004*, pp.142-143.
[02] *Symp. VLSI Circuits 2004*, pp. 246-249.
[03] *CICC 2004*, pp.99-102.
[04] *ISSCC 2005*, pp.264-265.
[05] *ISSCC 2006*, pp.424-425.
[06] *ESSCIRC 2006*, pp.3-6.
[07] *ISSCC 2007*, pp.264-265.
[08] *A-SSCC 2007*, pp.131-134.
[09] *ISSCC 2008*, pp.298-299.
[10] *ISSCC 2009*, pp.244-245.
[11] *ISSCC 2009*, pp.480-481.
[12] *Symp. on VLSI Circuits 2009*, pp. 256-257.
[13] *Symp. on VLSI Circuits 2009*, pp. 94-95.
[14] *Symp. on VLSI Circuits 2009*, pp. 92-93.
[15] *CICC 2009*, pp. 449-452.
[16] *A-SSCC 2009*, pp.305-308.
[17] *A-SSCC 2009*, pp.301-304.
[18] *ISSCC 2010*, pp.436-437.
[19] *ISSCC 2010*, pp.440-441.
[20] *ISSCC 2010*, ES3.
[21] *Symp. on VLSI Circuits 2010*, pp. 201-202.
[22] *A-SSCC 2010*, pp.81-84.
[23] *IEDM 2010*, p.17.1.1.
[24] *ISSCC 2011*, pp.490-491.
[25] *ISSCC 2013*, pp. 258-259.
[26] *Hot Chips 2014*.
[27] JST ACCEL project,
https://www.jst.go.jp/kisoken/accel/research_project/ongoing/h27_02.html
http://www.kuroda.elec.keio.ac.jp/accel_kuroda/
[28] *ISSCC 2007*, pp.266-267.
[29] *CICC 2007*, pp.13-2007.
[30] *A-SSCC 2008*, pp.113-116.
[31] *ISSCC 2009*, pp.470-472.
[32] *Symp. on VLSI Circuits 2009*, pp. 26-27.
[33] *ISSCC 2010*, pp.264-265.
[34] *ISSCC 2011*, pp. 492-493.
[35] *A-SSCC 2011*, pp. 145-148.
[36] *ISSCC 2012*, pp. 52-53.
[37] *CICC 2012*, pp. 7.9.1-7.9.4.
[38] *ISSCC 2013*, pp. 214-215.
[39] *ISSCC 2013*, pp. 200-201.
[40] *ISSCC 2014*, pp. 496-497.
[41] *ISSCC 2015*, pp. 176-177.
[42] *ISSCC 2015*, pp. 434-435.
[43] *Symp. on VLSI Circuits 2015*, pp. C128-129.
[44] JST CREST project,
http://www.jst.go.jp/kisoken/crest/project/41/41_01.html
[45] NEDO workshop, http://www.nedo.go.jp/events/CA_100115.html

978-1-5090-4661-4/17 $31.00 © 2017 IEEE

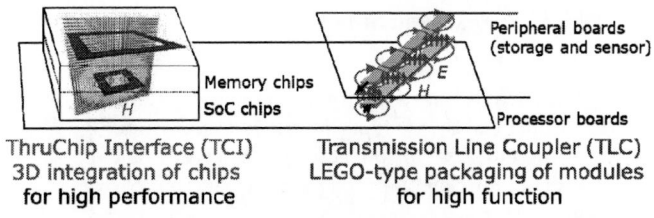

ThruChip Interface (TCI)
3D integration of chips
for high performance

Transmission Line Coupler (TLC)
LEGO-type packaging of modules
for high function

Fig.1 Near-field coupling integration technology.

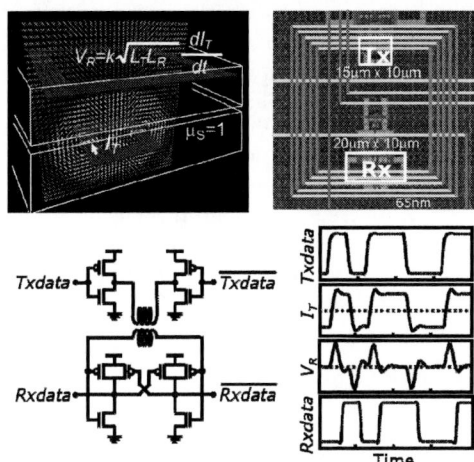

$V_R = k\sqrt{L_T \cdot L_R} \dfrac{dI_T}{dt}$

$\mu_S = 1$

15μm x 10μm Tx

20μm x 10μm Rx

65nm

Txdata $\overline{Txdata}$

Rxdata $\overline{Rxdata}$

Txdata I_T V_R Rxdata

Time

Fig.2 ThruChip Interface (TCI).

	TSV	TCI
Solution	Mechanical in package	Electrical on wafer
Wafer Technology	Additional steps needed	Standard CMOS
Package Technology	OSAT* involved	Conventional
Miniaturization	Difficult	Easy
Yield	Low, difficult to improve	High (~100%)
Eco-system	New model needed	Conventional model
Additional Cost	> 40%	A few %
Placement	Dedicated area w/KOZ**	Unconstrained
Speed	< 512 GB/s	> 512 GB/s
ESD Protection	Needed	No need
Power Dissipation	High	Low

OSAT*: Outsource Assembly and Test, KOZ**: Keep Out Zone

Fig.3 TSV vs. TCI.

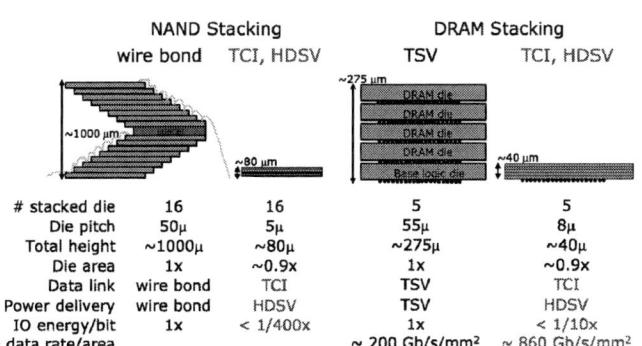

	NAND Stacking wire bond	NAND Stacking TCI, HDSV	DRAM Stacking TSV	DRAM Stacking TCI, HDSV
# stacked die	16	16	5	5
Die pitch	50μ	5μ	55μ	8μ
Total height	~1000μ	~80μ	~275μ	~40μ
Die area	1x	~0.9x	1x	~0.9x
Data link	wire bond	TCI	TSV	TCI
Power delivery	wire bond	HDSV	TSV	HDSV
IO energy/bit	1x	< 1/400x	1x	< 1/10x
IO data rate/area			~ 200 Gb/s/mm²	~ 860 Gb/s/mm²

Fig.4 Memory stacking with TCI and HDSV.

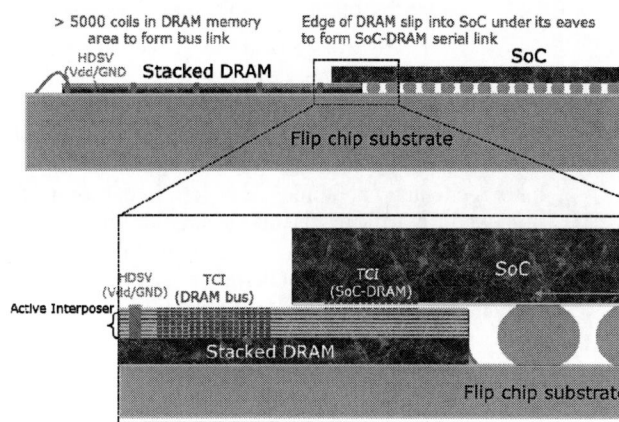

Fig.5 100 GFLOPS/W computer.

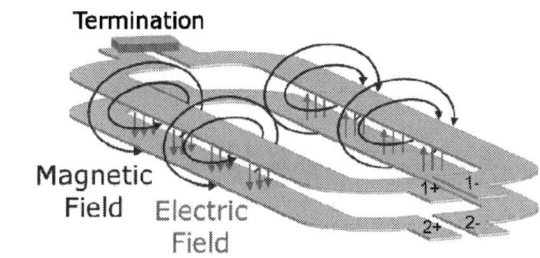

Termination

Magnetic Field Electric Field

1+ 1-
2+ 2-

Fig.6 Transmission Line Coupler (TLC).

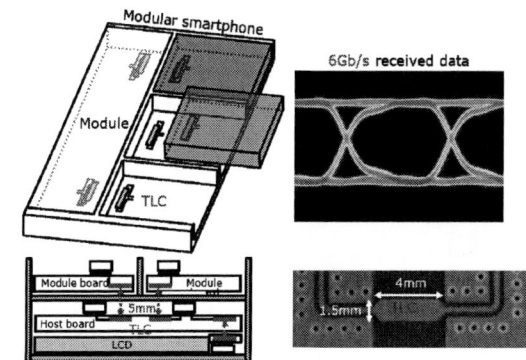

Memory Card
High-speed:50x(12Gb/s)
Low-power:1/500
Water proof
(pad-less, sealed)
ISSCC2013, pp.214-215

Display
High-speed:10x(6Gb/s)
Low-energy:1/10(16pJ/b)
Thin
(no mechanical structure)
ISSCC2013, pp.200-201

Smartphone
High-speed :5x(6Gb/s)
Low-energy :1/24(6pJ/b)
Modular design
(electrical connection)
ISSCC2015, pp.176-177

DIMM
High speed:5x(12.5Gb/s)
Multi-drop bus
(impedance controlled)
ISSCC2012, pp.52-53

In-vehicle LAN
Light: 30%
Strong EMC immunity
(wide band)
ISSCC2014, pp.496-497

Satellite
Light: 60%
Vibration immunity
(contactless connection)
ISSCC2015, pp.434-435

Fig.7 Applications of TLC.

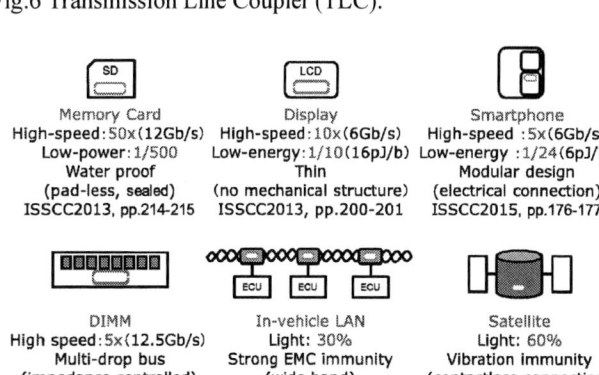

Fig.8 Modular design.

978-1-5090-4661-4/17 $31.00 © 2017 IEEE

4M-2 (Invited)

III-V/Si Low Temperature Direct Bonding Technology for Photonic Device Integration on SOI

Nobuhiko Nishiyama, Yusuke Hayashi, Junichi Suzuki, and Shigehisa Arai

Tokyo Institute of Technology, Tokyo, Japan, nishiyama@ee.e.titech.ac.jp

Abstract

Low temperature wafer scale direct bonding technology using plasma activated bonding (PAB) for heterogeneous photonic device integration is reviewed. Nitrogen plasma irradiation in a high vacuum chamber allows tight strength bonding between InP-based and SOI wafers with the bonding temperature of 150°C as well as low damage to GaInAsP quantum wells (QWs). In contrast Argon irradiation cause poor bonding strength for InP. Using this technology, lasing operations of 1.55-μm GaInAsP/SOI hybrid lasers were demonstrated.
(Keywords: Direct bonding, SOI, InP, laser)

Introduction

Si photonics technology is intensively studied and developed for Si-based high density photonic integrated circuits. Using CMOS fabrication technology, low loss wire waveguides, modulators, and many kinds of passive elements have been demonstrated [1,2]. For active elements such as lasers and amplifiers, in contrast, the integration of III-V semiconductor is crucial to realize high performance devices because Si is an indirect bandgap material. Direct bonding technologies have been demonstrated for such purpose and many types of active elements have been realized [3]. So far, major direct bonding technique is hydrophilic bonding which uses water molecular as glue between two wafers. This technique requires annealing process with several hundred °C. In this paper, direct bonding using nitrogen (N_2) plasma activated bonding (N_2-PAB), which enables low temperature bonding, and the performances of GaInAsP/SOI hybrid lasers using N_2-PAB are explained.

Wafer Bonding

Our N_2-PAB was done with the conditions summarized in Table 1. Usually, this kind of direct bonding uses Argon (Ar)-gas as plasma irradiation. However, Ar plasma causes surface roughness to InP wafer which is a relatively soft material. Therefore, bonding strength became worse as shown in Fig. 1. The bonding temperature is also an important parameter. Figure 2 shows bonding temperature dependence of bonding strength between InP and Si wafers. Higher bonding temperature made stronger bonding strength. In contrast, higher bonding temperature caused degradation of photoluminescence intensity of GaInAsP QWs after the wafer bonding on Si due to the difference of thermal expansion coefficient as shown in Fig. 3. Therefore, the bonding temperature of 150°C is good temperature for both bonding strength and maintain crystal quality. Figure 4 shows a photograph image of a bonded wafer of GaInAsP QWs/SOI. Similar bonding quality can be achievable even with a patterned SOI substrate for Si photonic circuits. A TEM image at the bonding interface is shown in Fig. 5. Although several nm amorphous layer could be observed, no defect propagating to the QWs was found.

Laser Characteristics

Figure 6 shows the schematic structure of our fabricated hybrid Fabry-perot laser. Five GaInAsP QWs emitting at 1.55 μm were grown on an InP substrate and bonded on a SOI substrate with Si wire waveguides. Then, the rest of fabrication process was done after wafer bonding process. I-L and I-V characteristics are shown in Fig. 7. A room-temperature CW operation of 750-μm-long cavity device was demonstrate with threshold current of 77 mA [4]. Our group also demonstrated hybrid lasers with Si-ring resonator toward single mode operation [5].

Conclusion

N_2-PAB bonding technology was explained for photonic device integration on SOI. The bonding at relatively low temperature was demonstrated for low crystal damage and a RT-CW operation of hybrid laser was realized by this bonding technology.

Acknowledgments

This work was supported by JSPS KAKENHI, NEDO, JST-CREST and JST-ACCEL.

References

[1] R. Soref, "Silicon-based optoelectronics," Proc. of IEEE, vol. 81, no. 12, pp. 1687-1706, 1993.

[2] C. Gunn, "CMOS photonics for high-speed interconnects," IEEE Micro, vol. 26, no. 2, pp. 58-66, 2006.

[3] A. W. Fang, H. Park, O. Cohen, R. Jones, M. J. Paniccia, and J. E. Bowers, "Electrically pumped hybrid AlGaInAs-silicon evanescent laser," Opt. Express, vol. 14, no. 20, pp. 9203-9210, 2006.

[4] Y. Hayashi, J. Suzuki, S. Inoue, T. Amemiya, N. Nishiyama, and S. Arai, "Room temperature continuous wave operation of GaInAsP/SOI hybrid lasers fabricated by N_2 plasma activated bonding," Soc. Meet. of IEICE 2016, C-4-17, 2016

[5] Y. Hayashi, J. Suzuki, S. Inoue, S. M. T. Hasan, Y. Kuno, K. Itoh, T. Amemiya, N. Nishiyama, and S. Arai, "GaInAsP/silicon-on-insulator hybrid laser with ring-resonator-type reflector fabricated by N2 plasma-activated bonding," Jpn. J. Appl. Phys., vol. 55, p. 082701, 2016.

Table 1: Bonding condition

Plasma Gas	N₂
Plasma Power	500 W
Irradiation Time	10 sec.
Bonding Temp.	150°C
Bonding Weight	0.25 MPa
Bonding Temp. Ramp Up	2 hour
Bonding Temp. Hold	0 min.
Bonding Temp. Cooling	Natural

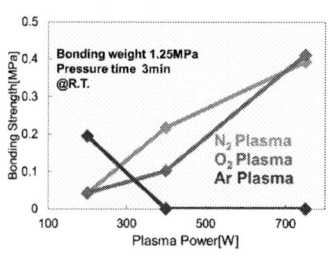

Fig. 1: Plasma power dependence of bonding strength.

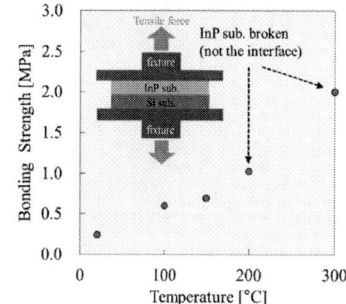

Fig. 2: Bonding temperature dependence of bonding strength.

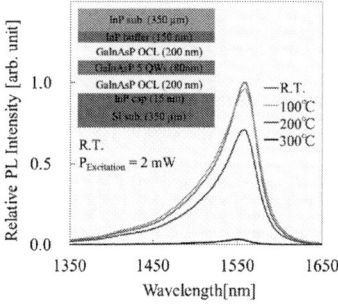

Fig. 3: Photoluminescence spectra of GaInAsP QWs on Si with various bonding temperatures.

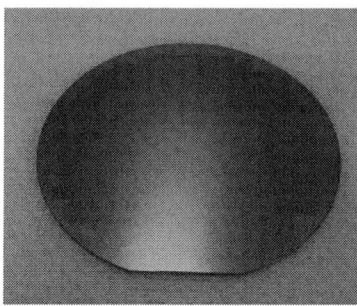

Fig. 4: GaInAsP QWs/SOI hybrid wafer after N₂-PAB.

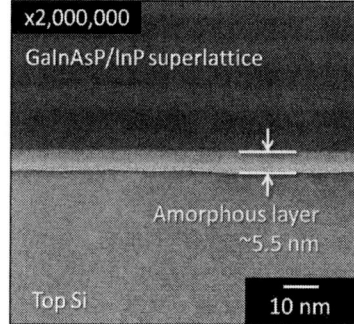

Fig. 5: A TEM image of the bonding interface between InP and Si.

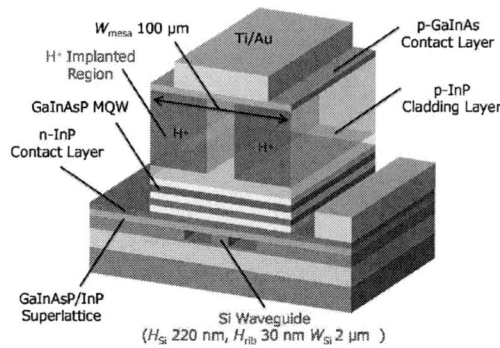

Fig. 6: The schematic structure of the fabricated hybrid GaInAsP/SOI laser.

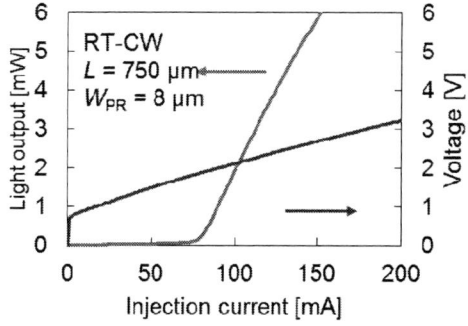

Fig. 7: I-L and I-V characteristics of a fabricated hybrid laser with the cavity length of 750 µm.

978-1-5090-4661-4/17 $31.00 © 2017 IEEE

4M-3 (Invited)

Focused technologies in near future from OSAT view point

Akio Katsumata[1], Akira Takashima[1], Kazuhiro Sawada[1], Hideki Sumihara[1],
and Norio Ito[1]

[1]J-DEVICES Corporation, Yokohama, Japan, akio.katsumata@j-devices.co.jp

Abstract

This paper explains indispensable "Key Words" to talk about future and J-DEVICES' approach.
(Keywords: High speed communication and Distributed processing for big data of IoT and Energy conversion efficiency)

Introduction

From relative stock valuation view point of 15 years cycle of electrical industry, 2017 is turning point to create the next application which will be formed in next 15 years. High speed communication, Distributed processing for big data and Energy conversion efficiency will be new key words to talk about next applications.

High speed communication like 5G and WiGig will move in millimeter wave domain. To decrease "Insertion loss" of wiring of semiconductor package at high frequency will be new challenge.

The distributed processing of "Smart Edge Devices", "Smartphone" & "Cloud Computer" will become mainstream. "Smart Edge Devices" will be big biz opportunity and down sizing of module will be continuous challenge.

Energy conversion efficiency consists of to reduce Conduction Loss and Switching Loss. Reduce Resistance and Inductance of semiconductor packaging will realize better Energy conversion efficiency of power devices.

Some OSATs and PCB companies are focusing on Embedded die in substrate technology. This paper explains the purpose of developing Embedded die in substrate technology and necessity of this technology for next application which will be formed in next 15 years.

Conclusion

The demand of Embedded Die in substrate technology like J-DEVICES' PLP will increase to satisfy the requirements from High speed communication, Distributed processing and Energy conversion efficiency.

Acknowledgments

The authors gratefully acknowledge the contributions of Mitsuru Oida, Fumihiko Taniguchi. Yuji Matsu, Atsuhiro Uratsuji for their developed PLP technology.

References

[1] http://news.panasonic.com/jp/topics/142678.html

[2] Takahashi, T, et al., "Present and Future of Panel Level Package (PLP) technology ", Mate, 2015, p. 335–338

[3] Mitsuru, O, et al., "Advanced Packaging Technologies supporting new semiconductor application" , IEEE CPMT, 2016, p125-128

Fig.1.Electrical industry cycle from relative stock valuation view point
Fig.2.Three Key Words of IoT (Internet of Things)
Fig.3.Demands of Sensing technology for Automotive
Fig.4.Distributed processing for big data
Fig.5.J-DEVICES' approach with PLP-Module
Fig.6.Energy conversion efficiency of power devices

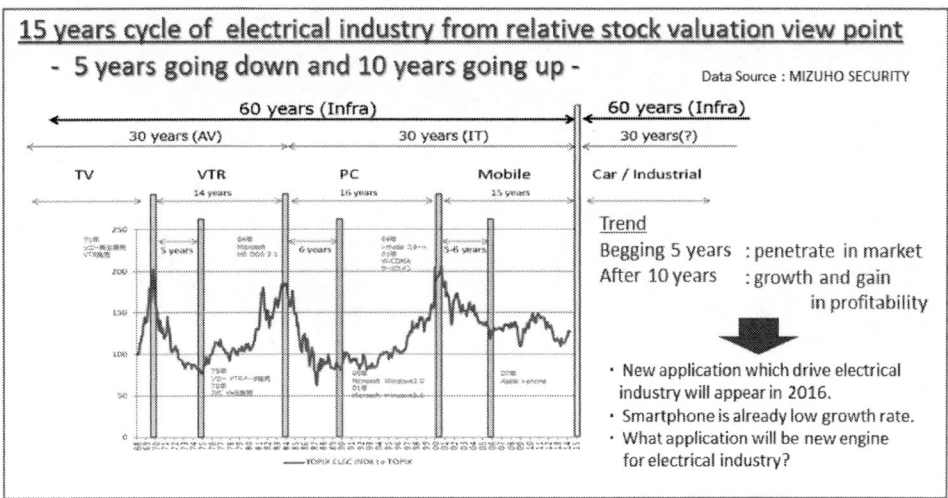

Fig. 1 Electrical industry cycle from relative stock valuation view point

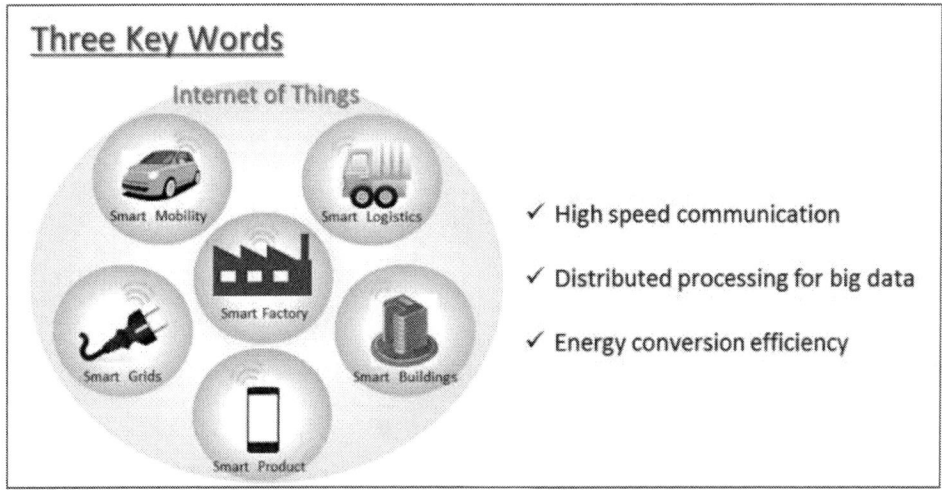

Fig. 2 Three Key Words of IoT (Internet of Things)

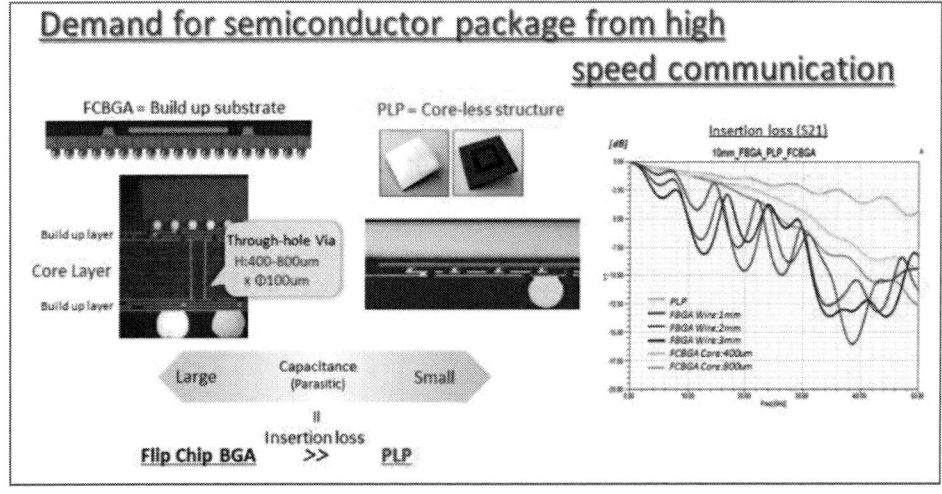

Fig. 3 Demands of Sensing technology for Automotive

978-1-5090-4661-4/17 $31.00 © 2017 IEEE

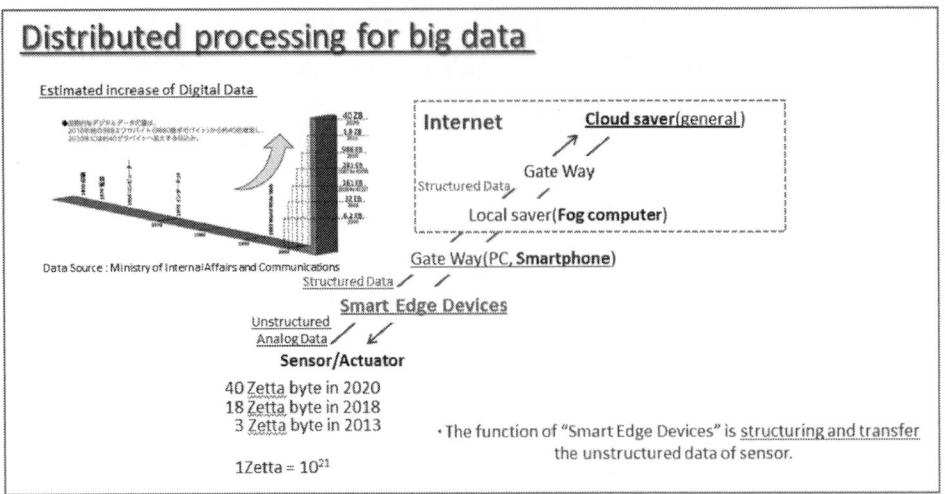

Fig. 4 Distributed processing for big data

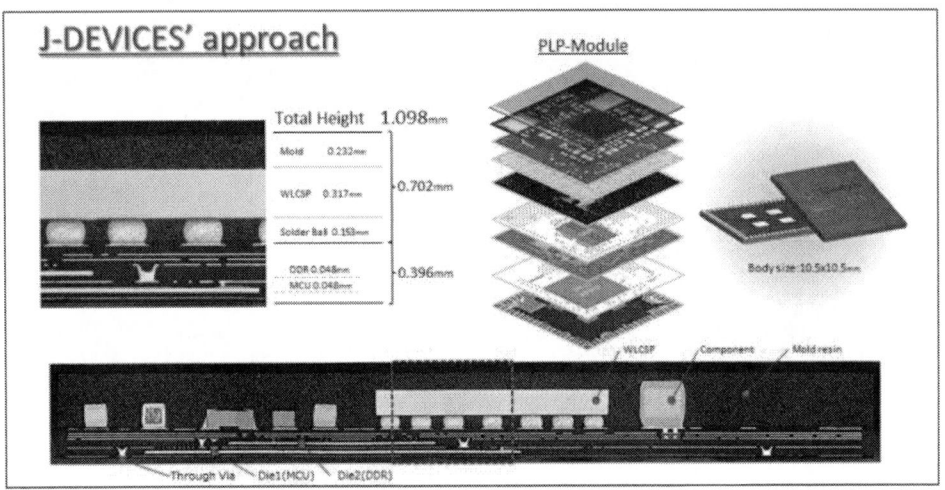

Fig. 5 J-DEVICES' approach with PLP-Module

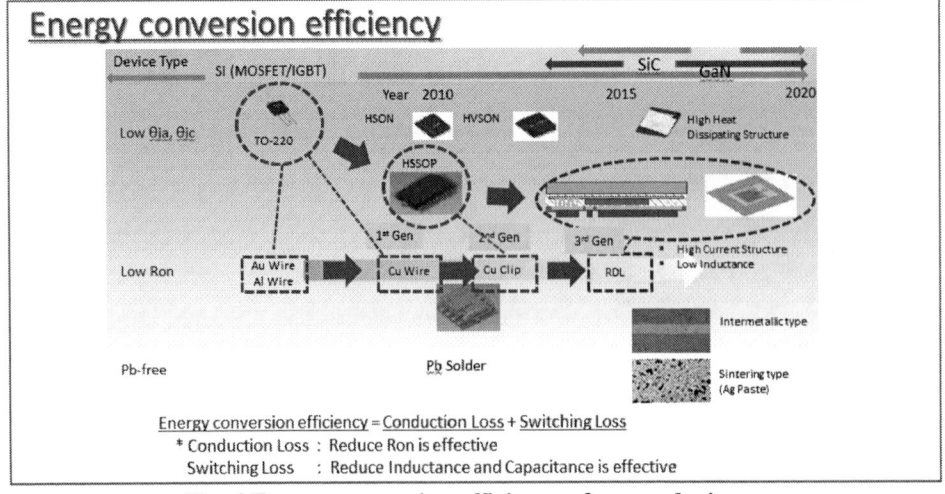

Fig. 6 Energy conversion efficiency of power devices

BGA Packaging process for a Device made by Minimal Fab

Sommawan Khumpuang[1,2], Fumito Imura[1,2], and Shiro Hara[1,2]

[1]AIST, [2]Minimal Fab, Tsukuba, Japan, sommawan.khumpuang@aist.go.jp

Abstract

We have developed a novel packaging tool-set of 13 machines to support continuous manufacturing process from a half-inch wafer process line until ready to be used. The packaging tools are made under minimal fab standard so that a half-inch wafer can be attached on a metal-substrate without dicing. The method that we employed is a BGA (Ball Grid Array)-type solder array which consists of following processes; a compression molding, a laser via, a copper redistribution layer (RDL) patterning, a solder-ball mounting, and a reflow. The development of each machine process and the total packaging process integration has been carried out. In this paper, we introduce a BGA packaging result using only our newly developed tools.

(Keywords: Minimal fab, Ball-Grid-Array and Packaging)

Introduction

In the minimal fab, a transistor process line on a half-inch wafer for custom device or low volume production is established [1]. Using our fab, transistors and MEMS sensor were successfully fabricated without a cleanroom circumstance. However, tools for packaging such devices on a half inch wafer were unavailable. Therefore we have developed a brand-new minimal packaging tools and integration process in order to complete the semiconductor manufacturing system for both front-end and back-end processes [2]. Wafer carriers and wafer transfer mechanism between a process chamber and the carrier are standardized for both the device process line and the packaging line.

Equipment and Process Development

Minimal packaging machines have been developed within our standard tool size of W30×D45×H144cm. Fig. 1 shows cross-sectional structure of our proposed BGA packaging. A half-inch (φ12.5 mm) wafer is die-bonded on a metal substrate (φ13.5 mm) and compression-molded by epoxy resin. Electrode-pads on the device wafer are connected to solder balls above where external path is formed by copper redistribution layer (RDL) between the solder resist and epoxy mold. Fig. 2 illustrated the minimal wafer carriers (minimal shuttles) for a packaging substrate and for a half-inch wafer. A schematic of the process flow is shown in Fig. 3. The process starts from (1) coating Ag paste on the φ13.5mm substrate using a dispenser and cured in our minimal die-bonder. (2) The half-inch wafer is covered with epoxy resin by a minimal compression molding machine. (3) Using our minimal laser ablation machine, laser via through the mold makes a contact path to an aluminum pad underneath. (4) After forming copper (Cu) film used as the seed layer by a minimal Cu sputtering machine, a Cu thick film is formed by a minimal electro-plating machine. (5) A copper redistribution layer is performed by minimal photolithography machines and a minimal Cu wet etcher. (6) To cover the Cu RDL, a solder resist pattern is printed using a minimal ink-jet. (7) Solder balls are aligned above the Cu pads by a minimal ball mounter. (8) Finally solder balls are melted by a minimal reflow machine. Fig. 3 shows the total BGA package flow according to the above-mentioned process. Fig. 4 shows the developed minimal BGA packaging line in AIST.

Experimental Results

Fig. 5 illustrates the completed BGA package using 13 newly-developed machines of minimal fab. The packaged wafer is ready to be mounted on a daughter board for a regular use as an IC chip.

Conclusion and Outcome

A novel packaging-line for seamlessly integrating to a device process-line for half-inch wafer is successfully established.

The machine designs of dimension, clean-localized units, user interfaces and wafer transfer system are successfully unified and standardized through the wafer process to packaging process. It has been suggested that the minimal integrated line from the wafer process to the packaging process will create a new world of semiconductor industry that can achieve as needed device or system in package with a minimum volume of 1 chip with reasonable production cost.

Acknowledgments

The authors would like to thank Minimal fab 3DIC development consortium for contribution in development of machines and system for minimal BGA packaging.

References

[1] S. Khumpuang *et al*., IEEE Transactions on Semiconductor Manufacturing, 28(3), 393-398 (2015).

[2] M. Inoue *et al*., Proc. of IEEE ICICDT & 4S symposium, Ho Chi Minh, Vietnam, (2016)

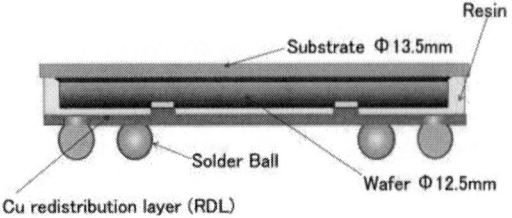

Fig. 1. A schematic diagram of cross-sectional view of BGA package for minimal fab.

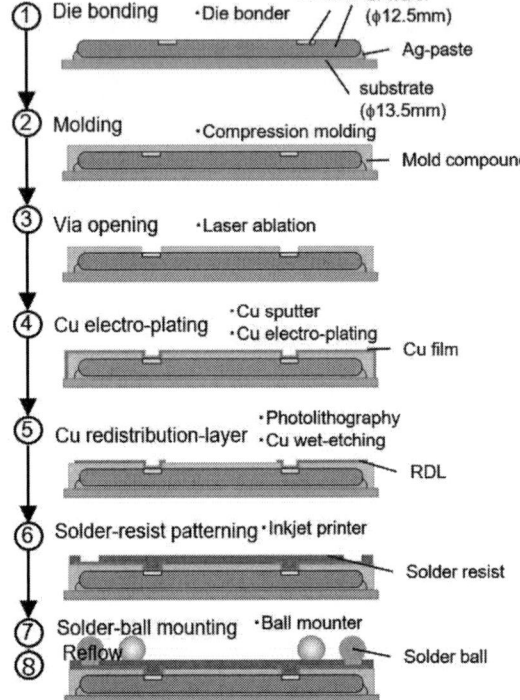

Fig. 3. A process flow for the BGA package fabricated by the minimal packaging process-line.

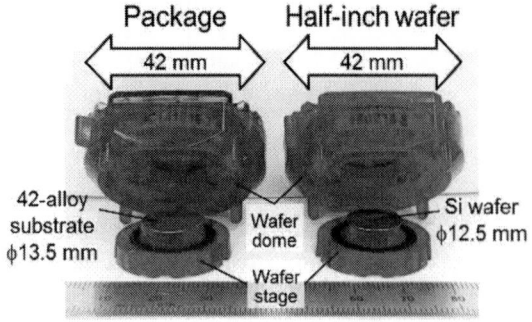

Fig. 2. A photograph of the wafer carrier (minimal shuttle) with wafer placed on the stage.

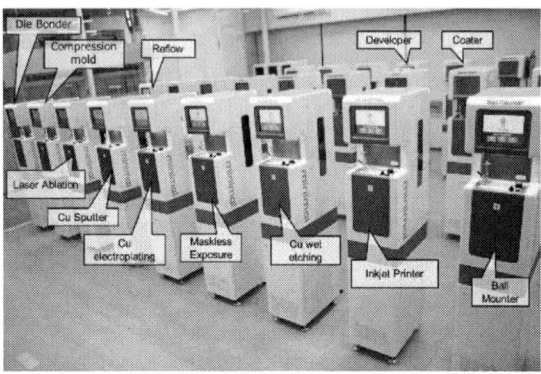

Fig. 4. A photograph of the minimal BGA packaging process-line in AIST.

Fig. 5. A photograph of fabricated BGA package placed on a minimal shuttle.

978-1-5090-4661-4/17 $31.00 © 2017 IEEE

Electrodeposited Cobalt for Advanced Packaging Applications

Bryan Buckalew, Justin Oberst, Thomas Ponnuswamy

Lam Research Corporation, Tualatin, OR USA, bryan.buckalew@lamresearch.com

Abstract

Electrodeposited cobalt has received significant attention in recent years as a suitable metallization alternative for many interconnect technologies. For instance, Co is being evaluated as back-end-of-line (BEOL) alternative for Cu in 10 nm node technologies due to challenges with scaling the diffusion barrier at small CDs and increasing electron scatter in small features [1]. A super-conformal Co plating bath was shown to fill a 5 x 56 μm through-silicon via (TSV) structure, thereby serving as a potential replacement for Cu TSV plating [2]. More related to the subject matter in this manuscript are applications where Co serves as an underbump metallization (UBM) alternative for conventional C4 applications [3] and, more recently, for fine pitch microbump applications [4]. Furthermore, this manuscript will discuss Co electrodeposition as an alternative to both nickel and copper electrodeposition for advanced packaging applications.

Introduction

Cobalt has been proposed as a viable alternative to Cu for the elimination of Kirkendall voids in Pb-free solder connections [5]. Further evaluation of Co-Sn intermetallic compounds (IMCs) reveals that this material is more compliant (lower Young's modulus) than traditional Cu-Sn IMCs and is therefore better suited for micropillar applications. Refer to Table 1.

Experimental Results

UBM layers of Cu, Ni, and Co were electrodeposited, on separate Cu seeds, to a 2 μm thickness. This was followed by 20 μm electrodeposition of Sn2.3Ag solder. The 3 samples were reflowed under standard SnAg reflow conditions. The intermetallic thickness between the three samples was analyzed. See Fig.1. The Cu UBM sample resulted in a much thicker, scalloped IMC, while the Ni and Co sample resulted in a thinner IMC. This is a critical factor since excessive IMC growth has been associated with mechanical stress failures in solder joints. One more noteworthy point is the fact that the Cu IMC resulted in Kirkendall voiding while the Co and Ni samples were void-free.

Additionally, a 1,000 Å Au (Au strike) layer was electrodeposited on a Ni and Co plated UBM. Slightly more Au deposited on the Co film suggesting slightly higher current efficiency for the plating process on Co than Ni. Close examination of the grain size distribution also revealed slightly finer grains on the Co film. See Fig. 2.

Lastly, samples were electrodeposited with a 3-metal stack: Co (or Ni) – Au strike – SnAg. The samples were reflowed under standard SnAg reflow conditions. The samples were cross-sectioned by focused ion beam (FIB) and analyzed by energy-dispersive x-ray spectroscopy (EDS). No appearance of deleterious Au IMCs were detected. See Fig. 3.

Conclusion

Cobalt is being widely evaluated for BEOL and advanced packaging applications. Herein, it was shown that Co resulted in minimal IMC thickness compared to Cu and was Kirkendall void-free. Au strike plating on Co resulted in slightly higher plating efficiency than on Ni. All in all, this study further supports using Co as an alternative to Cu and Ni for underbump metallization.

References

[1] Kelly, J., et al., "Experimental Study of Nanoscale Co Damascene BEOL Interconnect Structures," 2016 IEEE IITC/AMC Conference, pp. 40-42, 2016.

[2] Josell, D. et al., "Superconformal Bottom-up Cobalt Deposition in High Aspect Ratio Through Silicon Vias," ECS Trans., 2016, 75 (2): pp. 25-30.

[3] Humpston, G., "Cobalt: a universal barrier metal for solderable under bump metallisations," J. Mater. Sci.: Mater. Electron 21 (6): pp. 584-588, 2010.

[4] Derakhshandeh, J., et al., "Cobalt UBM for fine pitch microbump applications in 3DIC," 2015 IEEE IITC/MAM Conference, pp 221-224, 2015.

[5] Buckalew, B. "Enabling Advanced Packaging through Innovative Electrodeposition Technology," 16th International Conference on Electronic Packaging Technology (ICEPT), Changsha, China, August 2015.

[6] Vakanas, G., et al., "Formation, processing and characterization of Co-Sn intermetallic compounds for potential integration in 3D interconnects," Microelectronic Engineering, 140 (1), pp 72-80, 2015.

Table 1: Young's Modulus for Key Intermetallic Compounds

Intermetallic Compound	Young's Modulus (GPa)
$CoSn_3$	98
Cu_6Sn_5	120
Ni_3Sn_4	133
Cu_3Sn	134

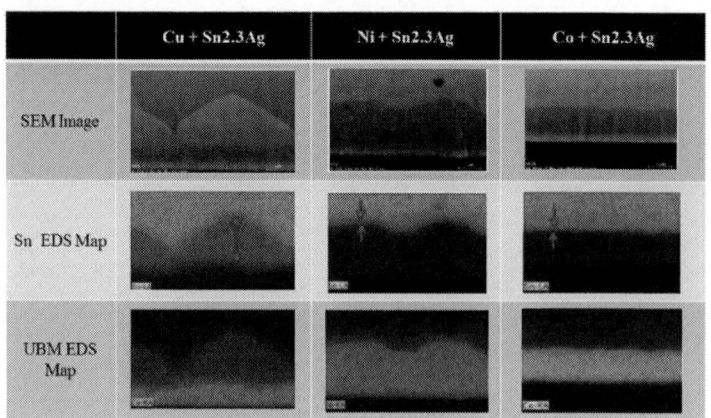

Fig. 1: 2 μm UBM (Cu, Ni, Co) followed by 20 μm Sn2.3Ag. Each sample was reflowed under standard SnAg reflow conditions.

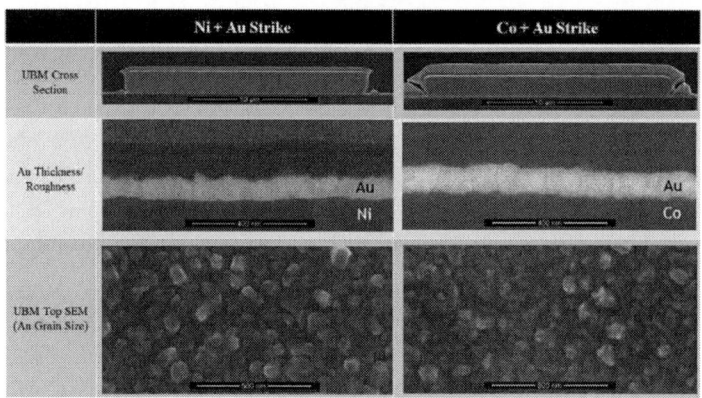

Fig. 2: Au strike (1,000Å) electrodeposition on both Ni and Co UBM.

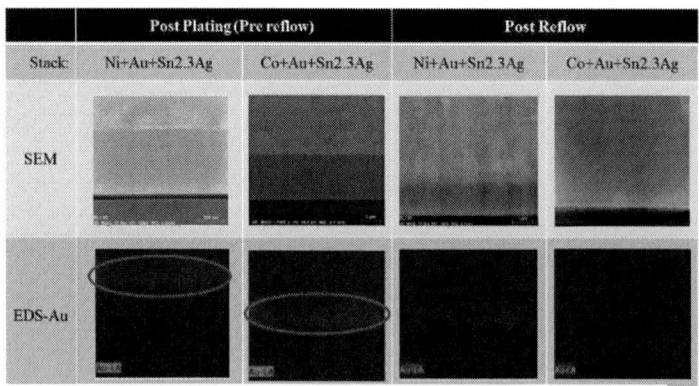

Fig. 3: Evaluation of pre- and post-reflow of Co (Ni) + Au + SnAg stack.

4M-6 (Invited)

Packaging design considerations for mobile and Internet of things (IOT)

Piyush Gupta,

Qualcomm Inc,

Abstract

This talk will go over the various considerations that lead into final selection of a package for a particular application and end form factor. These aspects not only include cost and performance requirements but also include die constraints and OEM PCB choices. It will show an example that for different tier of phones; these trade-offs are different which leads to unique package selection choices. Than the talk will segue into system in package. Essentially highlighting how two unrelated trends of node shrinkage and end device form factor shrinkage are affecting package choices.

(Keywords: Packaging, design)

Introduction

There are mobile phones that sell all the way from $50 to $800. How do you design packages for such a wide range of cost points? This ultimately boils down to optimizing each solution for its own design point. For example the packaging choices made for a $50 phone means prioritizing cost over performance as opposed to designing for a $800 phone where one prioritizes performance over cost. These differing constraints lead to very different decisions not only for the type of the package type, the package size, the package pitch, but also the type of package capacitors. Other big factors that play into deciding the package is what are the OEM's ID constraints and Surface mount technology (SMT) constraints. Increasingly the OEMs are asking for a thinner phone and that means thinner package. The IC node that goes in the package adds some unique tradeoffs in terms of the first level interconnect geometry and hence the package substrate technology that goes with the IC.

Processor package for mobile

This is normally the largest IC and the package in the system. Following are the key considerations when designing a processor package.

A. DDR performance requirements.

For a high tier phone, high DDR Bandwidth (BW) requirements mean high IO counts and unusually high DDR frequency. High frequency implies stringent Signal integrity requirements. This can usually only be met avoiding the PCB routing. Hence for this tier of phone POP becomes a natural choice. High IO requirements puts pressure on reducing the processor to the POP interconnect pitch. This leads to interposed based POP packages as

shown in Figure 1.

For low tier phones the cost of eMCP is cheaper than POP, hence POPs are avoided. Instead non-POP based packages are used.

B. Height

Twin trends of thinner phones and larger battery requirements are leading to pushing the package height reductions.

This is leading to some innovative package choices such as embedded die packages as shown in Figure 2.

C. OEM PCB technology choice

Higher tier phones pay a premium for a smaller PCB area by increasing the PCB micro-via layers. For a higher tier phone OEM uses typically 3 or more micro-via layers. Hence the package pitch can be as less as 0.35mm. For a low tier phone OEMs typically use 1-2 micro via layers and relaxed PCB design rules. And hence the package pitch typically is between 0.4 - 0.5mm pitch. So even though the high tier phones require more interconnects on the PCB the fact that their package pitches can be less, means the package sizes for high tier and low tier phones are typically same size ie between 12x12 mm to 15x15 mm.

C. Die Node

We see most advance die nodes first introduced with flagship high tier phones. As nodes are shrinking from $28 \rightarrow 16 \rightarrow 14 \rightarrow 10 \rightarrow 7nm$ the level 1 interconnect density between IC and the package is shrinking. This leads to interesting level 1 interconnect technology choices. The trend is to move from solder bump to Copper pillar based Level 1 interconnects. Finer Level 1 interconnects puts pressure on the package substrate technologies to support fine Line width and spacing to as small as 10um/10um.

C. Capacitors

Again the highest tier phones demand most processing power from their CPUs and GPUs, leading towards ever increasing # of cores or Fmax requirements. This means constant evolution of the Power delivery requirement (PDN). So highest tier processor packages see package capacitors inside the substrate. For the capacitor also the trend is moving from Multi-layerd ceramic capacitor (MLCC) to silicon based capacitors. Silicon based capacitors provide the lowest self-inductances. Low tier phones typically avoid any capacitors to save cost.

C. Other design considerations

Since the low tier phones usually use inexpensive PCBs with one or no HDI layers, they can-not afford to have any cross routing in the PCB. Thus many times in order to map the processor IC interfaces to the other components in the phone, the # of layers in the packages are increased. Leading to cost increases in the package. This can be avoided by co-designing the IC and package and PCB all together.

DRAM/Flash packages for mobile

These are typically multi stack WB dies. They come in the following combinations. a) POP DRAM (LP4,LP4x) + discrete eMMC, UFS b) eMCP/uMCP with LP4x . As memory density requirements are increasing, the number of memory dies in a stack is increasing. Hence a major trend is how to avoid increasing the memory package height. So Memory Fanout (FO) is a trend in the packaging space for this.

RF packages for mobile

The RF packages are typically chosen as a wafer level package (WLP), quad flat lead (QFN) or a laminate based package. For RF packages the main considerations are IOs and die size. For low IO requirements WLP is the first choice. If the IOs are large than the choice boils down to lead based packages or laminate based packages. For high tier phones typically board space is a constraint. And since laminate based packages are smaller than QFN based packages for similar IOs, hence for high tier phones laminate-based packages are chosen.

Power management packages

PMIC packages design is a function of the # of power domains that need to be supported. As power management becomes more and more important, each power domain needs to be controlled separately. This means multiple bucks and lots of unique trace routing on the OEM PCB. Hence the first decision that needs to be made is the # of PMIC packages and their position on the PCB to avoid crossing of routed traces. After that the decision to select the package type is made. The main package choices are WLP or a FC package. This decision between WLP and FC mainly is driven by cost and supply chain availability rather than any major requirements. For highest tier phones there is typically a Thermal interface material (TIM) at the back along with a heat sink. But for a cheap phone there is no heat sink needed.

IOT packages

Packages for IOT face some unique challenges depending on the end device. For example for wearable's, board space and the power are even more of a constraint than the performance requirements as compared to a mobile space.

Another trend here is of ePOP (Flash+ DRAM both in a POP) packages. ePOP is unique to IOT due to low DRAM density requirements space to put flash memory on the PCB.

Designing one chip and tooling many packages is also a prominent trend depending on the end user application. For example in wearable audio it may go to a ePOP sub 10x10 mm package vs in a home theatre it may go in a 21x21 mm 0.65mm pitch large package.

System on chip (SOC) vs system in package (SIP)vs discrete packages

The digital node shrink roadmap almost always precludes from SOC approach for digital and RF combined.

SIP approach is preferred when integration for miniaturization is desired over cost. When overall cost is optimized than usually discrete packages are cheaper. So for example smart watches SIP makes more sense, vs lets say a smart electric meter.

A. Digital die size reduction

As nodes shrink the IC size reduces, but the 2nd level of interconnect density does not reduce at the same scale. This implies there is free space inside the package. This could be used for putting some discrete chips as side by side in the package. Here the SIP becomes attractive as a means to save board space. Still SIP is desirable from cost point of view only if each of the components require same number of layers in the combined package.

B. Non-traditional IOT players

There is an accelerating tend where non traditional companies with not much background in traditional phone design are asking for a integrated and certified end to end solution so that they can focus on their strengths of IDs and fashion as a means of differentiation instead of investing in board design. This means that a system in package approach is very much needed. This means designing the RF+PMIC+processor packages together and selling it as a chipset.

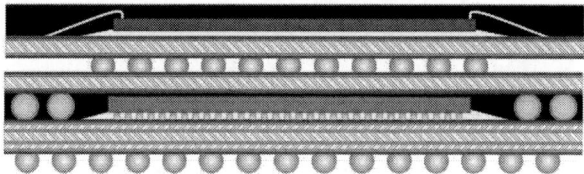

Fig. 1: Interposer based POP packages.

Fig. 2: Embedded die package

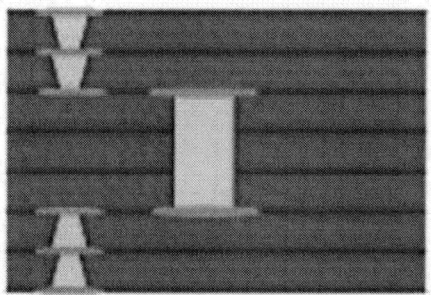

Fig. 3: 2 micro-via layer PCB

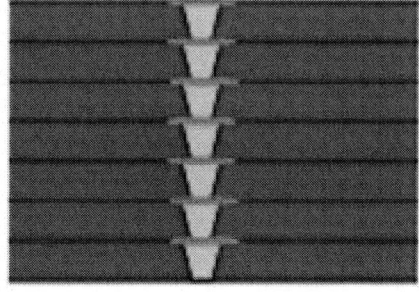

Fig. 4: 8 micro-via layer PCB

4A-1 (Invited)

Biocompatible ALD barrier coatings for medical devices

Mikko Matvejeff[1], Satu Ek[1], Riina Ritasalo[1], Jesse Kalliomäki[1], Päivi Järvinen[1], Oili Ylivaara[2], and Erik Östreng[1]

[1]Picosun, Masala, Finland, mikko.matvejeff@picosun.com

[2]VTT, Espoo, Finland

Abstract

Atomic layer deposition (ALD) is widely in use for depositing a variety of materials, such as metal oxides, metal nitrides and metals, in a conformal and defect-free form at low temperatures on high aspect-ratio substrates. These advantages make ALD uniquely powerful method for applications where sensitive substrate materials combine with extreme demands on coating quality and temperature/chemical resistance, such as those often seen in the medical applications.

(Keywords: ALD, medical device and barrier layer)

Introduction

Nanomaterials are gaining popularity in the coating of medical devices, due to their ability to meet the high demands of this application field. Nanomaterials can be used to apply uniform coating over challenging 3D shapes, and a wide range of biocompatible materials are already available. Additionally, nanomaterials can promote coating adhesion, strength and durability.

Atomic Layer Deposition (ALD) is in an important position to realize next generation of *in-vivo* medical applications to address the emerging health care challenges [1,2]. Defect-free ALD barrier layers enable perfectly hermetic nanoscale encapsulation of even the smallest, topologically most complicated surfaces enabling wide variety of applications from protective barriers to dielectrics [3] to biocompatible coatings in medical devices [4] to name just few.

The relatively low temperature of ALD processes allows deposition on, not only inorganic, but also temperature sensitive organic matrices such as polymers which form a key substrate group for many medical devices. ALD technology will therefore be an important building block in development of future integrated production lines for medical devices.

ALD barrier materials are often thin metal oxides or nitrides with a thickness of some tens of nanometers. Additionally, ALD technique enables tailoring of surface structures, i.e. deposition of well-defined laminates or mixtures of materials with excellent thickness control, where various material properties can be combined. Not only the ALD material and its thickness affect the barrier properties but also the

ALD process chemistry and parameters can be used to adjust the desired barrier properties. Mechanical properties of barrier ALD films [5,6] may play an important role in some medical devices which are exposed to bending, such as catheters. Thus, also the residual stress of ALD films need to be studied and taken into account when selecting the material for the device.

The current contribution focuses on ALD-TiO_2 as it exhibits one of the most interesting materials for medical applications. High temperature tolerance and chemical inertness [5] provide good compatibility with common sterilization methods and biological tissues.

Results and discussion

Low-temperature ALD deposition of TiO_2 was studied on 5x5 mm² chips with Au bumps on polymer sheets (20x20 mm²) (shown in Fig. 1). Titanium dioxide films with a thickness target of 50 nm were deposited on the samples using Picosun® R-200 Advanced ALD reactor. Focused ion beam (FIB)/scanning electron microscopy (SEM) cross sectioning was successfully performed on ALD-coated samples, where uniform 50 nm thick ALD layer was observed conformally covering the whole substrate. According to FIB/SEM study, adhesion on the horizontal and vertical surfaces of silicon was excellent. Slight delamination, on polymer materials, was observed, most likely caused by the conditions in FIB preparation.

In order to reap the full benefit of ALD, a good understanding of how to manipulate the structural and physical properties by tuning deposition parameters is required. In the medical field, this translates into surfaces which encourage cell growth (biomimetic) or which are inert to the body (bioinert). In this paper, we use phase control of TiO_2 deposited from $TiCl_4$ and water as an example. In the example, we will vary the deposition temperature and substrate to show how different structural properties arise. However, different chemistries can change other properties such as impurity profiles, grain sizes, crystallographic orientation and so on, which are all of paramount importance for medical applications.

In the following experiment, a series of samples

978-1-5090-4661-4/17 $31.00 © 2017 IEEE

deposited on Si (100) and Ru were analysed by x-ray diffractometry (XRD) and ellipsometry. There is a pronounced overall decrease in growth per cycle (GPC) with temperature consistent with loss of OH groups at increasing temperature (see Fig. 2). The interesting part is rather the discontinuous increase at $200^{\circ}C$. At the same temperature, the crystalline structure and the refractive index increases changes dramatically (see Figs. 3 and 4). We also observed a notable change in wafer uniformity (not shown here).

The change in refractive index is known to arise from film density and from different crystallinity. In this case the continuous increase from $50-200^{\circ}C$ can be explained by increasing film density, however the discontinuous increase from $200-225^{\circ}C$ has to be studied further.

The stress of TiO_2 films was also studied. Tensile residual stress was observed for ALD-TiO_2, in the scale of several hundred MPa. Observations were according to literature values for ALD-TiO_2 films [5,6].

The ALD-TiO_2 was deposited also on CeO_2 and Ru surfaces. Another degree of freedom was found, where crystalline rutile phase was deposited on oxidised Ru surfaces. Rutile is however out of scope for biological systems.

Conclusions

ALD has been utilized to deposit uniform, low temperature barrier films on polymers with good adhesion. We have also shown a study of the importance of good understanding of how the parameters are influencing the growth and properties of the film material for use in the final application. The stress of ALD-TiO_2 film is tensile, and in scale of several hundred MPa. TiO_2 has been used as a model system and as an example.

Acknowledgments

This work has been done in the ECSEL Joint Undertaking project InForMed (an integrated pilot line for micro-fabricated medical devices) coordinated by Philips Electronics Netherlands BV.

References

[1] S. A. Skoog, J. W. Elam and R. J. Narayan, "Atomic layer deposition: medical and biological applications", International Materials Review,

58(2), 2013, 113-129.

[2] S. Schröder, Ö. Dogan, J. Schneidewind, G. Bertotti, S. Keil and Hassan Gargouri, "Three Dimensional ALD of TiO2 for In-Vivo Biomedical Sensor Applications", Advances in Sensors and Interfaces (IWASI), 2015 6th IEEE International Workshop.

[3] G. Mestres, M. Espanol, W. Xia, M. Tenje and M. Ott, "Evaluation of Biocompatibility and Release of Reactive Oxygen Species of Aluminum Oxide-Coated Materials", ACS Omega, 1(4), 2016, 706 – 713.

[4] G. C. Correa, B. Bao, and N. C. Strandwitz, "Chemical Stability of Titania and Alumina Thin Films Formed by Atomic Layer Deposition", Applied Materials and Interfaces, 7, 2015, 14816 – 14821.

[5] J. Lyytinen, X. Liu, O.M.E. Ylivaara, S. Sintonen, A. Iyer, S. Ali, J. Julin, H. Lipsanen, T. Sajavaara, R. L. Puurunen, J. Koskinen, "Nanotribological, nanomechanical and interfacial characterization of atomic layer deposited TiO_2 on a silicon substrate", Wear, 342–343, 2015, 270-278.

[6] O.M.E. Ylivaara, L. Kilpi, X. Liu, S. Sintonen, S. Ali, M. Laitinen, J. Julin, E. Haimi, T. Sajavaara, H. Lipsanen, S.-P. Hannula, H. Ronkainen, R.L. Puurunen, Riikka L., R. L. (2017), "Aluminum oxide/titanium dioxide nanolaminates grown by atomic layer deposition: Growth and mechanical properties", J. of Vacuum Sci. & Techn. A: Vacuum, Surfaces, and Films, 35, 2017, 01B105.

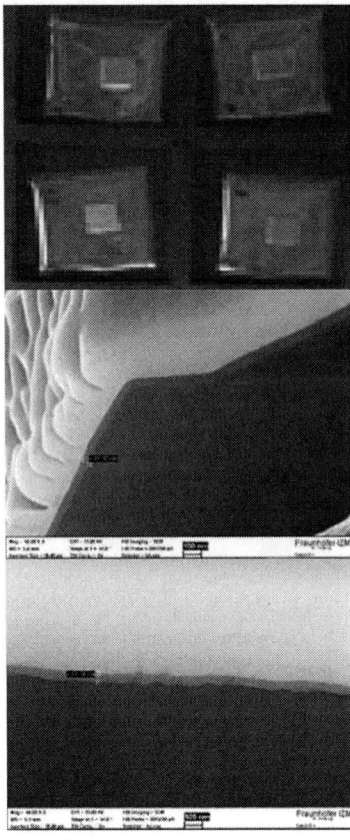

Figure 1: Glued chips on polymers (top) and pictures of the FIB/SEM study on interfaces between ALD films and various parts of the chip on polymer (middle and bottom).

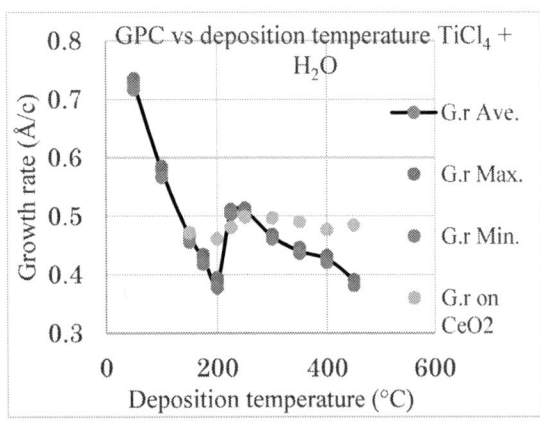

Figure 2: Growth per cycle in terms of deposition temperature of TiO_2

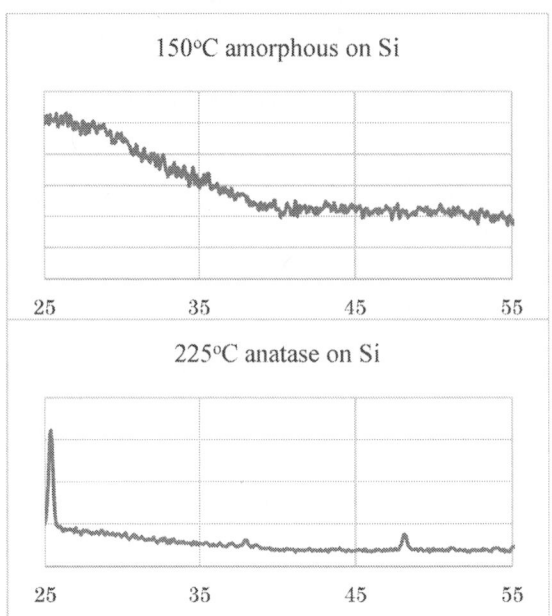

Figure 3: XRD patterns of TiO_2 samples deposited at 150°C (top) and 225°C (bottom). All the peaks are uniquely identified as anatase.

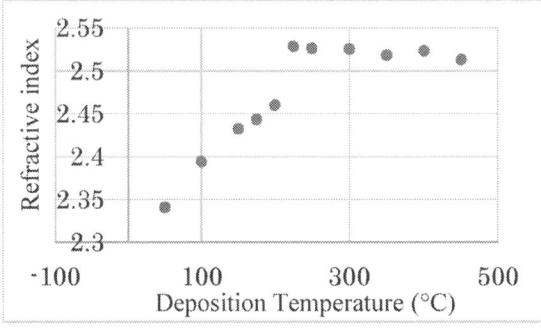

Figure 4: Refractive index of TiO_2 as function of deposition temperature.

978-1-5090-4661-4/17 $31.00 © 2017 IEEE   58

4A-2 (Invited)

Enablement of Cost Effective CVD/ALD Processing Through Precursor Design

Jean-Marc Girard

Air Liquide Advanced Materials, 3121 Route 22 East, Branchburg, New Jersey 08876, USA.

Abstract

CVD and more recently ALD have become methods of choice for the deposition of new materials, to deal with the 3D nature of new devices, and to meet the requirement for atomic level thickness control.

Several examples illustrate how precursor design can contribute to keeping the cost of new materials deposition in control, whether through easier facilitization thanks to better physical properties, or through process throughput enhancement.

(Keywords: ALD, CVD, Materials)

Introduction

The need new ALD or CVD thin films is leading the industry to adopt an increasing number of precursors. Specific constraints such as low thermal budgets, compatibility with underlying layers, or requirement for extremely pure films, are leading towards sub-optimal molecules from a physical property standpoint. Whether because they are solids, have a low volatily, or exhibit a low thermal stability, the facilitization of such compounds have a strong impact on the total cost of ownership, as they prevent the usage of bulk delivery to the tools ("auto-refill"), require non standard components, and limit the load per package. Typical examples of such issues are encountered with solid metal halides (F-free tungsten, Metal gate, etc.), with Rare Earth precursors or with the Cobalt precursors introduced as BEOL Logic adhesion/barrier layers (low thermal stability). The ALD process is also intrinsically slow owing to its self-limiting growth characteristic that is entirely molecule/co-reactant dependent. Here, high volatility (i.e. high dose despite a short exposure time) combined with a high intrinsic growth per cycle is strongly desired to accommodate relatively "thick" film such as in gapfill application or multiple patterning.

1- Turning solids into liquids

An example of turning solids into a liquid for metal halide can be exemplified by TAPEDIS™, a quasi-inorganic, volatile (1.4 mbar at 115°C) and versatile liquid Ta metal precursor, that compares very favorably to $TaCl_5$ (x3 in vapour pressure) and provides the capability to deposit TaN, TaC, TaSiN or TaSi [1]. As opposed to $TaCl_5$, that requires a complex sublimator, and limits loads to kg scale (hence large fixed cost per load), TAPEDIS™ can be supplied in bulk and fed to a refillable on-tool bubbler, hence drastically decreasing the CoO.

Similarly, Rare Earth precursors using the common ligand systems (β-diketonate, amidinates "AMD", or cyclopentadienyl "Cp" derivatives) are normally solids, especially for the smaller atoms (like Y, Yb, Sc, Lu). For these elements, a new Cp_2-AMD heteroleptic ligand system enables liquid state at room temperature and high thermal stability precursors [Fig. 1]. The performance of such precursors for ALD of oxide has been demonstrated both with O_3 and H_2O [2], and they are finding applications for instance as WF tuning layers or as mixed oxides for BEOL MIM applications.

2- The benefit of high thermal stability

The current most common Co BEOL precursor exhibit a limited thermal stability that prevents refillable bubblers, and limits the initial load of the canister owing to accumulation of less volatile degradation by product. A new precursor dubbed COSINE™ alleviates this issue, while exhibiting films with similar resistivity [3], [Fig. 2, 3]. A benefit of the higher stability is exemplified by the perfect step coverage in high aspect ratio structure [Pic. 1].

3- High Productivity Precursors

A new family of precursors having a high Si content has been developed to increase the throughput of SiO_2 (PE)ALD, showing a > 100% increase in growth per cycle (GPC) vs. usual aminosilanes [Fig. 4], leading to a tool productivity gains of up to 40% for a 30 nm film [4]. Combining high productivity equipment (spatial ALD) with such chemistry (dubbed ORTHRUS™) has enabled conformal growth up to 40 nm/min by PEALD. Similar high GPC benefits has been found for SiN PEALD [Fig. 5].

Conclusion

Through a few examples, we have shown how optimized precursor design can reduce the CoO of ALD/CVD processing, by improving physical properties or by increasing growth kinetics.

References

[1] C.Dussarrat et al., ECS Trans. 3 (2), 289-95, 2006

[2] Seppälä et al., Chem. Mater., 2016, 28 (15), pp 5440-5449.

[3] N. Blasco, 63rd AVS, Nashville, TN, 2016.

[4] A. Zauner, 16th ALD Conference, Dublin, 2016

978-1-5090-4661-4/17 $31.00 © 2017 IEEE

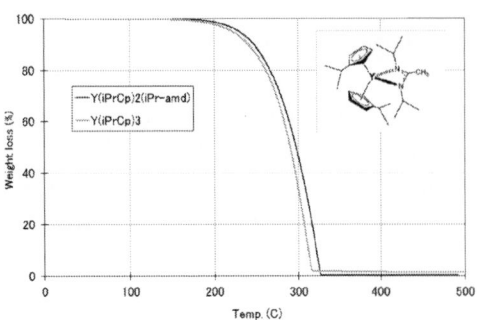

Figure 1:
Thermogravimetric analysis of liquid
Y(iPrCp)$_2$(iPrAMD) vs solid (iPrCp)$_3$ analog.

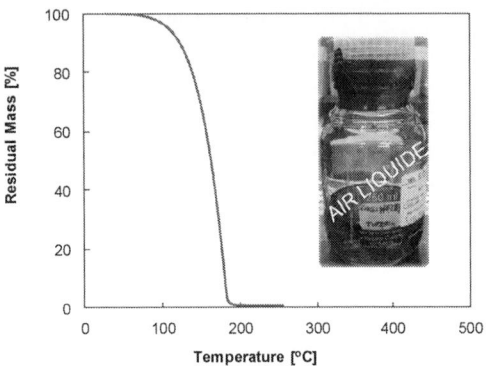

Figure 2:
Thermogravimetric analysis of COSINE™ showing a
clean evaporation.

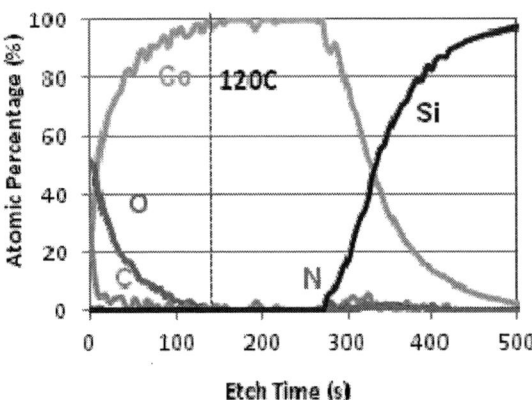

Figure 3:
COSINE™-based as deposited Cobalt film composition
by XPS (120°C, CVD, NH$_3$).

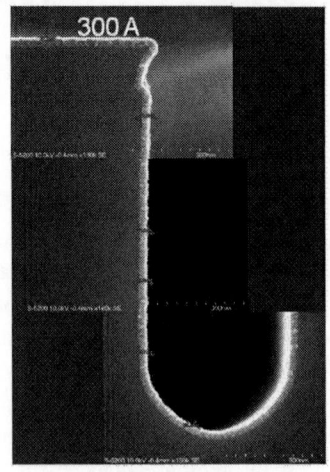

Picture 1:
Step Coverage of pure Co film obtained in 20:1 A/R
structure with COSINE/NH$_3$ at 170°C. The resistivity
of optimized film is in the range of ~15 μΩ.cm

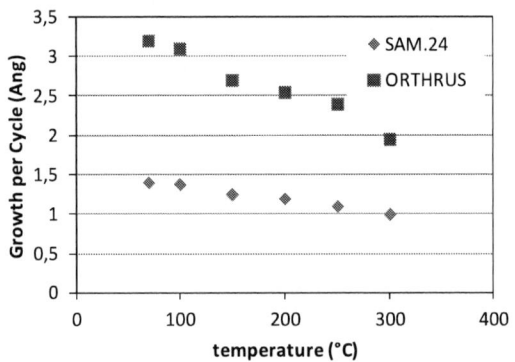

Figure 4:
GPC for SiO2 films by PE-ALD using SAM.24™ and
ORTHRUS™ at different temperatures for DPT
(50-100°C) and leave behind spacers (> 250°C).

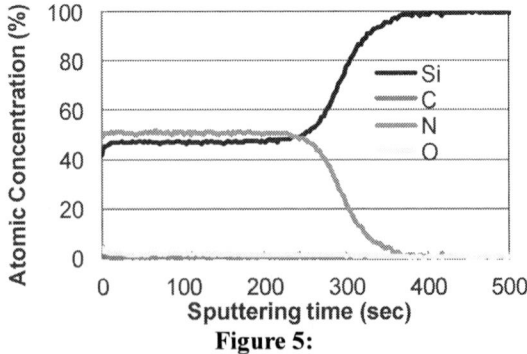

Figure 5:
Composition of SiN films deposited by PEALD using
ORTHRUS. GPC up to > 0.8 Ang/cycles were obtained,
with a dHF WER being < 50% of thermal oxide.

Impact of Hydrogen Annealing Behavior of C_3H_5 Carbon Cluster Ion Implanted Projection Range using Microwave heat treatment

Takeshi Kadono, Ryosuke Okuyama, Ayumi Masada, Ryou Hirose
Yoshihiro Koga, Hidehiko Okuda and Kazunari Kurita

SUMCO Corporation 1-52 Kubara, Yamashiro-cho, Imari-shi, Saga, 849-4256 Japan,
tkadono@sumcosi.com

Abstract

We describe the effect of microwave heating on C_3H_5 carbon cluster ion implanted epitaxial wafers using a high dose amount of carbon cluster ion implantation condition. A high dose amount condition of C_3H_5 carbon cluster ion implantation generates implantation-related defects, such as stacking faults, after epitaxial growth. Therefore, we investigated the control and reduction of stacking faults using the microwave heating technique. We revealed that this technique can control and reduce stacking faults by selectively heating of the carbon cluster ion implantation projection range.

(Keywords: Microwave heat, Gettering, Silicon)

Introduction

We developed a proximity gettering epitaxial wafer using the carbon cluster ion implantation technique to fabricate high-performance high-quality CMOS image sensors. In a previous study, we found that a carbon cluster ion implanted epitaxial silicon wafer has three characteristics for CMOS image sensors [1], as shown by Fig. 1. The first characteristic is that the carbon cluster ion implantation projection range has high gettering capability of metallic impurities. The second characteristic is that this range also has a barrier effect of oxygen impurity out-diffusing to the epitaxial layer from the silicon substrate. The third characteristic the passivation effect on process-induced defects is that expected due to the hydrogen of the carbon cluster ion trapped in the projection range diffusing during the CMOS image sensor device fabrication process.

These characteristics are enhanced by increasing the carbon cluster ion dose amount. However, a high dose amount generates stacking faults after epitaxial growth. Therefore, we considered the technical solution of controlling and reducing such ion implantation-related defects using the microwave heating technique to enable high dose carbon cluster ion implantation. The microwave heating technique can potentially be used to selectively heat only in the ion implantation projection range in the silicon crystal bulk. Moreover, this technique has been reported to result in the diffusion, activation, and re-crystallization of the amorphous layer formed by dopant ion implantation. [2] In this study, we investigated the re-crystallization of amorphous layer formed on C_3H_5 carbon cluster ion implantation projection range and reduction of stacking faults after epitaxial growth using microwave heat treatment.

Experiment

Figure 2 shows the experimental procedure for this study. The starting sample was subjected to C_3H_5 carbon cluster ion implantation at 80 keV with a dose amount of 8.3×10^{14} cluster/cm^2 (2.5×10^{15} Carbon atoms/cm^2). This sample was then heat treated by microwave heat treatment at a frequency of 2.45 GHz, power of 4.0 kW, and irradiation time of 300 sec. The carbon, hydrogen, and oxygen depth profiles were measured using SIMS and implanted defects were observed using cross sectional TEM before and after microwave heat treatment. We subjected the microwave heat treated sample to epitaxial growth of 5-μm in thickness and evaluated surface perfection using the laser scattering method.

Results and Discussion

Figure 3 shows the SIMS depth profiles of carbon, hydrogen, and oxygen before and after microwave heat treatment. We confirmed the formation of a unique hydrogen profile with peaks after microwave heating. On the other hand, the carbon concentration profile didn't diffuse after microwave heat treatment. These results indicate that hydrogen is selectively excited by microwaves. Figure 4 shows a cross sectional TEM image before and after microwave heat treatment. An amorphous layer formed by carbon cluster ion implantation was re-crystallized after microwave heat treatment. Therefore, it is considered that the amorphous layer can be selectively heat treated in the carbon cluster ion implantation projection range by microwave heat treatment. Figure 5 shows the surface perfection of carbon cluster ion implanted wafers after epitaxial growth with and without microwave heat treatment. Microwave heat treated sample decrease with surface defect density.

Conclusion

We achieved surface perfection after epitaxial growth using microwave heat treatment due to selective re-crystallization of C_3H_5 carbon cluster ion implantation projection range.

References

[1] K. Kurita et al., J. Surf. Sci. Soc. Jpa, Vol. 37, p. 104 (2016).

[2] Y-J. Lee et al., IEEE Transactions on Electron Devices, Vol. 61, p. 651-665 (2014).

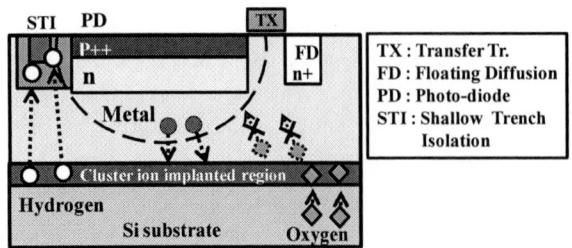

Fig. 1: Three characteristics of carbon cluster ion implanted epitaxial wafer

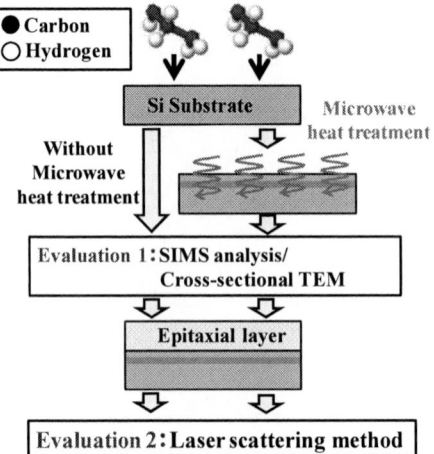

Fig. 2: Experimental procedure of this study

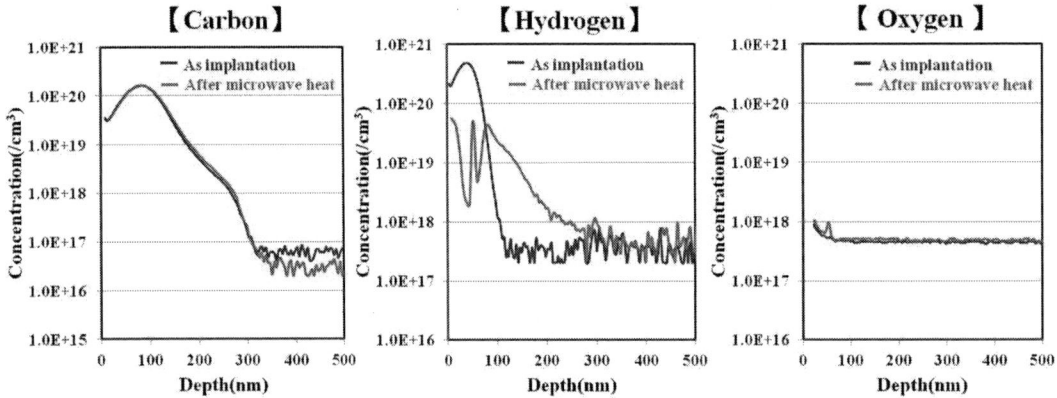

Fig. 3: Carbon, hydrogen, and oxygen concentration of carbon cluster ion implantation projection range before and after microwave heat treatment

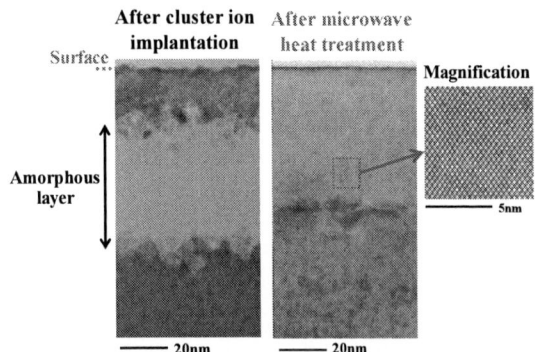

Fig. 4: Cross sectional TEM image of carbon cluster ion implantation projection range before and after microwave heat treatment

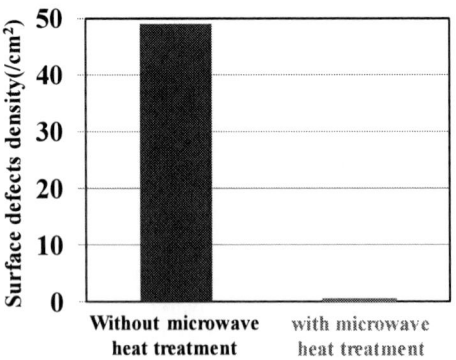

Fig. 5: Surface perfection of carbon cluster ion implanted epitaxial wafer with and without microwave heat treatment using laser scattering method

978-1-5090-4661-4/17 $31.00 © 2017 IEEE

New Opportunity of Ferroelectric Tunnel Junction Memory with Ultrathin HfO$_2$-based Oxides

Xuan Tian and Akira Toriumi

Department of Materials Engineering, The University of Tokyo, Tokyo, 113-8656, Japan
e-mail: xuan@adam.t.u-tokyo.ac.jp

Abstract

Ferroelectric tunnel junctions with ultra-thin Y-doped HfO$_2$ film was investigated for the low power, high speed, non-volatile memory. Polarization switching-induced resistance change amounts to ~100 in the on/off ratio. The electrode area-dependent current and no need of the forming process show a striking contrast to the filament type of ReRAM. FTJ memory with ferroelectric HfO$_2$ will provide us a new opportunity in the memory world using CMOS friendly materials and processes.

(Keywords: Non-volatile memory, FTJ, ferroelectric HfO$_2$)

Introduction

Magnetic tunnel junction (MTJ) is one of the attractive candidates for the "universal" memories in terms of fast, reliable and nonvolatile operation in high density memory arrays. The on/off ratio in MTJ, however, is typically ~5, the writing operation power is very high, and very complicated layered structure with sub-nm thickness control using exotic materials is needed [1]. Other resistive memories are so far not satisfied with requirements for universal memory.

Ferroelectric tunneling junction (FTJ) [2] has a simple structure with much larger on/off ratio (typically 10^2~10^3) and much lower writing current (~10^3 A/cm^2 in BTO-based FTJ [3]) than MTJ. However, there are two big concerns in FTJ. One is that CMOS unfriendly materials like BTO [3-4] and PZT [5] are used. The other is the poor thickness scalability of those ferroelectric films. Ferroelectric HfO$_2$ recently discovered by NamLab group [6] has a promising potential to change the outlook for ferroelectric application. This paper demonstrates opportunities of HfO$_2$-based FTJ memory application.

Experiment

3-nm-thick Y-doped HfO$_2$ (Y-HfO$_2$) film was deposited on 20-nm-thick TiN film on Si, by co-sputtering of HfO$_2$ and Y$_2$O$_3$ targets, followed by post-deposition annealing at 650°C in N$_2$ for 30s. Cation ratio of Y to Hf was estimated to be around 2.1% by XPS. Circular Ag electrodes with various areas were used for the top electrode. A schematic device structure of the FTJ is shown in **Fig. 1**.

Results and Discussion

Figure 2 briefly shows the operation principle of FTJ memory. Different from the normal dielectric film, the polarization charges are retained at the metal/insulator interface after removing the external electric field. Due to the finite screening of metal electrodes, the unscreened bound charges at ferroelectric layer establish a depolarization field inside the film, which alters the electrostatic potential of the ferroelectric layer. In asymmetric electrodes structure, the average potential seen by tunneling electrons is different for opposite polarization directions in the ferroelectric film. Hence, the tunneling current depends on the polarization direction coupled with asymmetric electrodes.

Since the polarization-electric field measurement is difficult for ultra-thin ferroelectric film, the piezo-response force microscope (PFM) measurement was carried out. **Fig. 3a, b** shows that the amplitude and phase images in the inner and outer flat areas where opposite writing voltages were applied. **Fig. 3c** shows amplitude and phase characteristics as a function of writing voltage at a certain point. The results clearly show ferroelectric properties of the 3-nm-thick Y-HfO$_2$.

Figure 4 shows memory effects in three different samples. The on and off voltages are -0.9V and +0.6V, respectively, and E$_c$ is around 2.5MV/cm in accordance with the local PFM phase hysteresis. The on/off resistance ratio is as high as 100. No forming process and no electrode area dependence of the current density were observed. They are quite different from the filament conduction type of ReRAM. **Fig. 5** shows the preliminary endurance properties using 100 μs pulses at 1.2V writing pulse. Considering the fast switching property of ferroelectricity, higher cycling endurance under faster pulse is expected. Since HfO$_2$ has almost twice the energy barrier height than the perovskite structure films [3-5]. The on-state resistance in Ag/Y-HfO$_2$/TiN FTJ at 0.5V is over 100kΩ, showing the current density at least two orders lower than FTJs with perovskite films as shown in **Fig. 6**. This may enable to considerably decrease the sneak current in the cross-bar type of memory arrays.

Conclusion

We have demonstrated FTJ memory operation with on/off ratio about 100 in ferroelectric HfO$_2$ FTJs. The results indicate the homogeneous current flow through the junction. The high on/off ratio and significantly low power consumption in the ferroelectric HfO$_2$ FTJs provide us a new opportunity in high-density nonvolatile memory world.

Acknowledgement

This work was supported by JST-CREST.

References

[1] J. Zhu, et al., *Mater. Today*, **9**, 36 (2006).
[2] H. Kohlstedt, et al., *Physical Review B*, **72**, 125341 (2005).
[3] A. Chanthbouala, et al. *Nat Nanotechnol*, 7, 101 (2012).
[4] Z. Wen, et al., *Nature Materials*, **12**, 617 (2013).
[5] D. Pantel, et al., *Appl Phys Lett*, **100**, 232902 (2012).
[6] T. S. Böscke, et al., *Appl Phys Lett*, **99**, 102903 (2011).

Fig. 1. Schematic device structure of the Y-doped HfO₂ FTJs with Ag and TiN electrodes. The probe on top Ag electrode was either μm-radius tungsten probe in the electrical testing stage or nm-radius Ti/Ir coated silicon tip in the conductive AFM measurement.

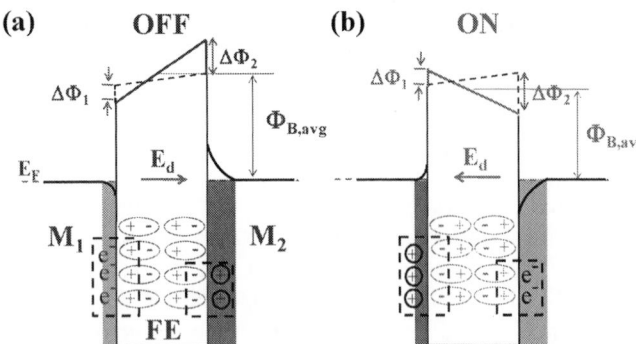

Fig. 2. Band diagram of M₁/FE/M₂ FTJ at V= 0 in **(a)** off and **(b)** on states. The dashed line in both band diagrams are original position of conduction band at V= 0 without polarizations in ferroelectric film. The different colored areas at the metal/ferroelectric interface represent the screening charges appearing on the metal side. The interfacial potential change is defined as ΔΦ, which is related to the finite band bending at the metal interface.

The dashed box means the total screened charges at metal/ferroelectric interfaces, and E_d denotes the depolarization field established by the unscreened bound charges. The $\Phi_{B,avg}$ represents the average barrier height, which determines the magnitude of tunneling current in the junction.

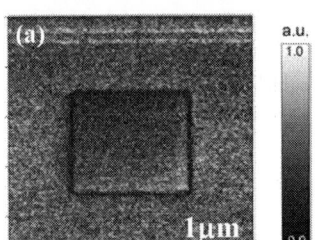

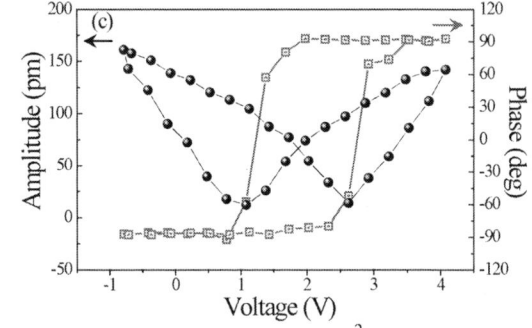

Fig. 3. PFM measurement results. Opposite writing voltages were applied to polarize the inner 2×2 μm² and the whole 5×5 μm² area on 3-nm-thick Y-HfO₂/TiN stack respectively. The surface state was subsequently read with zero bias. **(a)** PFM amplitude image shows clearly zero value at the boundary of two areas. **(b)** PFM phase image demonstrates opposite polarization direction with almost 180° phase difference. **(c)** The piezoelectric response of amplitude and phase collected from the local point presents typical ferroelectric characteristics with switching point occurring at same threshold voltage.

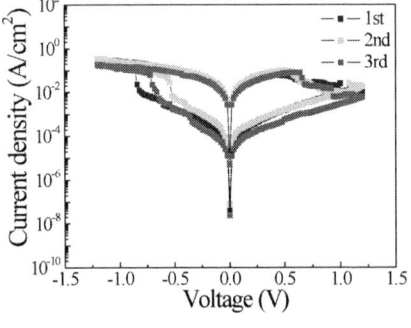
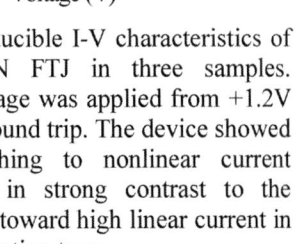
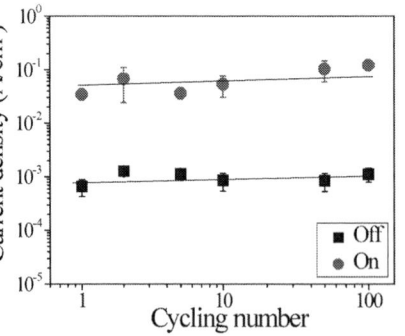
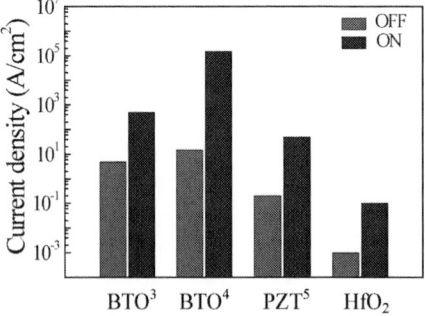

Fig. 4. Reproducible I-V characteristics of Ag/Y-HfO₂/TiN FTJ in three samples. Sweeping voltage was applied from +1.2V to -1.2V in a round trip. The device showed gradual switching to nonlinear current characteristic, in strong contrast to the sharp increase toward high linear current in filament conduction type.

Fig. 5. The cycling endurance test was performed with writing pulse voltage at 1.2V with 100 μs pulse width. After each cycling pulse, the current was read at 0.5V for 10 times. The average reading current was shown with the standard difference in the error bar.

Fig. 6. Comparison of on and off current density among FTJs on Y-doped HfO₂ and on perovskite structure films. All of the current densities are compared at the reading voltage of 0.5V.

978-1-5090-4661-4/17 $31.00 © 2017 IEEE

4B-2

Punch-Through Stop Doping Profile Control via Interstitial Trapping by Oxygen-Insertion Silicon Channel

Robert J. Mears[1a], Hideki Takeuchi[1], Robert J. Stephenson[1], Marek Hytha[1], Richard Burton[1],

Nyles W. Cody[1], Doran Weeks[1], Dmitri Choutov[1], Nidhi Agrawal[2b], and Suman Datta[2c]

[1]Atomera Inc., Los Gatos, CA, [2] Pennsylvania State Univ., PA

[a]e-mail: rjmears@atomera.com, [b]currently at Micron Technology Inc., [c]currently at Notre Dame Univ.

Abstract

Interstitial trapping by oxygen-inserted silicon channel results in blocking of boron and phosphorus transient enhanced diffusion as well as retention of channel boron profiles during the gate oxidation process. The enhanced doping profile control capability is applicable to punch-through stop of advanced CMOS devices and its benefits to 28nm planar CMOS and 20nm bulk FinFET devices projected by TCAD have been discussed.

(Keywords: oxygen-inserted silicon, transient-enhanced diffusion, CMOS, FinFET)

Introduction

OI (oxygen-inserted) silicon technology has been shown to achieve simultaneous e- and h+ mobility improvement, gate leakage reduction, and SSR (super-steep retrograde) channel formation leading to Vt variability reduction [1] – [5]. Here we report an important additional doping profile control capability. It is experimentally demonstrated that trapping of silicon interstitials by the OI layer leads to suppression of interstitial-mediated diffusion of boron and phosphorus. The observed feature is applicable to enhanced control of punch-through stop layer for planar and bulk FinFET CMOS devices.

Experimental and Results

A. Blocking TED (transient-enhanced diffusion)

Boron and phosphorus are well-known to form interstitial pairs to exhibit transient enhanced diffusion (TED) [6]. To investigate the interactions between interstitials and the OI layer, a high-energy implantation was performed through the OI layer, followed by a 1-hr anneal at 750°C to induce TED. Figs. 1 and 2 show boron and phosphorus profiles of the OI silicon and the control, respectively. It can be clearly seen that the 750°C anneal led to the upward diffusion of boron and phosphorus and that the diffusion is blocked by the OI layer thereby retaining low surface concentration whereas dopants diffuse to the top surface for the control.

B. Channel doping profile change by gate oxidation and subsequent anneals

TED is triggered by interstitial injection during thermal oxidation processes as well. Boron profiles after oxidation and an RTA were compared by growing an undoped epitaxial silicon film on top of boron implanted silicon substrates (Figs. 3 and 4). Whereas the non-OI epi film resulted in a reduction of boron peak concentration after thermal processing, almost no profile change near the peak was observed for the OI epi film. The difference is attributed to trapping of interstitials by the OI layer.

Application to Punch-Through Stop Layer

A. 28nm planar CMOS

Interstitial trapping by the OI layer enables the precise control of a PTS (punch-through stop) layer. A halo-free SSR channel for sub-65nm CMOS devices, advantageous for variability reduction and thus for low-voltage operation [7], can be realized. Fig. 6 compares (a) boron profiles at the channel center (b) transfer characteristics of Lg=28nm NMOS with and without TED by TCAD simulations (Fig. 5). The PTS layer defined by interstitial trapping is projected to improve Ion by 13% compared to regular Si epi with TED.

B. 20nm bulk FinFET

By taking advantage of the TED blocking effect, punch-through control of bulk FinFET devices can be achieved by implementing the OI layer on starting silicon substrates. The OI silicon layer blocks diffusion of not only PTS layer but also SD (source/drain) dopants so that SD and PTS can be vertically self-aligned (Fig.7). This allows aggressive diffusion anneals of SD dopants for series resistance reduction. Fig. 9 compares the performance benefits of 22nm bulk FinFET projected by TCAD calibrated to experimental data [8] (Fig. 8). Drain current is improved by mobility improvement by retaining an undoped fin channel as well as by series resistance reductions. Reduced RDF leads to variability reduction as well.

Conclusion

The OI silicon channel traps interstitials and thus suppresses boron and phosphorus TED. This feature is beneficial for controlling punch-through in planar and bulk-FinFET CMOS devices.

978-1-5090-4661-4/17 $31.00 © 2017 IEEE

References

[1] R. J. Mears et al., IEEE Intl SOI Conf., pp.23-24, 2007.

[2] R. J. Mears et al., IEEE Silicon Nanoelectronics Workshop, pp.33-34, 2012. [

[3] N. Xu et al., IEEE Intl Elect.n Dev. Meeting, pp.127-130, 2012.

[4] N. Xu et al., , IEEE Trans. Elect. Dev., pp.3345-3348, 2014

[5] N. Xu et al., Appl. Phys. Lett., 123502, 2015.

[6] P. A. Stolk et al., J. Appl. Phys., pp.6031-6050

[7] K. Fujita et al., IEEE Intl Elect. Dev. Meeting, pp.32.31-32.3.4, 2011

[8] C. Auth et al., Symp on, VLSI Tech. pp. 131-132, 2012.

[9] A. K. Kambham, Nanotechnology, vol.24, No. 27, 2013

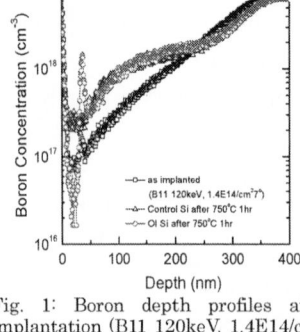

Fig. 1: Boron depth profiles after implantation (B11 120keV, 1.4E14/cm², 7°) and after 750C 1 hr anneal for the control and OI Si

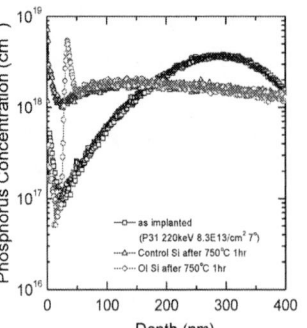

Fig. 2: Phosphorus depth profiles after implantation (P31 220keV, 8.3E13/cm², 7°) and after 750C 1 hr anneal for the control and OI Si.

- Screen Oxide (8nm)
- Well implantation: B 40keV 2.5E13 +BF2 6keV dose adjusted to meet Ioff=100pA/um
- Selective Si epi (10nm)
- Gate Oxidation (splits)
  - Control: Oxide removal → 800C 30min (6nm) → oxide removal→ 1000C 30s (1.4nm)
  - No TED split: Oxide removal → Si removal (2 nm) → Oxide deposition (1.4nm)
- Gate poly Si CVD and patterning
- Poly reox (850C 1min): skip for no TED split
- Offset spacer (13nm)
- Extension implant (As 2keV 1E15) and LSA (1050C 10ms)
- Sidewall spacer
- Deep SD implant (P 3keV 3E15) and RTA(1050C 0.1s)

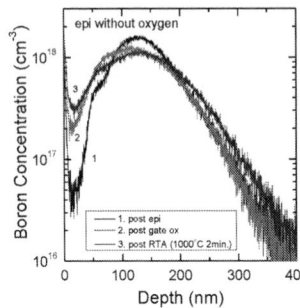

Fig. 3: Boron depth profiles: 1. after implantation (B11 25keV, 2E13/cm², 0°) and undoped epi Si growth(20nm); 2. after 800C 60min and 850C 30min. gate oxidations; 3. 1000°C 2min anneal.

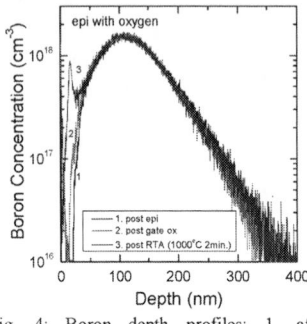

Fig. 4: Boron depth profiles: 1. after implantation (B11 25keV, 2E13/cm², 0°) and undoped epi Si growth with the OI layer (20nm); 2. after 800C 60min and 850C 30min. gate oxidations; 3. 1000°C 2min anneal.

Fig. 6: TCAD simulation flow of halo-free 28nm NMOS to assess impact of TED induced by oxidation processes.

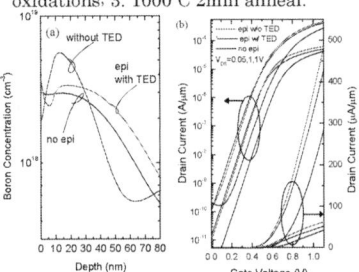

Fig. 5: Simulated (a) channel boron profiles at the channel center and (b) transfer characteristics of Lg=28nm NMOS for epi with and without TED during oxidation and processes and the Si control.

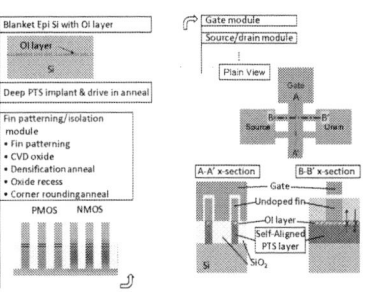

Fig. 7: Bulk FinFET process outline using the OI layer on starting substrate.

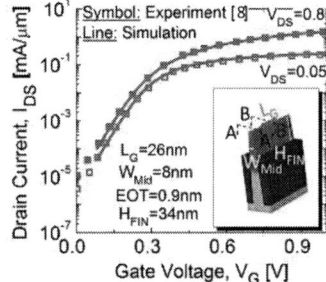

Fig. 8: 3D numerical simulations of 22nm FinFET in excellent agreement with experimental data [8]

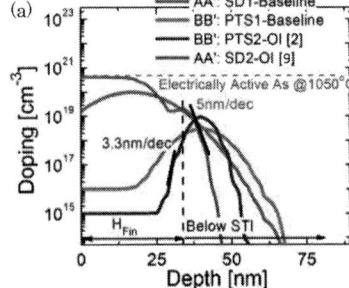

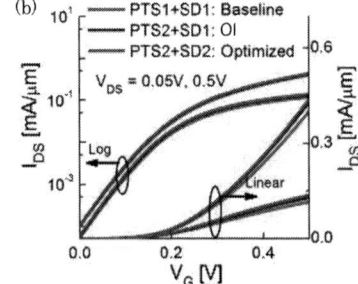

(c)

Electrical Parameter	(i) PTS1+SD1	(ii) PTS1+SD1 [(ii)/(i)]	(iii) PTS2+SD2 [(iii)/(i)]
Idsat (uA/um)	400	426 [1.07]	435 [1.10]
Idlin (uAum)	115	130 [1.13]	137 [1.21]
Vtlin (mV)	169	167 [0.99]	168 [1.06]
SSlin (mV/dec)	74	71 [0.99]	74 [0.99]
SSsat (mV/dec)	79	76 [0.99]	79 [0.99]
DIBL (mV/V)	55	51 [0.99]	47 [0.85]
σVt-RDF (mV)	19.98	15.24 [0.76]	14.22 [0.71]
σVt-Total (mV)	31.6	26.8 [0.85]	26.3 [0.83]

Fig. 9: (a) Source-drain doping profile optimization in 22nm FinFET using OI layers; OI layer (denoted by dashed vertical line) allows higher S/D doping concentration (green solid line) over the baseline S/D doping (red solid line) while reducing the junction depth in the sub-fin region below the STI at the same time. (b) Transfer characteristics of PTS2 + SD2 : OI FinFET benchmarked against the 22nm baseline FinFET and the PTS2 + SD1 : OI FinFET. OI. (c) FinFET with re-optimized source-drain doping profile and punch-through implant profile provides a combined performance boost of 21% IDLin and 10% IDSat improvement at VDD = 0.5V, due to reduction in external resistance and improvement in electron mobility from reduced impurity scattering. Considering RDF fluctuation as well as geometric fluctuation, OI FinFETs show 17% improvement in σV,Th-Total over baseline 22nm FinFET[8].

4B-3 (Invited)

(Invited) FinFET/Nanowire Design for 5nm/3nm Technology Nodes: Channel Cladding and Introducing a "Bottleneck" Shape to Remove Performance Bottleneck

Victor Moroz, Joanne Huang and Munkang Choi

Synopsys, Inc., 690 East Middlefield Road, Mountain View, CA, USA, victorm@synopsys.com

Abstract

Transition from planar MOSFETs to FinFETs enabled scaling beyond 28nm node. At 5nm/3nm design rules, a transition from FinFETs to nanowires has to be evaluated. We explore with rigorous NEGF (Non-Equilibrium Green's Functions) and sub-band Boltzmann transport models the impact of nanowire shape and SiGe/Si cladding layers on its performance and variability. Outside of the nanowire channel, a "bottleneck" shape of the source/drain extensions can either boost or ruin the performance, requiring NEGF-driven meticulous shape engineering.

(Keywords: Nanowire, FinFET, scaling, TCAD, NEGF, sub-band Boltzmann, SiGe channel cladding, band structure engineering)

Introduction

To avoid FinFET short-channel effects, the fin width W scaling has to be in sync with channel length L scaling. Scaling towards 5nm/3nm design rules requires key transistor specifications listed in Table 1. It will be difficult to reproducibly manufacture tall fins with the required fin widths. This is when nanowires can be introduced due to the better channel control by gate-all-around nanowire design [1]. Here, we benchmark FinFET and nanowire designs for 5nm/3nm nodes.

Modeling Methodology

We use 3D sub-band Boltzmann transport analysis for the PMOS FinFETs and nanowires, with 2D Schrödinger equation in the fin cross-section and 1D Boltzmann transport along the fin [2]. Considering that FinFETs have close to 80% ballistic transport [3], we do ballistic FinFET analysis. For the nanowires, ballistisity drops towards 50% [3], and therefore we include scattering mechanisms into nanowire analysis. The PMOS source/drain (S/D) extensions have $1 \cdot 10^{20}$ cm^{-3} doping and $3 \cdot 10^{-9}$ Ohm$\cdot$cm^2 contact resistance. For the NMOS FinFETs and nanowires, our analysis did not show any noticeable benefits by going to SiGe or cladded SiGe/Si channels, so we are not reporting such results in this work.

For the NMOS nanowire S/D extension engineering, we use 3D NEGF approach based on 3D Schrödinger transport [2] with explicit Coulomb scattering off atomistic dopants in S/D extensions.

PMOS FinFETs with Si, SiGe, and Cladded Fins

Si fin cladding with a 30% Ge SiGe layer for PMOS can be introduced to gain an equivalent of about 2 nm in terms of fin width scaling. Conformal and faceted fin cladding options are illustrated on Fig. 1. Conformal cladding exhibits slight advantage over faceted, so we focus on conformal cladding results here.

Figure 2 shows that holes tend to stay within the SiGe quantum well in the on-state, whereas homogeneous 30% Ge SiGe and Si fins tend to have most of the current flowing through the middle of the fin width. Somewhat similar patterns are observed for the off-state (Fig. 3). At all bias conditions, cladded fins bring the holes closer to the surface, where gate has a stronger control.

Benchmarking different fin designs for L ranging from 15 nm down to 11 nm, we see that cladded fin has the best Ion/Ioff trade-off (Fig. 4), whereas SiGe fin is the worst option at L=15 nm, but challenges Si at L=11 nm. This drastic improvement of SiGe channel performance happens due to the beneficial changes in SiGe band structure when fin width scales down to 5.5 nm. The source barrier height in the off-state increases by 137 mV (see Fig. 5, where different channel materials are compared in the off-state for a fixed on-state current). The flip side of such drastic performance improvement of SiGe fin is a strong performance variability due to inevitable fin width fluctuations.

PMOS Si, SiGe and Cladded Nanowires

The structure of a nanowire cross-section and the hole density distribution across the channel are depicted on Fig. 6, where most of the holes are located inside the SiGe cladding layer, similarly to the cladded fin. Cladding-induced stress patterns are very favorable for the PMOS performance (Fig. 7), with high compressive longitudinal and lateral stress values. Peak cladding stresses are: 1.4 GPa compressive longitudinal and lateral stress components and 843 MPa compressive vertical stress. Benchmarking of different nanowire designs is summarized on Fig. 8, where cladded channel is consistently the best, and Si channel is consistently the worst. As opposed to the FinFETs, nanowire performance is insensitive to scaling from 5nm to 3nm design rules.

NMOS Si Nanowire Access Resistance

One of the major issues for nanowire performance is source/drain extension resistance [1]. One seemingly obvious solution is to increase the extension cross-section (Fig. 9). However, rigorous NEGF analysis

978-1-5090-4661-4/17 $31.00 © 2017 IEEE

shows that it causes severe performance degradation due to the additional barrier when the band structure morphs according to geometric confinements of a wider extension and a narrow channel. Fortunately, moving the "bottleneck" shaped transition point 3 nm inside the extension solves the problem and provides additional performance boost (Fig. 10) by moving the band offset inside the heavily doped S/D where it has no effect on transistor behavior. Simultaneously, extension resistance reduces due to the wider cross-section area.

Conclusions

We have demonstrated a wide design space for the 5nm/3nm nodes that can be explored and optimized with rigorous physics-based modeling. One common observation for all described results is that behavior of transistors scaled down to 5nm/3nm design rules is determined by the band structure of underlying materials that is sensitive to specific shapes of the channel and S/D extensions. Therefore, accounting for band structure changes is critical and band structure

engineering becomes a key part of transistor optimization.

References

[1] H. Mertens, R. Ritzenthaler, A. Hikavyy, M. S. Kim, Z. Tao, K. Wostyn, S. A. Chew, A. De Keersgieter, G. Mannaert, E. Rosseel, T. Schram, K. Devriendt, D. Tsvetanova, H. Dekkers, S. Demuynck, A. Chasin, E. Van Besien, A. Dangol, S. Godny, B. Douhard, N. Bosman, O Richard, J Geypen, H Bender, K Barla, D. Mocuta, N. Horiguchi, and A. V-Y Thean, "Gate-all-around MOSFETs based on vertically stacked horizontal Si nanowires in a replacement metal gate process on bulk Si substrates" VLSI Technology Digest, (2016).

[2] Sentaurus Device QTX User's Guide, (2016).

[3] Munkang Choi, Victor Moroz, Lee Smith, and Joanne Huang, "Extending Drift-Diffusion Paradigm into the Era of FinFETs and Nanowires", SISPAD Proceedings, pp. 242 – 245, (2015).

Table 1. Key transistor design rules for 5nm, 4nm, and 3nm technology nodes

Key specifications	FinFET			Nano-wire		
	5nm node	4nm node	3nm node	5nm node	4nm node	3nm node
Channel length, nm	15	13	11	15	13	11
Fin/wire total width, nm	7.5	6.5	5.5	15	13	11
Fin/wire total height, nm	30	30	30	6	5.2	4.5
Cladding thickness, nm	2	1.75	1.5	1.5	1.25	1
Fin/wire pitch, nm	30	26	22	30	26	22
Access Resistance, Ohm	808	938	1145	2314	2879	3708

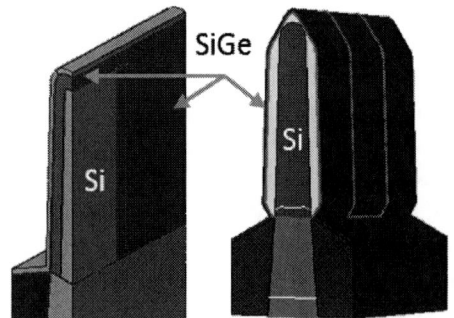

Fig. 1. FinFET with Si fin core and SiGe fin cladding. Conformal cladding (left) and faceted cladding (right).

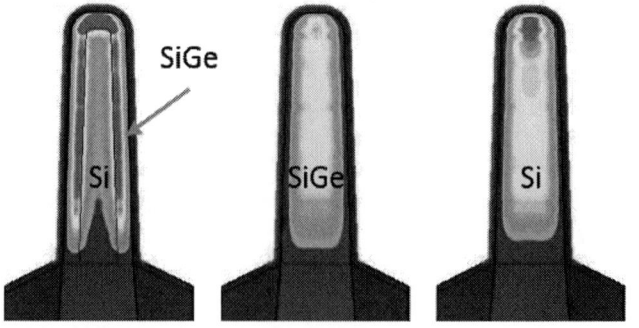

Fig. 2. Normalized hole distributions across the fin at on-state biases. FinFET with conformal SiGe cladding (left), SiGe fin (center), and Si fin (right).

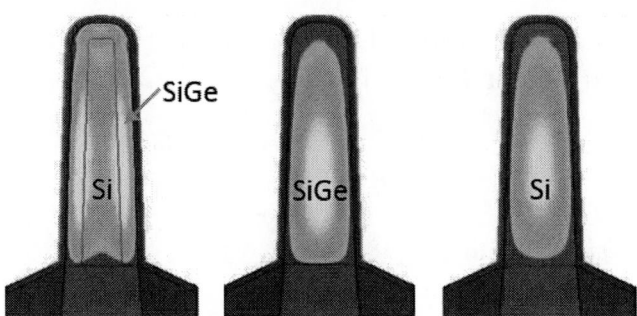

Fig. 3. Normalized hole distributions at off-state biases across the fin. FinFET with conformal SiGe cladding (left), SiGe fin (center), and Si fin (right).

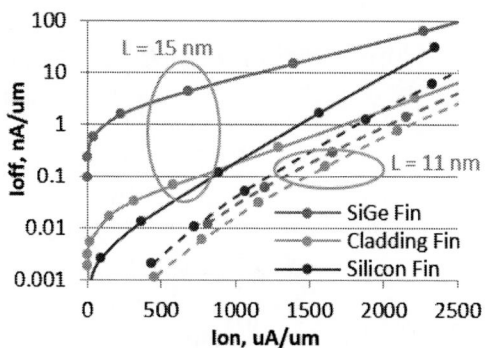

Fig. 4. FinFET performance benchmarking for L=15nm (solid lines) and L=11nm (dashed lines).

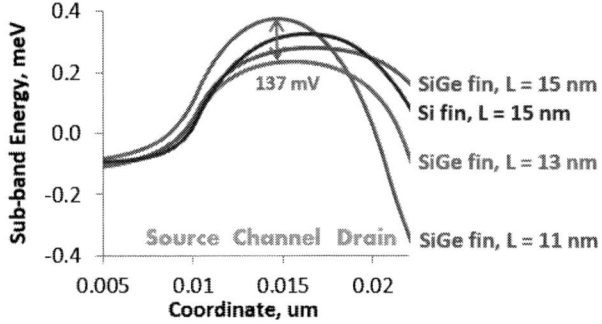

Fig. 5. Zeroth energy sub-band for SiGe & Si fins with different design rules under off-state bias conditions.

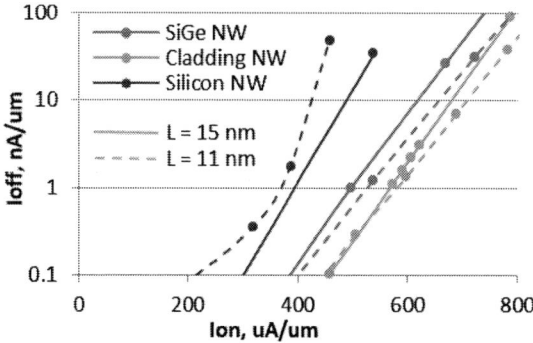

Fig. 6. Normalized hole density map in Si nanowire with SiGe cladding in the on-state.

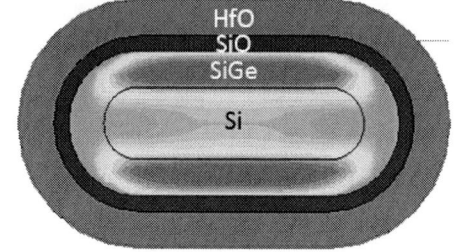

Fig. 8. Nanowire performance benchmarking for L=15nm (solid lines) and L=11nm (dashed lines).

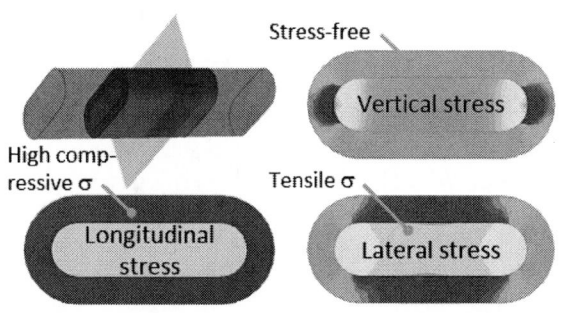

Fig. 7. A 3D nanowire (upper left) and nanowire stress components due to the SiGe cladding layer.

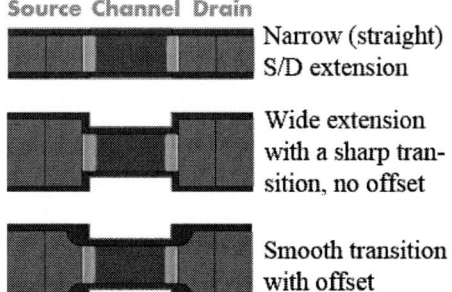

Fig. 9. Nanowire S/D extension engineering to reduce access resistance.

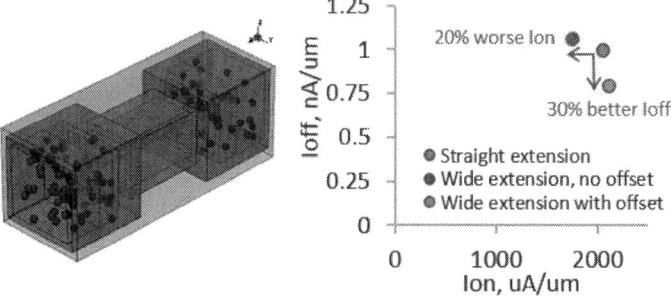

Fig. 10. Nanowire with a sharply wider source/drain extension (left) and nanowire on-state/off-state currents for different source/drain extension designs.

978-1-5090-4661-4/17 $31.00 © 2017 IEEE

4B-4

A Computational Study of Fundamentals and Design Considerations for Vertical Tunneling Field-Effect Transistor

Sheng Luo[*,a], Kain Lu Low[a], Xiaoyi Zhang[a], Qianyu Zhao[a], Hsin Lin[b,c], and Gengchiau Liang[a,c]

[a]Department of Electrical and Computer Engineering, National University of Singapore (NUS), 117583 Singapore
[b]Department of Physics, NUS, 117542 Singapore, [c]Centre for Advanced 2D Materials and Graphene Research
Centre, NUS, 117546 Singapore Email: a0123817@u.nus.edu

Abstract

A comprehensive and rigorous computational study at atomic level was performed for various vertical tunneling field-effect transistor (VTFET) structures based on *III-V* and two-dimensional (2D) materials. The key challenges of VTFETs were found to be induced by device structures and the channel materials' properties. An optimized VTFET structure was proposed to suppress the parasitic tunneling current and improve subthreshold region performance. A drive current ~421.6μA/μm is obtained based on the structural-optimized MoS_2-WSe_2 VTFET.

I. Introduction

Tunneling field-effect transistor (TFET) is a promising logic device that can achieve a steeper subthreshold swing (*SS*) than MOSFET [1], [2]. The vertical tunneling TFET (VTFET) was proposed to induce the drive current enhancement [3], [4]. In this work, we investigate the VTFET's performance based on group *III-V* and 2D materials, two potential candidates for next generation TFET. GaSb-InAs and MoS_2-WSe_2 are implemented as channel materials in this study. The challenges induced by material properties and device structures are then identified. Subsequently, an optimized VTFET structure is proposed in terms of suppressing the impact of parasitic tunneling.

II. Simulation Methodology

The cross-sectional view of a VTFET is shown in **Fig. 1(a)**. The key design parameters include the tunneling overlap length (L_{OV}), body thickness (T_{Body}), and underlap length (L_{UN}). T_{Body} is 4nm for III-V material heterojunction. A V_{DD} of 0.66 V and EOT of 0.31 nm are implemented. The simulation procedure is summarized in **Fig. 1(b)**. The device Hamiltonians for *III-V* materials are constructed based on $sp^3d^5s^*$ tight-binding model [5]. And a type II hetero-junction consists of monolayer MoS_2 and WSe_2 is implemented in an atomistic Hamiltonian derived from density functional theory calculation [6].

III. Results and Discussions

The I_{DS}–V_{GS} of GaSb-InAs VTFET with various L_{OV} is plotted in **Fig. 2(a)**. It notes that the characteristics show almost negligible dependence on L_{OV}. To explain this feature, **Fig. 3(b)–(d)** reveal that the majority of tunneling occur at the right edge of overlap region. Besides, in **Fig. 2(b)**, a drastic drop of local density of states (LDOS) across heterojunction with nearly **8 order** differences is observed. Due to the dominance of edge tunneling in total tunneling current and limited DOS, the drive current is independent of overlap length with the degradation of *SS* due to the parasitic tunneling at the edge. An optimized device structure shown in **Fig. 4(a)** is proposed to suppress the edge tunneling. The $I_{DS} - V_{GS}$ characteristics of the GaSb-InAs is shown in **Fig. 4(b)** with improved *SS* and drive-current scaling. The suppression of the leakage current is evidenced in the current contour comparing to the device without optimization (**Fig. 5(a), (b)**). For 2D material combination (MoS_2-WSe_2), similar scaling trend of drive current with the overlap length is also observed in **Fig. 4(d)**. In addition, comparing to the III-V heterostructure, MoS_2-WSe_2 possesses larger DOS in overlap region (**Fig. 5(d)**), enabling higher tunneling current (**~421.6μA/μm**).

IV. Summanry

In summary, we have identified the relatively low DOS and dominance of edge tunneling current contributed to the degradation of III-V material based VTFET. The proposed VTFET structure is verified to suppress the edge tunneling and *SS* improvement. For 2D material, providing relatively high DOS and short tunneling length, demonstrates similar scaling trend and achieves the drive current to ~421.6μA/μm.

Acknowledgement

We acknowledge support from the Singapore National Research Foundation (NRF) through a Competitive Research Program (Grant No: NRF-CRP6-2010-4) and the Ministry of Education under MOE2013-T2-2-125. S.L and HL. Acknowledge the NRF for the support under NRF Award No. NRF-NRFF2013-03. We also gratefully acknowledge the discussions with Dr. Kai-tak Lam.

References

[1] P. F. Wang *et al.*, *Solid State Electron.* 48, 2281, 2004
[2] Q. Zhang *et al.*, *IEEE Electron Dev. Lett.*, 27, 297, 2006. [3] L. De Michielis *et al.*, DRC Proceedings, 111, 2011. [4] Y. Lu *et al.*, *IEEE Elec. Dev. Lett.*, 33, 665, 2012.
[5] S. Datta, Quantum Transport: Atom to Transistor, Cambridge 2005
[6] Lam, Kai-Tak, *et al.*, *2014 IEDM* IEEE, 2014

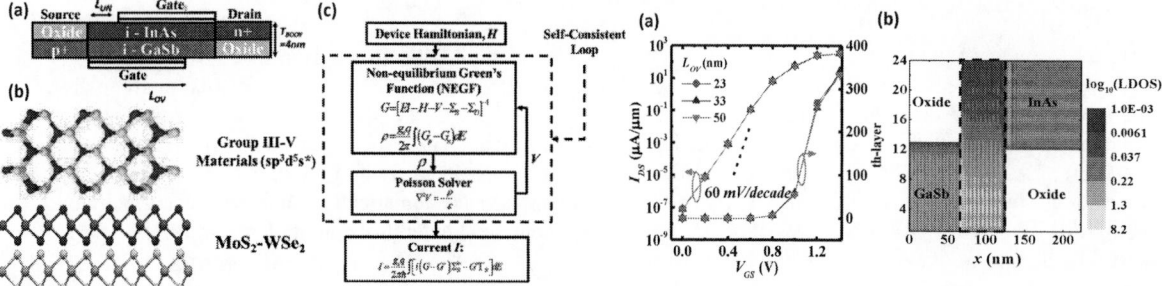

Fig. 1. (a) Schematic view of device. Blue/red denotes InAs/GaSb respectively (b) Crystal structures of the channel materials used in this work (c) Summary of simulation methodology adopted in this study.

Fig. 2. (a) I_{DS}-V_{GS} of hetero-structure with different L_{OV}. (b) 2D-local density of states (LDOS) for the energy level where the highest current is obtained. Black dash line denotes the overlap region.

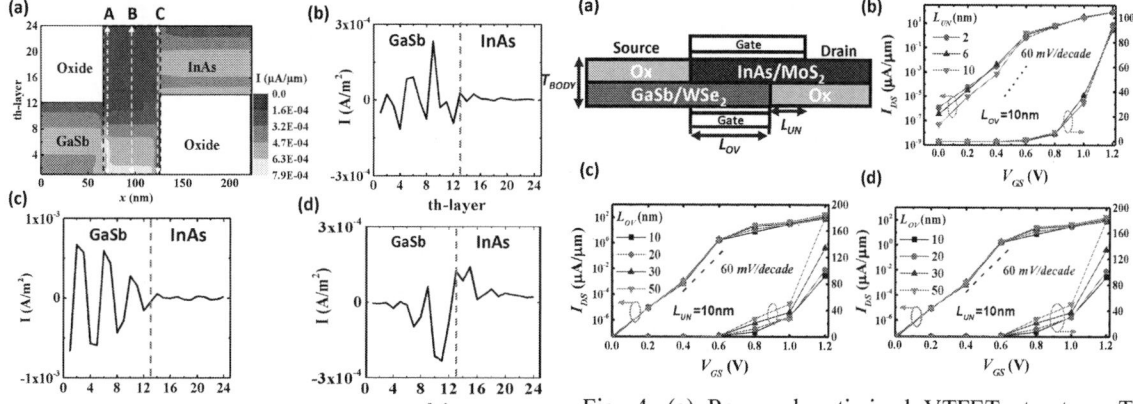

Fig. 3. (a) Current contour of GaSb-InAs VTFET with L_{OV} of 50 nm at V_{GS}=0.8 V. (b), (c), (d) Vertical tunneling current extracted from cutline A, B and C, respectively. Negative/Positive values at each layer indicate flow out/in at certain layer.

Fig. 4. (a) Proposed optimized VTFET structure. The oxide region extended to the edge of overlap region. An intrinsic underlap region controlled by gate is added at the top layer (b) , (c) and I_{DS}-V_{GS} characteristics of the optimized VTFET structure based on *III-V* materials (d) I_{DS}-V_{GS} characteristics of optimized 2D materials VTFET.

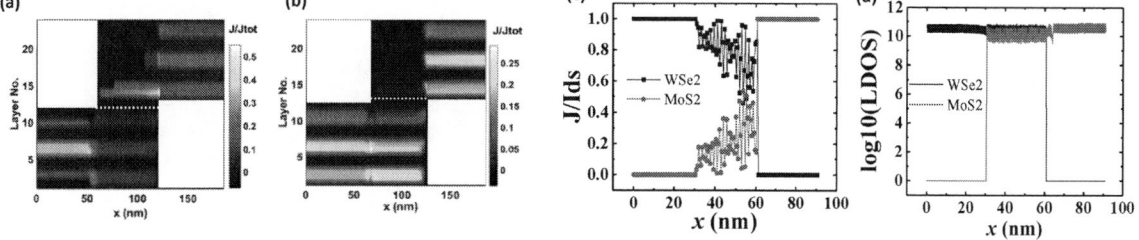

Fig. 5. (a), (b) The 2D- current contour of optimized *III-V* and non- optimized *III-V* VTFET, respectively. The current density has been normalized by I_{DS}. White dash line is the boundary between GaSb and InAs (c) Current distribution in 2D material VTFET. Current density flowing at each layer has been normalized by I_{DS}. (d) 2D-LDOS of MoS$_2$-WSe$_2$ VTFET. All figures are plotted under Vg=0.8V.

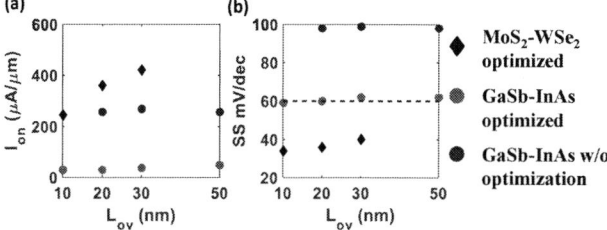

Fig. 6. (a) on-current and (b) *SS* of MoS$_2$-WSe$_2$ and GaSb-InAs VTFET under different overlap length. GaSb-InAs with edge-tunneling suppression demonstrates SS improvement and drive current-overlap length scaling. MoS$_2$-WSe$_2$ provides higher drive current comparing to the optimized III-V material due to large DOS in overlap region.

978-1-5090-4661-4/17 $31.00 © 2017 IEEE   71

4B-5

Analysis of break-even time for nonvolatile SRAM with SOTB technology

Daiki Kitagata, Yusuke Shuto, Shuu'ichirou Yamamoto, and Satoshi Sugahara
Tokyo Institute of Technology, Yokohama, Japan, kitagata.d@isl.titech.ac.jp

Abstract

Energy performance of nonvolatile power-gating (NVPG) that is a power-gating technique with nonvolatile state/data retention is demonstrated for nonvolatile SRAM (NV-SRAM) with spintronics retention and silicon-on-thin-BOX (SOTB) CMOS technologies. The NV-SRAM cell consists of an ordinary 6T cell and two magnetic tunnel junctions (for nonvolatile retention) with two pass-transistors. The cell and array architectures for leakage power reduction are developed, and the SOTB CMOS technology is used for the peripheral circuits to reduce the shut-down leakage. The break-even time (BET) that is an energy-performance index of NVPG is analytically formulated for a NV-SRAM array with its peripheral circuits. The BET behavior with respect to the cell/array architectures and the effect of the peripheral circuits are systematically analyzed using circuit parameters extracted from an implemented NV-SRAM.

(Keyword: Power-gating, Nonvolatile SRAM, SOTB, break-even time, cache, microprocessors, and SoCs)

Introduction

Reduction of static power dissipation is one of the most important issues for highly integrated CMOS logic systems, such as microprocessors and SoCs. Currently, power-gating (PG) becomes a standard technique to reduce static energy for these logic systems. In particular, logic systems based on a multi-core architecture come loaded with core-level PG. Although the core-level PG can effectively reduce static energy, the state and data retention in the cores restricts the energy performance of core-level PG, i.e., Back-up techniques for registers, register files, and caches in a core, such as data-transfer to a back-up memory and others [1], costs additional energy and latency. Therefore, nonvolatile power-gating (NVPG) that is a PG technique using nonvolatile bistable memory circuits is highly effective at achieving temporally fine granularity of PG executions, resulting in improvement of the energy reduction efficiency.

In this paper, energy performance of NVPG is investigated for nonvolatile SRAM (NV-SRAM) with spintronics retention and silicon-on-thin-BOX (SOTB)

CMOS technologies. Break-even time (BET) that is an energy-performance index of NVPG is analytically formulated for a NV-SRAM array with its peripheral circuits. The BET behavior with respect to the cell/array architectures and the effect of the peripheral circuits are systematically analyzed using circuit parameters extracted from an implemented NV-SRAM.

Cell and peripheral circuits

Figure 1 shows the circuit configuration of the NV-SRAM cell, which consists of magnetic tunnel junctions (MTJs) for nonvolatile retention and an ordinary 6T cell for normal SRAM operations. Since the MTJs are connected to the 6T cell through the pass transistors, the NV-SRAM cell can electrically separate the normal SRAM operations and the nonvolatile retention operation, i.e., the MTJs in the cell can be used only for the nonvolatile retention [2]. This is highly important to avoid degradation of the circuit performance during the normal SRAM operation mode and prevent an increase in the BET [3].

The details of the cell operations were described in Ref. [2,4]. When the cell enters the shutdown and wake-up modes, data on the storage nodes of the cell are transferred to the MTJs and the data in the MTJs are written back to the storage nodes, respectively. Hereafter, these are referred to as store and restore operations, respectively, for simplicity. In the cell array examined here, virtual-power-supply (VV$_{DD}$) architecture with a header power switch connected to the cell is employed.

Figure 2 shows the layout of the implemented 1kb NV-SRAM using the 65nm SOTB technology [4]. The

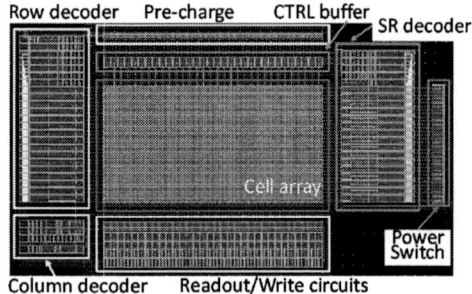

Fig. 2. Layout of a fabricated 1kb NV-SRAM chip.

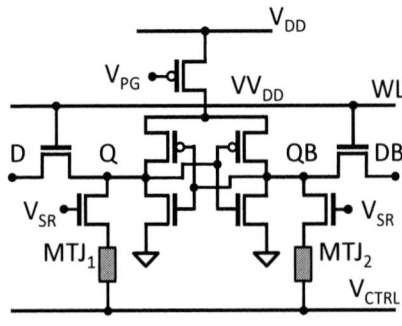

Fig. 1. Schematic of a NV-SRAM cell

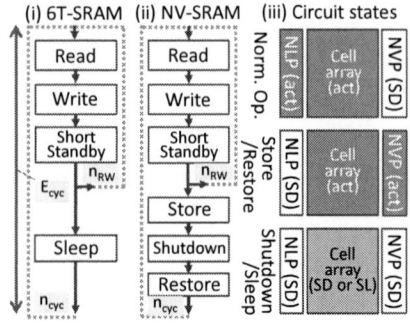

Fig. 3. Benchmark sequences for (i) 6T-SRAM, (ii) NV-SRAM, and (iii) circuit states.

978-1-5090-4661-4/17 $31.00 © 2017 IEEE   72

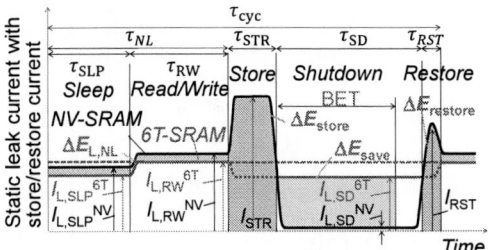

Fig. 4. Time chart of leak currents with store and restore currents for NV-SRAM and equivalent 6T-SRAM.

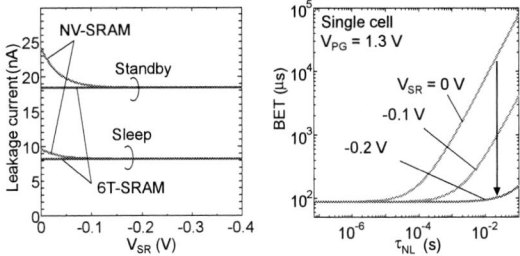

Fig. 6. (a) Standby and sleep leakage currents of the NV-SRAM cell as a function of V_{SR}. The leakage currents for the equivalent 6T cell are also indicated. (b) BET for the single NV-SRAM cell as a function of τ_{NL}, in which V_{SR} is varied.

NV-SRAM organization is basically the same as ordinary SRAM, and the SR decoder, CTRL buffer, and power switches are newly added for NVPG operations. The SOTB technology is used to reduce leakage currents of these peripheral circuits during the shut-down mode, employing a body bias. Hereafter, the peripheral circuits for the normal and NVPG operations are referred to as NLP and NVP, respectively.

BET formulation and analysis procedure

Figure 3 shows a benchmark sequence for BET evaluation. The circuit states of the cell array and peripheral circuits are also shown in the figure. Figure 4 schematically shows current components used for the following BET analysis. The BET of NV-SRAM is defined by the duration when the extra leakage energy during the normal operation mode and the store and restore energies for the NVPG operations are compensated by the saved energy due to the shutdown. When the shutdown duration (τ_{SD}) is longer than BET, the energy of the NV-SRAM can be effectively reduced in comparison with that of the equivalent 6T-SRAM. BET for a NV-SRAM cell array with the NLP and NVP circuits was formulated by [6]

$$BET = \frac{\begin{array}{c}(\Delta P_{L,NL}^{Array} + \Delta P_{L,SD}^{NVP})\tau_{NL} + \Delta E_{STR} + (\Delta P_{L,STR}^{Array} \\ + \Delta P_{L,STR}^{NVP} + \Delta P_{L,SD}^{NLP})\tau_{STR} + \Delta E_{RST} + (\Delta P_{L,RST}^{Array} \\ + \Delta P_{L,RES}^{NVP} + \Delta P_{L,SD}^{NLP})\tau_{RST}\end{array}}{\Delta P_{L,SD}^{Array} + \Delta P_{L,SD}^{NLP}} \quad (1)$$

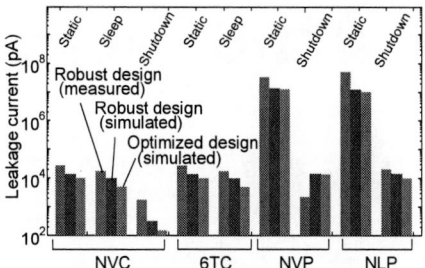

Fig. 5. Measured and simulated leakage currents of the NV-SRAM cell (NVC), 6T cell (6TC), NVP, and NLP.

where $\Delta P_{L,state}^{Circ}$ is the average leakage power difference during each state (that are normal; NL, store; STR, restore; RST, shutdown; SD operations) for Circ = cell array, NLP, and NVP, τ_{state} the duration of each state, and ΔE_{STR} and ΔE_{RST} the store and restore energies, respectively.

Figure 5 shows the measured leakage currents in the NV-SRAM cell and NLP/NVP peripherals of a fabricated 1kb NV-SRAM chip [4]. Simulated currents are also shown in the figure. The simulated values are not necessarily identical with the measured values. This is due to the imperfect HSPICE parameters and/or process variation for the SOTB CMOS technology under development. Nevertheless, the results can be considered to be in an acceptable range. Note that in this chip implementation, a slightly robust cell design was used. In the following analysis, the optimized design was used to extract the best performance of NV-SRAM. The differences between the measured and simulated results were corrected using the correction factors extracted from the results shown in Fig. 5.

Analysis results

Figure 6 (a) shows leakage currents of the NV-SRAM as a function of V_{SR} (see Fig.1) during the normal read/write and sleep (a low-voltage retention mode) operations. The leakage currents for the equivalent 6T cell are also shown in the figure. The leakage currents of the NV-SRAM cell during the normal read/write and sleep modes can be effectively reduced by negative V_{SR} and become the same levels as those of the 6T cell. Figure 6 (b) shows BET for the single NV-SRAM cell as a function of τ_{NL}. BET consists of two components. The plateau region results from the energies for the store/restore operations. The component proportional to τ_{NL} is governed by the leakage currents during the normal operation mode (that include the read/write and sleep operations). Hereafter, the former and latter components of BET are referred to as SR and NL components, respectively. The NL component shown in the fig. 6(b) is dramatically reduced by negatively increasing V_{SR} owing to the reduction of the cell leakage currents during the normal operation mode.

Figure 7 shows BET for cell arrays without the peripheral circuits, in which the array size are 16kB, 250kB, and 1MB (i.e., the array size is varied from typical L1$ to L2$ sizes). The arrays consist of 8kB ($256bit \times 256bit$) subarrays, and the bit widths for the

978-1-5090-4661-4/17 $31.00 © 2017 IEEE

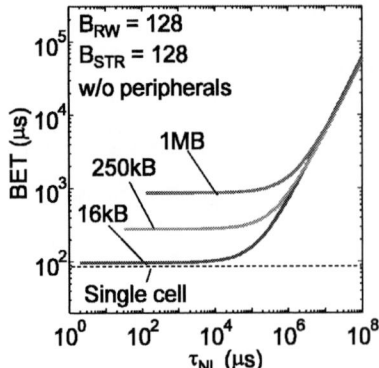

Fig. 7. BET as a function of τ_{NL} for the NV-SRAM cell arrays, in which array size is varied.

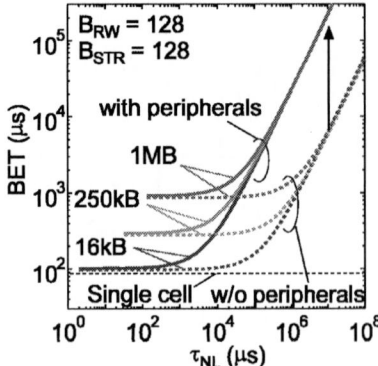

Fig. 8. BET as a function of τ_{NL} for the NV-SRAM cell arrays with and without their peripherals.

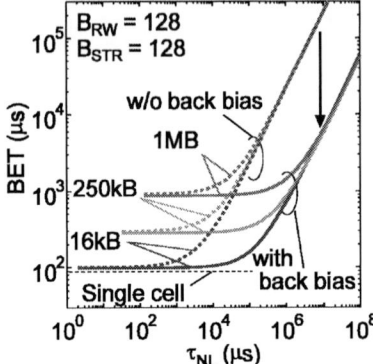

Fig. 9. BET of the NV-SRAM cell arrays with and without the body-bias control.

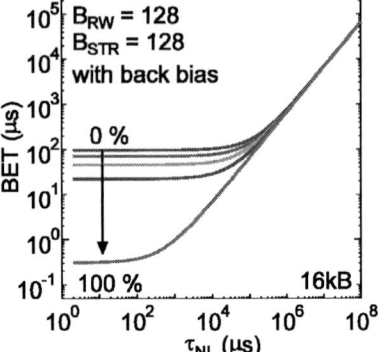

Fig. 10. BET for the 16kB NV-SRAM in which the proportion of the store-free region is varied.

normal read/write operations and the nonvolatile store operation are $B_{RW} = 128$bit and $B_{STR} = 128$bit, respectively. The SR component increased with increasing array size, since the store and leakage energies increase with increasing array size. The NL component is independent of the array size. This is because it only depends on the difference in the leakage current between the NV-SRAM and 6T cells.

Figure 8 shows BET as a function of τ_{NL} for the NV-SRAM arrays with and without peripheral circuits. The peripheral circuits (NLP and NVP) are allocated around each subarray. The peripheral circuits around an operating subarray are only active and the other peripheral circuits for inactive subarrays are shut down. The leakage currents of NLP and NVP circuits during the store operation mode play a minor role for the SR component of BET. This is because the subarray structure can effectively suppress an increase in the leakage currents of the peripheral circuits. However, the leakage current of the NVP circuits during the normal operation mode pushes up the NL component of BET, as shown in the figure.

To reduce the NL component of BET, the body-bias-induced leakage reduction using the SOTB technology [4] is effective, as shown in Fig. 9, since the shut-down leakage of the NVP circuits during the normal operation mode can be dramatically reduced [4]. The store-free shutdown architecture that can skip the store operation [5] can reduce the SR component of

BET. The SR component can be effectively reduced depending on the proportion of the store-free region of the array, as shown in Fig. 10.

Conclusion

The energy performance of NV-SRAM for the NVPG architecture is systematically analyzed using the circuit parameters extracted from an implemented 1kb NV-SRAM chip. The short BET of a few hundreds of microseconds or less for the 16kB NV-SRAM can be achieved, which would result in fine-grained core-level PG of multicore processors and SoCs.

Acknowledgement

This study was partly supported by JSPS KAKENHI Grant and JST. The VLSI chip in this study has been fabricated in the chip fabrication program of VDEC, the University of Tokyo in collaboration with Renesas Electronics Corp.

References

[1] Y. Kanno *et al.*, IEEE J. Solid-State Circuits, **42**, 1, pp. 74-83, 2007.
[2] Y. Shuto *et al.*, J. Appl. Phys., **105**, 07C933, 2009.
[3] S. Yamamoto *et al.*, Electron. Lett., **47**, 1027, 2011.
[4] Y. Shuto *et al.*, IEEE ESSCERC 2016
[5] Y. Shuto *et al.*, Jpn. J. Appl. Phys., **51**, 040212, 2012.
[6] D. Kitagata *et al.*, in preparation.

4B-6

Role of Floating Body Effect on Super Steep Subthreshold Slope PN-Body Tied SOI FET

Takahiro Yoshida, Jiro Ida, Syouta Inoue, Syougo Uchikura, Atsushi Hashimoto, Keisuke Hayashi,
Taichi Iwasaki, Masanori Kamako, Takashi Horii

Kanazawa Institute of Technology, Ishikawa, Japan, ida@neptune.kanazawa-it.ac.jp

Abstract

The SOI floating body effect and to keep floating on both the N layer and the body are important for appearance of the super steep SS on the PN-Body tied SOI FET. It was confirmed for the frist time with measuring the new test devices.

(Keywords: Steep SS, Floating body effect and SOI)

Introduction

The tunnel FET's [1] and the negative capacitance FET's [2] have been intensively researched as a steep Subthreshold Slope (SS) device for exploring ultralow power applications, in order to overcome the fundamental lower limit of SS on the conventional MOSFET. However, up to now, the steepness has not been good on both devices.

We have proposed and demonstrated the PN-Body Tied SOI FET's which show the super steep SS ($=35 \mu$V/dec) over 3 decades of the drain current with the ultralow drain voltage of 0.1V [3,4]. However, those needed the body bias over 5V. We have also reported recently that the body bias reduces from over 5V to below 1V when the impurity concentration of the N layer is redesigned from the N^+ (around 1E20/cm^3) to the N^- (around 1E17-1E18/cm^3) [5].

We have measured the device, to keep floating on both the N layer and the body because the SOI floating body effect is expected to contribute to the appearance of the super steep SS. In this study, we studied and confirmed it with the new test devices which have all terminals including the N layer and the body.

PN-body tied SOI FET

Fig.1 shows the cross sectional TEM view of 0.2um SOI [3], the 3D structure made by the 3D device simulator "HyENEXSS" [3] and also the plane view of the new test device, used in this study. The terminals from the N layer (Base of the bipolar transistor (BIP) along the body tied direction) and the body (MOS channel region, Corrector of BIP) are added. We also inserted the resistor (R_B or R_C) on measurements between the test device and the measurement instrument (Keysight B1500) in order to confirm the role of the floating.

Measurement Results

Fig. 2 shows the measured Id-Vg characteristics when the N layer is set to be floating and the body is set to be floating or connected with $R_C=0 \Omega$ (shorted)

or $R_C=1 G \Omega$. The results of the device with both the N^- and the N^+ on the N layer are shown. When $R_C=0$ Ω (shorted), the super steep (<1mV/dec) SS on the N^- and the steep (<60mV/dec) SS on the N^+ disappear and the SS becomes over 60mV/dec. It indicates that the hole is not accumulated in the body and discharged to the body terminal. When $R_C=1 G \Omega$, the super steep again appears. It should be noted that the super steep SS appear on only $R_C=1 G \Omega$ with the N^+. Fig. 3 shows the Id-Vg dependence on R_C. When R_C increases over 10MΩ on the N^- or 100MΩ on the N^+, the super steep SS appears. Those results indicate that the SOI floating body effect is a key for the appearance of the super steep SS.

Fig.4 shows the measured Id-Vg with the body floating and the N layer floating or connected with $R_B=1 G \Omega$ or 10MΩ or 0Ω (shorted). When $R_B=0 \Omega$, the super steep SS on the N^- and the steep SS on the N^+ disappear. The leakage current increases only with the N^- when R_B reduces. It may be because of the large forward PN junction current when the N^- layer is connected to the terminal with the low resistivity.

Conclusions

We confirmed that the SOI floating body effect and to keep floating on both the N layer and the body are important for appearance of the steep SS on the PN-Body tied SOI FET, which were confirmed with measuring the new test devices.

Acknowledgments

The authors gratefully acknowledge contributions of Lapis semiconductors Co. Ltd., and KEK to fabricating devices in this study.

References

[1] I. A Young, U. E. Avci, and D. H. Morris, "Tunneling Field Effect Transistors: Device and Circuit Considerations for Energy Efficient Logic Opportunities", IEDM Tech. Dig. pp.600, 2015.

[2] K.-S. Li, et al., "Sub-60mV-Swing Negative-Capacitance FinFET without Hysteresis", IEDM Tech. Dig. pp.620, 2015.

[3] J.Ida et.al., "Super Steep Subthreshold Slope PN-Body Tied SOI FET with Ultra Low Drain Voltage down to 0.1V", IEDM Tech. Dig. pp.624, 2015.

[4] T.Horii, J.Ida et.al., " Confirmation of SS=35µV/dec over 3 Decades of Drain Current and Hole Accumulation Effect on PN-Body Tied SOI Super Steep SS FET's ", Proc. of IEEE SNW, pp. 148, 2016.

[5] M. Yoshida, J. Ida, et.al., "Super Steep Subthreshold Slope PN-Body Tied SOI FET's of Ultra Low Drain Voltage=0.1V with Body Bias below 1.0V", IEEE S3S Conf., Session 6a.4, pp1-3, 2016

978-1-5090-4661-4/17 $31.00 © 2017 IEEE

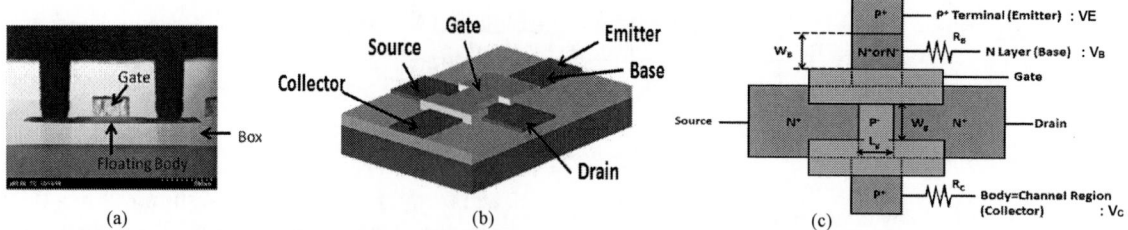

Fig.1 Cross sectional TEM photograph of 0.2um SOI (a), the 3D structure of PN-body tied SOI made by the 3D device simulator "HyENEXSS" (b), the plane view of the new test device (c), used in this study. Resistors are inserted between the test device and the measurements instrument. R_B: Resistor inserted on the N layer (Base of BIP) terminal, R_C: Resistor inserted on the body (Corrector of BIP) terminal.

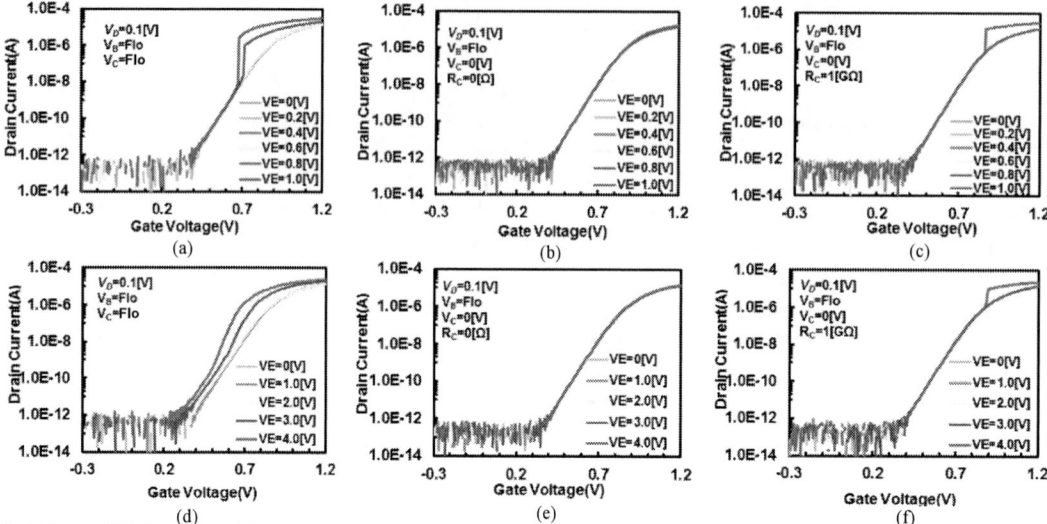

Fig.2 Measured Id-Vg characteristics on the new test device of Lg/Wg=0.2/1.0um, W_B=1.5um with V_D=0.1V and V_B=Floating. VE stands for the voltage of P$^+$ terminal (Emitter of BIP). Results of the devices with the N$^-$ are shown in (a) to (c). (a) : V_C is floating, (b) : R_C=0 Ω (shorted), (c) : R_C=1G Ω. Results of the devices with the N$^+$ are shown in (d) to (f). (d) : V_C is floating, (e) : R_C=0 Ω (shorted), (f): R_C=1G Ω. The terminal voltage (V_C) is set to be zero on (b), (c), (e), (f).

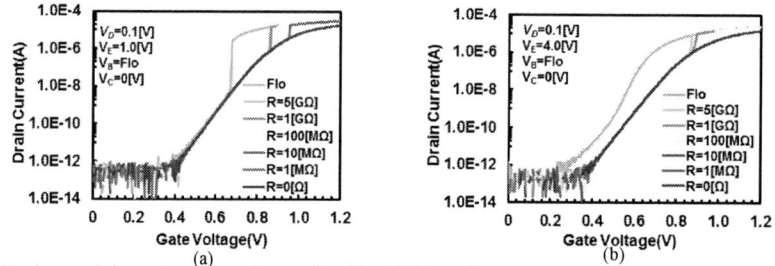

Fig.3 Measured Id-Vg characteristics on the new test device of Lg/Wg=0.2/1.0um, W_B=1.5um with V_D=0.1V and V_B=Floating. Results of the N$^-$ are shown in (a) when R_C changed from floating, 5G to 1M Ω, and 0 Ω (shorted). Results of the N$^+$ are shown in (b). V_C is set to be zero.

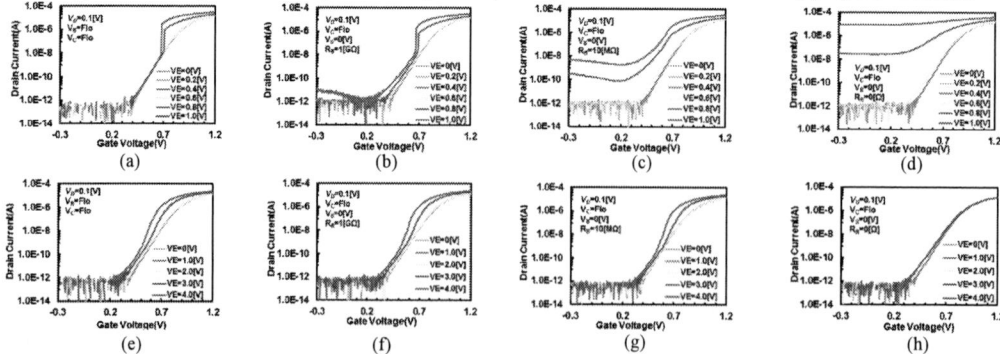

Fig.4 Measured Id-Vg characteristics on the new test device of Lg/Wg=0.2/1.0um, W_B=1.5um with V_D=0.1V and V_C=Floating. VE stands for the voltage of P$^+$ terminal (Emitter of BIP). Results of the devices with the N$^-$ are shown in (a) to (d). (a) : V_B is floating, (b) : R_B=1G Ω, (c) : R_B=10M Ω, (d) : R_B=0 Ω (Shorted). Results of the devices with the N$^+$ are shown in (e) to (h). (e) : V_B is floating, (f) : R_B=1G Ω, (g) : R_B=10M Ω, (h) : R_B=0 Ω (Shorted). The terminal voltage (V_B) is set to be zero on (b), (c), (d), (f), (g), (h).

978-1-5090-4661-4/17 $31.00 © 2017 IEEE

5M-1 (Invited)

EUV Lithography insertion for high volume manufacturing: status and outlook

Alek Chen[1], and Junji Miyazaki[2]

[1]ASML US Inc., 399 W Trimble Rd., San Jose, CA USA alek.chen@asml.com

[2]ASML Japan Co., Ltd., 4-7-35 Kita Shinagawa, Shinagawa, Tokyo, Japan

Abstract

We will present the current performance of the state-of-art EUV scanners, and an overview of EUV lithography process infrastructure status. The outlook for high volume manufacturing insertion will be also discussed.

(Keywords: Lithography and semiconductor manufacturing)

Introduction

EUV lithography has been in development for more than a decade. The complexity and difficulty of EUV source technology contributed to the delay of manufacturing insertion. ASML has introduced several generations of EUV scanners; from the NXE:3100 introduced in 2011 to NXE:3350B, which started shipping at the end of 2015. The NXE:3350B is using Master Oscillator Power Amplifier (MOPA) pre-pulse source technology, which enabled significant productivity scaling, demonstrated at ASML and at our customers.

This resulted in a growing consensus in the semiconductor industry that EUV insertion in volume manufacturing will start from 2018. EUV lithography not only introduces a significant resolution improvement but also enables the reduction of multiple patterning process complexity. This will offer end users increased design flexibility, shorter cycles of learning and as a consequence faster time to yield. An overview of NXE:3350B performance, in terms of productivity, critical dimension uniformity (CDU) and overlay, will be presented. ASML is committed to continue to improve EUV scanner capabilities, as shown in Figure 1.

EUV scanner performance overviews

A. Productivity

EUV source power continues to make improvement. TSMC recently achieved productivity champion results of 3-day average of >1500WPD.[1] ASML is in the process of delivering 125W source upgrade packages to our customers, and has demonstrated >200W in our source development facility. Using a combination of source power increases and wafer overhead time reduction, ASML is targeting to achieve >120 wafer per hour (WPH) at $20mj/cm^2$ targeting dose, as shown in Figure 2.

B. CDU control

Superior CDU performance of 16nm dense line/space

and 20nm isolated space patterns have been achieved on multiple systems, and on source power settings of 85W and 115W respectively, as shown in Figures 3 and 4.

C. Overlay and focus control

NXE:3350B has achieved dedicated chuck overlay near 1nm, and focus uniformity less than 10nm, as shown in Figure 5. Moreover, the EUV scanner is expected to work in a mix and match mode with ArFi immersion scanners as it is inserted in volume manufacturing in the beginning. NXE:3350B has achieved a mix and match overlay to an immersion scanner, NXT:1980, <3nm, shown in Figure 6.

EUV Lithography infrastructure status

An overview of the industrialization of EUV pellicle and the current status of EUV reticle and resist processes are briefly summarized.

A. Pellicle

ASML has demonstrated EUV pellicles to protect reticles from fall-on particles without impact to imaging and overlay performance.[2] Although ASML is committed to continue improve defectivity within the EUV scanner systems, many logic device end users, lacking sufficient circuit redundancy, would like to have this pellicle capability to prevent printing defects generated from particles falling on the reticle during exposures. ASML in collaboration with one of our customers has demonstrated hundreds of wafer exposures using a pellicled EUV mask without adding printing defects.[3]

B. Reticle

EUV mask blank quality and defectivity continue to show improvement towards volume manufacturing readiness. Defect free (defect size>=100nm) and <10 defects (size>=50nm) mask blanks are readily available.[4] EUV actinic inspection tools are also becoming available.[5]

C. Resist material and process

Resist material suppliers have shown significant progress in improving resolution and sensitivity while containing chemical stochastic effect for both chemical amplify resist (CAR)[6] and non-CAR materials.[7]

Conclusion

EUV lithography has made significant progress in the past couple of years. EUV scanner performance has

978-1-5090-4661-4/17 $31.00 © 2017 IEEE

steadily improved to meet volume manufacturing requirement of productivity, imaging and overlay. The EUV infrastructure has matured over the years, and its readiness to start volume manufacturing is gaining momentum for the anticipated insertion around 2018.

Acknowledgments

The authors gratefully acknowledge the contributions of EUV team members at ASML and Zeiss.

References

[1] T. Yen, Semicon Taiwan, Sep 2016.

[2] D. Brouns et al, 97761Y-1 SPIE 2016.

[3] B. Turkot et al, 9776-1 SPIE 2016.

[4] G. Zhang et al, SPIE BACUS 2016

[5] T. Weyh, EUVL Symposium, 2016

[6] JSR, SEMICON JAPAN 2016

[7] A. Grenville, EUVL Symposium, 2016

Figures

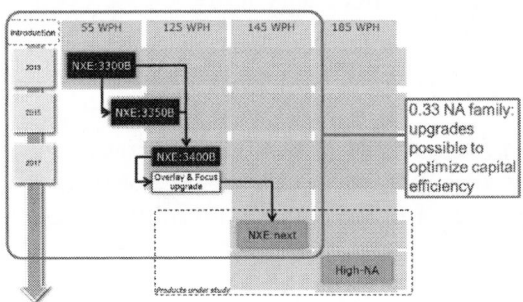

Fig. 1 ASML EUV scanner roadmap

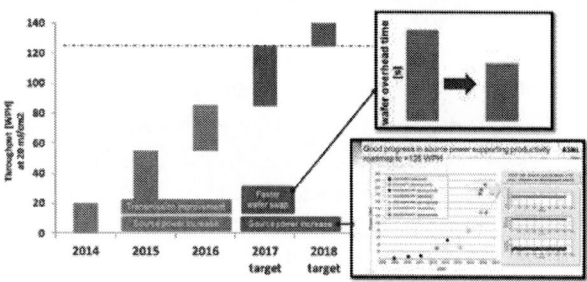

Fig. 2: EUV source power progress over the year

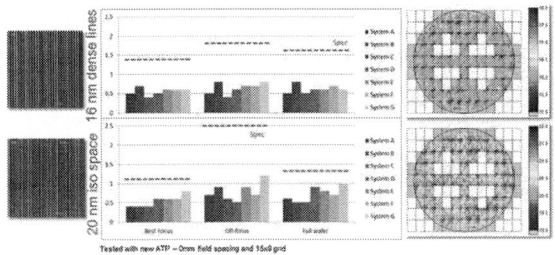

Fig. 3 NXE:3350B imaging: 16nm dense and 20nm iso space, consistently <1nm full wafer CDU

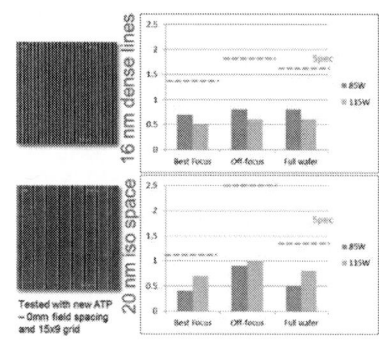

Fig. 4 NXE:3350B imaging test at 85 and 115W source power configurations respectively

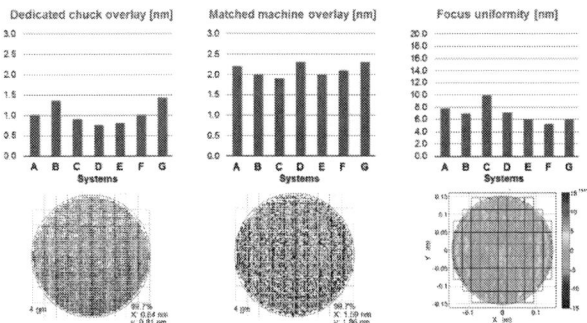

Fig. 5 Overlay and focus performance of NXE:3350B

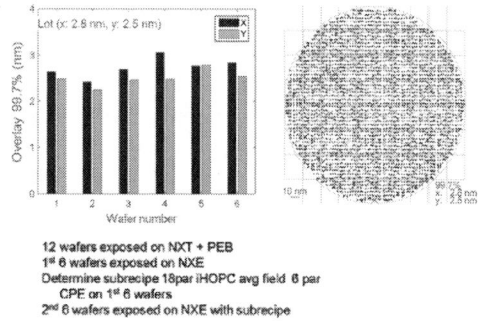

Fig. 6 Matching ArFi (NXT:1980D) to EUV (NXE:3350B) overlay performance

978-1-5090-4661-4/17 $31.00 © 2017 IEEE

Local, Isotropic, and Damageless Doping to Oxide Semiconductors by Using Electrochemistry

Takeaki Yajima[1], Tomonori Nishimura[1], and Akira Toriumi[1]

[1]The University of Tokyo, Tokyo, Japan, yajima@adam.t.u-tokyo.ac.jp

Abstract

Electrochemical doping, which introduces dopant impurities via isotropic electrochemical reaction, was proposed and demonstrated in an archetypical oxide semiconductor TiO_2. By controlling electrochemical potential of TiO_2 in electrolyte, the technique successfully introduced dopants to increase the conductivity of TiO_2 by ~5 orders. The technique was also applied through micro-patterned photo resist, verifying nano-scale compatibility. The dopants, which are hydrogen or oxygen vacancy in this case, were stabilized with an activation energy of 1.2 eV, making this technique attractive for oxide devices in displays, sensors, solar cells, and neuromorphic circuits.

(Keywords: Doping, electrochemistry, and oxide semiconductor)

Introduction

In the fabrication of electronic devices, dopant distribution in the semiconductor has to be controlled on the microscopic length scale. While ion implantation has widely been utilized for this purpose, it still poses challenges in isotropic and damageless doping to 3D structures. On the other hand, an alternative doping technique would be available for oxide semiconductors. It is the electrochemical doping, which takes advantage of the relatively high ionic diffusivity in these materials. The electrochemical doping will be performed in electrolyte as illustrated in **Fig. 1**, where the battery reaction between the metallic electrode and the oxide semiconductor introduces dopants to the unmasked area. Various electrochemistry-related techniques have already been utilized in oxides, including resistance switching for high-density memory [1] and electrochromism for smart windows [2], implying electrochemical doping would be a feasible alternative technique.

Experiments and Analyses

Polycrystalline single-phase anatase TiO_2 films was first fabricated on the SiO_2/Si substrate in a way described elsewhere [3]. By using this 35 nm TiO_2 film as the cathode, the electrochemical doping was performed by applying external voltage in the electrolyte as shown in **Fig. 2**. Under the external voltage, the TiO_2 cathode is reduced with the half reaction: $TiO_2 + D^+ + e^- \Rightarrow$ D-doped TiO_2,

where D denotes dopant such as hydrogen or oxygen vacancy. The counter half reaction occurs on the Pt anode: $4OH^- \Rightarrow O_2 + 2H_2O + 4e^-$. When the electrochemical potential of the TiO_2 film became lower than -0.9 V *vs.* standard hydrogen electrode (SHE), the sheet resistance of the TiO_2 films was decreased from ~1 $G\Omega$ to ~10 $k\Omega$ (**Fig. 3**). The Hall measurement in **Fig. 4** confirmed the decrease in the sheet resistance was mainly originated from the increased electron density rather than the change in their mobility. The structural analysis confirmed negligible damage in the crystallinity of TiO_2. Electrochemical doping was further performed through the micro-patterned photo resist (5 μm line and space) on the TiO_2 surface. In this case, instead of applying external voltage on each patterned region, a base metal electrode in each region was used as the anode as illustrated in **Fig. 1**. The conductive atomic force microscope image in **Fig. 5** revealed the electrochemical doping occurred only in the uncovered region with the edge sharpness of 100 nm, which is almost the measurement resolution. The time decay analysis confirmed the dopants are stabilized by the activation energy of 1.2 eV (**Fig. 6**), which could be exploited to ensure the dopant stability with the help of surface passivation.

Conclusion

Electrochemical doping was demonstrated in an archetypical oxide semiconductor TiO_2. The technique caused negligible damage in the crystallinity of TiO_2, and was successfully applied to photolithography patterns with sub-micrometer resolution. The dopants, which are hydrogen or oxygen vacancy, were stabilized with an activation energy of 1.2 eV, making this technique attractive for fabricating oxide devices in displays, sensors, solar cells, and neuromorphic circuits.

Acknowledgments

The authors acknowledge the experimental support by S. Yamaguchi and S. Miyoshi.

References

[1] R. Waser, R. Dittmann, G. Staikov, K. Szot, *Adv. Mater.* **21**, 2632 (2009).

[2] C. G. Granqvist, Sol. Energy Mater. Sol. Cells **60**, 201 (2000).

[3] T. Yajima, T. Nishimura, A. Toriumi, *Physica Status Solidi A* **213**, 2196 (2016).

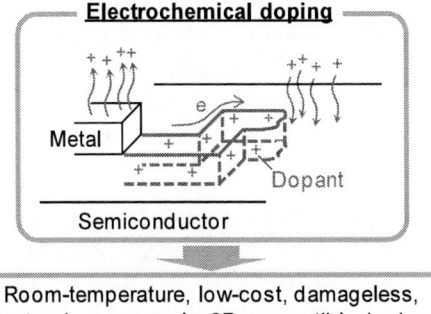

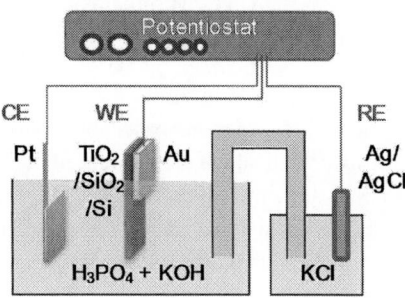

Fig. 1: A schematic illustration of electrochemical doping. When the base metal electrode and the oxide semiconductor are dipped in the electrolyte, the battery reaction occurs in between. The positively charged dopant impurities are introduced to the oxide semiconductor at the cathode, and the counter charges of electrons are provided by the oxidization of metallic electrodes at the anode. Due to the convenience of electrochemical reactions, this doping technique has various advantages as listed, compared with other techniques such as ion implantation.

Fig. 2: The three terminal experiment for electrochemical doping, which is a proof-of-concept experiment before utilizing the battery reaction in Fig. 1. In this experiment, the negative external voltage was applied on the TiO_2 film by the potentiostat to tune the electrochemical potential. Then, the TiO_2 is reduced as $TiO_2 + D^+ + e^- \Rightarrow$ D-doped TiO_2, and the counter half reaction occurs on Pt electrode: $4OH^- \Rightarrow O_2 + 2H_2O + 4e^-$. Here, the TiO_2 film is the working electrode (WE), the Ag/AgCl is the reference electrode (RE), and the Pt is the counter electrode (CE).

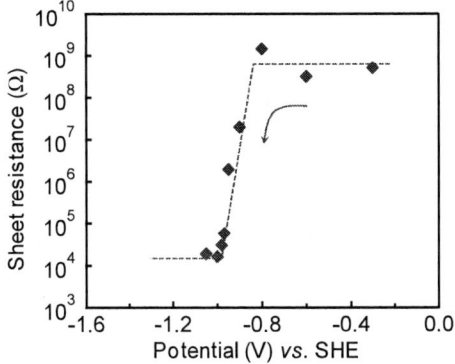

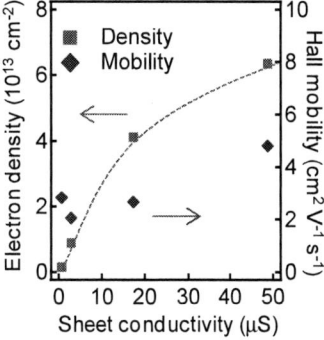

Fig. 3: Sheet resistance of the TiO_2 film as a function of the electrochemical potential in the electrolyte. When the potential was set below -0.9 V *vs.* SHE by external voltage, the insulating TiO_2 became conductive.

Fig. 4: Electron density and Hall mobility of electrochemically doped TiO_2 films with different sheet conductivity. The results indicate the conductivity of the doped TiO_2 films mainly originates from the increased electron density rather than the increased mobilities.

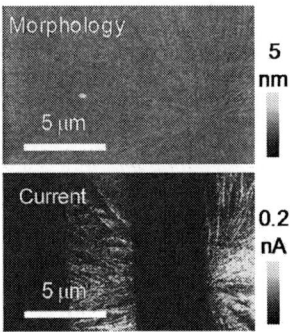

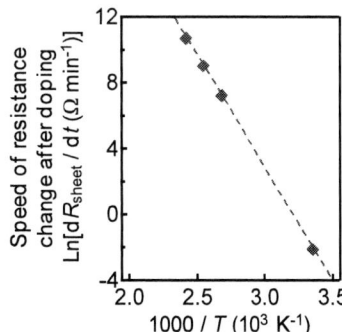

Fig. 5: Surface morphology (top) and the current map (bottom) of the TiO_2 film which was electrochemically doped through a micro patterned photo resist (5 μm line and space). There is a clear conductivity contrast between the doped and the non-doped regions although the surface morphologies are the same.

Fig. 6: The logarithmic speed of resistance change in oxygen after the electrochemical doping, which was plotted as a function of the inverse temperature 1000 / T. The linear extrapolation (red dashed line) gives the activation energy 1.2 eV in this dopant-release process.

978-1-5090-4661-4/17 $31.00 © 2017 IEEE

5M-3 (Invited)

Process Development for CMOS fabrication using Minimal Fab

Sommawan Khumpuang[1,2], Kazuhiro Koga[1,2], Yongxun Liu[1], and Shiro Hara[1,2]

[1]AIST, [2]Minimal Fab, Tsukuba, Japan, sommawan.khumpuang@aist.go.jp

Abstract

CMOS fabrication processes based on clean-localized technology of Minimal fab are introduced in this work. Without a cleanroom, the particle and impurities are locally controlled at each machine and wafer carrier during the fabrication process. Two methods of CMOS inverter fabrication are performed, 1) using only equipment of a minimal fab for entire process on Si bulk wafer and 2) hybridizing the minimal fab with a conventional fab on SOI wafer. Both methods employ thermal diffusion for doping impurities. We have confirmed that both CMOS have good electrical-properties including interface state density.

(Keywords: Minimal fab, CMOS and diffusion doping)

Introduction

We have developed a novel semiconductor manufacturing system for electronic device realized on a half-inch wafer namely "minimal fab". Due to its distinct features of cleanroom-free and simplicity, a huge wafer transfer mechanism and complicated functions to uniform a large wafer process are eliminated [1]. The system is suitable for customizing device designs in a low volume production where the use of conventional fab takes time and cost unreasonably. In a typical CMOS fabrication process, an ion-implantation and a CVD are usually employed for impurity doping and polysilicon gate deposition. However, the equipment for minimal fab was still partly under development. Thus, we introduce a simple impurity doping process by SOD coating following by thermal diffusion and sputtering system to form Al gate for full minimal process [2], while making hybrid process with the conventional fab possible to increase variation of gate metal selection. In this hybrid process, we employ PVD of conventional fab to form TiN-gate of a CMOS [3].

CMOS Process using Full Minimal Fab

Fig. 1 shows the transistor fabrication line of minimal fab installed in an office atmosphere. A fundamental process for CMOS fabrication using the minimal fab is explained as follows. A half-inch Si (100) wafer is used. The fabrication process is started with forming p-well of NMOS by coating BSOD agent, followed by diffusion in O_2/N_2 atmosphere. At the PMOS side, the same BSOD is doped and diffused. The source and drain doping at NMOS is performed with PSOD coating and following diffusing at in N_2 atmosphere. A 40-nm SiO_2 is then formed as the gate oxide. Finally, Al contacts and gate-metals are formed at the same time. Fig. 2 illustrates the schematic structure of the CMOS with the process flow.

CMOS Process using Hybrid Fab

In this work, n-type (100) silicon-on-insulator (SOI) wafers are used. The fabrication process started with boron doping (p+) for source and drain formation of PMOS with the same tradition as the one fabricated using full-minimal fab. Then a 158 nm-thick TEOS layer is deposited using conventional PE-CVD. With the same method as p+ doping, n+ doping for source and drain of NMOS is formed. After device isolation layer is etched, PMOS and NMOS are separated. A 5.8 nm-thick gate oxide layer is formed by conventional thermal oxidation followed by a 30-nm-thick conventional PVD-TiN deposition as a gate and a patterned structure. Contact holes and Al contacts are then realized. Finally, a post-metallization annealing is performed. Fig. 3(a) and 3(b) illustrate the schematic structure of the CMOS and the process flow, respectively. The rate of utilization of minimal-fab is about 60 %.

Experimental Results

Fig. 4(a) and 4(b) shows photograph of a resulting CMOS-inverter fabricated by the full minimal fab and the hybrid fab, respectively. The I-V characteristic of that of the full minimal fab and that of the hybrid fab are shown in Fig. 5(a) and 5(b), respectively. The time required for whole fabrication was 3 days for the full minimal fab while at least one week was required for the hybrid fab process. The interface states density of both CMOS is in the order of 10^{10} states/cm^2.

Conclusion

Fundamental processes for CMOS fabrication using the minimal fab and the hybrid fab are developed. It is assured that using clean-localized technology of minimal, a fabrication is speedy and in an equivalent level of electrical characteristic to a conventional fab using a cleanroom.

References

[1] S. Khumpuang et al., IEEE Trans Semiconductor Manufacturing, 28(3), 393-398 (2015).

[2] S. Khumpuang et al., IEEE Trans Semiconductor Manufacturing, 28(4), 551-556 (2015).

[3] Y. Liu et al., proc. of IEEE NANO 2016, Sendai, Japan (2016)

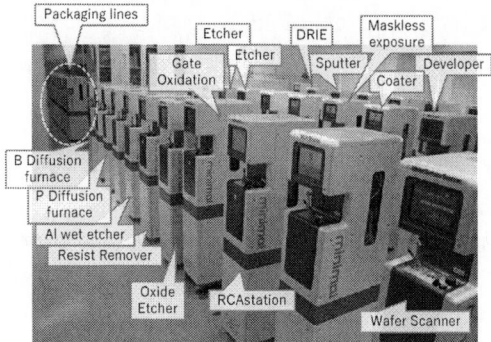

Fig. 1 Minimal fab line employed for CMOS fabrication which is established in an office

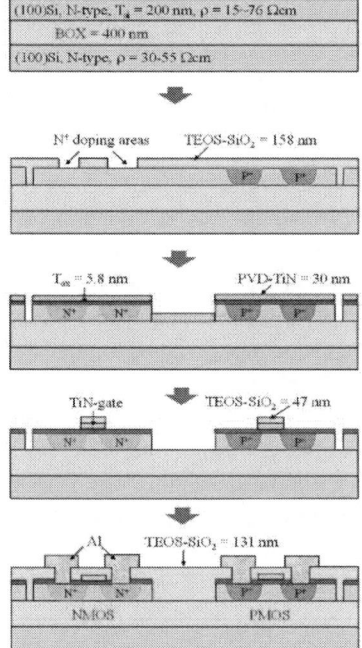

Fig. 3 Schematic cross-sectional structure of CMOS and process flow for hybrid fab fabrication

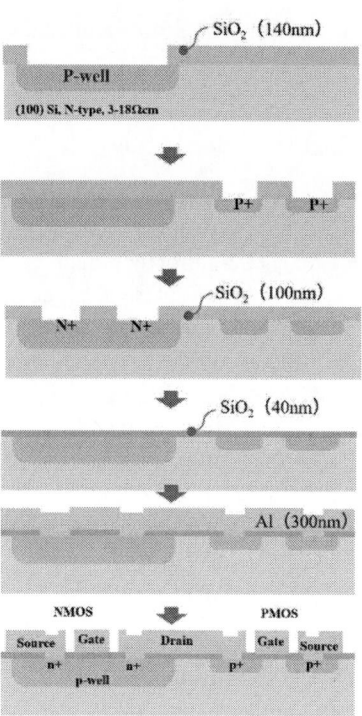

Fig. 2 Schematic cross-sectional structure of CMOS and process flow for full minimal fab

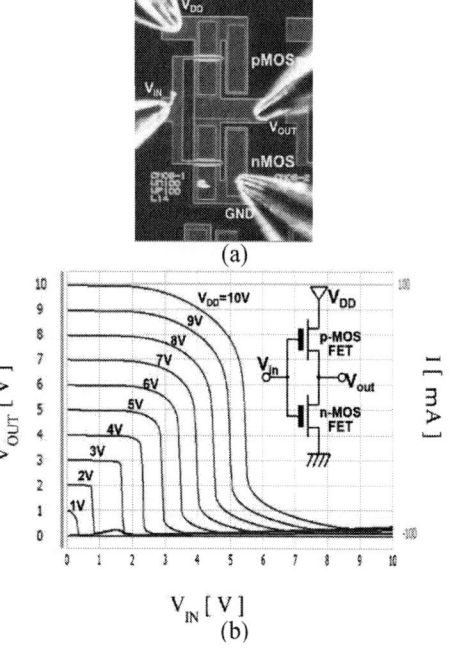

Fig. 4 Resulting CMOS fabricated using full minimal fab (a) photograph image and (b) V_{in}-V_{out} characteristic of the inverter.

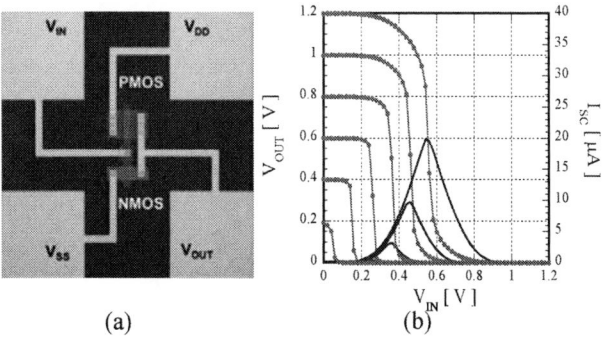

Fig. 5 Resulting CMOS fabricated using hybrid fab (a) photograph image and (b) V_{in}-V_{out} characteristic

978-1-5090-4661-4/17 $31.00 © 2017 IEEE

5M-4

New Compact ECR Plasma Source for Silicon Nitride Film Formation in Minimal Fab System

Tetsuya Goto[1], Kei-ichiro Sato[2], Yuki Yabuta[3], Shigetoshi Sugawa[1], and

Shiro Hara[4]

[1]Tohoku University, Sendai, Japan, tetsuya.goto.b2@tohoku.ac.jp

[2] Kotec Co., LTD., [3] Seinan Industries Co., LTD., [4] The National Institute of Advanced Industrial Science and Technology

Abstract

A compact magnetic-mirror confined ECR plasma source for low-damage plasma processings was developed, especially aiming for the realization of high-quality silicon nitride film formation for the sub-micron CMOS device processes in the minimal fab system. Magnetically-confined plasma could be produced, and the wet-etch resistance against HF solution for the silicon nitride film formed at 400 °C was the same level as that of the film formed by the conventional LPCVD at 750 °C.

(Keywords: Minimal fab system, Silicon nitride, Plasma damage)

Introduction

Recently, we are developing a new compact plasma source used for the minimal fab system [1]. Here, the minimal fab system is the new IC fabrication system using the half-inch wafer and the clean-localized small process machines which can eliminate a need of clean room. The investment cost can be reduced by 1/1000 compared to the current 300-mm wafer mega fab system. Our plasma source was developed for high-quality plasma CVD of silicon nitride films, to realize high-performance sub-micron CMOS devices in the minimal fab system. In this paper, plasma measurement results obtained by Langmuir probe and preliminary results of silicon nitride film depositions are demonstrated.

Plasma Equipment

In the new plasma system, the magnetic-mirror plasma confinement, which is the well-known technique in a field of fusion research [2], was incorporated. Fig. 1 shows a schematic view of the plasma source, a magnetic field strength distribution at the axis, and a picture of typical Ar plasma. A 5.8-GHz microwave was introduced to excite plasma using the ECR method. As can be seen in the photograph of plasma in Fig. 1, the magnetically confined mirror plasma was successfully produced. The wafer is set at the neighborhood of the confined plasma (not inside of the plasma), to supply high-density neutral reactive species produced by the high-density plasma (neutral species can escape from the plasma), while ion flux can be suppressed because the wafer was set at the outside of plasma, resulting in the reduction in ion-bombardment induced plasma damages. Fig. 2 shows photographs of the chamber, the magnet system, and the waveguides installed in the main frame of the minimal fab system. The components were successfully minimized to be installed in the frame for the minimal fab system. We have already a good prospect to install the other components such as electrical control system, microwave power generator, gas flow control system, evacuation system and detoxifier.

Experimental Results and Discussions

A. Plasma Properties

Figs. 3(a) and 3(b) show (a) electron density and electron temperature as a function of applied microwave power (Ar, 0.27 Pa), and (b) electron density as a function of pressure (microwave power of 40 W) for Ar plasma. The electron density larger than 10^{11} cm^{-3} could be obtained for the wide pressure range. For the electron temperature, value is ranged between 3 eV to 4 eV. Fig. 3(c) shows radial distribution of ion saturation current for various pressures. Core plasma radius was 15 mm. For the outside region of core plasma where the wafer was set, ion current was decreased rapidly, which can reduce ion bombardment induced damages.

B. Silicon nitride Film Formations

Fig. 4(a) shows a photograph of Ar/SiH$_4$/N$_2$/H$_2$ plasma and the Si wafer for the silicon nitride film formations. Fig. 4(b) shows a refractive index of the silicon nitride film, obtained at 400 °C, as a function of N$_2$ flow rate. The refractive index could be controlled by N$_2$ flow rate, and ideal value of approximately two could be obtained. Fig. 4(c) shows etching rates of the film when dipped in the 5% diluted HF solution. In the optimized condition, the wet-etch resistance against HF solution for the silicon nitride film formed at 400 °C was the same level as that of the film formed by the conventional LPCVD at 750 °C.

References

[1] S. Khumpuang et al., IEEE Transactions on Semiconductor Manufacturing, 28(3), 393-398 (2015).

[2] T. C. Simonen, Proc. of the IEEE, **69** 935 (1981).

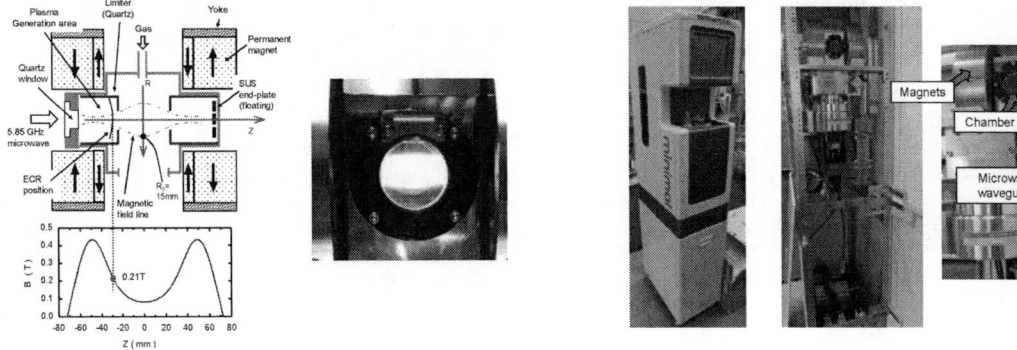

Fig. 1: Schematic view of new plasma source, axial distribution of magnetic field strength, and the photograph of typical Ar plasma.

Fig. 2: Chamber, magnet system, and waveguides installed in the main frame of the minimal fab system.

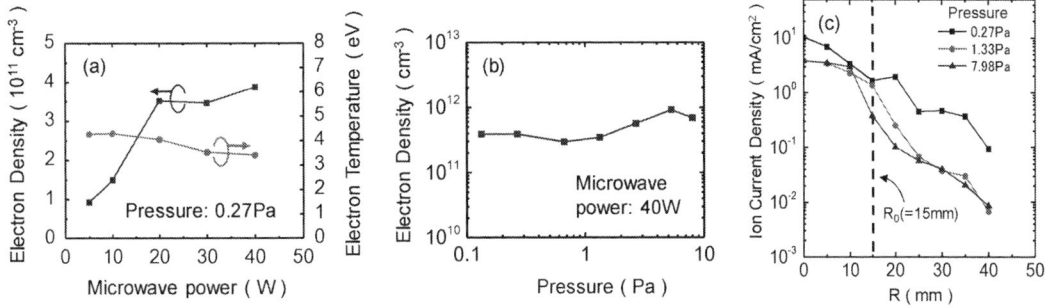

Fig.3: Ar plasma measurement results obtained from the Langmuir probe. (a) Electron density and electron temperature as a function of microwave power (Ar, 0.27 Pa). (b) Electron density as a function of pressure (microwave power of 40 W). (c) Radial distributions of ion current density for various pressures.

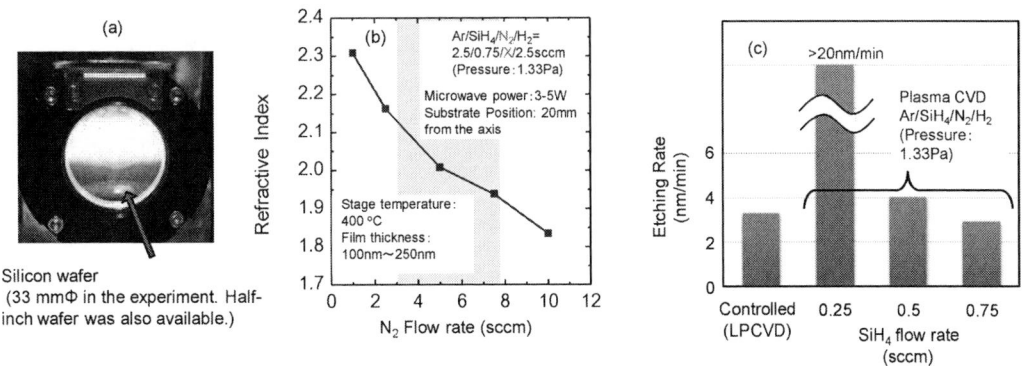

Fig.4: (a) Photograph of $Ar/SiH_4/N_2/H_2$ plasma and Si wafere for Si_3N_4 film formation. (b) Refractive index of silicon nitride film as a function of N_2 flow rate. (c) Etching rate of silicon nitride films formed at 400 °C when dipped in the 5% HF solution for various SiH_4 flow rate conditions. The controlled sample of LPCVD silicon nitride film formed at 750 °C was also shown.

978-1-5090-4661-4/17 $31.00 © 2017 IEEE

5A-1

A Scalable Si-based Micro Thermoelectric Generator

Takanobu Watanabe[1], Shuhei Asada[1], Taiyu Xu[1], Shuichiro Hashimoto[1], Shunsuke Ohba[1]

Yuya Himeda[1], Ryo Yamato[1], Hui Zhang[1], Motohiro Tomita[1], Takashi Matsukawa[2],

Yoshinari Kamakura[3] and Hiroya Ikeda[4]

[1]Waseda University, Tokyo, Japan, watanabe-t@waseda.jp

[2]AIST, [3]Osaka University, [4]Shizuoka University

Abstract

A new device architecture of micro thermoelectric generator (μ-TEG) is proposed. The μ-TEG utilizes silicon nanowires as the thermoelectric (TE) material, and it can be fabricated by the CMOS-compatible process. It is driven by an "evanescent thermal field" exuding around a heat flow perpendicular to the substrate. We demonstrate experimentally that the TE power increases in the shorter TE leg lengths. The results show that the TE power density is scalable by miniaturizing and integrating the proposed structure.
(Keywords: Energy harvester, thermoelectrics, Si nanowire)

Introduction

A μ-TEG is anticipated as a key device for realizing a trillion sensor network. The recent discovery of the superior TE property of silicon nanowires (Si-NWs) [1] opens the way for Si-based μ-TEGs, which can be fabricated by Si-CMOS processes. However, miniaturized structures have been considered to be unfavourable for TEGs, because of the relative increase in parasitic thermal resistances which decreases the temperature gradient in the TE leg.

In this paper, we propose a new μ-TEG architecture which utilizes a very steep temperature gradient formed in the proximity of a main heat flow. It is found that the TE power is enhanced by shortening the TE leg, so that the power generation density can be increased by miniaturizing and integrating the device structure.

Proposed μ-TEG Architecture

Figure 1 illustrates the proposed μ-TEG structure together with a conventional planer TEG device [2]. Conventional device is fabricated on a thin insulating membrane suspended in air, in order to block the shortcut of the heat flow and to ensure the temperature gradient in the TE leg. Contrary, in the proposed device, the leg is placed on the substrate without forming any cavity underneath the insulating layer, and the heat current flows perpendicularly to the substrate. Figure 2 shows a result of the thermal analysis [3], showing that a very steep temperature gradient is formed in the vicinity of about few hundred nm. We call this "evanescent thermal field" hereafter. The new structure is advantageous in terms of fabrication cost and simplicity. Furthermore, the evanescent thermal field can be enhanced by shortening the TE leg, so that the power generator is scalable as well as the Si-CMOS technology.

Experimental Demonstration

We have confirmed experimentally that the TE power is enhanced by shortening the leg length. Figure 3 shows a schematic of the μ-TEG, which is fabricated by electron beam (EB) lithography and dry etching. It comprises 400 Si-NW legs with a P dose of $1.0 \times 10^{15} cm^{-2}$. The definition width of Si-NW in EB lithography is 100 nm, and the length (L) is varied from 8 μm to 90 μm. Both ends of NW bundle are connected to NiSi pads, on one of which an AlN film is deposited as the thermal conductive layer. The TE characteristic of the μ-TEG is measured with a prober system equipped with a micro IR thermography camera and a custom made micro thermostat (Fig.4). The micro thermostat is used as the heat source, which is contacted on the AlN film and is kept at about 45 °C. The bottom of the substrate is kept at R.T. The TE current is measured with applying loading voltage between two NiSi pads.

Figure 5 shows an I-V characteristic and TE power P curves of the μ-TEG. The maximum power is obtained in the shortest leg device (L = 8 μm). Figure 6 shows the temperature distribution captured by the thermography camera. A steep temperature gradient is formed in Si-NW, especially in the vicinity of the NiSi pad of the hot side. The mean temperature gradient in the Si-NW increases as the NW length decreases. Figure 7 shows the L dependences of the maximum TE power and the internal resistance of μ-TEG. The TE power in the shortest leg device is much more enhanced than that expected from the internal resistance. The TE power enhancement can be attributed to the increase in temperature gradient in the shorter leg devices.

Discussion on Scalability

Figure 8 shows scaling laws for the proposed μ-TEG and the conventional TEG with vertical leg architecture. The areal power generation density of the proposed μ-TEG is inversely proportional to the square of leg length L, so that the TE performance is improved by shrinking the device footprint. Conversely, the conventional vertical structure, which has been considered the most efficient TEG structure, is not scalable by shrinking the horizontal dimensions. Thus the scalability of μ-TEG for high TE power density is attained, for the first time, by the proposed device structure.

Conclusion

We have designed a scalable Si-based planer μ-TEG, which can be fabricated by CMOS-compatible processes. Unlike the conventional TEGs, the proposed μ-TEG is driven by the steep temperature gradient formed in the very proximity of a heat flow perpendicular to the substrate, thereby it does not require any cavity structure underneath the TE legs. The experimental results show that the TE power increases by shortening the TE legs, demonstrating the scalability for high TE power density.

Acknowledgments

This work was supported by the CREST project of Japan Science and Technology Corporation (JST).

References

[1] A. I. Boukai, et al., Nature, 451.7175 (2008) 168.

[2] Gao Min, et al., Electron. Lett., 34, 222 (1998)

[3] T. Xu et al., Book of Abstract of ECT 2016, 296, (2016).

978-1-5090-4661-4/17 $31.00 © 2017 IEEE

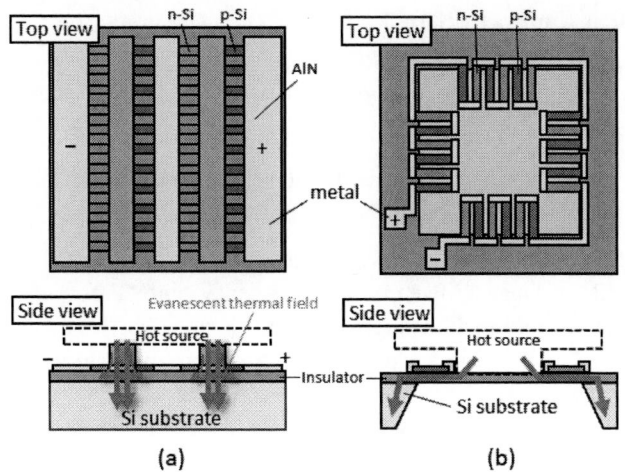

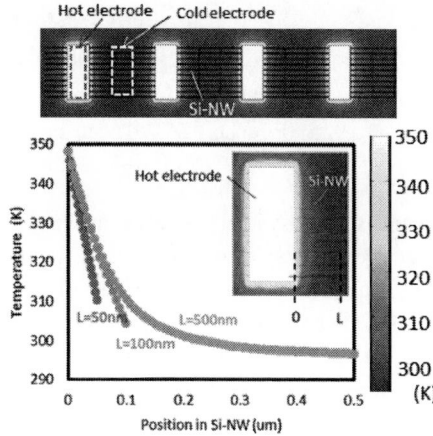

Fig. 1 Schematic μ-TEGs. (a) Proposed structure. (b) Conventional planer structure (modified from ref. 2). Red arrows shows the heat flow.

Fig. 2 Thermal analysis of the proposed μ-TEG using COMSOL Multiphysics ver.5.2a. Steep temperature gradient is formed near the hot electrode.

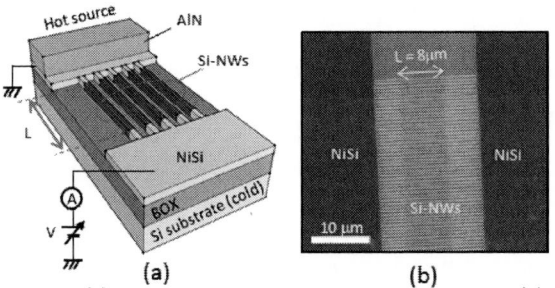

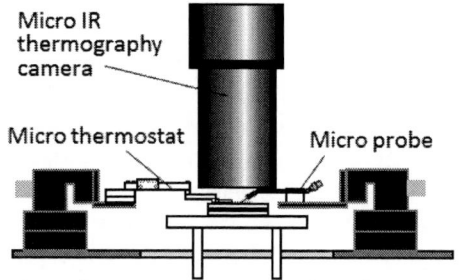

Fig.3 (a) Schematic of test μ-TEG structure. (b) SEM image of Si-NWs connected to NiSi pads.

Fig.4 prober system equipped with a micro thermography camera and micro thermostat

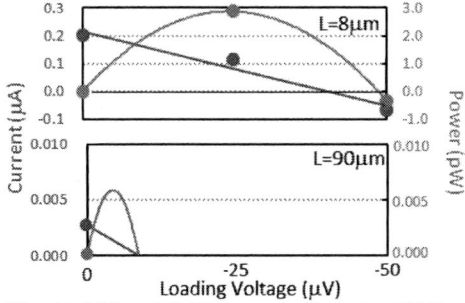

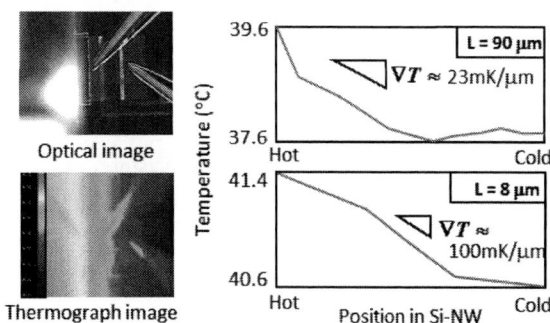

Fig. 5 I-V and P-V curves of test μ-TEGs.

Fig. 6 Temperature distribution in TE legs.

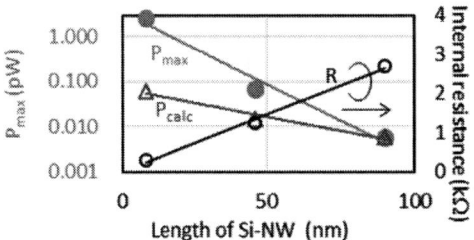

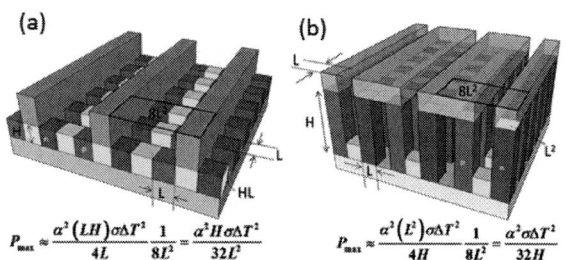

Fig. 7 Leg length dependency of maximum power and internal resistance R. P_{calc} is the expected value from the resistance R.

Fig. 8 Scaling laws of (a) proposed μ-TEG and (b) vertical leg TEG. α is the Seebeck coefficient of TE material.

978-1-5090-4661-4/17 $31.00 © 2017 IEEE

5A-2 (Invited)

Health Monitoring of Houses and Communities
by Recording Earthquake Response of Buildings

Minoru Yoshida[1], Yoichi Tanaka1[1], Shoichi Ikeda[1],
Soichiro Murata[2], Yoshiya Oda[3], and Masatake Ushiro[4]

[1]Hakusan Corporation, Fuchu City, Tokyo, Japan, yoshida@hakusan.co.jp
[2]SAP Japan, [3]Tokyo Metropolitan University, [4]Business Break-Through University

Abstract

A cloud server system used as a repository for earthquake response records can also provide for the health monitoring and diagnostics of residential buildings at low cost. With the free application iJishin, mobile phones fixed to the floor are used to collect shake data from houses.

(Keywords: Mobile phone seismometer, Real-time earthquake info, Health monitoring of house, Seismic capacity)

Introduction

The seismic capacity of a building is legally defined and technically tested by engineers when it is designed. After construction, however, no verification of its capacity is conducted. The owner of a building, usually a nonspecialist, has no way to evaluate seismic capacity unless the individual makes great effort to conduct his or her own testing.

Recent technologies such as IoT and cloud computing allow us to attempt preliminary solutions for the public.

We have developed a cloud server system that makes a real-time connection with the shake response data of individual houses using various seismic disaster information issued by the government and research institutes (Fig. 1). After an earthquake of any strength occurs, users of this system can determine the damage and state of deterioration of their houses and community. The system has provided services for house owners in the Tokyo metropolitan area.

Network

A. Seismic Response

An iOS application named iJishin was developed to use the MEMS acceleration sensor installed in the iPhone and iPod Touch as a seismometer [1]. Compared with commercial seismometers, iJishin can measure a tremor with an intensity of 1 or higher on the Japanese seismic scale with a precision of 0.1 of the seismic intensity. 100 Hz sampling time is synchronized to Coordinated Universal Time within 5 ms by Network Time Protocol.

iPods installed in each house are connected to the cloud system at all times so that data acquired by iJishin are uploaded to the cloud server immediately after seismic events.

B. Earthquake Early Warning

Japan Meteorological Agency provides public information regarding the hypocenter, estimated time, and shake intensity anywhere in Japan within 5 s of the outbreak of an earthquake by using approximately 1,000 seismic observation sites.

iPods with the iJishin application receive an early warning about the earthquake, and measure quake shaking in synchronization with each other. Each iPod's data for an individual house are uploaded to an archive of the earthquake responses of buildings.

C. Cloud Computing Platform

Our cloud server system consists of a communication section, data storage section, database, and web service section. The spatial distribution of seismic responses and earthquake events are archived in a time-line window. The server provides wave analysis tools such as fast Fourier transform for the calculation of integrations and response spectra. Data are transferred to the in-memory database platform to process big data.

D. Social Media

The scope of the system includes the connection of iPod seismometers to social network services on individual smartphones to access information for urgent and daily use.

Project

iPod seismometers have been deployed in 300 houses in Tokyo since October 2016, as part of the project "Community Development with Information of Disaster Prevention" funded by the New Enterprise Sector Creation Project of the Tokyo Metropolitan Government.

Users can confirm the shake data for their houses, and a history of responses to every earthquake, on the website (Fig. 2). In response to nonspecialist users' requests, the project plans to distribute diagnostics of the present states of houses in order to evaluate the seismic capacity.

Conclusion

The iJishin network and a cloud server system were integrated to collect earthquake response data for residential houses. This contributes to the

978-1-5090-4661-4/17 $31.00 © 2017 IEEE

development of an earthquake-resilient community.

Acknowledgments

The authors gratefully acknowledge the contributions of the Tokyo Metropolitan Government and the National Research Institute for Earth Science and Disaster Resilience.

References

[1] Hiroyuki Fujiwara, Hiroki Azuma, Shohei Naito, Shgeki Senna, Hiromitsu Nakamura, Hao Ken Xiansheng, Minoru Yoshida, Noboru Yuki, Yoshiharu Hirayama, "A Sharing System on Earthquake Response Information of Buildings Using a Sensor Cloud Technology", Journal of JAEE, Vol.13, No.5, 2013.

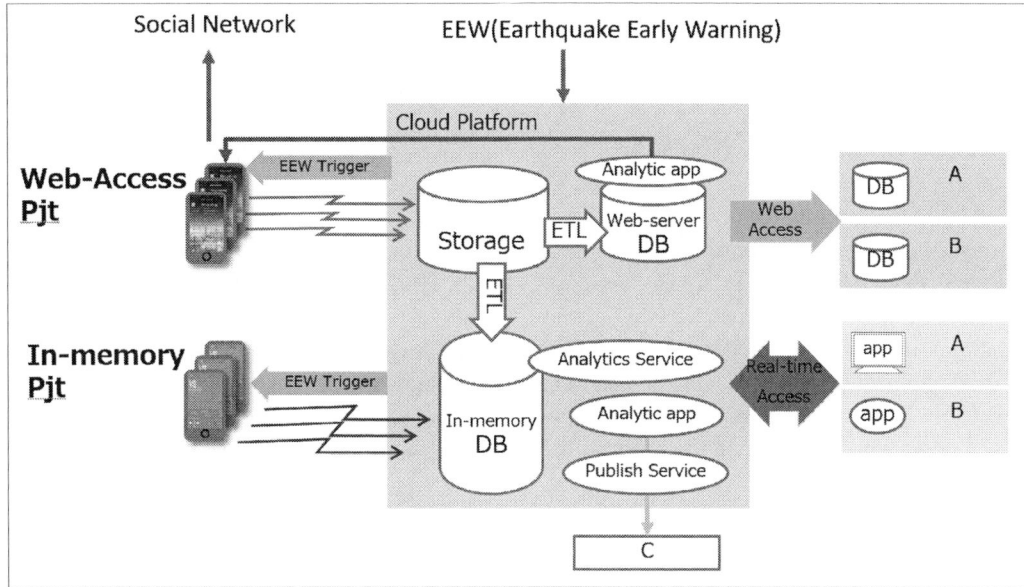

Fig. 1: The cloud server system of the smartphone seismometer network.

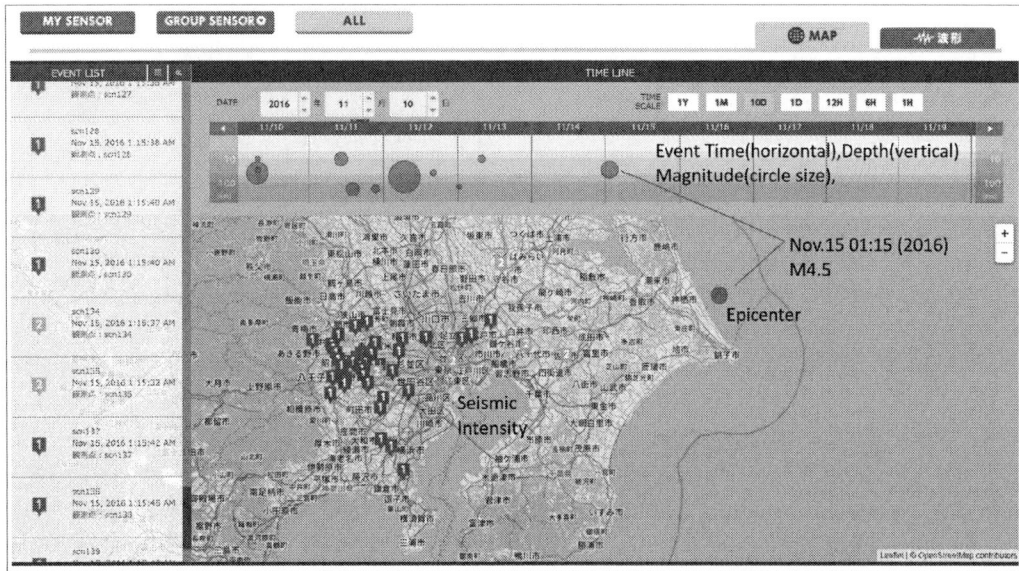

Fig. 2: Spatial distribution of seismic responses and earthquake events in time-line window.

5A-3 (Invited)

Smart Biosensing Technologies
to Detect Single Bacteria and Viruses

Masateru Taniguchi

The Institute of Scientific and Industrial Research, Osaka University, Osaka, Japan,

taniguti@sanken.osaka-u.ac.jp

Abstract

Biosensing systems containing micro- and nano-diameter pores on a Si substrate can detect and discriminate biomolecules with different molecular volumes from the changes in ionic current passing through the pores. Pattern recognition using machine learning to analyze a large number of ionic current-time profiles allows us to discriminate bacteria with similar molecular volumes.

(Keywords: Nanopores, Biosensors and AI)

Introduction

Technologies to diagnose and prevent infections from bacteria or viruses and to control the health management of foods are fundamental technologies for the realization of safe and prosperous societies. Technologies should be designed to be low-cost, portable, highly sensitive, and have low-power consumption sensors that ideally detect bacteria and viruses at single analyte levels because the sensors are expected to be used on-site at hospitals, in food factories, and at airports. For example, sensors that detect an infinitesimal influenza virus would allow us to prescribe medicines before the influenza develops. The on-site information about bacteria and viruses can allow for on the spot response and treatment, while the information obtained from analysis and integration of big data will enable to understand and predict the generation statuses of bacteria and viruses, and to perform pandemic prevention and countermeasures when necessary.

Working towards this goal, we have been developing micro and nanopore devices to detect single bacteria and viruses. A biosensing system would analyze measurement data using machine learning/artificial intelligence. In the near future, devices would provide information about the generation status and predictions of bacteria and viruses using our biosensing system with on-site AI and AI in the data center.

Results and Discussions

Micro and nanopore devices can be applied to biosensors, classified into either bio-nanopores based on channel proteins or solid-state nanopores made on Si substrates [1]. Reservoirs of ionic solutions and electrodes are installed on the both sides of the substrate with nanopores. Appling voltage between the electrodes results in an ionic current that flows through the nanopore. Negatively charged biomolecules move to the electrode with positive voltage due to electrophoretic flow. When a biomolecule enters into a nanopore, the ionic current decreases because the biomolecule interrupts ionic flow. The electric signal is characterized by the maximum current (I_p) and duration of the ionic current (t_d). When the thicknesses of the nanopores are larger than the diameters of the biomolecules, changes in ionic currents provide information about the molecular volume of the biomolecules. In the case of substrates with a smaller thickness than the diameters of the biomolecules, changes in the ionic currents include information about the molecular structures of the biomolecules.

We can discriminate particles and biomolecules with different diameters due to changes in ionic currents, using thick and thin nanopores [2]. When using analytes with similar diameters, the discriminating accuracies based on I_p-t_d analyses are low even if thinner nanopores are used for measuring the ionic current-time profiles. Since ionic current-time profiles include structural information about the analytes, pattern recognition of the profiles using machine learning can extract feature values that largely reflect structural information, resulting in a discriminating accuracy of approximately 90%.

Conclusion

Biosensors based on nanopore devices made by microfabrication technologies can detect bacteria due to changes in ionic currents, and succeeded in discriminating between single bacteria using machine learning that analyzes a large amount of ionic current-time profiles.

Acknowledgments

This work was supported by ImPACT Program of Council for Science, Technology, and Innovation (Cabinet Office, Government of Japan).

References

[1] M. Taniguchi, "Selective Multidetection Using Nanopores," Anal. Chem., vol. 87, 188 (2015).

[2] M. Tsutsui, Y. He, K. Yokota, A. Arima, S. Hongo, M. Taniguchi, T. Washio, and T. Kawai, "Particle Trajectory-Dependent Ionic Current Blockade in Low-Aspect-Ratio Pores," ACS Nano, vol. 10, 803 (2016).

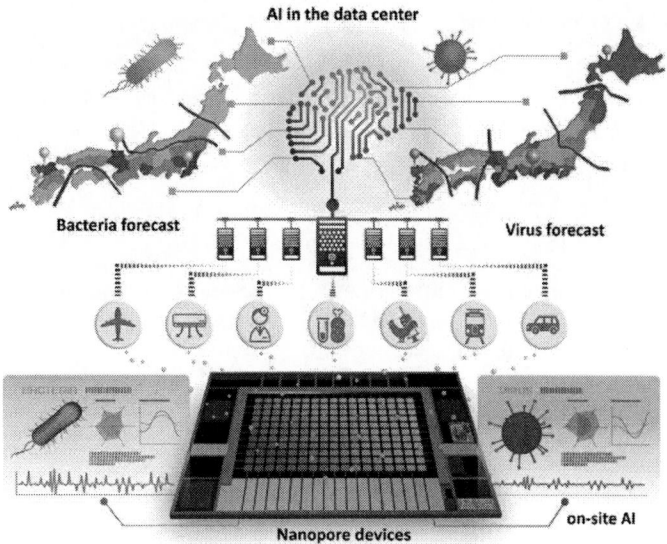

Fig. 1: Schematic figure of smart biosensing technologies using micro and nanopore devices. Ionic current-time profiles are analyzed using machine learning on a sensor system, resulting in diagnosis, detection, and identification of bacteria and viruses. The information will be integrated in a data center for analysis of big data. Artificial intelligence in the data center will report generation statuses and predictions of bacteria and virus development in the high resolution for pandemic prevention and countermeasures when necessary.

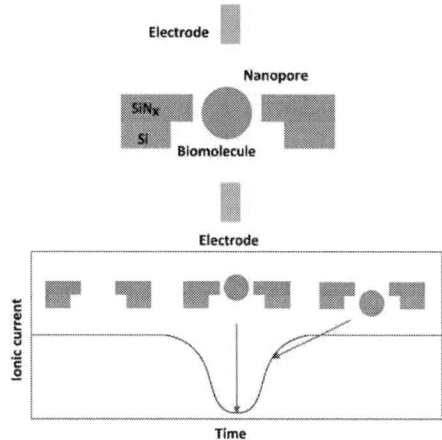

Fig. 2: Operating principle of micro and nanopore devices. The devices discriminate biomolecules passing through a nanopore by detecting changes in the ionic current flowing parallel to the nanopore when a voltage is applied across the SiN_x membrane.

978-1-5090-4661-4/17 $31.00 © 2017 IEEE

5A-4

Strain-Engineering in Germanium Membranes towards Light Sources on Silicon

D. Burt[1], A. Z. Al-Attili[1], Z. Li[1], F. Liu[1], K. Oda[2], N. Higashitarumizu[3], Y. Ishikawa[3],

O.M. Querin[4], F. Gardès [1], R. W. Kelsall[4], and S. Saito[1]

[1]Nanoelectronics & Nanotechnology Research Group, ZI, ECS, FPSE, Univ. of Southampton, UK. email: S.Saito@soton.ac.uk

[2]Research & Development Group, Hitachi, Ltd., 1-280 Higashi-Koigakubo, Kokubunji, Tokyo 185-8601, Japan.

[3]Department of Materials Engineering, Graduate School of Engineering, The University of Tokyo, Tokyo, Japan.

[4]Institute of Microwaves and Photonics, School of Electronic & Electrical Engineering, University of Leeds, UK.

Abstract

Bi-axially strained Germanium (Ge) is an ideal material for Silicon (Si) compatible light sources, offering exciting applications in optical interconnect technology. By employing a novel suspended architecture with an optimum design on the curvature, we applied a biaxial tensile strain as large as 0.85% to the central region of the membrane.
(Keywords: Si Photonics, Germanium, Light source)

Introduction

Bi-axially suspended germanium membranes exhibit extremely high strain values reducing the direct band-gap and thus enhancing light emission [1, 2]. Previous works have demonstrated high biaxial strain up to 1.9% [1] and 2.4 times PL enhancements [2] by using Ge-On-Insulator (GeOI) wafers. Despite the successful demonstrations by GeOI wafers [1, 2], the cost might be increased. We have utilized Ge on bulk Si wafers to manufacture suspended membranes.

Design, Simulation and Fabrication

2D Simulations were performed in COMSOL Multiphysics 5.2. The "Structural Mechanics" module was used with elasticity matrix defined in Voigt notation to simulate the strain values. The "dilute species" module was also used to simulate the diffusion of the solution during the wet etching to identify the boundaries of the etched Si substrate.

The design steps are shown in Fig. 1 with the curvature ratio, C, defined in terms of the comprising primitives.

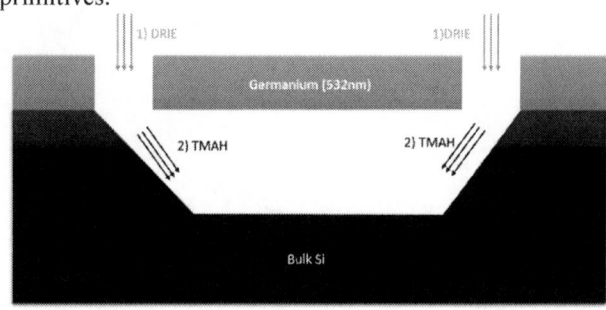

Fig 2: Schematic of the wafer stack and the main etching steps.

DRIE defines the etch windows followed by an anisotropic wet

TMAH under etch.

An intrinsic Ge on Si wafer was patterned using electron-beam lithography. The etching windows were patterned using deep reactive ion etching (DRIE) with 38 sccm Ar and 12 sccm CHF$_3$ at 200W RF power. Finally, the suspended Ge membranes were made using a Tetra-Methyl-Ammonium-Hydroxide (TMAH) wet etching process. Fig. 2 shows the wafer and main etching steps.

Results

Fig. 3 shows an optical 3-Dimentional ($3D$) micrograph of a structure with C of 3.0. The etching profile can be seen with etched slopes corresponding to the <111> crystallographic plane.

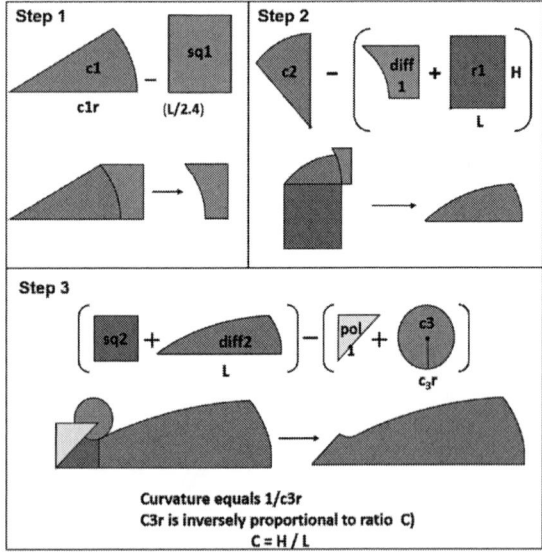

Fig 1: Schematic of design process for half an arm with

curvature ratio C.

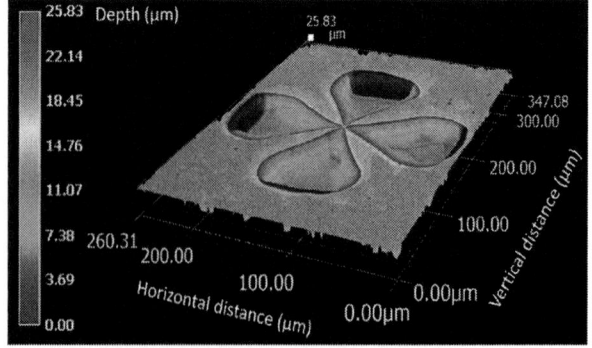

Fig 3: 3D micrograph of structure with C of 3.0

978-1-5090-4661-4/17 $31.00 © 2017 IEEE   92

Fig. 4 shows a titled optical micrograph of a structure with C of 3.0 and the etched window, in which the shadows show successful suspension.

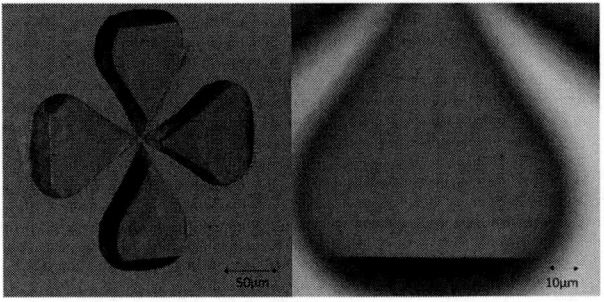

Fig 4: a) Tilted micrograph of structure at C of 3.0. b) Normal micrograph of etch window with etched Si crystallographic planes visible.

Fig. 5 shows strain distribution at the extremities of C of 1.6 and 3.0. These simulations show improved homogeneity in the membrane with C of 3.0 relative to that with C of 1.6.

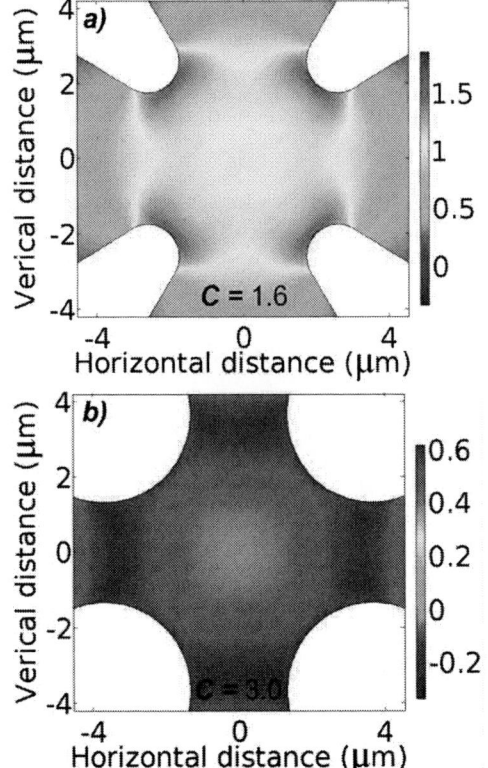

Fig 5:   Simulated strain distributions with C of a) 1.6 and b) 3.0. As C increases the strain homogeneity increases.

Fig. 6 shows the Raman spectra at various C with Lorentzian fittings. The peak widths and peak positions was then extracted from this data.

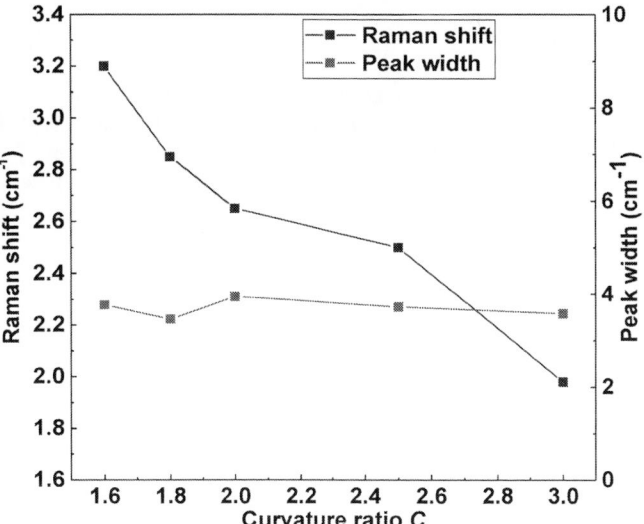

Fig 6: Raman spectra with Lorentzian fitting plotted over with increasing c

Figure 7 shows the effect of C on the peak widths and the Raman shifts. The peak widths remained relatively constant with increasing C, suggesting no major change in crystalline qualities. The decrease of the Raman shift on increasing C corresponds to the decrease of strain.

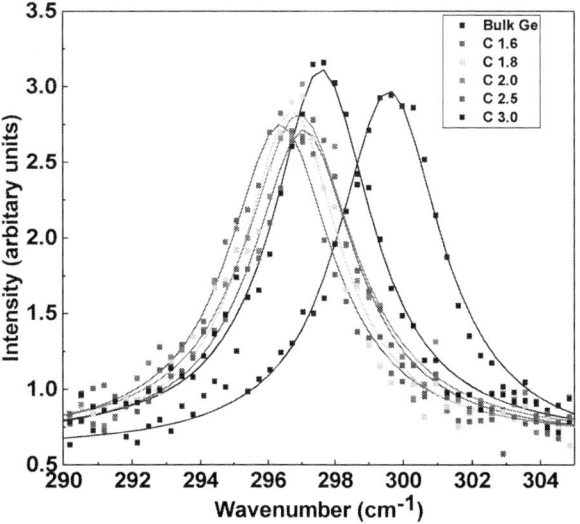

Fig 7: Raman shift and peak width with increasing C. Peak width remains relatively constant suggesting no significant change in crystalline quality. Raman shift decreases corresponding to decreasing central strain.

Fig. 8 shows the strain estimated from the Raman shift in comparison with the simulated values. Both the measured and simulated strain values decreased with increasing C, with a maximum strain of 0.85% observed at C of 1.6. The small differences between experimental and simulated values would be attributed to the simulated boundaries assuming $2D$ isotropic wet etching, while actual wet etching was highly anisotropic. This highlights the need for a more sophisticated simulation for defining the profile taking into account the anisotropic etching of the silicon crystal in three dimensions.

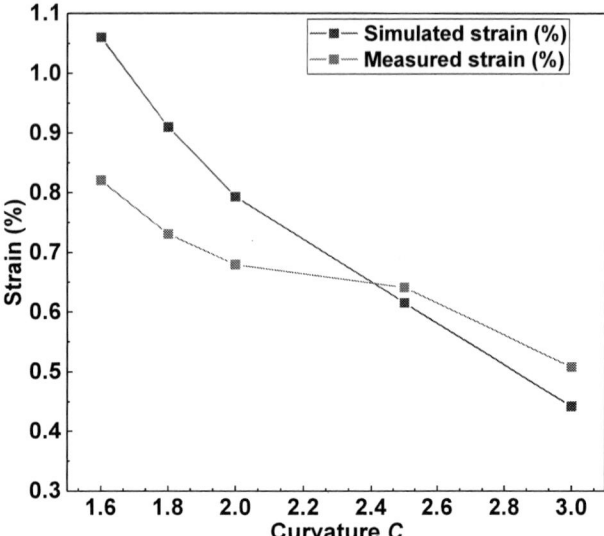

Fig 8: Simulated and measured strain at various C. Both simulated and measured strain decreases with increasing C. There is a discrepancy between the two sets of values is due to a simple isotropic etching profile being assumed.

Fig. 9. shows the micro-PL spectra at the center of structures with increasing C. Surprisingly, the devices with smaller C and thus higher central strain values exhibit smaller PL intensities. One explanation of this trend could be increased dislocations and thus increased non-radiative recombination due to higher strain values. In fact, we found significant local increase of the tensile strain at the edges of the membrane with C of 1.6 (Fig. 5 (a)). The steeper strain gradient resulted in the local band-gap distribution, and carriers might be recombined at the edges [2].

The peaks identified in the spectrum in Fig. 9 from the membrane with C of 3.0 would be attributed from the splitting of Light-Hole (LH) and Heavy-Hole (HH) bands with the presence of biaxial strain.

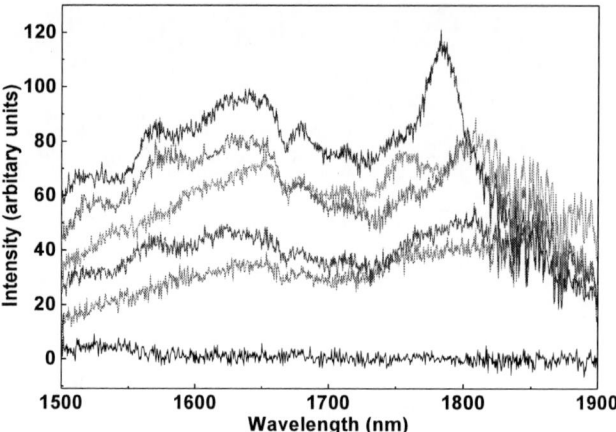

Fig 9: PL spectra at various C and bulk Ge. HH and LH splitting can be visualized as well as a decrease in intensity with increasing C despite the central strain value increasing.

Conclusion

We have successfully fabricated bi-axially suspended membranes using Ge on Si wafers with the tensile strain up to 0.85%. We found that the inhomogeneity of the strain reduces the PL intensities, so that the homogeneous strain-engineering would be critical towards developing monolithic Ge light sources on Si.

Acknowledgments

This work is supported by EPSRC Manufacturing Fellowship (EP/M008975/1), EU FP7 Marie-Curie Carrier-Integration-Grant (PCIG13-GA-2013-61811), and EPSRC Institutional Sponsorship Research Collaboration. The data from the paper can be obtained from the University of Southampton e-Print research repository: http://doi.org/10.5258/SOTON/403064.

References

[1] A. Gassenq, Appl. Phys. Lett. **107**, 191904 (2016).
[2] D. Sukhdeo, Opt. Express **23**, 16740 (2015).

5B-1

Transient Characterization of Graphene NEMS Switch ESD Protection Structures

Qi Chen[1], Cheng Li[1], Jimmy Ng[2], Fei Lu[1], Chenkun Wang[1], Feilong Zhang[1], Rui Ma[3],

Ya-Hong Xie [2], and Albert Wang[1]

[1]Dept. of ECE, University of California, Riverside, CA, USA, aw@ece.ucr.edu

[2]Dept. of MSE, University of California, Los Angeles, CA, USA; [3]Intel, USA

Abstract

Above-IC graphene NEMS (gNEMS) switch electrostatic discharge (ESD) protection structures are designed to replace traditional in-Si PN-based ESD structures. Built in CMOS back-end without PN junctions, gNEMS eliminates ESD-induced parasitics (leakage, capacitance, noise) inherent to PN junctions. Transient characterization of gNEMS ESD structures were conducted by transmission line pulsing (TLP) measurement to understand impacts of zapping waveforms on ESD behaviors. gNEMS ESD switch is a potential ESD protection solution to advanced ICs at nano nodes.

(Keywords: Graphene, TLP, gNEMS, ESD)

Introduction

ESD protection is required for all ICs, which becomes a design challenge for advanced ICs at sub-30nm nodes. ESD-critical parameters, including triggering voltage (V_{t1}), discharging resistance (R_{ON}), and failure current (I_{t2}), must be carefully designed to ensure full-chip ESD protection [1-2]. For decades, ESD protection relies on in-Si PN-junction-based active devices to discharge the fast and strong ESD surges. Unfortunately, the ESD-induced parasitic effects that are inherent to PN junctions, including leakage current, capacitance and noise, can seriously affect IC performance, especially for advanced ICs at sub-30nm nodes. Additionally, robust ESD protection requires large ESD device size and full-chip ESD protection needs multiple ESD devices par pad, which not only consume Si area, but also make IC layout difficult [3-5]. Using the unique mechanical and thermal properties of graphene [6], we devised a new mechanical switch based gNEMS ESD protection mechanism and demonstrated gNEMS ESD switches as an above-IC ESD protection structure [7]. Without any PN junction, gNEMS ESD structures can resolve all the problems associated with in-Si PN-type ESD structures. In this paper, we discuss comprehensive transient characterization of gNEMS switches by TLP testing for human body model (HBM) ESD protection.

gNEMS Switch ESD Protection Structure

Fig. 1a illustrates the fabrication process for gNEMS ESD switches. First, a 100nm layer of Si_3N_4 was grown on Si/SiO_2 (300nm) substrate by LPCVD and etched by RIE to create an opening which is covered by a CVD-grown monolayer graphene (Fig. 1b). Next, the graphene membrane was patterned by oxygen plasma. Ti/Pd/Au (5/30/50nm) was deposited as top electrodes. Then, HF steam was used to etch the chambers and release the graphene membrane. Fig. 1c shows the SEM image of gNEMS ESD switch, which has two terminals to be connected to IC pads.

TLP Characterization of gNEMS Switches

Table 1 outlines dimension splits of gNEMS devices including chamber depth (d), graphene ribbon length (L) and width (W). Complete transient characterization was conducted by TLP testing using varying pulse waveforms, i.e., pulse duration (t_d) and rise time (t_r) as shown in Table 1. Under TLP zapping, a fast ESD pulse is applied across the two gNMES terminals and the transient electrostatic force will pull the suspend graphene ribbon toward the bottom. When the graphene membrane touches the bottom, the gNEMS switch is turned on to discharge the ESD pulses for ESD protection. gNEMS switch has ultra-low leakage (~pA) and capacitance. Fig. 2 depicts a transient I-V curve for a gNEMS switch (L=10µm, W=7µm, d=350nm) by TLP testing (t_r=200ps, t_d=100ns), showing V_{t1}~8.9V and fast response down to 200ps. The measured I_{t2} (J_{t2} ~1.8×10^8 A/cm^2) shows very robust ESD protection. Fig. 3 shows the impact of varying t_d (75/150ns for t_r=10ns) on ESD I-V behaviors. It is observed that as TLP pulse t_d increases, ESD V_{t1} decreases, which is attributed to that a shorter pulse duration allows less response time, hence requires high electrostatic force to trigger the gNEMS switch. For similar reason, as t_d increases, ESD I_{t2} decreases due to the extra transient energy accumulation for a longer ESD pulse. Fig. 4 depicts the I-V curves for varying t_r (0.2/10ns, at t_d=100ns). Due to the much smaller scale of $\triangle t_r$ versus $\triangle t_d$, much weaker impacts on V_{t1} and I_{t2} were observed. Fig. 5 shows that the gNEMS device works at up to T=110ºC. Fig. 6 shows a SEM image for a failed gNMES where the ESD damage is observed in the graphene membrane, possibly due to combined thermal and mechanical failures [8].

Conclusion

We report a study of transient characterization of gNEMS ESD switches, showing strong impacts TLP pulse waveforms on ESD behaviors. The insight observed is critical to practical ESD designs.

978-1-5090-4661-4/17 $31.00 © 2017 IEEE

References

[1] A. Wang, *On-Chip ESD Protection for Integrated Circuits,* Kluwer, 2002.

[2] S. Voldman, *ESD: RF Technology and Circuits,* Wiley, 2006.

[3] A. Wang, et al, *IEEE TED,* pp.1304, July 2005.

[4] X. S. Wang, et al, *IEEE JSSC,* pp.1927, Sept. 2014.

[5] F. Lu, et al, *IEEE TCASI,* pp.1746, Oct. 2016.

[6] A. K. Geim, *Nature Materials 6.3 (2007)*: 183-191.

[7] R. Ma, et al, *IEEE EDL,* pp.674, May 2016.

[8] Q. Chen, et al, *IEEE TED,* pp.3205, July 2016.

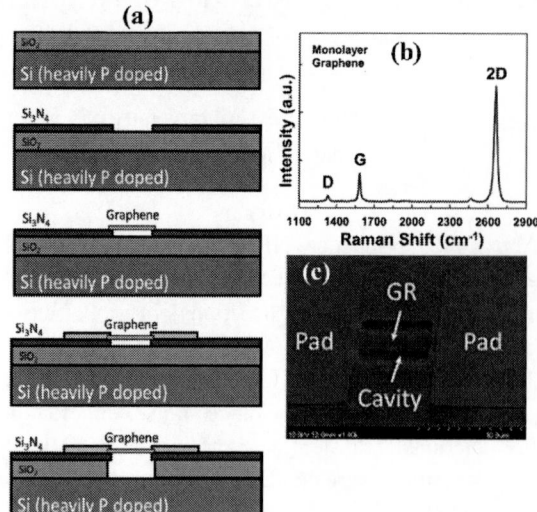

Table 1 gNEMS splits and TLP conditions

d (nm)	L (μm)	W (μm)	t_d (ns)	t_r (ns)
350	5/7/10/15	3/5/7/10/15	75/100/150	0.2/2/10

Fig. 1 (a) Process flow of gNEMS. (b) Raman spectrum shows monolayer graphene. (c) SEM image of gNEMS shows a suspended graphene ribbon (GR).

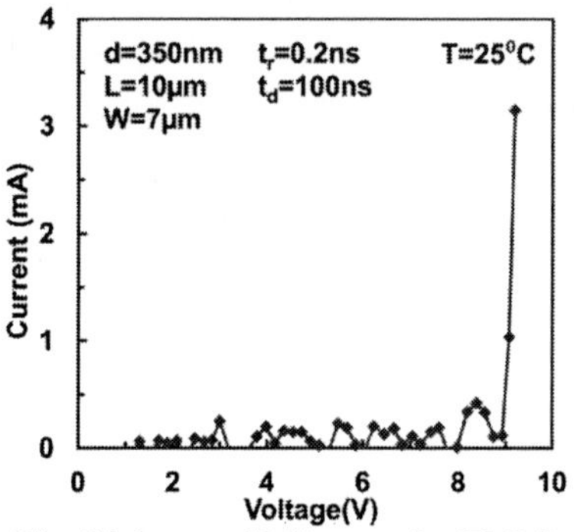

Fig. 2 Measured I-V curve by TLP for a gNEMS shows ESD discharging.

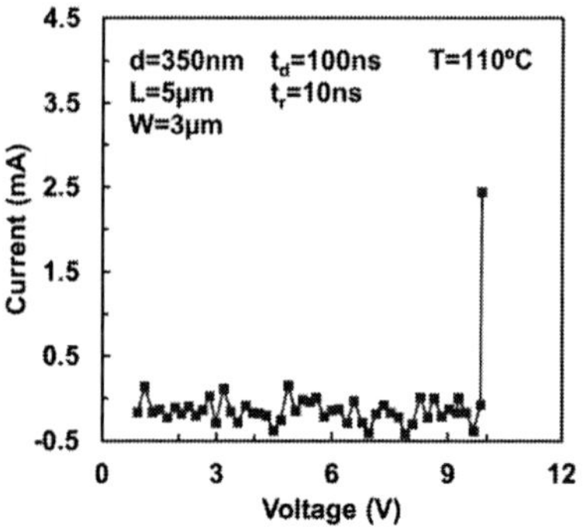

Fig. 5 I-V curve at T=110ºC shows temperature effect.

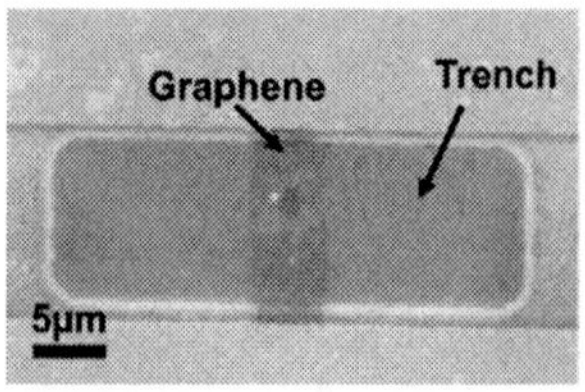

Fig. 6 SEM image for a failed gNEMS shows ESD failure signature.

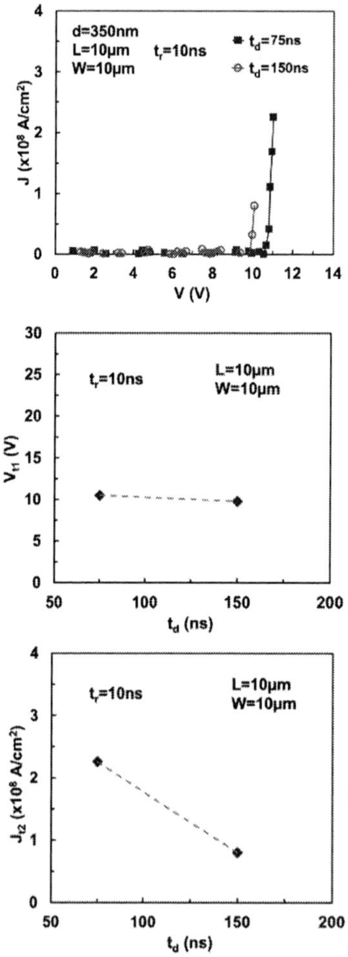

Fig. 3 Transient I-V curves shows impacts of TLP pulse durations (t_d).

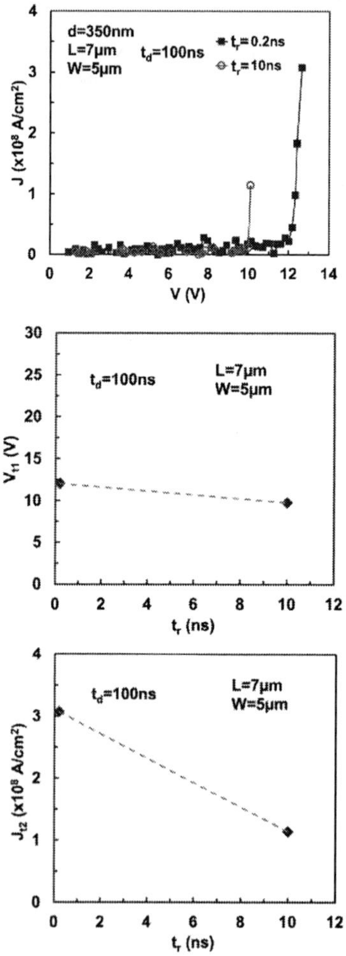

Fig. 4 Transient I-V curves shows impacts of TLP pulse rise time (t_r).

Gap in pagination due to unavailable paper.

Pages 97-99

5B-3 (Invited)

Multi-scale and Multi-domain Simulation of Electrical Power System

Shimeng Huang[1], Xiao Li[1], Tinghao Yeh[2], Shanghsun Mao[2],

Takayuki Sekisue[3], Vel Ambalavanar[1] and Sameer Kher[1]

[1]ANSYS Inc., Canonsburg, PA, USA, shimeng.huang@ansys.com

[2]ANSYS Taiwan, [3]ANSYS Japan

Abstract

An integrated simulation solution in ANSYS Simplorer is presented for multi-scale and multi-domain simulation of electrical power systems. An accurate device-level power semiconductor modeling tool is introduced and verified. Reduced order model (ROM) techniques are then demonstrated to import detailed models from multiple ANSYS finite element analysis (FEA) tools, so they can be interconnected and integrated to system level simulation.

(Keywords: Simulation, Compact model, ROM)

Introduction

With rising scale and complexity, electrical power systems are becoming more challenging to design and analyze [1][2]. This paper presents an integrated simulation solution in Simplorer, which can interconnect various types of physical and behavioral models to simulate complex systems. Two modeling techniques are introduced: 1) compact model and characterization of power semiconductors; 2) ROM extraction of cable, and device packaging with parasitic and cooling effects from ANSYS FEA tools. To showcase combined simulation of these models, two demonstrative examples of electromagnetic interference (EMI) and electro-thermal simulation are described.

Compact Model of Power Semiconductors

To simulate switching characteristics of power electronics, a library of device level power semiconductors and a model characterization tool are provided in Simplorer [3][4]. It enables accurate simulation of off-the-shelf devices, including IGBT, power MOSFET and diode. All information needed to parameterize these models can be found in datasheet.

Fig.1 shows the equivalent circuit inside the IGBT model. It consists of a static core of a MOSFET driven bipolar transistor, around which a set of lumped passive elements and current sources are built to model the dynamic behavior. All capacitances between the terminals are modeled with a capacitor that includes both a depletion capacitance behavior and an enhancement capacitance behavior. It is a function of voltage across the junction V_J:

$$C(V_J) = C_0 \left[1 + (\beta - 1) \left(1 - e^{\frac{-V_J \cdot \alpha \cdot (1-\delta)}{(\beta - 1) \cdot V_D}} \right) \right]$$

when $V_J > 0$

$$C(V_J) = C_0 \left[\delta + \frac{1-\delta}{\left(1 - \frac{V_J}{V_D}\right)^\alpha} \right] \text{ when } V_J \leq 0$$

where α, β, V_D are constant 0.1, 0.5, and 0.6 respectively. And C_0 and δ are parameters to be fitted according to dynamic characteristics of the device. The curves remain differentiable at the transition. In addition, the model also includes operating condition dependency, built-in power loss calculation, and junction to case thermal model for junction temperature simulation.

For demonstration, a model is extracted for Infineon FS400R12A2T4 IGBT. Using the automatic device characterization tool, static and thermal parameters are fitted to datasheet curves, and dynamic parameters are automatically tuned to match switching time and energy data. Fig.2 and Fig.3 plot the simulated output and thermal characteristics of this model against points sampled from datasheet. And Fig.4 verifies the model's switching loss at multiple levels of nominal current.

EMI Simulation with HF Parasitic ROM

Fig.7 is an example of PMSM speed control system, in which the FS400R12A2T4 IGBT model characterized in last section is used to build the inverter.

For this system, the parasitic effects (RLCG) are important for noise spectrum analysis, and are mainly determined by 3D geometry of the printed circuit board (PCB), device packaging, etc. These 3D structures can be modeled in ANSYS Q3D, which extracts frequency dependent RLCG parameters using FEA. To study the influence of the parasitics to the power electronic system, Q3D can generate highly accurate ROM in form of scattering parameter, which can be used in circuit simulation. Fig.5 shows the Q3D model of the example IGBT's package. Its RLCG matrices are calculated up to 100 MHz, and the resulting parasitic ROM is integrated to the IGBTs in Simplorer, as shown in upper right of Fig.7. The same ROM extraction technique is also applied to model a 1m long 3-phase power cable between inverter and electric machine in Fig.7.

The close-loop PMSM speed control is simulated with a reference change at 0.75s. Fig.8 shows the simulated current, and Fig.6 is the junction temperature rise on one of the IGBTs under ideal cooling condi-

978-1-5090-4661-4/17 $31.00 © 2017 IEEE

tion. LISN (CISPR25) is used to measure the conductive EMC between the DC source and the inverter (EUT). The spectrum plot of the voltage output from LISN EMI receiver port is shown in Fig.9, with red represents the result without packaging and cable ROMs and blue represents the result with the ROMs.

Electro-thermal Simulation with Thermal ROM
To simulate realistic thermal performance, 3D geometry of package/heat sink, as well as air flow condition need to be considered. In this example, the 3D package model in Fig.5 is used again in ANSYS Icepak to perform FEA analysis for cooling. Based on the step response file from Icepak, a state space ROM can be generated to use in Simplorer. Fig.10 shows the ROM simulated in Simplorer with matching step response as in Icepak. And Fig.11 demonstrate how the thermal ROM can be connected to the characterized IGBT by exposing its thermal pins.

Conclusion
This paper demonstrates the capability of Simplorer to interconnect accurate power device model with ROMs extracted from various FEA tools. Combination of these modeling techniques within an integrated simulation environment enables investigation on different level details for a complete electrical power system/sub-system. A demonstrative example with integrated components is discussed with reasonable simulation results.

References
[1] S. Meliopoulos, J. Meisel, G. Cokkinides, and T. Overbye, "Power system level impacts of plug-in hybrid vehicles," Power Systems Engineering Research Center (PSERC), Tech. Rep, pp. 09–12, 2009.

[2] A. Elhafez and A. Forsyth, "A review of more-electric aircraft," 13th International Conference on Aerospace Sciences & Aviation Technology, 2009.

[3] F. Wang, S. Kher, T. Fichtner, and J. Aurich, "A new power mosfet model and an easy to use characterization tool using device datasheet," in IEEE 14th Workshop on Control and Modeling for Power Electronics, June 2013, pp. 1–5.

[4] J. Aurich and T. Barucki, "Fast dynamic model family of semiconductor switches," in IEEE 32nd Annual Power Electronics Specialists Conference, vol. 1, 2001, pp. 67–74 vol. 1.

978-1-5090-4661-4/17 $31.00 © 2017 IEEE

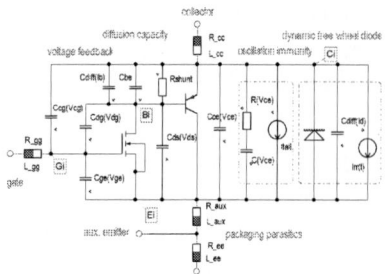

Fig. 1: Equivalent circuit of dynamic IGBT model.

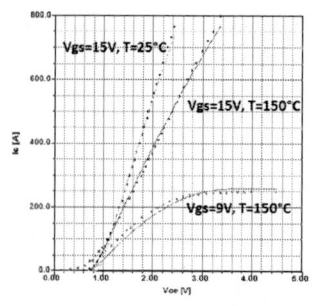

Fig. 2: Output characteristics of extracted IGBT model

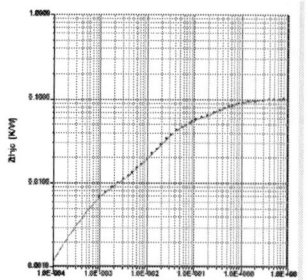

Fig. 3: Transient thermal resistance junction to case

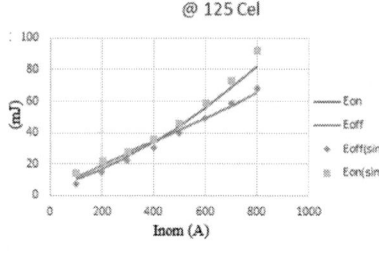

Fig. 4: Switching energy of extracted IGBT model.

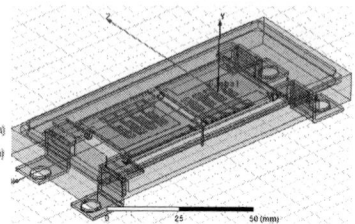

Fig. 5: Package structure of the IGBT device

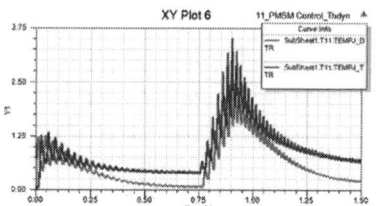

Fig. 6: Simulation of junction temperature rise

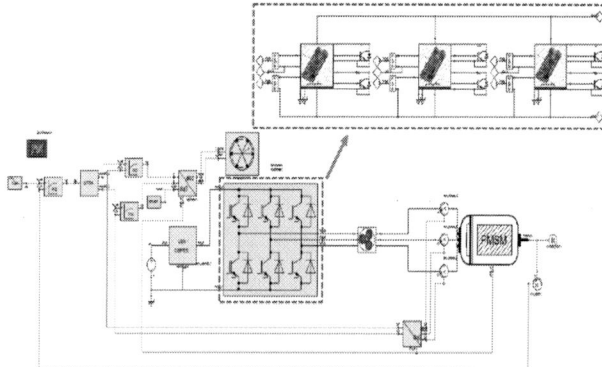

Fig. 7: System level PMSM speed control schematic with packaging and cable.

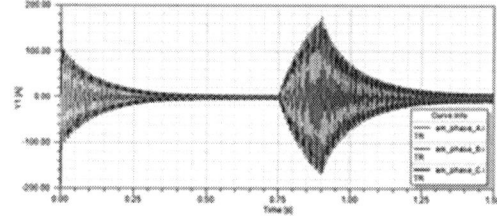

Fig. 8: PMSM speed control three phase currents

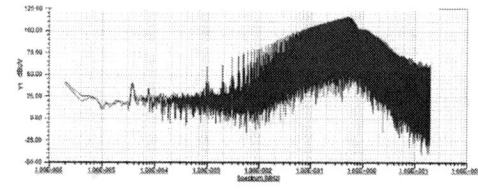

Fig. 9: Noise spectrum plot from LISN

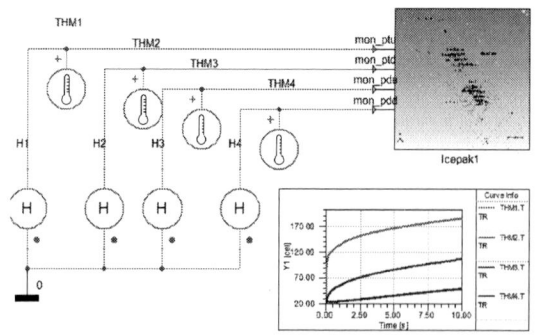

Fig. 10: Step response of thermal ROM for IGBT package

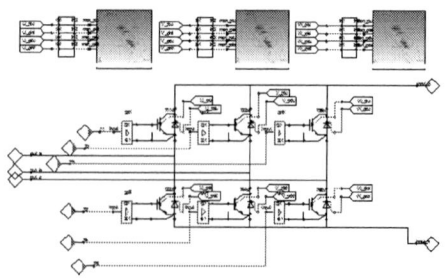

Fig. 11: Electro-thermal simulation of a 3-phase IGBT inverter

978-1-5090-4661-4/17 $31.00 © 2017 IEEE

6M-1 (Invited)

Nano-Structure-Controlled Very Low Resistivity Cu Wires Formed by High Purity Electrolyte and Optimized Additives

Jin Onuki[1], Kunihiro Tamahashi[1], Takashi Inami[1], Takatoshi Nagano[1],

Yasushi Sasajima [1], and Shuji Ikeda[2]

[1]Graduate School of Science and Engineering, Ibaraki University Ibaraki-ken, Japan,

jin.ohnuki.jn@vc.ibaraki.ac.jp, [2]Tei Solutions

Abstract

Resistivity increase in nano-level Cu wires is becoming a critical issue for high speed ULSIs. We have established the new manufacturing process utilizing very high-purity 9N electrolyte and optimized additives to control nano-structures of Cu wires, and we realized Cu wires with resistivity 50% lower than that of wires made by a conventional process. We also have ascertained the reason for getting very low resistivity Cu wires by STEM analyses and first-principle simulation.

(Keywords: High Purity Electrolyte, Optimized Additives, Nano-Structure Controlled Cu Wire, Very Low Resistivity)

Introduction

Cu wire resistivity increase due to miniaturization is the crucial common issue to be solved among leading LSI manufacturing companies. Researchers have worked on the thinning of high resistivity barrier metals [1]-[3]. Replacement of high-resistivity barrier metals with low resistivity barrier metal like Ru has also been investigated [4]. However, the area percentage of barrier metals to the entire area of the Cu wires is less than 20%. This means that thinning barrier metals is not particularly effective against resistivity increase. Hence, the reduction of resistivity in Cu wires (Cu core) is much more important. We have been insisting from 2007 that the resistivity increase of very narrow Cu wires is due to electron scattering at very small grain boundaries of around 50nm [5]. Recently, grain boundary scattering in the Cu core was shown to be the dominant factor for resistivity increase [6]. Hence, the objective for our study was to lower electron scattering at a grain boundary by nano-structure control of the Cu core, i.e., enlargement and homogenization, leading to very narrow Cu wires with much lower resistivity than that of conventional Cu wires.

Results and Discussion

A. Lowering resistivity

Fig.1 schematically explains our objective of lowering resistivity. To lower the resistivity, we thought that lowering of electron scattering at grain boundaries by enlargement and homogenization is mandatory.

B. Microstructural analyses of conventional Cu wire

Fig.2 shows the dark field STEM image around a grain boundary of 60nm wide conventional Cu wires formed using 6N Cu electrolyte (in current use) and standard additives. Many impurities, indicated by arrows, of about 5nm length and 2nm width were found along the grain boundary. We found these impurities included O, Cl, and Fe. We clarified these impurities existed at the grain boundary as very stable Fe (ClO) compounds and our ab-initio calculation results showed these compounds pinned grain growth during annealing.

C. Investigation of uniform grain size enlargement process

We thought there were two key points to getting large and uniform grains based on the above results. The first was to identify the grain size miniaturization process to utilize the high grain boundary energy for grain growth [7]. We found that very small grains can be obtained when using 1/10 the leveler in addition to the standard accelerator and suppressor. (We call these the optimized additives hereafter). The second key point was to purify the Cu electrolyte, raising it from 6N to 9N. We developed high purity 9N Cu electrolyte from 6N metal through an electrolytic method.

D. Resistivity and microstructure of Cu wires formed by the newly established method

By using the newly established plating process with high purity 9N Cu electrolyte and optimized additives, we formed Cu wires and evaluated grain sizes of Cu wires. Fig.3 shows longitudinal cross-sectional micrographs of 60nm wide Cu wires obtained by (a) 6N electrolyte with standard additives and (b) 9N electrolyte with optimized additives. The newly developed plating process gave Cu wires with much larger and more homogeneous grains than those formed by the conventional process. We also evaluated resistivity of 50nm Cu wires obtained by (a) and (b) and found the values were $5.8 \mu \Omega \cdot cm$ and $3.7 \mu \Omega \cdot cm$, respectively.

Conclusion

By combining 9N Cu electrolyte with optimized additives, we realized Cu wires with 1.6 times larger grain sizes and much lower resistivity ($3.7 \mu \Omega \cdot cm$) than the sizes and resistivity ($5.8 \mu \Omega \cdot cm$) of Cu wires made by conventional 6N electrolyte.

978-1-5090-4661-4/17 $31.00 © 2017 IEEE

Acknowledgments

We gratefully acknowledge the contributions of Prof. K. Kondo of Osaka Prefectural University, Mr. T. Ishigami of Kusaka Rare Metal Products, Mr. S. Yoshida of Toshiba Corp., Dr. M. Uchikoshi of Tohoku University, Dr. N. Ishikawa of NIMS, and Mr. M. Majima of JST for their encouragement and valuable discussions.

References

[1] James S. Clarke, Christopher George, Christopher Jezewski, Arantxa Maestre Caro, David Michalak, Jessica Torres, "Process Technology Scaling in an Increasingly Interconnect Dominated World", Digest of Technical Papers, Symposium on VLSI Technology, 2014. Session 16-2, pp. 142-143

[2] J. Koike and M. Wada, "Self-forming diffusion barrier layer in Cu-Mn alloy metallization", Appl. Phys. Lett., Vol.87, 041911, 2005

[3] Takeshi Nogami, Benjamin D Briggs, Sevim Korkmaz, Moosung Chae, Christopher Penny, Juntao Li, Wei Wang, Paul S McLaughlin, Terence Kane, Christopher Parks, Anita Madan, Stephan Cohen, Thomas Shaw, Deepika Priyadarshini, Hosadurga Shobha, Son Nguyen, Raghuveer Patlolla, James Kelly, Xunyuan Zhang, Terry Spooner, Donald Canaperi, Theodorus Standaert,Elbert Huang, Vamsi Paruchuri, and Daniel Edelstein, "Through-Cobalt Self Forming Barrier (tCoSFB) for Cu/ULK BEOL: A Novel Concept for Advanced Technology Nodes", Technical Digest of IEDM 2015. Session 8-1, pp. 181-184

[4] R.-H. Kim, B.H. Kim, T. Matsuda, J.N. Kim, J.M. Baek, J.J. Lee, J.O. Cha, J.H. Hwang, S.Y. Yoo, K.-M. Chung, K.H. Park, J.K. Choi, E.B. Lee, S.D. Nam, Y.W. Cho, H.J. Choi, J.S. Kim, S.Y. Jung, D.H. Lee, I.S. Kim, D.W. Park, H.B. Lee, S. H. Ahn, S.H. Park, M.-C. Kim, B.U. Yoon , S.S. Paak, N.-I. Lee, J.-H. Ku, J.S. Yoon, H.-K. Kang, and E.S. Jung, " Highly Reliable Cu Inter Connect Strategy for 10nm Node Logic Technology and Beyond", Technical Digest of IEDM 2014. Session 32-2, pp. 768-771

[5] Khyoupin Khoo, Jin Onuki, Takahiro Nagano, Yasunori Chonan, Haruo Akahoshi, Masahiro Chiba, Tatsuyuki Saito and Kensuke Ishikawa, "Influence of Grain Size Distributions on the Resistivity of 80nm Wide Cu Interconnects", Mater. Trans., Vol.48, pp. 622-624, 2007

[6] A. Pyzyna, R. Bruce, M. Lofaro, H. Tsai, C. Witt, L. Gignac, M. Brink, M. Guillorn, G. Fritz, H. Miyazoe, D. Klaus, E. Joseph, K. P. Rodbell, C. Lavoie, D.-G. Park, "Resistivity of copper interconnects beyond the 7nm node", Digest of Technical Papers, Symposium on VLSI Technology, 2015. Session 8-3, pp. 120-121

[7] Yasushi Sasajima,Junpei Kageyama,Khyoupin Khoo and Jin Onuki,"Grain coarseninh mechanism of Cu thin films by rapid annealing", Thin Solid Films,Vol.518,pp.6883-6890,2010

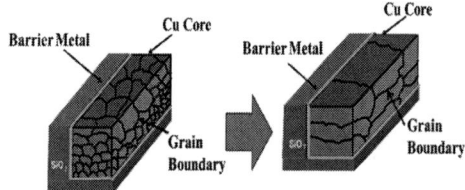

Fig.1 Objective of our study to lower Cu wire resistivity

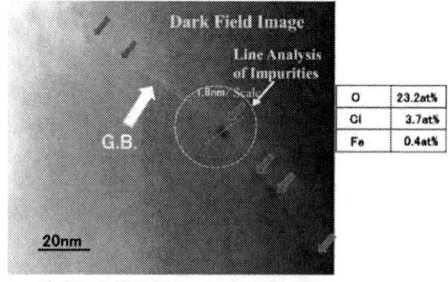

Fig.2 Dark Field Image of a Grain Boundary Red arrows indicate impurity compounds.

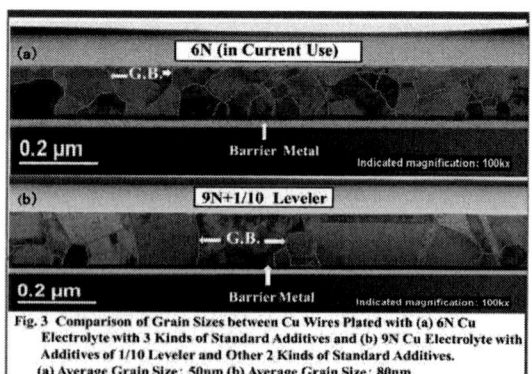

Fig. 3 Comparison of Grain Sizes between Cu Wires Plated with (a) 6N Cu Electrolyte with 3 Kinds of Standard Additives and (b) 9N Cu Electrolyte with Additives of 1/10 Leveler and Other 2 Kinds of Standard Additives.
(a) Average Grain Size：50nm (b) Average Grain Size：80nm

6M-4 (Invited)

Proximity Gettering Technology for Advanced CMOS Image Sensors Using C_3H_5 Carbon Cluster Ion Implantation Techniques

Kazunari Kurita, Takeshi Kadono, Ryosuke Okuyama, Ayumi Masada, Ryou Hirose
Yoshihiro Koga, and Hidehiko Okuda

SUMCO Corporation 1-52 Kubara, Yamashiro-cho, Imari-shi, Saga, 849-4256 Japan,
k-kurita@sumcosi.com

Abstract

CMOS image sensor has been widely used for ubiquitous devices. However, CMOS image sensor devices performance is dramatically influenced by process induced defects such as metallic impurities contamination at devices active region during CMOS devices process. Thus, it is extremely important to study metallic impurities influence on CMOS image sensor devices performance and to develop effectiveness metallic impurities gettering techniques. We introduce our new proximity gettering technique for advanced CMOS image sensor using carbon cluster ion implantation technique. Furthermore, we demonstrated that the carbon cluster ion implanted silicon wafer has high gettering capability of metal, oxygen and hydrogen impurity during CMOS image sensor devices fabrication process.

(Keywords: Cluster, Gettering, Silicon wafer)

Introduction

The process for fabricating advanced CMOS image sensor devices, such as thermal annealing and plasma etching treatment, substantially influence metallic impurity contamination [1-2]. Since, CMOS image sensor manufacturers strongly require silicon wafers with the highest metallic impurity gettering ability, we have developed a new gettering wafer production concept using a carbon cluster ion implantation technique for advanced CMOS image sensors [3]. This technique can implant a silicon wafer surface simultaneously with carbon and hydrogen elements that form the projection range by using a hydrocarbon compound as shown in Fig.1. In our previous study, it was found that a carbon cluster ion implanted silicon wafer had three characteristics for high performance of advanced CMOS image sensors [4-5]. First, a carbon cluster ion projection range has high gettering ability of metallic impurities as shown in Fig.2. Second, this projection range also has a diffusion barrier effect for oxygen impurities out-diffusing to the device active region from the silicon wafer substrate as shown in Fig.3. Third, it is expected that diffusing hydrogen to the device active region from the hydrogen of carbon cluster ions gettered in the projection range during the device fabrication process will have a passivation effect on process induced defects such as those in the Si/SiO_2 interface state as shown in Fig.4. In this article, we demonstrated that a carbon cluster ion implanted

silicon wafer has characteristics of gettering ability for metallic impurities, oxygen, and hydrogen during the device fabrication process.

Experiment

We evaluated the gettering ability of a carbon cluster ion implanted silicon wafer contaminated by iron metallic impurities after diffusion heat treatment. We measured the depth profile of nickel metal, oxygen and hydrogen impurity concentration using secondary ion mass spectroscopy (SIMS) analysis.

Results and Discussion

Figs.2-4 shows a SIMS depth profile of metal, carbon, oxygen, and hydrogen impurity after diffusion heat treatment. It confirms that nickel is gettered in the carbon cluster projection range. We confirmed that the oxygen peak that diffused out from a silicon wafer substrate and implanted hydrogen was formed in a projection range. These obtained results demonstrate three characteristics about the gettering ability of a carbon cluster ion implanted silicon wafer using nickel metallic impurity contamination diffusion heat treatment.

Conclusion

We provide the first ever demonstration of the gettering capability of C_3H_5 carbon cluster ion implanted epitaxial growth silicon wafers such as metal, oxygen, and hydrogen impurity after CMOS image sensors fabrication processes. It was found that the C_3H_5 carbon cluster ion implantation range could getter the metal, oxygen and hydrogen impurities after CMOS image sensor heat treatment. In conclusion, we assume that the carbon cluster ion implanted silicon wafers are an effective mean of decreasing dark current and white spot defects for CMOS image sensors. We believe that such wafers will be beneficial for advanced CMOS image sensor fabrication processes.

References

[1] T. Kuroda,"Essential Principle of Image sensor"(CRC press,2014).

[2] S.Inoue, Oyo Buturi **73**,1207(2014).

[3] K. Kurita, Oyo Buturi, Vol. 84, p628 (2016)

[4] K. Kurita et al., J. Surf. Sci. Soc. Jpa, Vol. **37**, p. 104 (2016)

[5] T. Kadono and K. Kurita , Japan Patent 5673811 (2015).

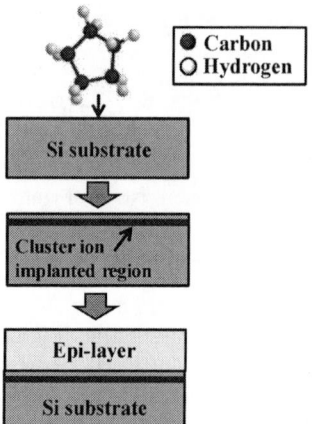

Fig. 1: Production concept of carbon cluster ion implanted silicon wafers

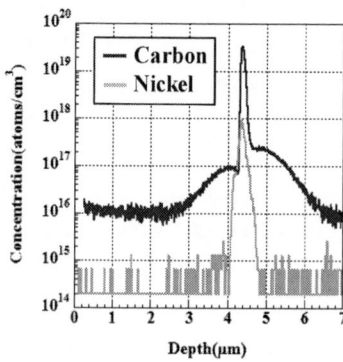

Fig. 2: SIMS depth profile measured on a carbon cluster ion implanted epitaxial growth silicon wafer after nickel contamination subject to heat treatment

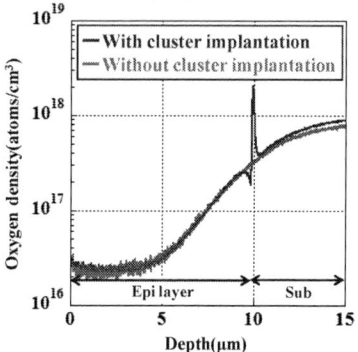

Fig. 3: SIMS depth profile measured on oxygen impurity in carbon cluster ion projection range after epitaxial growth

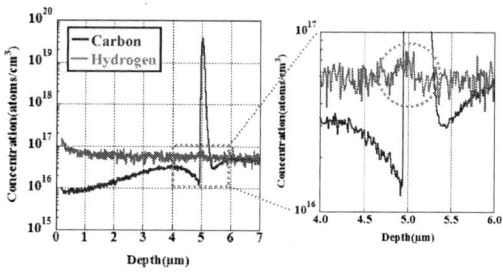

Fig. 4: SIMS depth profile measured on hydrogen impurity in carbon cluster ion projection range after epitaxial growth and after CMOS simulation heat treatment

978-1-5090-4661-4/17 $31.00 © 2017 IEEE 106

6A-1 (Invited)

New Visions for IC Yield Detractor Detection

Bill Nehrer, Kelvin Doong, Dennis Ciplickas
PDF Solutions, San Jose , CA

Abstract

The observability of conventional electrical test site and imaging techniques needs to be extended and coupled with all of the actual product layout attributes in order to reflect the relevant yield detractors of the current technologies in production and development.

This paper discusses new electrical test site strategies that have been recently developed and deployed developed for parametric yield detection and systematic hard defect detection by layout attribute. Such test structures are derived from the actual product and in some key cases also embedded in the product utilizing available space between the active circuitry and detected in-line with non-contact techniques.

(Keywords: SoC Yield, electrical test structures, yield prediction, parametric defects, systematic pattern defects, Manufacturing, CMOS and SOI)

Introduction

Modeling and yield prediction of the semiconductor product has been an integral part of the IC manufacturing process from the very early stages of the industry. For accuracy, the detection and measurement of the yield detractors must be precise. The capability to image and measure yield detractors before the completion of the lengthy serial process flow to manufacture a product has be a key tool for successful technology development and the effective monitor and control of the mass production process. However, as key features and dimensions continue to shrink, the layout environments need to be better reflected in the electrical test structures used to breakdown, measure, and determine the root cause of parametric and systematic pattern defects.

Parametric Yield Detractor Detection

The design methodology of turning a standard cell library or product chip into a test chip will be presented. Experimental data in terms of gap analysis of Silicon-to-SPICE/RC of advanced logic process are reported for the various applications including technology development, standard cell library design, and product production yield .

The use is depicted in Figure 1, which shows that some patterns (as identified by pattern classification in x-axis groups) are measured at the extreme edge of the SPICE model prediction used for SoC design. These patterns has been shown, Fig 2., to accurately relate to the product level parametric bin yield, in this example Iddq.

Systematic Pattern Yield Detractor Detection

The key objective of yield detractor detection is to quantitatively measure and control the process to minimize the adverse effects, and maximize the yield. Typically, this involves a design to process co-optimization. However, as process dimension have decreased to today's levels of 7 and 5 nm, and the structures have become three-dimensional in active configurations, and the ability to image the defective sites has become limited. This paper discusses a technique in recent use of an embedded electrical test structure in the product area which is measured electrically for open and shorts by using ebeam voltage contract techniques. The resultant technology is shown to be a non-contact electrical test structure which provides the quantitative and accurate fail rates. Figure 3 illustrates the system concept and application.

Conclusion

Two new yield detection technologies have been shown that extend the Yield "vision" of the technology development and yield enhancement engineer. The incorporation of the actual layout attributes of the product layout to the electrical test site has increased the accuracy to enable the precise prediction and measurement on process layer failure rates as demonstrated by the examples shown.

Acknowledgments To be Added

References   To Be added

978-1-5090-4661-4/17 $31.00 © 2017 IEEE

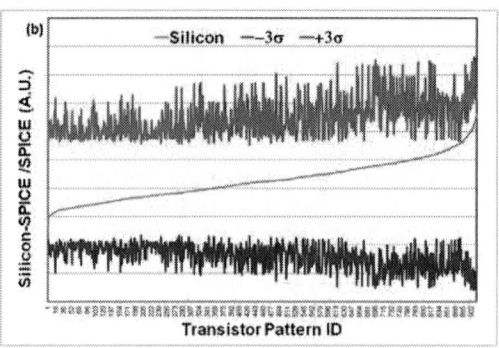

Figure 1: Si-to-SPICE model gap. NMOS devices measured on test site sorted as a function of pattern class vs. the SPICE model predictions.

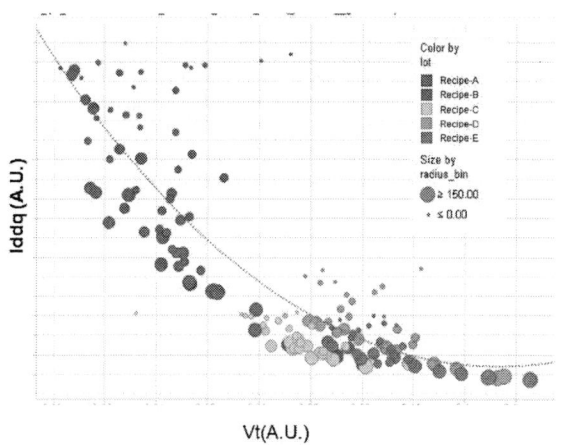

Figure 2: Correlation of Iddq and the use of the classified pattern parameters in the product / device targeting.

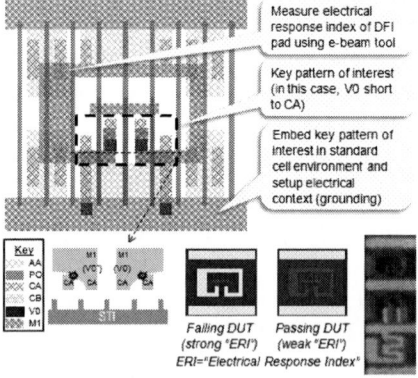

Figure 3: System for embedding electrical test sites using the actual product layouts to measure the failure rates (in terms of ERI) of patterns by process.

978-1-5090-4661-4/17 $31.00 © 2017 IEEE   108

Comparative Study on RTN Amplitude in Planar and FinFET Devices

Zexuan Zhang[1], Zhe Zhang[1], Shaofeng Guo[1], Runsheng Wang[1*],
Xingsheng Wang[2], Binjie Cheng[2], Asen Asenov[2,3], Ru Huang[1]

[1]Institute of Microelectronics, Peking University, Beijing 100871, China. (*Email: r.wang@pku.edu.cn)
[2]Synopsys, Glasgow G3 7JT, U.K. [3]School of Engineering, University of Glasgow, Glasgow G12 8LT, U.K.

Abstract

In this paper, the amplitude of random telegraph noise (RTN) in FinFET is studied, comparing with RTN in planar devices. The impacts of intrinsic characteristics in FinFET (channel non-uniformity and quantum confinement) on its RTN amplitude are comprehensively studied, based on the framework of "hole in the inversion layer" (HIL) model and the 3D device simulations. The results indicate that, the conventional HIL model for planar device can be extended to FinFET, if taking into account the non-uniform Fin current density. It is also found that, the RTN-induced "hole" in FinFET is smaller than that in planar device under the same inversion carrier density per gate, due to strong quantum confinement in FinFET. These results are helpful for accurate RTN amplitude modeling in FinFET.

Introduction

Random telegraph noise (RTN) is one of the greatest reliability concerns for 16/14nm node and beyond as its amplitude increases rapidly with device scaling. Typical RTN signal observed in drain current is shown in Fig.1(a). Accurate modeling on RTN amplitude is therefore urgently needed for robust circuit design. However, unlike planar devices, model for RTN amplitude in FinFET is still lacking. One natural thought is to extend the RTN amplitude model for planar device (i.e. "hole in the inversion layer" (HIL) model [1-11], as shown in Fig.1(b)) to FinFET. Therefore, a comparative study of RTN amplitude will be helpful for RTN amplitude modeling in FinFET.

In this paper, the impact of intrinsic channel non-uniformity in FinFET on RTN amplitude is theoretically analyzed based on HIL model, and confirmed by 3D 'atomistic' device simulations. Besides, the underlying physics for the difference of "hole" radius in planar and FinFET device is thoroughly studied considering the stronger quantum confinement effect of FinFET. The results are essential for accurate modeling of RTN amplitudes in FinFET.

Device Simulation

Planar MOSFET ($W=L=25$nm) and FinFET ($L=20$nm, $W_{Fin}=8$nm and $H_{Fin}=40$nm) are studied in this work (as shown in Fig.2) using carefully calibrated 3D 'atomistic' device simulator [12]. EOT is 0.8nm in both devices and V_d is set at 0.05V.

Impact of Channel Non-Uniformity on RTN Amplitude

In FinFET, due to its non-planar nature, the current density shows an intrinsic non-uniformity with increasing V_G (shown in Fig.3(a)) [13]. The highest current density (J_{max}) position for the FinFET studied is around $y=6$nm at $V_G=0$V (definition of zero point throughout this paper is illustrated in Fig.2) and transfers upward with increasing V_G as shown in Fig.3(b). In HIL model framework, it is assumed that the trapped carrier creates a "hole" in the inversion layer beneath the trap. The higher local current density around the trap, the more current will be "blocked" by the "hole", implying larger RTN amplitude. Therefore, unlike in planar MOSFET (with an intrinsic uniform channel), RTN amplitude and its V_G dependence in FinFET will rely on the trap position in y direction and $y=6$nm will be a turning point. The simulation results are well consistent with expectations. As can be seen in Fig.4, though both devices present a rapid decrease in RTN amplitude with increasing V_G above threshold, unlike in planar device where RTN amplitude shows a plateau, V_G dependence of RTN amplitude in FinFET is rather complicated with a turning point (6nm in this case) below threshold. RTN amplitude shows a monotonic decrease with V_G for traps with y larger and a peak for traps with y smaller than the turning point. Based on HIL model, as the channel is uniform in planar device, RTN amplitude depends solely on "hole" radius. Below threshold, the "hole" radius is a constant [6,10-11]. Therefore, RTN amplitude shows a plateau below threshold in planar device. In FinFET, for a trap with $y>6$nm, it is farther away from the J_{max} position with increasing V_G, resulting in a monotonic reduce in RTN amplitude. And for a trap with $y<6$nm, a peak of RTN amplitude will appear when the trap is around J_{max} position.

RTN amplitude dependence on trap position along y direction in planar device is shown in Fig.5. As expected, RTN amplitude has weak dependence on trap location due to the uniform channel. For FinFET, considering the evolution of intrinsic channel non-uniformity and carrier distribution in y direction with increasing V_G, three regions (i.e. below threshold, near-threshold volume inversion and near-surface strong inversion) are studied respectively. Below threshold, as can be seen in Fig.6, a "killer" trap position (where trap induces largest RTN amplitude) exists following J_{max} position (shown in Fig.7). This phenomenon is consistent with HIL model as is discussed earlier. In volume inversion region, RTN amplitude changes slightly with trap location along y direction (as can be seen in Fig.8) due to the fact that the channel is uniform between $y=-15$nm to $y=10$nm just like a planar device, as shown in Fig.9. In near-surface inversion region, as is shown in Fig.10, RTN amplitude shows a monotonic increase with y. This finding can be explained based on HIL model considering carrier becoming nearer to surface with increasing y (shown in Fig.11), resulting in a larger "hole" blocking more drain current and causing larger RTN amplitude.

It is well known RTN amplitudes show a "bell shape" with trap position along channel length direction for planar devices [10]. This is also the case with FinFET [14]. Normalized RTN amplitude with x is shown in Fig.12 to characterize the decreasing speed of RTN amplitudes when trap moves from the middle of the channel to source and drain. The decreasing speed is same regardless of device structure as the non-planar nature in FinFET does not introduce channel non-uniformity in length direction.

Impact of Quantum Confinement on "Hole" Radius

The differences between planar and FinFET RTN amplitude have been well explained in the framework of the HIL model, suggesting the possibility of extending HIL model to FinFET. According to HIL model, inversion carrier can be treated as a charge sheet. For a uniform channel, RTN amplitude can be gained by solving a 2D resistor network:

$$\frac{\Delta I_d}{I_d} = \frac{4R^2}{(L-2R)(W-2R)+2WR} \quad (1)$$

The strong inversion for both FinFET and planar devices are considered here, because the channels are almost uniform in these regions (Fig.9). Based on (1) and the simulation results, the "hole" radius can be extracted. In the strong inversion regions, effect of the trapped carrier is screened mainly by the high-density inversion layers [2-3,5-7,10]. Therefore, "hole" radius is determined by local inversion carrier density. The dependences of "hole" radius on 2D carrier density per gate in planar and FinFET are shown in Fig.13. "Hole" radius is about 23% smaller in FinFET than that in planar device under the same 2D carrier density per gate. This difference can be partly explained by the fact that t_{inv} is slightly larger in FinFET (shown in Fig.14). More importantly, HIL model is a 2D model, where carrier is regarded as a sheet through a line integral. For FinFET, the carrier density drops much more slowly compared to planar device due to the stronger quantum confinement effect (Fig.14). A large portion of carriers farther away from the surface than t_{inv} are in fact not so largely affected by the trapped carrier as are assumed through charge sheet approximation. Therefore, "hole" radius is smaller in FinFET than planar device under same 2D carrier density per gate. For accurate modeling of RTN amplitudes in FinFET, carrier distribution vertical to the surface (which relies on Fin width) has to be taken into account.

Summary

Based on framework of HIL model, the impacts of intrinsic channel non-uniformity and strong quantum confinement in FinFET on its RTN amplitude are theoretically discussed, and confirmed by 3D device simulations.

Acknowledgements: This work was partly supported by NSFC (61522402, 61421005).

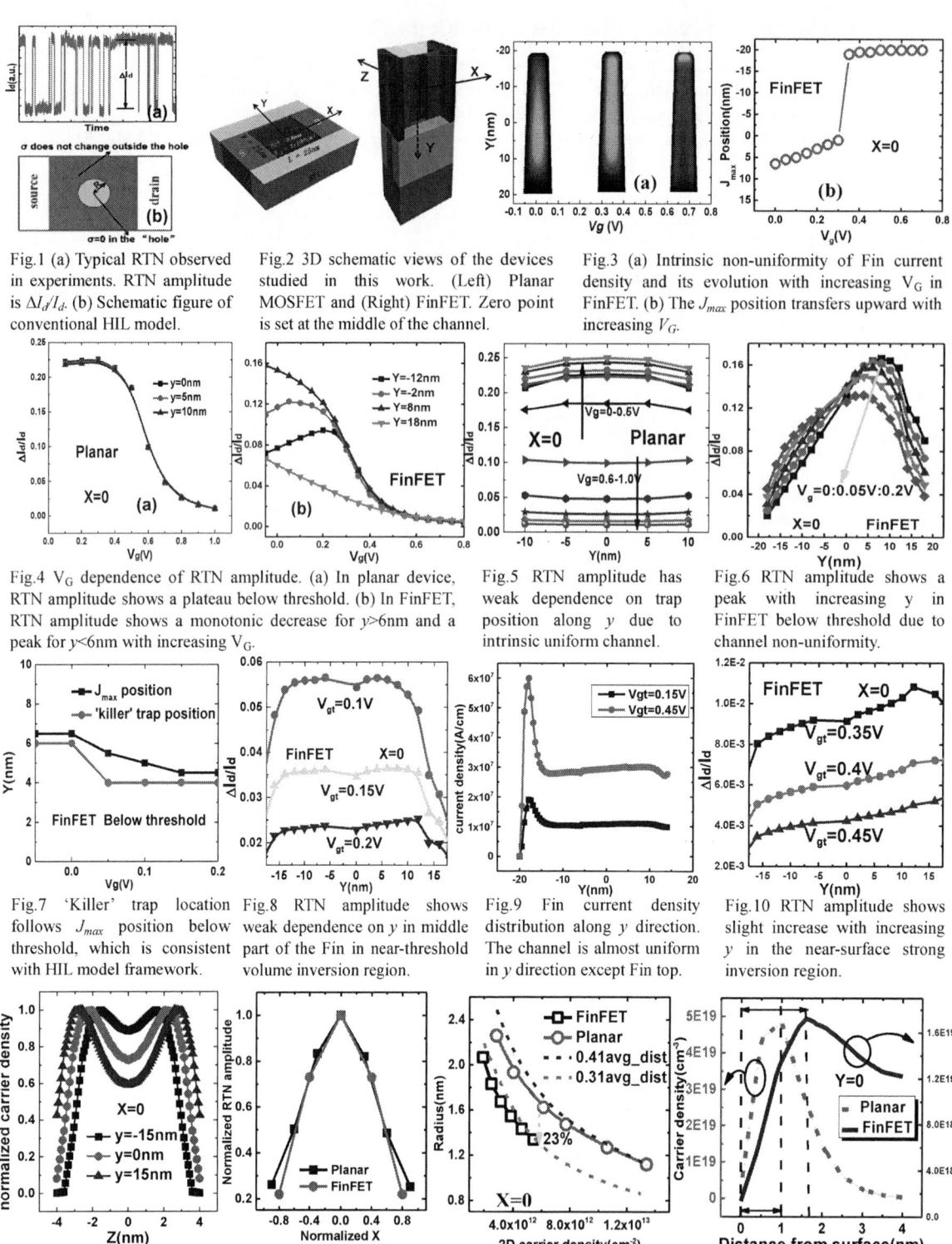

Fig.1 (a) Typical RTN observed in experiments. RTN amplitude is $\Delta I_d/I_d$. (b) Schematic figure of conventional HIL model.

Fig.2 3D schematic views of the devices studied in this work. (Left) Planar MOSFET and (Right) FinFET. Zero point is set at the middle of the channel.

Fig.3 (a) Intrinsic non-uniformity of Fin current density and its evolution with increasing V_G in FinFET. (b) The J_{max} position transfers upward with increasing V_G.

Fig.4 V_G dependence of RTN amplitude. (a) In planar device, RTN amplitude shows a plateau below threshold. (b) In FinFET, RTN amplitude shows a monotonic decrease for $y>6$nm and a peak for $y<6$nm with increasing V_G.

Fig.5 RTN amplitude has weak dependence on trap position along y due to intrinsic uniform channel.

Fig.6 RTN amplitude shows a peak with increasing y in FinFET below threshold due to channel non-uniformity.

Fig.7 'Killer' trap location follows J_{max} position below threshold, which is consistent with HIL model framework.

Fig.8 RTN amplitude shows weak dependence on y in middle part of the Fin in near-threshold volume inversion region.

Fig.9 Fin current density distribution along y direction. The channel is almost uniform in y direction except Fin top.

Fig.10 RTN amplitude shows slight increase with increasing y in the near-surface strong inversion region.

Fig.11 Normalized Fin carrier density at V_{gt}=0.55V. Carriers are nearer to the surface with increasing y.

Fig.12 Normalized RTN amplitudes in planar and FinFET devices, which is almost the same under a given x position regardless of device structure.

Fig.13 Extracted "hole" radius in FinFET and planar devices. "Hole" radius is smaller in FinFET under the same 2D carrier density per gate.

Fig.14 Carrier density distribution vertical to surface at V_{gt}=0.55V in FinFET and planar devices. Carrier density drops more slowly in FinFET than planar device.

References: [1] L. D. Yau, *TED* p.170 (1969). [2] K. P. Cheung, *ICICDT* p.1 (2011). [3] K. P. Cheung, *IRPS* p.GD1 (2012). [4] R. G. Southwick III, *SNW* p.147 (2012). [5] K. P. Cheung, *ICSICT* p.531 (2012). [6] E. Simoen, *TED* p. 422 (1992). [7] C. Liu, *IRPS* p.XT.17 (2014). [8] H. H. Mueller, *JAP* p.1734 (1997). [9] M. J. Chen, *TED* p.2495 (2014). [10] A. Asenov, *TED* p.839 (2003). [11] K. Sonoda, *TED* p.1918 (2007). [12] GARAND. http://www.goldstandardsimulations.com. [13] Z. Zhang, *IEDM* session #7.2 (2016) to be published. [14] X. Wang, *SNW* p.77 (2012).

978-1-5090-4661-4/17 $31.00 © 2017 IEEE

6A-3 (Invited)

A BTI Analysis Tool (BAT) to Simulate p-MOSFET Ageing Under Diverse Experimental Conditions

Souvik Mahapatra, Narendra Parihar, Subrat Mishra, Beryl Fernandez and Ankush Chaudhary

Department of Electrical Engineering, Indian Institute of Technology Bombay, Mumbai, India

Abstract

A physical modeling framework is demonstrated for Negative Bias Temperature Instability (NBTI). It can simulate temporal kinetics of threshold voltage shift (ΔV_T) during and after DC and AC stress and mixed DC-AC stress for dynamic voltage, frequency and activity conditions. It can predict gate insulator process dependence and is consistent with large and small area devices. The framework is included in a commercial TCAD software to simulate degradation of FinFETs and GAA NWFETs.

(Keywords: NBTI, Reaction-Diffusion (RD) model, trap generation, hole trapping, TCAD)

Introduction

NBTI is a crucial p-MOSFET reliability issue and impacts Silicon Oxynitride [1] and High-K Metal Gate [2] devices. It causes shift in device parameters due to buildup of positive gate insulator charges. However, it relaxes after the removal of stress, giving relief for AC as compared to DC stress [3]. Prediction of NBTI time evolution during and after stress is needed to calculate ageing at end-of-life for technology qualification.

NBTI physical mechanism was a topic of debate. It was proposed that NBTI is due to only interface trap generation (ΔN_{IT}) [4], only trapping of holes in pre-existing traps (ΔN_{HT}) [5], strongly correlated ΔN_{IT} and ΔN_{HT} [6] and uncorrelated ΔN_{IT} and ΔN_{HT} [1], [7], refer to [8] for a recent review. It is shown that although ΔN_{IT} dominates ΔV_T for industrial quality devices [9], uncorrelated mix of ΔN_{IT} and ΔN_{HT} is needed to explain ultra-fast measurements and gate insulator process impact [1], [7]-[9]. Only ΔN_{HT} and correlated ΔN_{IT} and ΔN_{HT} models are not consistent with long-time stress experiments as shown in [8].

A framework consisting of uncorrelated ΔN_{IT} and ΔN_{HT} and bulk trap generation (ΔN_{OT}) was proposed to explain NBTI [10]-[12]. It can predict temporal ΔV_T kinetics during and after DC and AC stress for different stress (V_{GSTR}) and recovery (V_{GREC}) biases, temperature (T), frequency (f) and pulse duty cycle (PDC), and for mixed DC-AC stress under dynamic voltage, frequency and activity conditions. It is fully capable to explain gate insulator process dependence. It is consistent with data from large and multiple small area devices. The framework was incorporated in commercial TCAD [13] and used for simulation of 3D FinFETs and GAA NWFETs. In this paper, a few selected results from earlier publications [10]-[13] will be reviewed.

NBTI Modeling Framework

Fig.1 explains the NBTI framework [10], [11]. The time evolution of ΔN_{IT} is calculated using the double interface H/H_2 Reaction-Diffusion (RD) model, and contribution from charged traps (ΔV_{IT}) is calculated by Transient Trap Occupancy Model (TTOM). The RD model calculates dissociation and passivation of interfacial Si-H bonds during and after stress and diffusion of Hydrogen (H and H_2) species. Breaking of Si-H bonds create donor like interface traps, and those above Fermi level contribute to ΔV_{IT}. Kinetics of ΔV_{IT} for long-time stress is dictated by molecular H_2 diffusion. ΔV_{IT} kinetics during recovery is due to trap passivation ($\Delta V_{IT-Slow}$, RD) and electron capture in traps ($\Delta V_{IT-Fast}$, TTOM) that go below the Fermi level. Contributions due to hole trapping (ΔV_{HT}) and bulk trap generation (ΔV_{OT}) are calculated by using empirical expressions. The kinetics of empirically calculated ΔV_{HT} is verified by physical models [14].

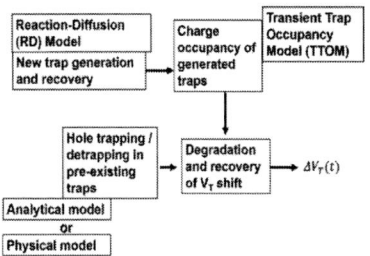

Fig.1 Schematic of NBTI framework [10]-[12].

Prediction of Measured Data

The time evolution of ultra-fast measured ΔV_T and model calculations are shown in Fig.2. During stress, early degradation is decided by ΔV_{HT} that saturates fast, and long-time degradation exclusively depends on ΔV_{IT} that dominates the end of life degradation, Fig.2 (a). Early part of recovery is due to ΔV_{HT} and $\Delta V_{IT-Fast}$, while $\Delta V_{IT-Slow}$ dictates long-time recovery, Figs.2 (b)-(d). The model can predict time evolution of ΔV_T during AC stress.

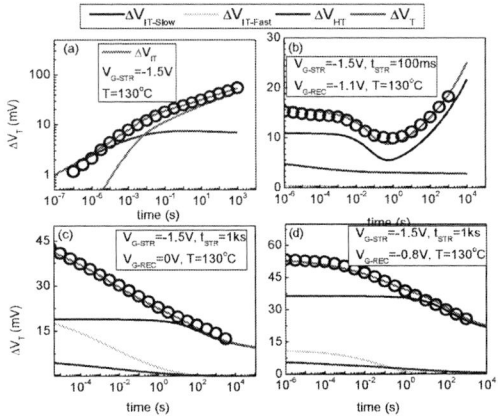

Fig.2 Time evolution of ΔV_T from measurements and TTOM enabled RD model with all the sub components for (a) DC stress, (b) recovery at -1.1V after 100ms stress, (c) recovery at 0V after 1ks stress and (d) recovery at -0.8V after 1k stress, from [11].

The f and PDC dependence of ΔV_T for Mode-A

978-1-5090-4661-4/17 $31.00 © 2017 IEEE

and Mode-B AC stress is shown in Fig.3. The kink near DC for PDC dependence and f (in) dependence can be readily predicted. Figs.4, 5 respectively show measured multi-cycle DC stress at different bias and AC stress at different bias and f. Model prediction and subcomponents are shown as well. Note, all model calculations are done by using consistent parameters for DC and AC stress. More examples of model prediction for DC and AC experiments under diverse conditions are shown in [10]-[12].

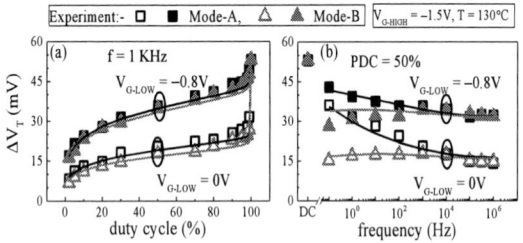

Fig.3 Measured data and model prediction of (a) PDC and (b) frequency dependence for different pulse low bias for Mode-A and Mode-B AC stress, from [10].

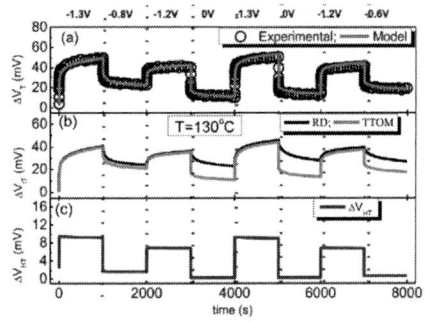

Fig.4 (a) Experimental (symbols) and model prediction of ΔV_T time evolution with constant stress time and varying V_G. (b) ΔV_{IT} from the RD and TTOM enabled RD model. (c) ΔV_{HT} time evolution; from [11].

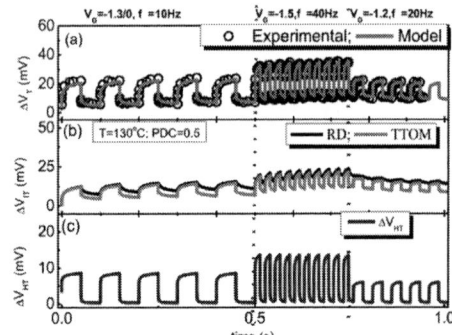

Fig.5 (a) Experimental (symbols) and model prediction of ΔV_T time evolution under AC with varying voltage and frequency. (b) ΔV_{IT} from RD and TTOM enabled RD model. (c) ΔV_{HT} time evolution; from [11].

Stochastic NBTI Framework

A Kinetic Monte Carlo based stochastic framework is also developed for small area devices [12], [15]. It calculates interface trap kinetics using stochastic RD model; electron capture in generated traps and hole trapping/detrapping in pre-existing traps using time constant dispersion. Mean of multiple simulations follow deterministic model for stress and recovery [12]. Fig.6 (a) shows stochastic recovery simulations for small area devices. Mean of measured recovery matches with mean simulation, Fig.6 (b). Note, RD only simulation does not capture the full physics and cannot be compared with measurements. Electron capture and hole detrapping processes are needed to predict measured data.

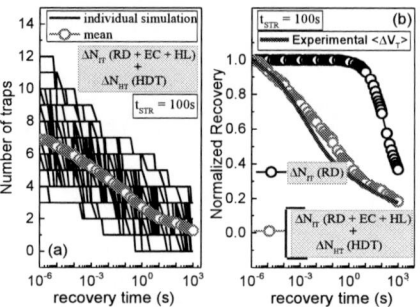

Fig.6 (a) Recovery simulations using KMC stochastic framework for V_{IT} and ΔV_{HT}. (b) Experimental data compared to complete stochastic simulaiton framework and basic RD only framework, from [12].

TCAD Implementation

The Multi State Configuration (MSC) - Hydrogen Transport degradation model in Sentaurus™ TCAD [15] is used to calculate ΔN_{IT} kinetics during stress and recovery. It self consistently solves the 3D H/H_2 RD equations with Poisson and quantum corrected inversion hole density to simulate ΔN_{IT} kinetics. The isometric views of devices used are shown in Fig.7. 3D simulations were done using realistic Back End of Line (BEOL) structure for Bulk and SOI FinFETs and GAA NWFETs [13]. The Low-κ material that surrounds the gate metal vias for isolation are not shown for better viewing.

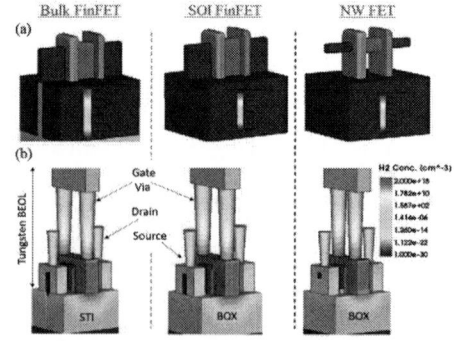

Fig. 7(a) 3D isometric view of various structures for which NBTI degradation was simulated (b) Snapshot of H_2 diffusion profile after stressing the devices for 1000s/130C. At long-term stress, H_2 diffuses through via and finally gets absorbed at the contacts, from [13].

978-1-5090-4661-4/17 $31.00 © 2017 IEEE

Fig.8 shows TCAD model calibration of measured data (ΔV_T is dominated by ΔV_{IT}) in bulk FinFETs at different stress overdrives and T. The device details can be found in [13]. Scaling of device dimensions in advanced nodes requires the fins/nanowire widths to be smaller for better sub-threshold characteristics. In such ultra-scaled dimensions (below 10nm), both electrostatic and geometrical quantum confinements become important. Fig.9 shows the implications of width scaling on NBTI. Results show that reducing fin and NW width increases NBTI degradation [13]. More TCAD results can be found in [13], [16].

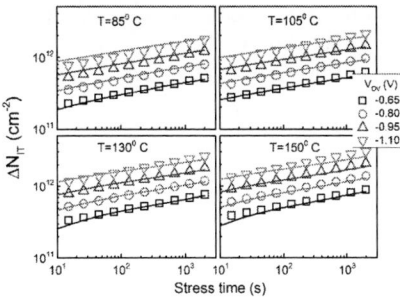

Fig.8 Calibration of TCAD parameters. Simulation of NBTI degradation (lines) and measured bulk FinFET degradation data (symbols) for different V_{GSTR} and T, from [13].

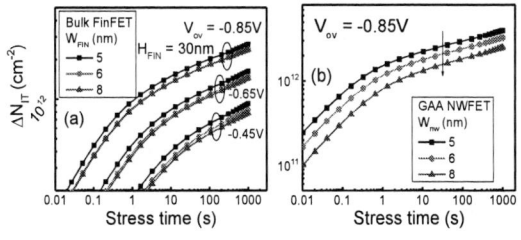

Fig.9 Simulated time evolution of (a) Bulk FinFET and (b) GAA NWFET with different widths stressed under different overdrive conditions, from [13].

Conclusion

A physics-based NBTI framework is demonstrated. It can predict diverse results for DC and AC stress and recovery experiments. It is fully consistent for large area deterministic and small area stochastic simulations. It is implemented in commercial TCAD for simulations of advanced technologies.

References

[1] S. Mahapatra *et al.*, "On the Physical Mechanism of NBTI in Silicon Oxynitride p-MOSFETs: Can Differences in Insulator Processing Conditions Resolve the Interface Trap Generation versus Hole Trapping Controversy?", in *Proc., Int. Rel. Phys. Symp (IRPS)*, p.1, 2007.

[2] K. Joshi *et al.*, "HKMG process impact on N, P BTI: Role of thermal IL scaling, IL/HK integration and post HK nitridation", in *Proc., Int. Rel. Phys. Symp (IRPS)*, p. 4C.2.1, 2013.

[3] G. Chen et al., "Dynamic NBTI of PMOS transistors and its impact on device lifetime", in *Proc., Int. Rel. Phys. Symp (IRPS)*, p. 196, 2003.

[4] D. Varghese *et al.*, "On the dispersive versus Arrhenius temperature activation of NBTI time evolution in plasma nitrided gate oxides: Measurements, theory and implications", in *Proc. IEEE Int. Electron Device Meeting (IEDM)*, p. 684, 2005.

[5] T. Grasser, "Stochastic charge trapping in oxides: From random telegraph noise to bias temperature instabilities", in *Microelectronics Reliability (MR)*, Vol. 52, Issue 1, Jan. 2012.

[6] T. Grasser *et al.*, "A two-stage model for negative bias temperature instability", in *Proc., Int. Rel. Phys. Symp (IRPS)*, p. 33, 2009.

[7] S. Mahapatra *et al.*, "A critical re-evaluation of the usefulness of R-D framework in predicting NBTI stress and recovery", in *Proc., Int. Rel. Phys. Symp (IRPS)*, p. 6A.3.1, 2011.

[8] S. Mahapatra *et al.*, "A comparative study of different physics-based NBTI models", in *IEEE Trans. Electron Devices (TED)*, vol. 60, no. 3, p. 901, Mar. 2013.

[9] S. Mahapatra *et al.*, "Universality of NBTI - From devices to circuits and products", in *Proc., Int. Rel. Phys. Symp (IRPS)*, p.3B.1.1, 2014.

[10] N. Goel *et al.*, "Combined trap generation and transient trap occupancy model for time evolution of NBTI during DC multi-cycle and AC stress," in *Proc., Int. Rel. Phys. Symp (IRPS)*, p. 4A.3.1, 2015.

[11] N. Parihar *et al.*, "A Modeling Framework for NBTI Degradation Under Dynamic Voltage and Frequency Scaling," in *IEEE Trans. Electron Devices (TED)*, vol. 63, no. 3, p. 946, Mar. 2016.

[12] A. Chaudhary *et al.*, "Consistency of the Two Component Composite Modeling Framework for NBTI in Large and Small Area p-MOSFETs," in *IEEE Trans. Electron Devices (TED)*, vol. 64, no. 1, p. 256, Jan. 2017.

[13] S. Mishra *et al.*, "TCAD-based Predictive NBTI Framework for Sub-20nm node Device Design Considerations," in *IEEE Trans. Electron Devices (TED)*, vol. 63, no. 12, p. 4624, 2016.

[14] N. Parihar *et al.*, "Resolution of Disputes Concerning the Physical Mechanism and DC-AC Stress/Recovery Modeling of Negative Bias Temperature Instability (NBTI) in p-MOSFETs," in *Proc., Int. Rel. Phys. Symp (IRPS)*, Apr. 2017.

[15] Synopsys Inc., CA, USA, Sentaurus™ Device User Guide (2016).

[16] S. Mishra *et al.*, "Predictive TCAD for NBTI Stress-Recovery for Different Device Architectures and Channel Materials", in *Proc., Int. Rel. Phys. Symp (IRPS)*, Apr. 2017.

6A-4

Accurate Mapping of Oxide Traps in Highly-Stable Black Phosphorus FETs

Yu.Yu. Illarionov,[1,2] G. Rzepa,[1] M. Waltl,[1] T. Knobloch,[1] J.-S. Kim,[3] D. Akinwande,[3] T. Grasser[1]

[1]TU Wien, Austria   [2]Ioffe Physical-Technical Institute, Russia   [3]The University of Texas at Austin, USA

Abstract – We examine *highly-stable black phosphorus field-effect transistors* and demonstrate that they can exhibit *reproducible characteristics for over ten months*. Nevertheless, we show that the performance of these devices is affected by thermally activated charge trapping in oxide traps. In order to characterize these important traps, we introduce a universal experimental technique which allows for *an accurate mapping of the defects with different time constants*. At room temperature the extracted oxide trap densities are close to those reported for more mature Si/SiO$_2$ devices.

Introduction

Black phosphorus (BP) is a crystalline 2D semiconductor which is now considered for applications in next-generation electronic devices [1]. In particular, a few successful attempts at fabricating black phosphorus FETs (BPFETs) have been reported recently [1–3]. However, so far the poor air-stability of these devices has not allowed any analysis of their reliability, which has only become possible due to the recent introduction of conformal capping [3].

The reliability of all 2D transistors investigated so far is reduced by charge trapping in oxide traps [4,5], which typically results in a hysteresis of the gate transfer characteristics [6,7] and threshold voltage shifts [8,9]. As such, the information about density and energetic alignment of these defects is of utmost importance, especially for such unexplored system as BP/SiO$_2$.

Devices and Experimental Technique

Our devices are few-layer BPFETs with an 80 nm thick SiO$_2$ insulator and conformal Al$_2$O$_3$ encapsulation (Fig. 1a). During ten months after fabrication, we have either performed intensive measurements or stored the devices in ambient conditions (Fig. 1b). The I_d-V_g characteristics measured at different time intervals remain similar (Fig. 1c). Remarkably, some drifts are observed only after several months of intensive measurements, while long storage in ambient conditions does not have any significant impact on the device performance. Furthermore, the consequently measured I_d-V_g characteristics in a vacuum and in ambient conditions are similar. All this confirms the high stability of our devices.

Aiming to map the oxide traps with widely distributed time constants and different energy levels, we suggest the following experimental technique: An elementary loop consists of measurements of the I_d-V_g characteristics in a vacuum in both forward (V^+) and reversed (V^-) sweep directions using a fixed sweep range V_{gmin} to V_{gmax} and different step voltages V_{step} and sampling times t_{step} [8]. As shown in Fig. 2, the full measurement procedure consists of repeating these loops using either, 1) a fixed $V_{gmin} = -20$ V and V_{gmax} varied from 0 to 20 V in 1 V steps or 2) a fixed $V_{gmax} = 20$ V and V_{gmin} varied from 0 to -20 V in -1 V steps. We express the results in terms of the hysteresis width in the electron/hole branches $\Delta V_{Hn|p}$ versus the measurement frequency $f = 1/(Nt_{step})$ [8] curves, which contain the information about spatial and energy distribution of the density of charged oxide traps.

Results and Discussion

In Fig. 3a we show the I_d-V_g characteristics measured at $T = 165\,°C$ using different sweep rates $S = V_{step}/t_{step}$. As S is decreased, the charge neutrality point V_{NP}^+ becomes more negative, which means that some defects become charged while passing through the hole conduction region. At the same time, hole emission which takes place while approaching V_{gmax} leads to a more

positive V_{NP}^-. As a result, the observed hysteresis is due to both capture and emission. Since both processes are thermally activated, the hysteresis width ΔV_{Hn} increases versus temperature (Fig. 3b). Furthermore, the total hysteresis width ΔV_{Hn} is a superposition of the threshold voltage shifts ΔV_{Tn}^+ and ΔV_{Tn}^- with respect to the fast sweep reference curve (Fig. 3c).

Fig. 4a shows the $\Delta V_{Tn}^+(f)$ and $\Delta V_{Tn}^-(f)$ curves measured using $V_{gmin} = -20$ V and V_{gmax} between 1 and 20 V. Clearly, the $\Delta V_{Tn}^-(f)$ curves follow the increase of V_{gmax} along which the number of traps which can emit a hole becomes larger. As such, the concentration of oxide traps which contribute between V_{gmax}^i and V_{gmax}^{i+1} is

$$N_{ot}^i(f) = \left(\Delta V_{Tn}^-(f, V_{gmax}^i) - \Delta V_{Tn}^-(f, V_{gmax}^{i+1})\right)\frac{C_{ox}}{q} \quad (1)$$

To contribute to the charge trapping processes, these traps should be situated within $d \approx 3$ nm from the BP/SiO$_2$ interface. Thus, the oxide trap density within the electron conduction region is

$$D_{ot}\left(\frac{V_{gmax}^i + V_{gmax}^{i+1}}{2}, f\right) = \frac{N_{ot}^i(f)}{d \mid V_{gmax}^{i+1} - V_{gmax}^i \mid} \quad (2)$$

The results for $V_{gmax} = 20$ V and V_{gmin} between 0 and -20 V are shown in Fig. 4b. Since V_{gmin} impacts the number of traps which can capture a hole, we use the $\Delta V_{Tp|n}^+(f)$ dependences for an analogous extraction of D_{ot} within the hole conduction region.

Taking into account the previously extracted position of the trap level E_T [9], we next evaluate the energy distributions $D_{ot}(E)$, see Fig. 5. At higher temperatures these distributions are more sensitive to the measurement frequency, while the trap density is larger. As such, our results are fully consistent with thermally activated charge trapping. At the same time, the shape of the $D_{ot}(E)$ curves is well reproducible at different temperatures. This confirms both the accuracy of our technique and the high stability of the BPFETs.

Finally, in Fig. 6 we compare the typical D_{ot} extracted for our BPFETs with literature values for different technologies [10–17]. At room temperature the density of active oxide traps in our devices is $\sim 10^{17}$ cm^{-3}/eV, which is close to Si/SiO$_2$ FETs. At the same time, this is considerably lower than for MoS$_2$/SiO$_2$ and Si/high-k devices (10^{19}–10^{20} cm^{-3}/eV). Although at $T = 165\,°C$ D_{ot} increases, the obtained values ($\sim 10^{19}$ cm^{-3}/eV) remain reasonable.

Conclusions

We have performed an accurate mapping of oxide traps with widely distributed time constants for our highly-stable BPFETs. Although the thermal activation of charge trapping is very important, we found that at room temperature the obtained values of the oxide trap density can be *comparable to Si technologies*. As such, we conclude that *a considerable advancement in the manufacturing and technology of next-generation 2D FETs has been obtained*.

Acknowledgements: MoRV project n° 619234 and FWF grant n° I2606-N30. D.A. acknowledges the support of the NSF and the ONR.

[1] H. Liu *et al.*, Nano **8**, 4033 (2014). [2] S. Das *et al.*, Nano **8**, 11730 (2014). [3] J.-S. Kim *et al.*, Sci. Rep. **5**, 1 (2015). [4] Y. Illarionov *et al.*, APL **105**, 143507 (2014). [5] Y. Guo *et al.*, APL **106**, 103109 (2015). [6] D. Late *et al.*, Nano **6**, 5635 (2012). [7] A.-J. Cho *et al.*, SSL **3**, Q67 (2014). [8] Y. Illarionov *et al.*, 2DM **3**, 035004 (2016). [9] Y. Illarionov *et al.*, ACS Nano (2016). [10] F. Wang *et al.*, SSE **45**, 351 (2001). [11] M. von Haartman *et al.*, ICNF (2003), p. 381. [12] E. Simoen *et al.*, T-ED **51**, 780 (2004). [13] B. Min *et al.*, APL **86**, 2102 (2005). [14] E. Simoen *et al.*, T-ED **60**, 3849 (2013). [15] J. Renteria *et al.*, APL **104**, 153104 (2014). [16] Z. Çelik Butler *et al.*, SSE **111**, 141 (2015). [17] L. Yuan *et al.*, Chin. Phys. B **24**, 088503 (2015).

978-1-5090-4661-4/17 $31.00 © 2017 IEEE

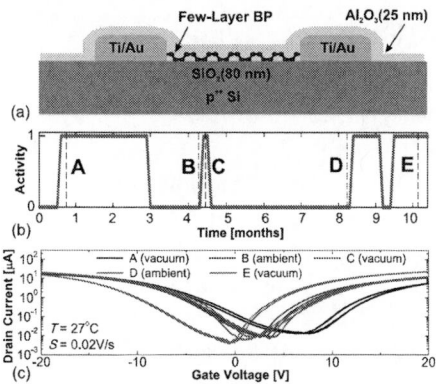

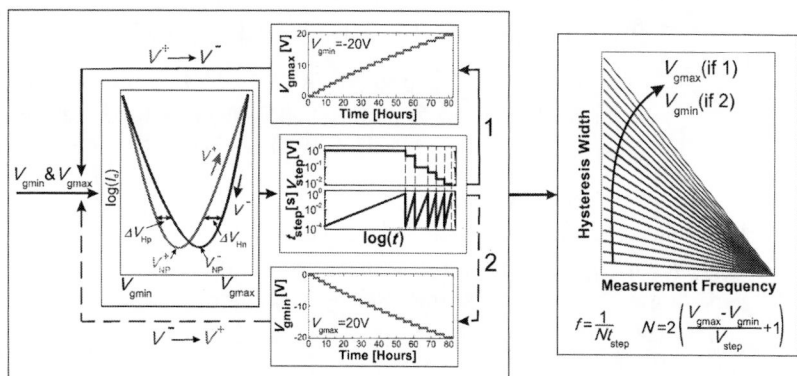

Fig. 1: (a) Schematic layout of our BPFETs ($L = 500$ nm) with Al$_2$O$_3$ encapsulation. (b) Measurement activity versus time since fabrication. "0" means that the devices have been stored in ambient conditions, while "1" expresses intensive measurements in vacuum (mostly stressing) up to $T = 165$°C. (c) The I_d-V_g characteristics measured for the same device at different stages of our long-term study (as marked in (b)).

Fig. 2: Schematic illustration of our experimental technique. We measure the I_d-V_g characteristics at $V_d = 0.1$ V in both sweep directions using the step voltage V_{step} range [1 V ... 0.01 V] and sampling time t_{step} varied between 0.2 ms and 0.5 s for each V_{step}, i.e. the sweep rate $S = V_{step}/t_{step}$ between 0.02 and 5000 V/s. The measurements are repeated with either $V_{gmin} = -20$ V and V_{gmax} varied between 0 and 20 V (1) or with $V_{gmax} = 20$ V and V_{gmin} varied between 0 and -20 V in -1 V steps (2). As a result we obtain two sets of the hysteresis width versus measurement frequency ($\Delta V_{Hn|p}(f)$) characteristics which contain the information about the distribution of the charged oxide traps with different time constants. Interestingly, although each measurement loop takes over 80 hours, our BPFETs remain highly stable.

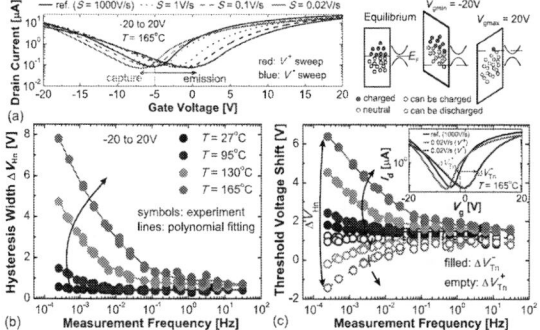

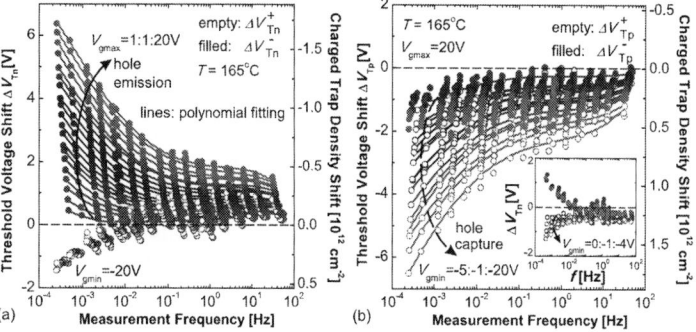

Fig. 3: (a) The I_d-V_g characteristics measured using different sweep rates. For slow sweeps, many defects can become charged within the hole conduction region. This makes V_{NP}^+ more negative than it was for the reference curve ($S = 1000$ V/s). At the same time, large V_{gmax} leads to a more positive V_{NP}^- due to a hole emission. (b) The $\Delta V_{Hn}(f)$ curves are consistent with thermally activated charge trapping. (c) The total hysteresis width can be split into the threshold voltage shifts ΔV_{Tn}^+ and ΔV_{Tn}^- (inset) with respect to the reference I_d-V_g curve.

Fig. 4: (a) The $\Delta V_{Tn}^+(f)$ and $\Delta V_{Tn}^-(f)$ characteristics obtained at $T = 165$°C for $V_{gmin} = -20$ V and V_{gmax} from 1 to 20 V. While for larger V_{gmax} the number of defects which can emit a hole becomes larger, the distances between the $\Delta V_{Tn}(f)$ curves are proportional to the concentrations of oxide traps which discharge within the corresponding V_g interval. (b) The $\Delta V_{Tp}^-(f)$ and $\Delta V_{Tp}^+(f)$ characteristics obtained for $V_{gmax} = 20$ V and V_{gmin} from -5 to -20 V. Since V_{gmin} determines the number of traps which can become charged, we analyze the distances between the $\Delta V_{Tp}^+(f)$ curves. However, since for V_{gmin} from 0 to -4 V ΔV_{Tp}^+ is not accessible, we have to use the $\Delta V_{Tn}^+(f)$ curves (inset).

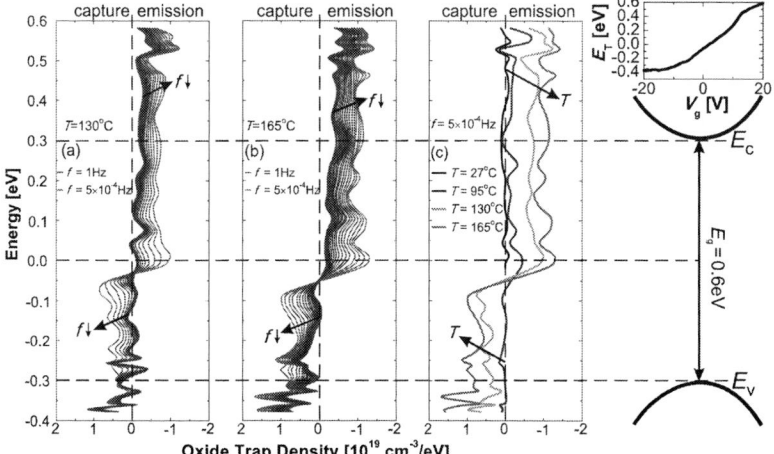

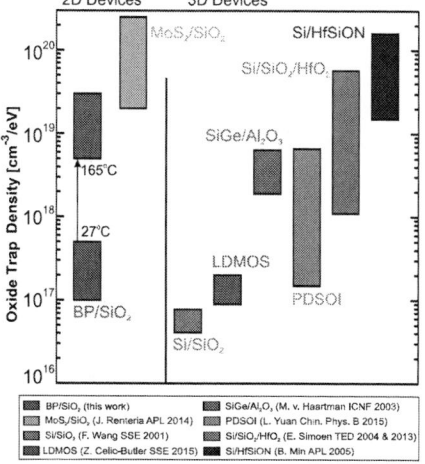

Fig. 5: The differential energy distributions $D_{ot}(E)$ for (a) $T = 130$°C and (b) $T = 165$°C; f is spaced logarithmically between 5×10^{-4} and 1 Hz. Since the originally extracted D_{ot} is given in cm^{-3}/V, we recalculated the gate voltage into the trap level E_T (top right plot) by considering the band-bending within 3 nm from the interface [9]. We can clearly discern the regions with dominating hole capture and emission. Both processes become more efficient for smaller f, while emission is more sensitive to f than capture, i.e. the emission times are more widely distributed. Also, at higher temperature the impact of f becomes more pronounced, which is consistent with thermal activation. (c) The energy distributions of slow oxide traps ($f = 5 \times 10^{-4}$ Hz) obtained for different temperatures. Due to thermal activation, the trap density becomes larger for higher temperatures, while the positions of the spikes on the $D_{ot}(E)$ curves remain fixed.

Fig. 6: Comparison of typical D_{ot} values for our BPFETs with literature reports for different technologies [10–17]. At room temperature the obtained densities of active oxide traps can be $\sim 10^{17}$ cm^{-3}/eV. This is lower than in MoS$_2$/SiO$_2$ and Si/high-k FETs, while being close to Si/SiO$_2$ devices. Taking into account the novelty of BP and recent issues with its stability, we found these values to be unexpectedly low. Nevertheless, thermal activation of oxide traps presents a crucial issue for BPFETs.

978-1-5090-4661-4/17 $31.00 © 2017 IEEE   115

6B-1 (Invited)

ESD Performance Enhancement Methodologies for CMOS Power Transistors

Mahadeva Iyer Natarajan, Jian-Hsing Lee

GLOBALFOUNDRIES Inc., 400 Stonebreak Rd. Extension, Malta, NY, 12020 USA,
mahadevaiyer.natarajan@globalfoundries.com

Abstract

Key challenges in providing ESD protection for High Voltage CMOS technology is presented in this paper. Based on that, various methodologies to make the high voltage power transistor ESD self-protecting without changing the device IV characteristics and dimension, for different HV technologies is outlined.
(Keywords: LNDMOS, ESD, HBM and NBL)

Introduction

Enabling ESD protection in semiconductor devices has been one of the challenges faced with ever changing application needs and technology scaling. While ESD protection challenges in logic low voltage nodes are well understood and managed, making high voltage (HV) laterally diffused NMOSFET (LNDMOS) electrostatic-discharge (ESD) robust without any special design rules is one of the biggest challenges today in the semiconductor device designs. Conventional schemes to enhance the ESD performances of the low-voltage device all cannot be used for the HV LNDMOS such as the gate-coupling circuit, substrate-triggering circuit, wide salicide-blocking region and embedded SCR. The HV LNDMOS designed with the special rule often cannot be used as the power transistor since it will increase the chip size and thus economically not a viable solution [1]. However, the very huge power transistors still might be unable to achieve the 2kV Human-body model (HBM) specification if it is designed with the minimum design rules (Fig. 1).

Fig. 1: Hot spot caused by the damage of the power transistor after a 2kV HBM event.

ESD Enhanced Methodologies

The drain of LDNMOS is composed of two N-type implants, heavily doped implant (N+) and lightly doped implant (N_{LD}), as shown in Fig. 2 [2]. If the stress current I_{ESD} is higher than the maximum current I_{NLD} of the N_{LD} at the pre-snapback region

(Fig. 3), the LDMOS will fail at the low-voltage HBM stress. On the contrary, the LDNMOS can pass high-voltage HBM stress. So, reducing the pre-snapback current I_{PRE-SB} by decreasing the device trigger voltage (Vt1) or increasing the I_{NLD} both can enhance the ESD performance of HV LDMOS.

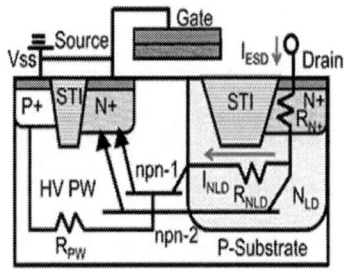

Fig. 2 Equivalent circuit of LDNMOS.

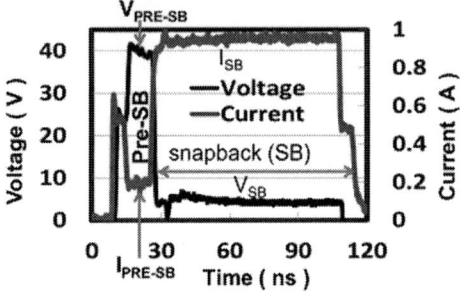

Fig. 3 Current and voltage waveforms of LDNMOS under 100ns TLP stress.

A. Additional NBL for drain

For silicon, the temperature to change from the positive coefficient resistor to negative coefficient resistor increases with the doping concentration. This implies that the I_{NLD} is enlarged with the doping concentration [3]. For some applications, the HV technologies often need to provide the n-type buried layer (NBL) to isolate the HV LDNMOS from the p-substrate. So, this layer can be used to change the doping concentration of the N_{LD} and provide the additional junction to increase its I_{NLD}, resulting in the device ESD performance improvement [4]. Fig. 4 shows the simulated currents of the 40V LDNMOS's with and without NBL. The current of the 40V LDNMOS without the NBL only can flows through the junction of the N_{LD}, while the current of the 40V with the NBL can flow not only through the junction of the N_{LD} but also through the junction of the NBL junction. This significantly decreases the current density of the 40V LDNMOS to make it that can sink more current. So, the 40V LDNMOS is damaged after the

978-1-5090-4661-4/17 $31.00 © 2017 IEEE   116

snapback if it is without the NBL, while it can sustain ~1.8A TLP current if it has the NBL on the drain (Fig. 5).

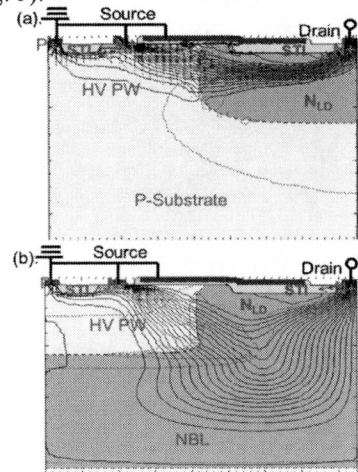

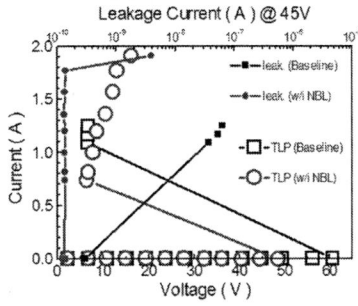

Fig. 4: Simulated currents for the baseline devices of 40V LDMOS's (a) without NBL, (b) with NBL.

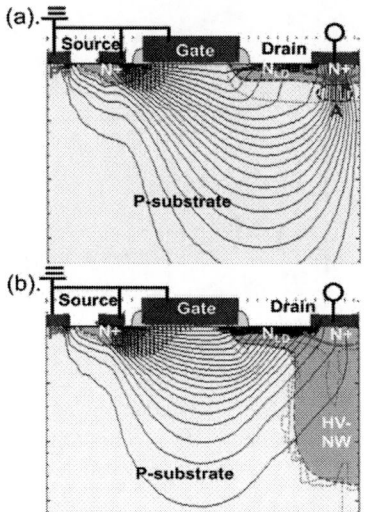

Fig. 6: Simulated currents for the baseline devices of 18V LDMOS's(a) without HV-NW, (b) with HV-NW.

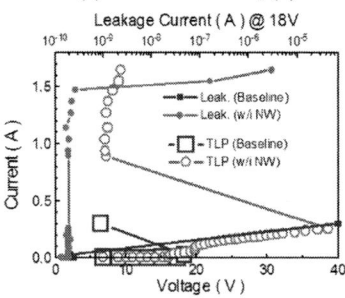

Fig. 7 TLP IV curves for the two devices shown in Fig. 6.

Fig. 5 TLP IV curves for the two devices shown in Fig. 4.

B. Additional HV-NW for drain

For lower cost processes, HV-NWell for HV LDPMOS can be used to enhance the ESD performance of HV LDNMOS. Fig. 6 shows the cross-sections of the 18V LDNMOS's with and without HV-NWell [5]. Different from the NBL, the HV-NWell cannot cover whole N_{LD} to prevent the device IV characteristics changing. Similar to the 40V LDNMOS, the 18V LDNMOS without the HV-NWell is damaged after the snapback (Fig. 7). This is caused by that the junction depth of the N_{LD} is too shallow and close to the N+ junction depth. The two implants make the electrical field of this region higher than other region without the N+ implant, resulting in the current crowded at a small region 'A' as shown in in Fig. 6a. With the HV-NWell, the current can uniformly distribute the junctions of the N_{LD} as well as HV-NWell (Fig. 6b). So, it is not damaged at the 1st snapback and can sustain ~1.1A TLP current (Fig. 7).

C. Active area optimization methodologies

In technologies where the LDNMOS fails at the low voltage ESD stress even it is buried inside the HV-NWell, drain design can be optimized with the many small N+ active area instead of a long N+ strap to build the ballast resistance. Alternately, the well tap can be designed with many small P+ active area instead of a long P+ strap to reduce the device Vt1 (Fig. 8) [1]. The two layout changes do not alter the device IV characteristics since the R_{NLD} is much larger than the R_{N+}. Fig. 9 shows that the new LDNMOS has the smaller Vt1 (~28V) and much higher It2 (2.8A) compared to that for the baseline LDNMOS (38V Vt1 and 0.5A It2).

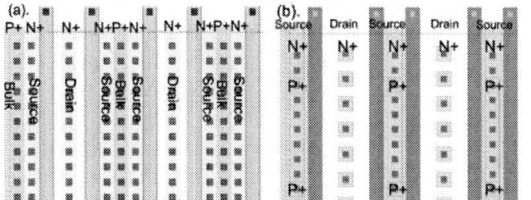

Fig. 8: Layouts for 15V LDMOS's (a) baseline device, (b) new device.

978-1-5090-4661-4/17 $31.00 © 2017 IEEE   117

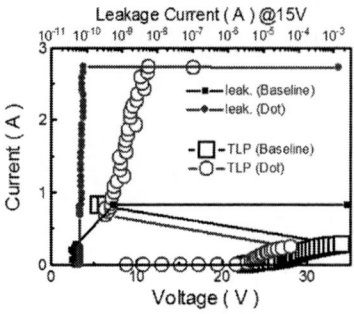

Fig. 9 TLP IV curves for the two devices in Fig. 8.

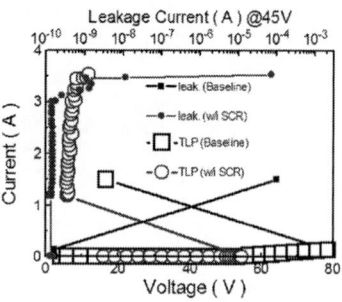

Fig. 11 TLP IV curves for the two devices in Fig. 10.

D. Building the SCR on the guard-rings

Emergence of many HV technologies using the double reduced-surface field (RESURF) structure, to reduce the R_{DS} of LDNMOS, introduces additional challenges to enable 2kV HBM performance, even designed with the optimized structure as shown in Fig. 8b. The double RESURF LDNMOS is often designed with many guard-rings (Fig. 10) and buried inside the NBL except the outermost P+ guard-ring. Careful layout optimization with different types of implants added into the two inner guard-rings (Fig. 10b), SCRs can be realized. Different from the embedded SCR which inserting P+ implant into the drain, this modification does not increase the LDNMOS dimension since the widths of the two guard-ring are large enough to insert another type implants in them [6]. Fig. 11 shows that the LDNMOS can be improved from ~0.1A to 3A once the SCR is built on the guard-rings.

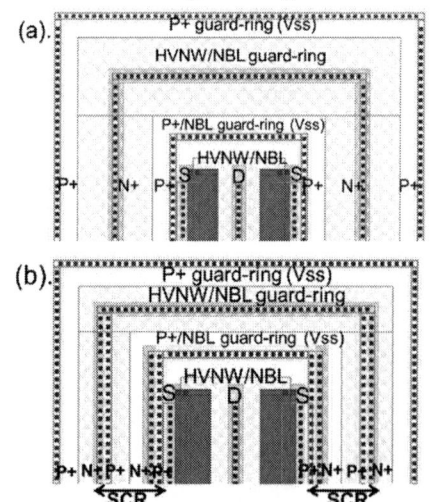

Fig. 10: Layouts for the baseline devices of double RESURF 40V LDMOS(a) without SCR, (b) with SCR on the guard-rings.

Conclusion

Methodologies involving prevention of the LDNMOS damage at the pre-snapback region is the key to make self-protecting power transistor devices. This can be implemented by increasing the I_{NLD} or decreasing the ESD stress current of the LDNMOS at the pre-snapback region such as adding the NBL to the drain, adding HV-NWell to the drain, using many distributed N+ implants instead of a long N+ implant for the drain, and building the SCR on the existing guard-rings. These methodologies all can enhance the device ESD performance without changing the device dimension and IV

References

[1] J. H. Lee, H. D. Su, C. L. Chan, D. H. Yang, J. F. Chen, and K. M. Wu, "The Influence of the Layout on the ESD Performance of HV-LDMOS," *Proc. 22th* ISPSD, 2010, pp. 303-306.

[2] J. H. Lee, Natarajan M. Iyer, Ruchil Jain and M. Prabhu, "Predictive High Voltage ESD Design Methodology," Proc. EOS/ESD Symp., 2016, 1A.3

[3] V. De Heyn, G. Groeseneken, B. Keppens, M. Natarajan, L. Vacaresse, G. Gallopyn, "Design and Analysis of New Protection Structures for Smart Power Technology and Controlled Trigger and Holding Voltage ," Proc. IRPS 2001, pp. 253-258.

[4] J. H. Lee, S. H. Chen, Y. T. Tsai, D. B. Lee, F. H. Chen, W. C. Liu, C. M. Chung, S. L. Hsu, J. R. Shih, Alan Y. Liang, and K. Wu, "The Influence of NBL Layout and LOCOS Space on Component ESD and System Level ESD for HV-LDMOS," *Proc. 19th* ISPSD, 2007, pp. 173-176.

[5] J. H. Lee, J. R. Shih, T. C. Ong, and K. Wu, "Low-Side Driver's Failure Mechanism in a Class-D Amplifier under Short-Circuit Test and a Robust Driver Device," *Proc.* IRPS, 2010, pp. 182-187.

[6] C. H. Wu, J. H. Lee, and C. H. Lien, "A New Low-Voltage Triggering SCR for the Protection of a Double RESURF HV-LDMOS," *IEEE Electron Device Lett.*, 2016, pp. 1201-1203.

6B-2

Aging Simulation of SiC-MOSFET in DC-AC Converter

Kenshiro Sato[1], Shinya Sekizaki[2], Dondee Navarro[3], Yoshifumi Zoka[2], Naoto Yorino[2],

Hiroshi Zenitani[2], Hans Jürgen Mattausch[1,3], and Mitiko Miura-Mattausch[3]

[1]Graduate School of Advanced Sciences of Matter, Hiroshima University, Japan,   m155449@hiroshima-u.ac.jp

[2]Graduate School of Engineering, Hiroshima University, Japan

[3]HiSIM Research Center, Hiroshima University, Japan

Abstract

This investigation focuses on the aging simulation of a DC-AC converter during the stress situation of a lightning strike. The newly developed compact model HiSIM_HSiC for high-voltage SiC MOSFETs, which considers the carrier-trap increase, is applied for the simulation. Simulation results reproduce the measured converter characteristics during the conventional earth fault protection of the DC-AC converter. It is verified that the device aging occurs in spite of this protection. The modeled device aging after the converter has endured several lightning strikes predicts enhanced efficiency reduction by more than 10% in addition to the ordinary usage.
(Keywords: SiC-MOSFET, DC-AC converter, Lightning strike, Device aging, Efficiency)

Introduction

The diversity of DC-AC converter applications is increasing due to the development of different kinds of power supply systems. These systems are usually protected against failure under unexpected stress situations, such as protection by earth-fault connection in the case of lightning strikes. Lightning is an electrostatic discharge during a thunder-storm weather condition, which is known to severely affect electrical and electronic devices. Here the question arises, whether the devices are sufficiently protected from degradation due to a lightning strike by such earth-fault connections. To clarify the device reliability, we have developed an aging model enabling to simulate the device degradation under prolonged lightning-strike conditions [1] and evaluated the effect on converter efficiency.

Circuit Simulation and Device Aging Prediction Results

The studied three phase DC-AC converter, shown in Fig. 1, consists of SiC-MOSFETs and SiC free-wheeling diodes with the bias condition and parasitic component values shown in Table 1. To investigate the reliability of the circuit performance, HiSIM_HSiC [2], a newly developed compact SiC-MOSFET model, has been applied. The model solves the Poisson equation explicitly to allow inclusion of MOSFET-internal changes by the trapped carrier

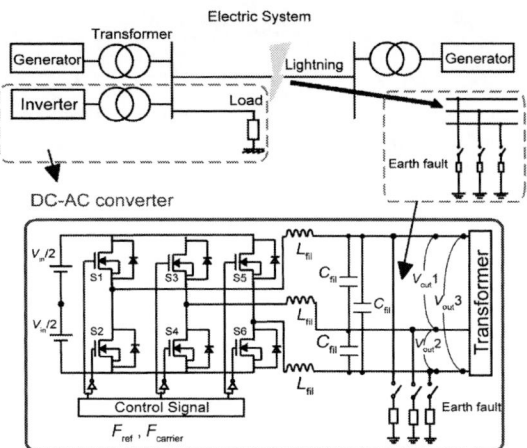

Fig. 1. DC-AC converter circuit with earth fault protection.

Table. 1. Operation conditions and the passive component.

V_{in}	450V
V_{out}	200V$_{rms}$
V_{gs}	10V
C_{fill}	100nF
L_{fill}	39mH
F_{ref}	60Hz
$F_{carrier}$	20kHz

density increase during stress applied. The top of Fig. 1 shows the experimental environment to obtain the measurement data reproducing the lightning strike accident. The lower part is the studied DC-AC converter applied for the experiment.

Fig. 2a shows the measured characteristics of the DC-AC converter, where the lightning-strike conditions are applied for 70ms at t=0.1s. It is seen that the converter recovers after the removal of this lightning stress at t=0.16s. The lightning strength and duration give a large effect on device aging. Simulation study is performed with HiSIM_HSiC. Fig. 3a shows the studied UMOS device structure. Fig. 3b compares the measured drain current I_{ds} as a function of drain voltage V_{ds} and gate voltage V_{gs} to the model calculation results with HiSIM_HSiC. Fig. 2b shows the simulation result of the circuit shown in Fig. 1, where measurement characteristics are well reproduced.

978-1-5090-4661-4/17 $31.00 © 2017 IEEE

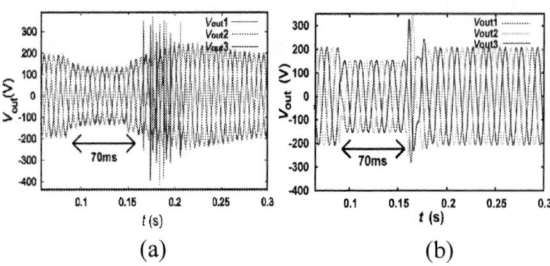

Fig. 2. (a) Measured DC-AC characteristics during earth fault. (b) Simulated DC-AC characteristics using HiSIM_HSiC model.

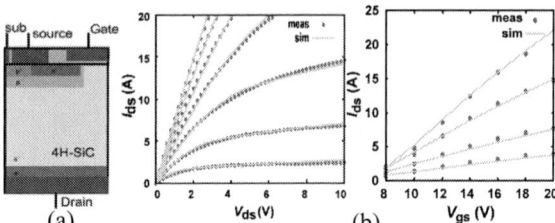

Fig. 3. (a) SiC-MOSFET device structure. (b) HiSIM_HSiC model results fitted to measurements.

Table. 2. Comparison of operation condition under normal and lightning condition

	Vd (V)	Id (A)
normal	0.5	2.3
lightning	139	37

Fig. 3. Simulated switching characteristics during earth fault.

To predict device aging during the lightning stress, we investigate each individual SiC MOSFET within the circuit shown in Fig. 1. Fig. 3 depicts the simulated effective bias-stress conditions of the SiC MOSFET during the earth fault around t=0.12s (see Fig. 2b). During the lightning, V_{ds} and I_{ds} reach up to 139V and 37A, respectively. These values are much higher than those of the normal operation condition, where only up to 0.5V for V_{ds} and 2.3A for I_{ds} occur (see Table. 2). The lightning-strike stress usually continues up to 100ms order, which can be expected to cause severe device aging [1].

The origin of the device aging is hot electrons

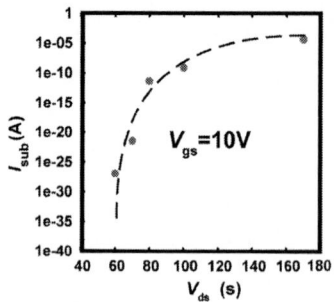

Fig. 4. I_{sub} characteristics extracted from 2D device simulation.

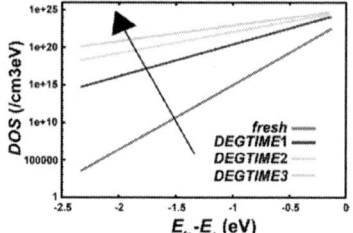

Fig. 5. DOS characteristics as I_{subt} increases.

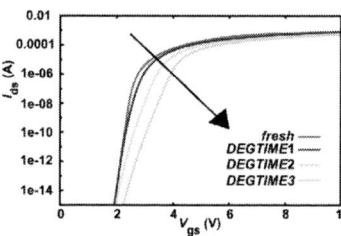

Fig. 6. Effect of I_{subt} on SiC MOSFET I_{ds}-V_{gs} degradation.

induced by high fields, which causes carrier trapping. Aging investigation is usually performed under enhanced stress conditions with different stress durations, DEGTIME. To model the device aging, the substrate current I_{sub} is utilized as the measure to describe the trap increase [3]. The integrated I_{sub} over the operation time enables to consider the stress dynamic condition change, which can be written as

$$I_{subt} = \int_{t0}^{DEGTIME} I_{sub}(V_{gs}(t), V_{ds}(t), V_{bs}(t))dt \quad (1)$$

where I_{subt} is the integrated I_{sub}.

To illustrate the capability of the aging model, I_{sub} characteristics is extracted from 2D-device simulation as shown in Fig. 4. The resulting trap density increase for different DEGTIME values is shown in Fig. 5. Fig. 6 depicts I_{ds}-V_{gs} aging characteristics for different DEGTIME values with

978-1-5090-4661-4/17 $31.00 © 2017 IEEE   120

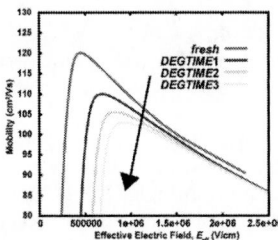

Fig. 7. Carrier mobility characteristics as a function of the vertical electric field.

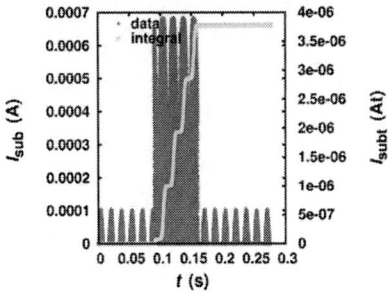

Fig. 8. I_{sub} and its integration I_{subt} during earth fault.

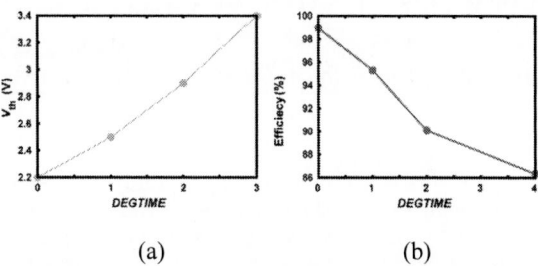

(a)   (b)

Fig. 9. (a) V_{th} shift as a function of lightning-stress duration. (b) DC-AC efficiency degradation as a function of lightning-stress duration.

While such aging simulation becomes possible with the developed aging model for HSiC, enhanced substrate current is expected for SiC-based devices because of the high-field channel which accelerates carriers and lead to greater impact ionization in the substrate, in addition to the contribution from junction breakdown for UMOS structure. As such, accurate I_{sub} measurement is necessary.

Conclusion

We have successfully simulated the efficiency aging of a SiC-MOSFET-based DC-AC converter during normal condition and earth faults due to lightning strikes by modeling the SiC-MOSFET aging based on the hot-carrier-induced trap-density increase.

Acknowledgment

The authors are grateful to the Core Research for Evolutional Science and Technology (CREST) program of the Japan Science and Technology Agency (JST) for supporting this research.

References

[1] S. B. Naderi et al., "Efficient fault ride-through scheme for three phase voltage source inverter-interfaced distributed generation using DC link adjustable resistive type fault current limiter," Renewable Energy. Vol.92, July 2016.

[2] Y. Tanimoto et al., "Power-Loss Prediction of High-Voltage SiC-MOSFET Circuits with Compact Model Including Carrier-Trap Influences," IEEE Trans. Power Electron., vol.31, No.6, June 2016.

[3] H. Tanoue et al., "Universal Properties and Compact Modeling of Dynamics Hot-Electron Degradation in n-MOSFETs," IEEE IRPS, Montrey, 2013.

[4] A. Saito et al., "Efficiency Analysis of SiC-MOSFET-Based Bidirectional Isolated DC/DC Converters" IEICE Trans. Electron., vol.E990C, No.9, September 2016.

[5] C. Ma et al., "Compact Reliability Model for Degradation of Advanced p-MOSFETs Due to NBTI and Hot-Carrier Effects in the Circuit Simulation" IEEE IRPS, Montrey, 2013.

increased densities shown in Fig. 5. It is seen that the increase of trap densities caused subthreshold shift as well as on-current reduction. The main reason for the current reduction is attributed to the mobility reduction shown in Fig. 7, showing the carrier mobility as a function of effective electric field E_{eff}, which makes V_{th} and R_{on} larger [4].

Fig. 8 depicts an example of I_{sub} and its integrated value I_{subt} during the first 0.3s shown in Fig. 2b. I_{subt} drastically increased during the earth fault. Under normal converter operation, devices are stressed by hot carriers as well. Even though this normal stress is not drastic, it can also introduce a non-negligible effect, e.g. a threshold-voltage shift.

The threshold-voltage shift as a function of DEGTIME is depicted in Fig. 9a. Fig. 9b shows the corresponding decline of the DC-AC converter efficiency by 12% due to device aging [5]. Simulated converter efficiency is defined as

$$efficiency = \frac{P_{out}}{P_{in}} = \frac{V_{out} \times I_{out}}{V_{in} \times I_{in}} \qquad (2)$$

where V_{out}, I_{out}, and V_{in}, I_{in} are output and input powers as indicated in Fig. 1, respectively [4]. Long term accumulation of I_{subt} make the threshold-voltage shift and worse the circuit performance. It can be expected that this decline would be further enhanced, if the lightning stress occurs many time.

6B-3

Degradation Caused by Negative Bias Temperature Instability Depending on Body Bias on NMOS or PMOS in 65 nm Bulk and Thin-BOX FDSOI Processes

Ryo Kishida and Kazutoshi Kobayashi

Department of Electronics, Graduate School of Science and Technology, Kyoto Institute of Technology

Abstract

Reverse Body Bias (RBB) control on Fully Depleted Silicon On Insulator (FDSOI) with thin Buried OXide (BOX) layer mitigates power consumption on the standby mode. However, Degradation caused by Negative Bias Temperature Instability (NBTI) is changed by RBB. We measure aging degradation of ring oscillators by applying RBB to NMOS or PMOS. In bulk, RBB to PMOS suppresses NBTI-induced degradation because increasing threshold voltage reduces carriers in channel. However, RBB to NMOS does not suppress NBTI-induced degradation because Positive BTI (PBTI) is not dominant in NMOS. In FDSOI, RBB to not only PMOS but also NMOS suppresses NBTI-induced degradation because BOX layer intercepts carriers to flow to substrate.

I. Introduction

In recent years, low power consumption is mandatory to reduce power density and to prolong battery life. Thin Buried OXide (BOX) Fully-Depleted Silicon On Insulator (FDSOI) is one possible candidate to operate in the low voltage of 0.4 V [1]. Reverse Body Bias (RBB) can reduce power consumption further more. However, degradation caused by Negativec Bias Temperature Instability (NBTI) is changed by RBB [2], [3]. NBTI causes aging degradation and become significant concern in nano-scaled devices [4]. The same amount of RBB is generally applied to NMOS and PMOS. However, it requires additional voltage source for RBB to NMOS. Our purpose is to optimize RBB voltage in NMOS and PMOS to suppress NBTI-induced degradation. We measure aging degradation of Ring Oscillators (ROs) by periodically counting the number of oscillation during stress conditions.

II. BTI-induced Degradation by RBB

Fig. 1 shows how to control body bias. N-well in PMOS and P-well in NMOS are connected to source terminals on the standard mode as shown in Fig. 1 (a). Fig. 1 (b) shows the RBB conditions. N-well and P-well are biased to positive and negative for source terminals of PMOS and NMOS respectively. Fig. 2 shows an example how to use RBB. RBB suppresses power consumption on the stand-by mode because threshold voltage (V_{th}) increases and leakage current decreases. Body bias cannot be controlled in SOI with conventional thick BOX. However, thin-BOX FDSOI can control the body bias as shown in Fig. 3 [1]. We evaluate BTI-induced degradation by changing RBB in bulk and thin-BOX FDSOI with 10 nm BOX layers. RBB suppresses aging degradations and V_{th} increases with time as voltage and temperature increase [4]. The atomistic trap-based BTI (ATB) model [5] is one of proposed theories of BTI as shown in Fig. 4. BTI occurs and V_{th} increases since defects in a gate oxide trap carriers as shown in Fig. 4 (a). The number of carriers in channel decreases by applying RBB as shown in Fig. 4 (b). Probability to trap carriers decreases because the number of carriers induced in the channel decreases. Therefore, it becomes harder to trap carriers to the defects as RBB increases. There are Negative BTI (NBTI) and Positive BTI (PBTI). NBTI occurs on PMOS when gate-source voltage is negative. Likewise, PBTI is observed in NMOS especially in technologies with high-k gate dielectrics [6].

III. Measurement Setup

We fabricated a chip including 11-stage ROs composed of NOR gates to dominate NBTI as shown in Fig. 5. PMOS of NORs are stressed by NBTI when ENB is high to stop oscillation. Fig. 6 shows our measurement flow. During frequency measurement, ENB becomes low only for 28 μs and RBB is restored to 0 V to oscillate in the same condition. RBB is applied while ROs stop oscillation for 50 s. ROs suffer from NBTI in the low-power mode. NBTI measurement condition is at 1.6 V power supply voltage and 120 °C temperature to accelerate NBTI-induced degradation. Fig. 7 shows the test chip fabricated in 65 nm bulk and thin-BOX FDSOI processes. Layout patterns are same in both processes.

IV. Measurement Results

Fig. 8 shows measurement results by the body bias. X-axis is stress time and Y-axis is degradation rate based on initial frequencies. Dots represent average of measurement data and curves are drawn by the fitting function along $S_{NBTI} \log(t + 1)$. S_{NBTI} is the fitting parameter indicating degradations caused by NBTI. This function comes from the ATB model since defects have a time constant distributed uniformly on the log scale from 10^{-9} s to 10^9 s [7]. Fig. 8 (a) and (b) show the measurement results in bulk when RBB is applied to NMOS or PMOS. V_{BN} and V_{BP} are defined body bias of NMOS and PMOS respectively. Difference of degradation rate at 0 and 1 V of V_{BN} is only less than 0.05%. While in PMOS, degradation rate decreases by 38% when V_{BP} of 1 V is applied. NBTI-induced degradation is suppressed by applying V_{BP} in bulk because NBTI is more dominant than PBTI in 65 nm process. Fig. 8 (c) and (d) show the measurement results in FDSOI when V_{BN} or V_{BP} are applied. Degradation rate decreases as RBB increases in RBB of both MOSFETs. We assume oscillation frequencies increase because carriers are captured in channel by RBB. BOX layer intercepts carriers to flow to substrate. We also evaluate degradation factor S_{NBTI} from fitting functions. Fig. 9 (a) and (b) show S_{NBTI} in bulk by applying V_{BN} or V_{BP}. Fig. 9 (c) and (d) show those in FDSOI. S_{NBTI} are almost constant in any V_{BN} of bulk, which means NBTI is not suppressed by V_{BN} in bulk. Likewise, S_{NBTI} decreases by 40% at V_{BP} of 1 V. S_{NBTI} in FDSOI is about twice as large as that in bulk because V_{th} of FDSOI is lower than that of bulk. Electric field in the gate oxide is higher in lower V_{th} and aging degradation is accelerated. In FDSOI, S_{NBTI} decrease as RBB on both MOSFETs and are almost equivalent. Applying RBB on PMOS is enough to suppress NBTI-induced degradation without RBB on NMOS and negative voltage sources are not required.

V. Conclusion

We fabricated ring oscillators in 65 nm bulk and thin-BOX FDSOI processes and measured aging degradation applied RBB to NMOS or PMOS. RBB on NMOS cannot suppress degradation in bulk, while RBB on PMOS, the degradation is suppressed. The degradation factor decreases by 40% from 0 to 1 V of RBB on PMOS. In FDSOI, the degradation factor decreases by RBB on both NMOS and PMOS. Applying RBB to PMOS is enough to suppress NBTI-induced degradation.

978-1-5090-4661-4/17 $31.00 © 2017 IEEE

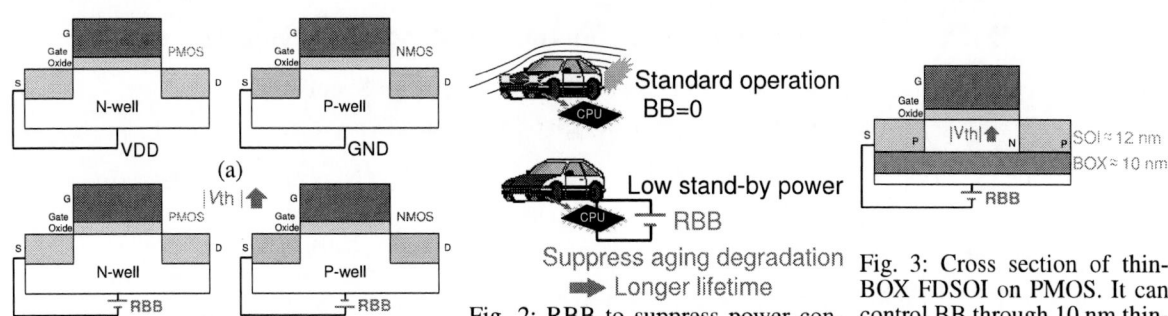

Fig. 1: Body bias (BB) control. (a) no BB and (b) reverse BB (RBB).

Fig. 2: RBB to suppress power consumption and aging degradation for chips on cars.

Fig. 3: Cross section of thin-BOX FDSOI on PMOS. It can control BB through 10 nm thin-BOX layers.

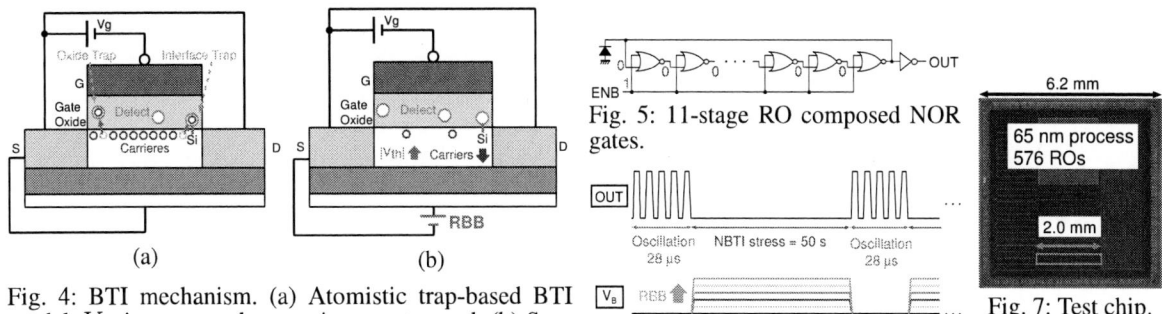

Fig. 4: BTI mechanism. (a) Atomistic trap-based BTI model. V_{th} increases when carriers are trapped. (b) Suppression of NBTI by RBB. V_{th} increases and the number of carries decrease.

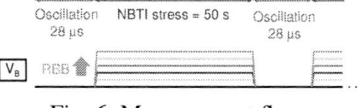

Fig. 5: 11-stage RO composed NOR gates.

Fig. 6: Measurement flow.

Fig. 7: Test chip.

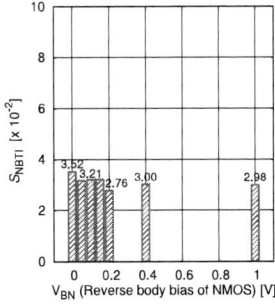

(a) NMOS RBB in bulk.

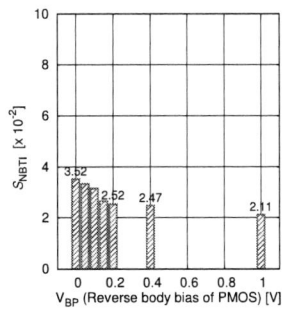

(b) PMOS RBB in bulk.

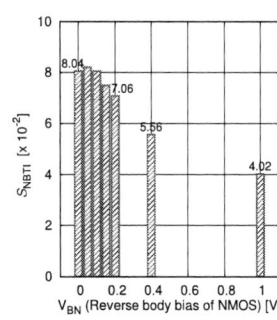

(c) NMOS RBB in FDSOI.

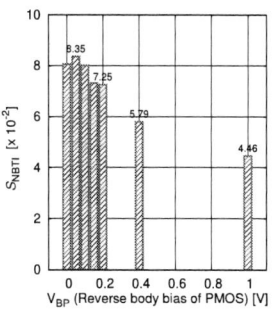

(d) PMOS RBB in FDSOI.

Fig. 8: Measurement results of NBTI.

(a) NMOS RBB in bulk.

(b) PMOS RBB in bulk.

(c) NMOS RBB in FDSOI.

(d) PMOS RBB in FDSOI.

Fig. 9: Degradation factor S_{NBTI}.

Acknowledgments

We thank Prof. Kumashiro for valuable comments. The chip for this work was fabricated by Renesas Electronics and designed by utilizing the EDA system supported by the VLSI Design and Education Center (VDEC), the University of Tokyo in collaboration with Synopsys Inc., Cadence Design System and Mentor Graphics Inc.

References

[1] R. Tsuchiya et al., *IEDM*, pp. 631-634, 2004.
[2] R. Kishida et al., *SSDM*, pp. 711-712, 2016.
[3] J. Franco et al., *IEDM*, pp 18.5.1-18.5.4, 2011.
[4] V. Huard et al., *IRPS*, pp. 289-300, 2008.
[5] H. Kukner et al., *IEEE Trans. on Dev. & Mat. Rel.*, pp. 182-193, 2014.
[6] S. Zafar et al., *VLSI Tech.*, pp. 23-25, 2006.
[7] B. Kaczer et al., *IRPS*, pp XT.3.1-XT.3.5, 2011.

978-1-5090-4661-4/17 $31.00 © 2017 IEEE

6B-4

JFETIDG: A Compact Model for Independent Dual-Gate JFETs

Kejun Xia, Colin C. McAndrew, and Hanyu Sheng

NXP Semiconductors, Tempe AZ, USA (kejun.xia@nxp.com)

Abstract

This paper presents a new compact model, JFETIDG, for independent dual-gate JFETs. The model is applicable to JFETs with any combination of p-n junction or MOS gates, captures geometry and temperature dependencies. As a special case, it can model junctionless MOSFETs. The model is verified by comparison to experimental and TCAD data. Verilog-A code for the model is available in the public domain.

(Keywords: JFET, device modeling, SPICE)

Introduction

Recently, we generalized the dual-gate JFET concept to include MOS as well as p-n junction gates, and for such devices derived a core I_{ds} model that is valid in all regions of operation [1-2]. Based on this we developed a complete large-signal model, JFETIDG, which we present here. The model improves on existing work in that it is physically based, in contrast to the empirical second gate model of [3], and does not approximate the depletion pinching effect using linearization, as is done in [4]. It also includes self-heating, which is important but is ignored in [3-4], and has a non-singular velocity saturation model.

Our model has application beyond JFETs, it can be used to model: the collector resistance of 4-terminal vertical bipolar transistors; the drift region of high voltage MOS transistors with field plates; resistors (diffused or polysilicon) with metal shields; and recently developed junctionless MOSFETs [5].

Details of the Model

JFETIDG follows the structure of the R3 model for single gate JFETs [6], with the addition of a second gate. Fig. 1 shows the large-signal equivalent circuit. Self-heating is modeled via an RC thermal network. The core I_{ds} model uses the exact solution for depletion pinching modulation of the channel conductance [1-2]. Two changes to the formalism were made. First, we modified $V_{s,e}$ (eq. (16) in [2]) to add fitting flexibility in pinchoff and transition regions as

$$V_{s,e} = V_{sp} - n\phi_t \ln\left[1 + \exp\left(\frac{V_{sp} - V_s}{n\phi_t} M_{OFF}\right)\right] f_{voff} \tag{1}$$

where

$$f_{voff} = \frac{1 + \exp\left[\frac{(V_S - V_{sp})M_{OFF} + V_{OFF}}{n\phi_t}\right]}{M_{OFF} + \exp\left(\frac{V_S - V_{sp}}{n\phi_t}\right)} \tag{2}$$

V_{OFF} and M_{OFF} are model parameters. Second, we replaced V_{ds} in the channel length modulation (CLM) model (eq. (17) in [2]) with a "smoothed" V_{dsx}

$$V_{dsx} = \sqrt{V_{ds}^2 + 0.01} - 0.1. \tag{3}$$

The key parameters for I_{ds} are dfb, dft, psirb, psirt, and rsh. We provide two ways to calculate these, as Fig. 2 shows. If the model parameters dfbinf, dftinf, psirbinf, psirtinf, and rsh are given they are used directly; otherwise they are calculated based on specified physical parameters (layer doping levels, oxide thickness and flatband voltage for MOS gates, and channel mobility).

Model switches, parameters and their nomenclature, parasitic gate diode formulation, and geometry and temperature dependencies closely follow those of R3. Unlike R3, MKS units are used exclusively.

To verify the model, we created p-channel dual p-n junction gate JFETs in an SOI technology. Deep trench isolation was used to decouple the gates.

Figs. 3 and 4 show I_d vs. V_{gt} (for fixed V_{gb}) at low V_{ds} on log and linear ordinate axis scales, respectively, for a device with W/L=50μm/20μm. Fig. 5 shows g_m vs. V_{gt} for the same data. The model fits the measured data well.

Fig. 6 shows I_d vs. V_{gt} (for $V_{gb}=V_{gt}$) at different V_d, for a W/L=20μm/5μm device. Clearly, DIBL is also modeled well. Figs. 7 and 8 show I_d vs. V_d and g_o vs. V_d, respectively, for the same W/L=20μm/5μm device. The accuracy of modeling CLM, and the smooth transition between non-pinchoff and drain-pinchoff behavior, is apparent.

Conclusion

We have developed a complete compact model for independent dual-gate JFETs and have verified the model against experimental data. Verilog-A code for JFETIDG, under the NEEDS license, is publically available.

Acknowledgments

The authors gratefully acknowledge the contributions of Bernhard Grote for design of the JFETs.

References

[1] K. Xia, C. C. McAndrew, and B. Grote, *IEEE T-ED*, vol. 63, no. 4, pp. 1408-1415, Apr. 2016.

[2] K. Xia, C. C. McAndrew, and B. Grote, *IEEE T-ED*, vol. 63, no. 4, pp. 1416-1422, Apr. 2016.

[3] S. Banas et al., *Solid-State Electron.*, vol. 123, pp. 133-142, Sep. 2016.

978-1-5090-4661-4/17 $31.00 © 2017 IEEE

[4] W. Wu, S. Banerjee, and K. Joardar, *Proc. IEEE ICMTS*, pp. 37-41, 2015.

[5] X. Lin et al., *IEEE T-ED*, vol. 63, no. 3, pp. 959-965, Mar. 2016.

[6] C. C. McAndrew, in *Compact Modeling:*

Principles, Techniques and Applications, G. Gildenblat, Ed., Springer, pp. 271-297, 2010.

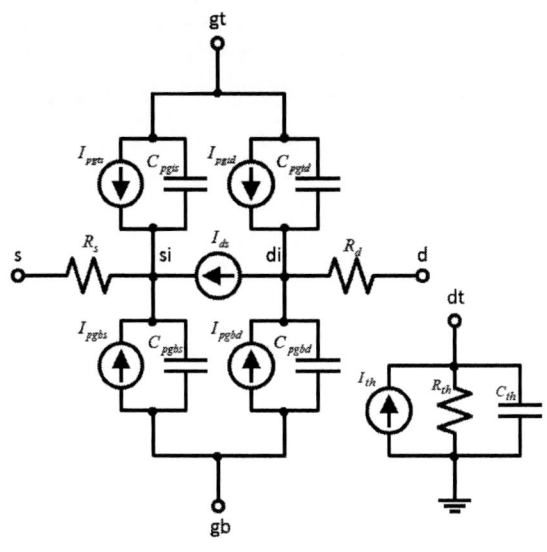

Fig. 1: JFETIDG model equivalent circuit.

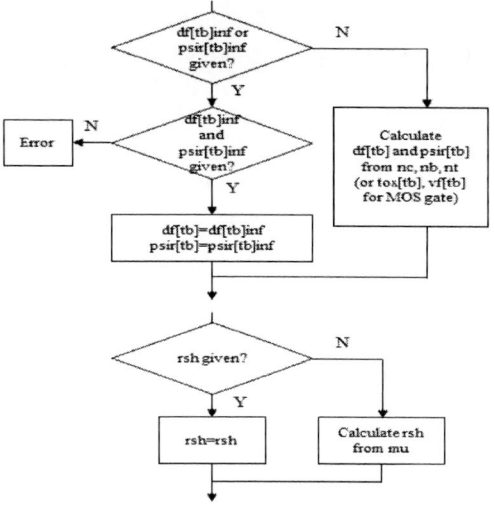

Fig. 2: Key channel parameter calculation flow.

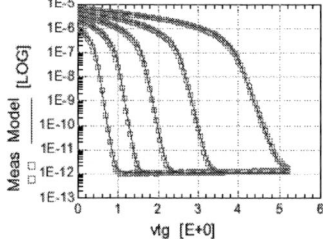

Fig. 3: I_d (log scale) vs. V_{gt}: V_{bg}=0, 0.875, 1.75, 2.625, 3.5, V_d=-0.1, V_s=0, W/L=50μm/20μm, T=27 C.

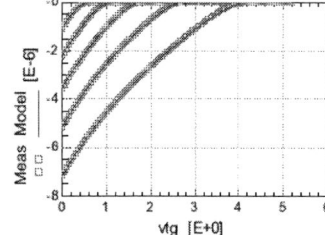

Fig. 4: I_d (linear scale) vs. V_{gt}: same device and biasing as Fig. 3.

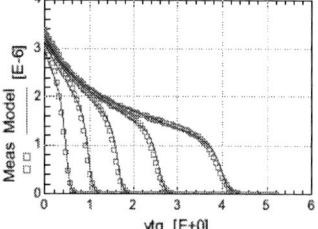

Fig. 5: g_m vs. V_{gt}: same device and biasing as Fig. 3.

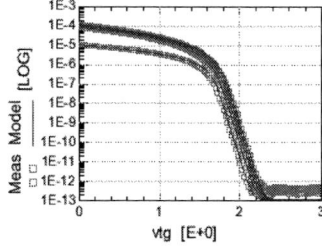

Fig. 6: I_d vs. V_{gt}: V_{gb}=V_{gt}, V_d=-0.1, -1.325, -2.55, -3.775, -5, V_s=0, W/L=20μm/5μm, T=27 C.

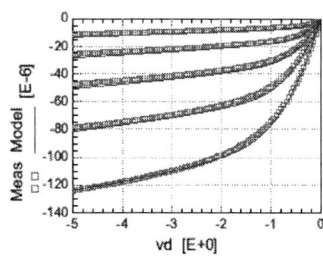

Fig. 7: I_d vs. V_d: V_{gt}=0, 0.75, 1.5, 2.25, 3, V_{gb}=0, V_s=0, W/L=20μm/5μm, T=27 C.

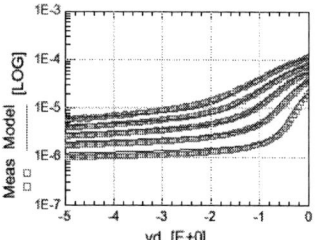

Fig. 8: g_o vs. V_d: same device and biasing as Fig. 7.

978-1-5090-4661-4/17 $31.00 © 2017 IEEE

7M-1 (Invited)

How Non-ideality Effects Deteriorate the Performance of Tunnel FETs

Andreas Schenk[1], Saurabh Sant[1], Kirsten Moselund[2], and Heike Riel[2]

[1]Integrated Systems Laboratory, ETH Zurich, CH-8092 Zürich, Switzerland, schenk@iis.ee.ethz.ch

[2]IBM Research-Zurich, Säumerstrasse 4, CH-8803 Rüschlikon, Switzerland

Abstract

Physics-based TCAD simulations of measured vertical and lateral InAs/Si hetero nanowire tunnel FETs are presented to demonstrate the effect of major non-idealities on slope and ON-current. The D_{it} limit for sub-thermal TFET operation is predicted, and it is shown that a high defect density at the InAs/Si interface can result in a slope close to 60 mV/dec due to thermionic emission in an arising MOSFET with the intrinsic Si region as gated channel.

(Keywords: TFETs, steep slope, non-idealities)

Introduction

Non-ideality effects as trap-assisted tunneling (TAT) at interface and bulk traps, roughness of material interfaces, channel quantization, and density-of-state (DOS) tails degrade the performance of TFETs [1]. It is shown by physics-based TCAD simulations how sub-threshold swing (SS) and ON-current are influenced by crucial physical parameters like D_{it}, capture cross sections, relaxation energies, and roughness amplitude. Temperature-dependent measurements of hetero nanowire (NW) TFETs (Fig. 1) provide the input for a quantitative assessment of the D_{it} limit which still guarantees a sub-60 mV/dec slope averaged over 3-4 decades. A new model for the combined effect of channel quantization and interface roughness on line tunneling quantifies the expected smoothening of the sharp onset of tunneling, which otherwise is typical for the 2D-3D DOS matching.

Effect of trap-assisted tunneling

Fig. 2 compares the theoretical band-to-band tunneling (BTBT) current with TAT currents due to generation centers at the InAs/oxide interface (measured parameters used) and at the InAs/Si hetero interface (D_{it} fitted) in one of the lateral devices shown in Fig. 1. The latter current matches the experimental curve at 300 K which has a SS ~ 60 mV/dec [2]. The band diagram in Fig. 3 reveals that the strong TAT generation at the hetero-interface acts like a current source in an internal MOSFET that arises from self-consistent electrostatics where the interface charges in conjunction with the gate voltage create a thermionic barrier to the flow of holes towards the drain. The intrinsic Si channel provides an almost perfect gate control which explains the measured slope. The comparison for two temperatures in Fig. 4 shows three distinct segments of the transfer curve - thermionic emission, TAT, and BTBT – being the bottleneck at low, intermediate, and high V_{GS}, respectively. It is found that $D_{it} < 5e11 cm^{-2}eV^{-1}$ at both interfaces is required to reach sub-thermal SS (still caused by TAT) [3].

Effect of surface roughness

Line tunneling (vertical tunnel paths under the gate) is affected by surface roughness because of the effective smoothening of the 2D DOS in the TFET channel. Within Ando's perturbation approach one can model the combined effect of channel quantization and roughness [1]. Fig. 5 presents a test device with predominant line tunneling used to quantify the roughness effect. Simulations performed with different roughness amplitudes Δ and an auto-correlation length $L = 1.9$ nm are shown in Fig. 6. The DOS smoothening degrades the SS with rising Δ and decreasing L (not shown).

Conclusion

Low defect densities, in particular at interfaces, and smooth semiconductor-oxide interfaces are necessary to achieve sub-thermal slope in InAs/Si NW TFETs. By combining defect characterization, temperature-dependent IV-measurements, and physics-based TCAD simulations one can predict the size of both effects.

Acknowledgments

Funding from the European Community's Seventh Frame-work Program under Grant Agreement No. 619509 (Project E2SWITCH) is acknowledged.

References

[1] A. Schenk, S. Sant, K. Moselund, and H. Riel, "III-V-based Hetero Tunnel FETs: A Simulation Study with Focus on Non-ideality Effects" Proc. EUROSOI-ULIS, Jan 25-27, 2016, pp. 9-12, doi: 10.1109/ULIS.2016.7440039

[2] K. E. Moselund, D. Cutaia, H. Schmid, M. Borg, S. Sant, A. Schenk, and H. Riel, "Lateral InAs/Si p-Type Tunnel FETs Integrated on Si - Part 1: Experimental Devices", IEEE Trans. Electron Devices, Vol. 63 (11), p. 4233, Nov. 2016.

[3] S. Sant, K. E. Moselund, D. Cutaia, H. Schmid, M. Borg, H. Riel, and A. Schenk, "Lateral InAs/Si p-Type Tunnel FETs Integrated on Si - Part 2: Simulation Study of the Impact of Interface Traps", IEEE Trans. Electron Devices, Vol. 63 (11), p. 4240, Nov. 2016.

978-1-5090-4661-4/17 $31.00 © 2017 IEEE

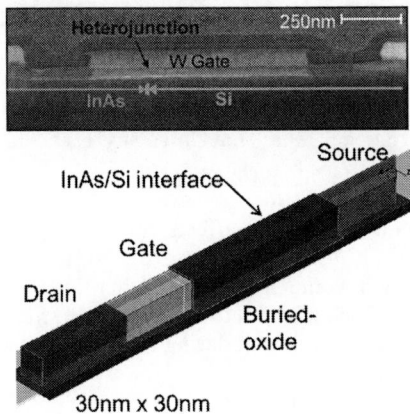

Fig. 1: SEM image of the lateral InAs/Si NW TFET with inclined hetero-interface (top) and 3D simulation domain (bottom).

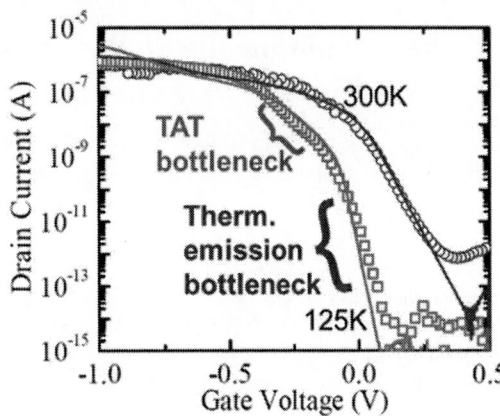

Fig. 4: Transfer characteristics at two different temperatures reveal three distinct intervals of the transfer curve, where the current originates from thermionic emission, TAT, or BTBT, respectively.

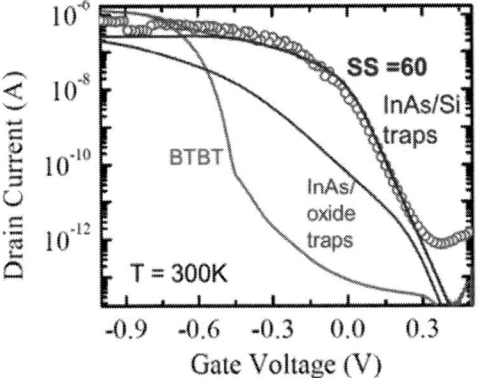

Fig. 2: Individual contributions of BTBT and the two TAT processes to the transfer characteristics of the lateral InAs/Si NW TFET at 300 K. Solid lines: simulation, circles: measurement.

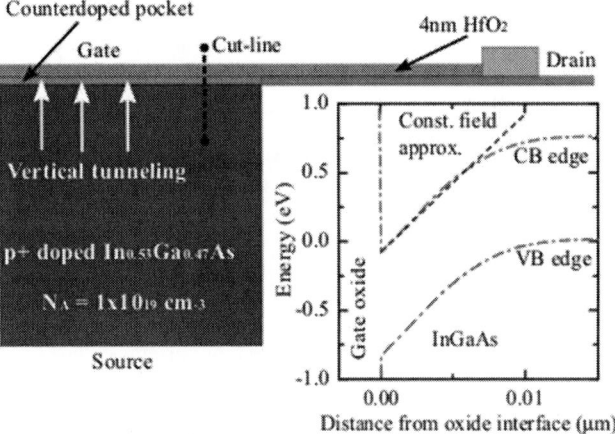

Fig. 5: Left: Simulation domain of an InGaAs vertical TFET with counter-doped pocket. The special geometry suppresses point tunneling and is used to study the combined effect of channel quantization and surface roughness. Right: Band edge diagram at $V_{GS} = 0.625V$ along the cut line perpendicular to the channel.

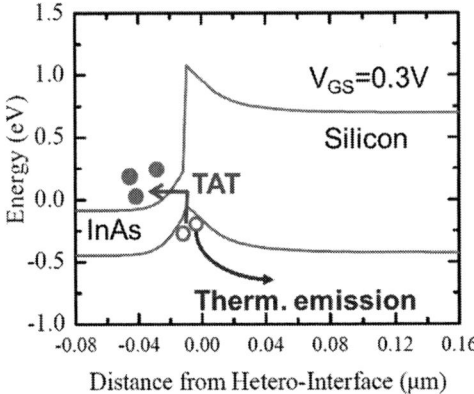

Fig. 3: Band diagram along the channel for three different gate voltages near the onset of the TFET. In case of a high defect density at the InAs/Si interface, transport of carriers happens via two subsequent steps, viz., TAT and subsequent thermionic emission.

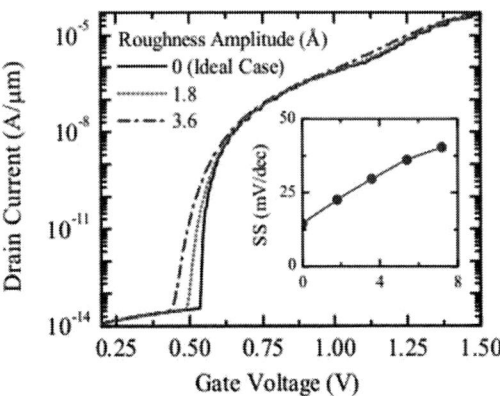

Fig. 6: Transfer characteristics of the vertical TFET in Fig. 5 for different values of the roughness amplitude Δ. Increasing Δ degrades the SS which is plotted in the inset.

978-1-5090-4661-4/17 $31.00 © 2017 IEEE

Charge Splitting In-situ Recorder (CSIR) for Monitoring Plasma Damage in FinFET BEOL Processes

Ting-Huan Hsieh, Yi-Pei Tsai, Chrong Jung Lin and Ya-Chin King

Microelectronics Laboratory, Institute of Electronics Engineering, National Tsing Hua University, Hsinchu, Taiwan

Phone/Fax: +886-3-5162219/+886-3-5721804, E-mail: ycking@ee.nthu.edu.tw

Abstract

A new monitoring device records positive/ negative charging effects in FinFET BEOL plasma processes through polarity direction scheme is first-time proposed. This charge splitting in-situ recorder (CSIR) is composed of separate floating gates to independently detect ion charging and electron charging effects. Through this on-chip charge directing structure, the separation of positive or negative plasma charging in FinFET BEOL processes has been successfully demonstrated. The independent charge level in both polarities reflects more truthfully what a transistor experiences under plasma treatments in fabrication. This new test scheme provides powerful assistance for process optimization and reliability evaluations as CMOS technologies push for finer metal lines in the future.

I. Introduction

Plasma induced damage (PID) is one of the critical challenges in advanced CMOS process[1]. High-selectivity plasma etching processes are used in etching of fine contact, via and Cu metal lines. In FinFET technologies, plasma charging effect is expected to be worsen as the discharging path includes narrow- fin type substrate and finer and more complex metal stacks are more frequently present[2-3]. Plasma charging effect is the most common issues to raised gate leakage current and/or early dielectric breakdown in FinFETs. A PID recorder with a floating gate coupled by antenna structure has been proposed with in-situ detection capability in our previous study [4-5]. In this work, we proposed a revised PID recorder with charge splitting feature. Positive and negative charging path are directed through a forward diode and a reverse diode connected to a common antenna structure. Moreover, it still provides real-time feedback of plasma charging levels by sensing the stored charge right after BEOL process. In addition, this new charge splitting in-situ recorder with small antenna enables future study of plasma charging effect in middle-end of the line processes.

II. Plasma Charging Polarity

During BEOL process for forming metal layers, the plasma induced charging rate varies drastically not only spatially, but also timely. Namely, its magnitude and polarity changes within the process time of a single etching step. As shown in figure 1, the distribution of plasma charging rate, $J_P(x,y,t)$, across the wafer during etching process at different stages change in both magnitude as well as in polarities. The plasma charging recorder proposed in our prior study, as shown in figure 2(a). The raised potential on the antenna with Q_P from plasma charging can induce electron to inject to or eject out of the floating gate. A transistor may be affected by complex plasma charging sequences when forming the multiple metal layers, see Figure 2(b). Processing of different metal layer causes different polarity of Q_P when

the dominant charging effects shift form ion to electron, as reveal by the data figure 2(c). Positive and negative charge will compensate each other in a floating gate recorder, which may fail to reveal the real stress conditions of a device experienced during BEOL processes.

III. Test Pattern for Charge Separation

In order to record positive ion charging and negative electron charging effect independently, a new Charge Splitting In-situ Recorder (CSIR) with two floating gates, FG1 and FG2 is proposed, as illustrated in figure 3. The simulated 2D potential distribution on cross-section a CSIR with positive Q_P and negative Q_P on the antenna are shown in figure 4 (a) and (b), respectively.

IV. Measured Data and Device Optimization

Figure 5 compares threshold voltage shift (ΔV_T) on devices under FG1 (with forward diode), FG2 (with reverse diode) and that without diode along the center line of a wafer. Figure 6 summarized the collected charge on FG1 and FG2 across a wafer, separately.

In CSIR, it does not respond to antenna effect, as shown in figure 7. High antenna ratio (AR) on traditional PID monitoring devices have a more severe T_{BD} degradation, see figure 8. As antenna area increases, the corresponding capacitance, C_{Ant}, increases accordingly, as shown in figure 9(a). The capacitance of the antenna structure dominates the total capacitance on the coupling node, hence, the Q_{FG} on the recorder with large antenna is independent of AR, see figure 9 (b). This result suggests that the new CSIR can detect the same plasma charging rates with a much smaller antenna structure. Data retention characteristics of PID recorder is shown in figure 10. Both charge lost and charge gain may occurs after wafer fab out.

V. Conclusions

A novel Charge Splitting In-situ Recorder (CSIR) for monitoring plasma induced damage is first-time proposed and demonstrated. The CSIR provides a tool for future understanding charging polarity effects and its correlation to device reliability. CSIRs with small antennas area further allow for future investigate of PID in MOEL.

Acknowledgments

The authors would like to thank the support from the Ministry of Science and Technology (MOST), Taiwan, and Taiwan Semiconductor Manufacturing Company (TSMC).

References

[1] C.Y. Chang, et al. SNW, pp. 1-2, Dec. 2014.
[2] K. Eriguchi, et al. ICICDT pp. 1-5, May. 2014.
[3] A. Matsuda, et al. ICICDT, pp. 191-194.
[4] C. H. Wu, et al. IEDM, pp. 7.1.1 - 7.1.4, Dec. 2015.
[5] Y. P. Tsai, et al. TED, vol. 63, pp. 2497-2502, Jun. 2016.

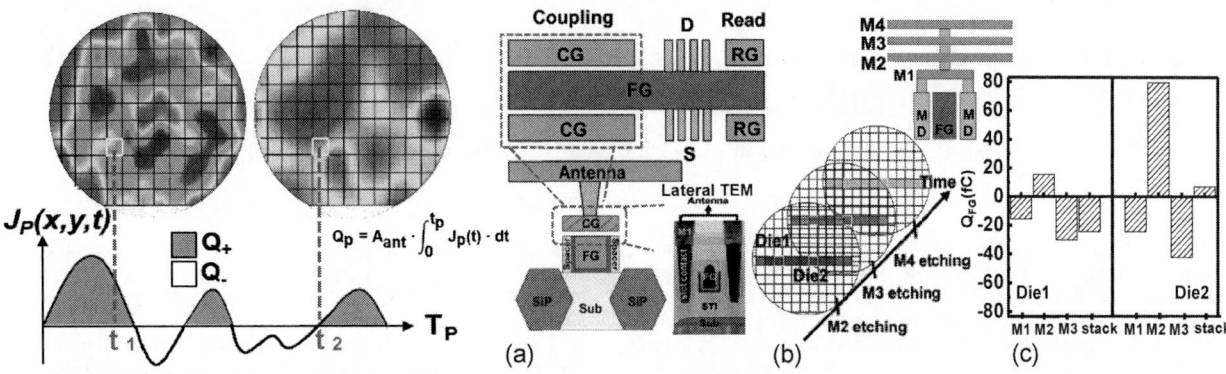

Figure 1 Distribution of plasma induced charging rate across the wafer during metal etching process at different stages. Charging polarity on a particular site may change within one step.

Figure 2 (a) Illustration of a contact coupled plasma charging recorder with stack-metal structures. (b) Plasma charging effect during the different metal layer. (c) As positive and negative charging effects compensate each other, Q_{FG} fails to indicate the real stresses a device experience.

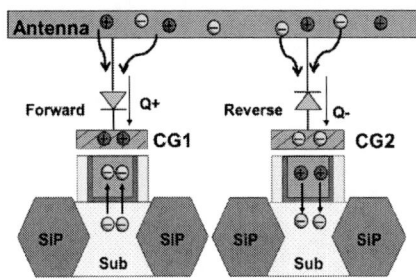

Figure 3 Charge splitting in-situ recorder with two separate floating gates, connecting to a forward and a reverse diode, respectively.

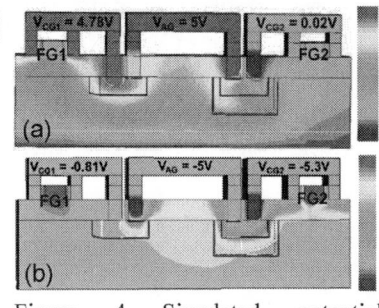

Figure 4 Simulated potential distribution for a charge splitting recorder with (a) positive and (b) negative charge on the antenna, respectively.

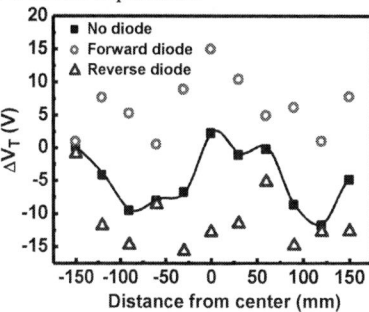

Figure 5 Comparison of ΔV_T on FG1 (forward diode), FG2 (reverse diode) and devices without diode, for CSIR along the center line of a wafer.

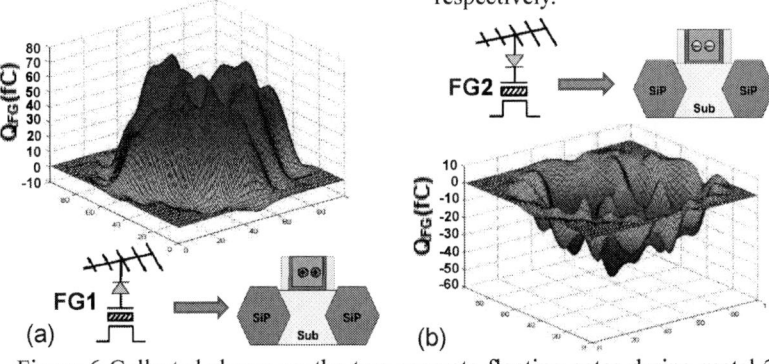

Figure 6 Collected charge on the two separate floating gates during metal 2 etching across a wafer from (a) electron and (b) ion charging effects.

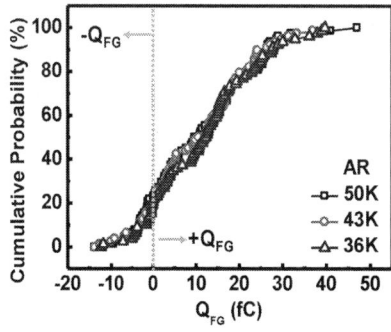

Figure 7 No antenna effect are found on FG recorders, namely, Q_{FG} does not respond to increasing AR.

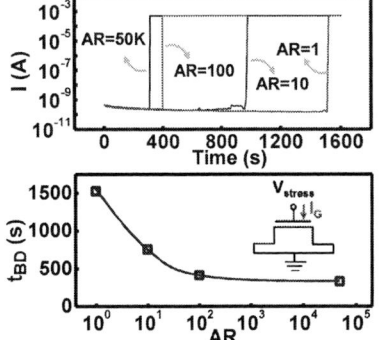

Figure 8 Antenna effect on conventional FinFETs reveals strong T_{BD} degradation on devices with increasing antenna ratio (AR).

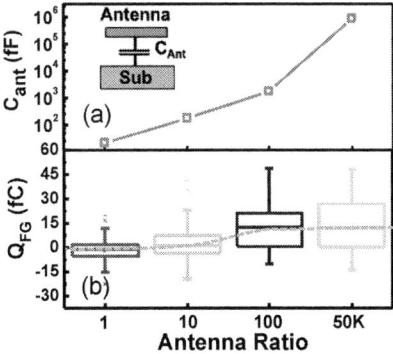

Figure 9 (a) As the capacitance of the antenna increases, V_{CG} becomes independent of the AR. (b) Q_{FG} saturates as the AR exceeds 100X.

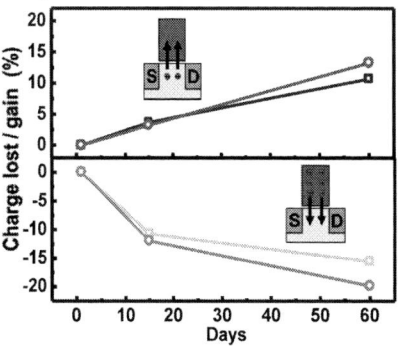

Fig. 10 Monitoring the charge lost and charge gain to/from FG of PID recorders after wafer fab out.

978-1-5090-4661-4/17 $31.00 © 2017 IEEE

7M-3

Geometric Variation: A Novel Approach to Examine the Surface Roughness and the Line Roughness Effects in Trigate FinFETs

E. R. Hsieh[1], Y. C. Fan[2], C. H. Liu[2], Steve S. Chung[1], R. M. Huang[3], C. T. Tsai[3], and T. R. Yew[3]

[1]Department of Electronics Engineering & Institute of Electronics, National Chiao Tung University, Hsinchu, Taiwan,
[2] Department of Mechatronic Engineering, National Taiwan Normal University, Taipei,Taiwan   [3]UMC, Hsinchu, Taiwan

Abstract- A new theory has been developed for geometric variation of trigate FinFETs. This geometric variation includes both *line roughness induced variation* and *oxide-thickness variation*, which can be measured from gate capacitance and Ig current variations, respectively. Experimental results show that trigate devices are subject to serious *line variations* as the fin height scales up and the fin-width scales down, leading to large I_{on} current variation, i.e., as we increase the fin aspect-ratio, line variation becomes worse which shows an increase of the *active power* consumption. On the other hand, *oxide-thickness variation* reveals significant impacts on the off-state leakage, i.e., a rough gate oxide yields to larger *static power*. These valuable results provide us important guideline for the design and manufacturing of high quality 3D gate FinFETs.

1. Introduction

To improve performance and maintain good gate-electrostatic, FinFET device scaling favors taller fin-height so as to extend the Moore's law [1]. The 3D gate with good gate controllability and lightly-doped channel also allows a better suppression of variation sources, such as random-dopant fluctuation (RDF) [2], gate work function fluctuation (WFF) [3], random-telegraph-noise (RTN) [4], and random-trap-fluctuation (RTF) [5] etc. However, scale-up of fin-height and scale-down of fin-width, i.e., high-aspect-ratio of the fin, exhibits a worse oxide roughness (that we studied in [6]) as well as potential line variations [7]. So far, previous studies are not sufficient to provide a systematic approach to distinguish the differences for various geometry-dependent variation sources.

In this work, we will focus on developing a new theory and experimental verifications of geometric variations and demonstrate the impacts on the performance of trigate devices. Finally, the inverters transfer characteristics will be demonstrated as a benchmark to compare the power consumption between trigate and planar ones.

2. Preparation of Experimental Devices

28nm poly-Silicon gate bulk-trigate devices, with SiON insulator, were fabricated, Fig. 1(a), on a pilot foundry platform. Devices with different channel lengths and fin numbers were prepared for experimental measurements. Also, planar devices on the same process were used as control samples for the comparison.

3. Results and Discussion
A. The Method to Decouple the Geometric Variations

Fig. 1(b) shows the cross section of trigate devices in Fig. 1(a). Its zoom-in, Fig. 1(c), shows a rough oxide film caused by the etching process to form STI between fins. The degree of the surface roughness can be represented by the so-called *oxide variation*, σT_{ox}, as we defined in [6], Fig. 1(d), along the z-direction. On the other hand, Fig. 2 illustrates a second geometric variation related to the length and width fluctuations which can be derived from σA (*area variation*). As a consequence, σI_g and σC_g as functions of σT_{ox} and σA, are derived in Table 1. In practice, from the I_g and C_g measurement results (Figs. 3 and 4), followed by calculation of σI_g and σC_g(Fig. 5), we may complete the evaluations of oxide thickness variation, σT_{ox}, and gate area variation, σA, as shown in Figs. 6 and 7. The results revealed that trigate device exhibits much larger oxide thickness variation, i.e., surface roughness, and also larger area variation as a result of the 3D gate structure. In other words, σT_{ox} and σA become larger as fin-height is scaled-up, which is caused by the device process, such as damage of the fin-sidewall by high-energy radicals during STI, oxide uniformity, and gate patterning.

B. Line-edge Roughness versus Line-width Roughness

In order to study why and how geometric variations affect device variations, a tree path is illustrated in Fig. 8. The geometric variation can be divided into the oxide-thickness variation and the area

variation, σA. The surface-roughness can be deduced from the oxide variation, σT_{ox} (e.g. see [6]), while area variation (σA_g) can be represented by Line-Edge Roughness (LER) and Line Width Roughness (LWR) because $\sigma^2 C_g = \sigma^2 C_{gc} + \sigma^2 C_{gd}$, where LER indicates the variation of the gate width vertical to the current flow; LWR is defined as the variation of gate length along the channel direction (Fig. 9(a)). The former can be measured by the gate-to-drain capacitance, C_{gd} while the latter can be measured by the gate-to-channel capacitance, C_{gc}. In Fig. 9(b), as the channel length reduces, LER (shaded area) dominates the line roughness, in comparison to LWR, and is even more significant with an increasing fin-height.

C. Impact of Variations on I_{off} and I_{on}

Fig. 10 shows that at the near threshold, e.g., I_{off} leakage at **LP**(low power) operation, the normalized variance of measured gate current is decoupled into normalized $\sigma^2 A_g$ and $\sigma^2 T_{ox}$, and the latter(shaded area) dominates the gate leakage. Since the I_{off} leakage is the major source of standby power consumption at the off state ($V_{gs}= 0V$), the *oxide-thickness variation* becomes the dominant factor of the standby power consumption. In contrast, Fig. 11 shows the normalized $\sigma^2 C_g$, and the line-roughness variation, in terms of σA_g (the blank area), dominating at high performance (**HP**) regime, i.e., at $V_{gs}= V_{ds}= 1V$. In order to understand why σT_{ox} dominates at low electric field and σA_g in high field, Fig. 12 shows the results of σT_{ox} and σA_g. It was found that σT_{ox} decays more rapidly against the field than that of σA, which indicates that σA should dominate the electric parameters in the high field regime. In other words, σA has direct influence on the transistor $I_{d,sat}$. Fig. 13 shows that the variation of I_{on} current ($I_{dsat}@V_{ds}= V_{gs}= 1V$), and σI_{dsat} of trigate devices (with larger fluctuation) is larger than those of planar ones. To understand the reasons, σI_{dsat} is further decomposed into the terms of σA, σT_{ox}, σRDF, and $\sigma DIBL$, Table 2, with results given in Fig. 14. It was found that σT_{ox} is weak in planar devices; however, σT_{ox} and σA are two dominant factors in trigate devices, especially line roughness is the dominant variation source. In other words, considering the impact of line roughness on I_{dsat}, this effect increases with the increasing fin height. Also, σI_{dsat} of trigate devices is larger than planar ones because of *more serious line roughness*. Finally, the voltage transfer curves of CMOS inverters have been evaluated as a benchmark (Fig. 15), in which the overall variation of output voltage (pink color) shows an increase of 11%, compared to the planar one. It reveals that the influence of the geometric variations on the trigate inverter is really significant.

In summary, geometric variations have been experimentally and systematically studied, including σA (representing the line roughness), and σT_{ox} (representing the surface roughness). The results show that line-roughness dominates in the high-field, due to the Line-edge roughness (LER). In addition, the line roughness has significantly caused an increase of $\sigma I_{d,sat}$ in trigate devices, critical to the active power. In contrast, σT_{ox} has become more significant at off-state for high aspect ratio of the fin, critical to the static power. In other words, both geometric factors affect the power consumption of the trigate significantly. These findings and methodology are extremely useful for the design and manufacturing of high quality 3D gate FinFETs.

Acknowledgments   This work was support in part by the Ministry of Science & Technology, Taiwan, under contract, *MOST-104-2221-E-009 -005*, and *NCTU-UC Berkeley* I-RiCE program, under *MOST-105-2911-E-009-301*.

References: [1] G. Moore, *Electronics*, p. 114, 1965. [2] E. R. Hsieh, et al., in *Symp. on VLSI*, p. 194, 2011.[3] E. R. Hsieh, et al., in *Symp. on VLSI*, p. 204, 2016. [4] E. R. Hsieh, et al., in *IEDM Tech. Dig.*, p. 19.2.1, 2012. [5] H. M. Tsai, et al., in *Symp. VLSI*, p. 189, 2012. [6] E. R. Hsieh et al., in *IEDM Tech. Dig.*, p. 31.2.1, 2013. [7] R. Wang, et al., in *IEEE T-ED*, vol.60, p. 3676, 2013.

978-1-5090-4661-4/17 $31.00 © 2017 IEEE

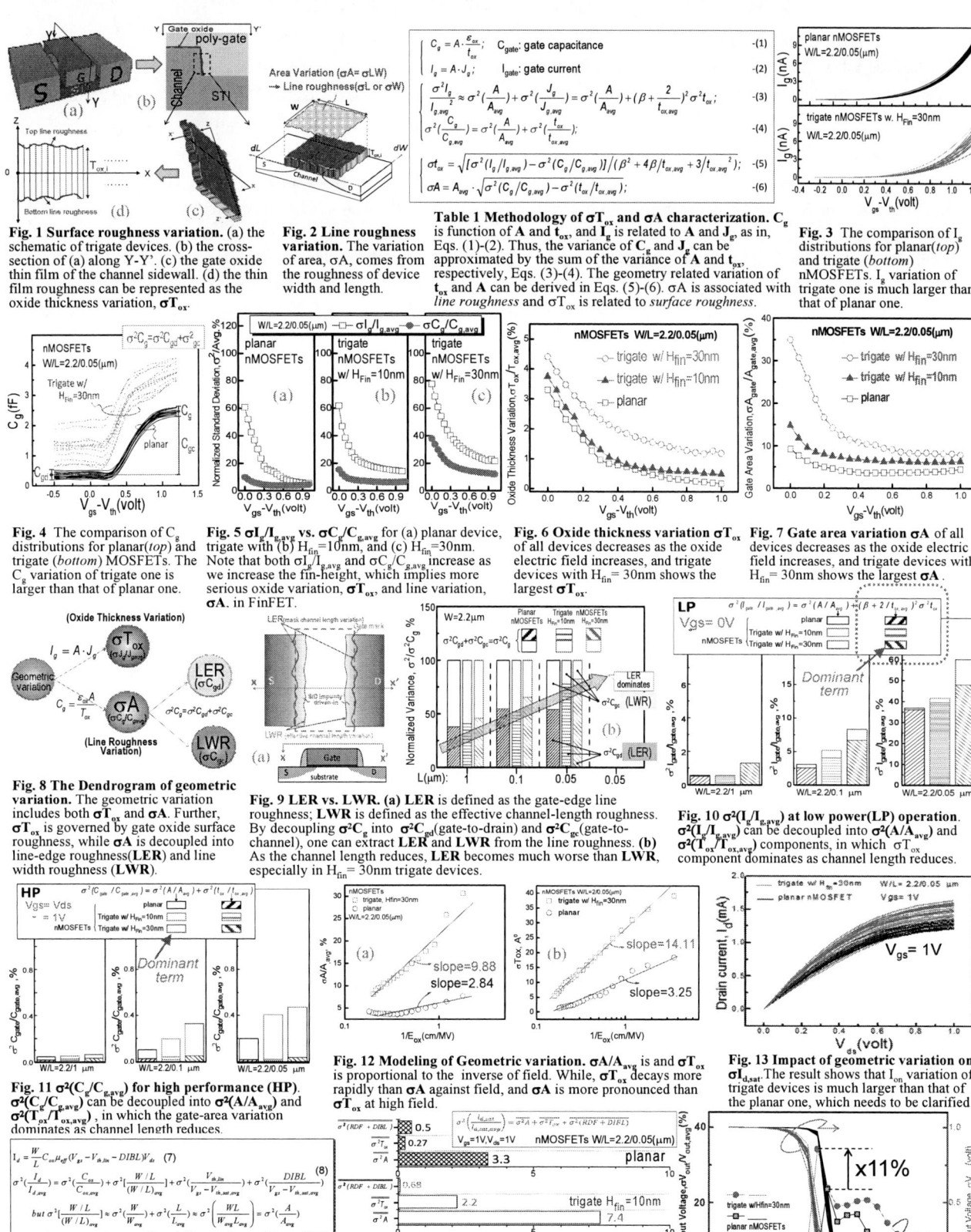

Table 1 Methodology of σT$_{ox}$ and σA characterization. C$_g$ is function of A and t$_{ox}$, and I$_g$ is related to A and J$_g$, as in, Eqs. (1)-(2). Thus, the variance of C$_g$ and J$_g$ can be approximated by the sum of the variance of A and t$_{ox}$, respectively, Eqs. (3)-(4). The geometry related variation of t$_{ox}$ and A can be derived in Eqs. (5)-(6). σA is associated with *line roughness* and σT$_{ox}$ is related to *surface roughness*.

$$C_g = A \cdot \frac{\varepsilon_{ox}}{t_{ox}}; \quad C_{gate}: \text{gate capacitance} \quad \text{-(1)}$$

$$I_g = A \cdot J_g; \quad I_{gate}: \text{gate current} \quad \text{-(2)}$$

$$\frac{\sigma^2 I_g}{2} \approx \sigma^2 \left(\frac{A}{A_{avg}}\right) + \sigma^2 \left(\frac{J_g}{J_{g,avg}}\right) = \sigma^2 \left(\frac{A}{A_{avg}}\right) + (\beta + \frac{2}{t_{ox,avg}})^2 \sigma^2 t_{ox}; \quad \text{-(3)}$$

$$\sigma^2 \left(\frac{C_g}{C_{g,avg}}\right) = \sigma^2 \left(\frac{A}{A_{avg}}\right) + \sigma^2 \left(\frac{t_{ox}}{t_{ox,avg}}\right); \quad \text{-(4)}$$

$$\sigma t_{ox} = \sqrt{[\sigma^2 (I_g/I_{g,avg}) - \sigma^2 (C_g/C_{g,avg})]/(\beta^2 + 4\beta/t_{ox,avg} + 3/t_{ox,avg}^2)}; \quad \text{-(5)}$$

$$\sigma A = A_{avg} \cdot \sqrt{\sigma^2 (C_g/C_{g,avg}) - \sigma^2 (t_{ox}/t_{ox,avg})}; \quad \text{-(6)}$$

Fig. 1 Surface roughness variation. (a) the schematic of trigate devices. (b) the cross-section of (a) along Y-Y'. (c) the gate oxide thin film of the channel sidewall. (d) the thin film roughness can be represented as the oxide thickness variation, **σT$_{ox}$**.

Fig. 2 Line roughness variation. The variation of area, σA, comes from the roughness of device width and length.

Fig. 3 The comparison of I$_g$ distributions for planar(*top*) and trigate (*bottom*) nMOSFETs. I$_g$ variation of trigate one is much larger than that of planar one.

Fig. 4 The comparison of C$_g$ distributions for planar(*top*) and trigate (*bottom*) MOSFETs. The C$_g$ variation of trigate one is larger than that of planar one.

Fig. 5 σI$_g$/I$_{g,avg}$ vs. σC$_g$/C$_{g,avg}$ for (a) planar device, trigate with (b) H$_{fin}$=10nm, and (c) H$_{fin}$=30nm. Note that both σI$_g$/I$_{g,avg}$ and σC$_g$/C$_{g,avg}$ increase as we increase the fin-height, which implies more serious oxide variation, σT$_{ox}$, and line variation, σA. in FinFET.

Fig. 6 Oxide thickness variation σT$_{ox}$ of all devices decreases as the oxide electric field increases, and trigate devices with H$_{fin}$= 30nm shows the largest σT$_{ox}$.

Fig. 7 Gate area variation σA of all devices decreases as the oxide electric field increases, and trigate devices with H$_{fin}$= 30nm shows the largest σA.

Fig. 8 The Dendrogram of geometric variation. The geometric variation includes both **σT$_{ox}$** and **σA**. Further, **σT$_{ox}$** is governed by gate oxide surface roughness, while **σA** is decoupled into line-edge roughness(**LER**) and line width roughness (**LWR**).

Fig. 9 LER vs. LWR. (a) **LER** is defined as the gate-edge line roughness; **LWR** is defined as the effective channel-length roughness. By decoupling **σ²C$_g$** into **σ²C$_{gd}$**(gate-to-drain) and **σ²C$_{gc}$**(gate-to-channel), one can extract **LER** and **LWR** from the line roughness. (b) As the channel length reduces, **LER** becomes much worse than **LWR**, especially in H$_{fin}$= 30nm trigate devices.

Fig. 10 σ²(I$_g$/I$_{g,avg}$) at low power(LP) operation. σ²(I$_g$/I$_{g,avg}$) can be decoupled into σ²(A/A$_{avg}$) and σ²(T$_{ox}$/T$_{ox,avg}$) components, in which σT$_{ox}$ component dominates as channel length reduces.

Fig. 11 σ²(C$_g$/C$_{g,avg}$) for high performance (HP). σ²(C$_g$/C$_{g,avg}$) can be decoupled into σ²(A/A$_{avg}$) and σ²(T$_{ox}$/T$_{ox,avg}$), in which the gate-area variation dominates as channel length reduces.

Fig. 12 Modeling of Geometric variation. σA/A$_{avg}$ is and σT$_{ox}$ is proportional to the inverse of field. While, σT$_{ox}$ decays more rapidly than σA against field, and σA is more pronounced than σT$_{ox}$ at high field.

Fig. 13 Impact of geometric variation on σI$_{d,sat}$. The result shows that I$_{on}$ variation of trigate devices is much larger than that of the planar one, which needs to be clarified.

Table 2 Derivations of the composition of σ²I$_{d,sat}$: From the approximation of multi-variable Taylor expansion, σ²(I$_{d,sat}$/I$_{d,sat,avg}$) can be decomposed into the terms of geometric variances(σ²A & σ²T$_{ox}$) and the variances of **RDF** and **DIBL**, Eq. (9).

$$I_d = \frac{W}{L} C_{ox} \mu_{eff} (V_{gs} - V_{th,lin} - DIBL) V_{ds} \quad (7)$$

$$\sigma^2 (\frac{I_d}{I_{d,avg}}) = \sigma^2 (\frac{C_{ox}}{C_{ox,avg}}) + \sigma^2 [\frac{W/L}{(W/L)_{avg}}] + \sigma^2 (\frac{V_{th,lin}}{V_{gs} - V_{th,sat,avg}}) + \sigma^2 (\frac{DIBL}{V_{gs} - V_{th,sat,avg}}) \quad (8)$$

$$but \; \sigma^2 [\frac{W/L}{(W/L)_{avg}}] \approx \sigma^2 (\frac{W}{W_{avg}}) + \sigma^2 (\frac{L}{L_{avg}}) \approx \sigma^2 (\frac{WL}{W_{avg}L_{avg}}) = \sigma^2 (\frac{A}{A_{avg}})$$

$$\Rightarrow \sigma^2 (\frac{I_d}{I_{d,avg}}) = \underbrace{2 \cdot \sigma^2 (\frac{A}{A_{avg}})}_{\sigma^2 A} + \underbrace{[1 + (\frac{V_{gs} + 0.1V}{V_{gs} - V_{th,sat,avg}})]^2 \sigma^2 (\frac{T_{ox}}{T_{ox,avg}})}_{\sigma^2 T_{ox}} + \underbrace{\sigma^2 RDF + \sigma^2 DIBL}_{\sigma^2 RDF + \sigma^2 DIBL} (9)$$

Fig. 14 The line roughness(σ²A) dominates σ²I$_{d,sat}$, and surface roughness (σT$_{ox}$) is increased from the planar device to the taller fin-height device. Note that **RDF** variance is minor because **RDF** is dominant only at the near threshold region.

Fig. 15 The DC inverter transfer curves and the variation. The trigate inverter shows larger variation than the planar ones. An increase of active power in trigate during the switching reflects the importance to minimize the geometric variation.

978-1-5090-4661-4/17 $31.00 © 2017 IEEE

A Novel Method to Characterize DRAM Process Variation
by the Analyzing Stochastic Properties of Retention Time Distribution

Min Hee Cho[1], Namho Jeon[1], Moonyoung Jeong[1], Sungsam Lee[1], Satoru Yamada[1], and Hyeongsun Hong[1]

[1]Semiconductor R&D center, Samsung Electronics Co., Gyeonggi-do, Korea, cmh12.cho@samsung.com

Abstract

This study proposes an innovative method to measure the variation of cell leakage current. Extreme cell leakage determines DRAM refresh time (tREF). Although the average leakage current from the test element group (TEG) has been the only index for determining cell leakage, it does not provide the distribution of unit cell leakage. We find that cell leakage distribution can be calculated from the slope at the retention time-fail bit plot. A steep slope indicates a small cell leakage distribution that corresponds to a long tREF. It is proved with statistical models and experimental results with mass data. (Keywords: DRAM, Cell leakage, time–FBT slope, and Process variation)

Introduction

DRAM developers have struggled to find the correct index to analyze leakage current that defines the boundary of tREF. Usually, gate-induced drain leakage current (GIDL) is the major cell leakage current. To monitor GIDL, a TEG is used as shown in Fig. 1. However, GIDL of a unit cell is so small (0.01–100 fA/cell) that it is impossible to measure each cell leakage current. Currently, GIDL from TEG is the only indicator for predicting tREF; however, this only provides the average GIDL, not the leakage distribution. Mismatching present between average GIDL and tREF is a serious obstacle for DRAM research, because GIDL from TEG cannot always predict tREF. We analyze the reason why discordance occurs between GIDL and tREF for the first time using statistical models and experimental results.

Cell Leakage Mechanism and Mismatching

Hurkx' model [1] explains the cell leakage mechanism related to trap energy level, trap density, and electric field (e-field). The source and drain in the cell (S/D) ion implantation (iip) process (Reference: Process A) can make the S/D doping profile broad, which causes a weak e-field when voltage is applied. Therefore, the GIDL of Process A becomes smaller than that of the no S/D iip process (Process B) because the e-field at S/D of Process B increases due to the abrupt doping profile. However, contrary to the expectation, the tREF of Process A is shorter than that of Process B (Fig. 2). This phenomenon can be explained by the distribution of cell leakage.

Leakage Distribution

The cell leakage has a lognormal distribution [2] and ranges from very small current (<0.1 fA) to large leakage current (hundreds of fA) in DRAM cell arrays (several Giga bits). If a unit cell loses its data because of leakage, its results can be counted as fail bit (FBT) with retention time (i.e., how long the cell can hold data). Some cells with excessive leakage current must be FBT. This case is known as tail bit. Leakage current is exponentially related to e-field, and defect and e-field are assumed to have normal Gaussian distributions in the DRAM chip. Therefore, the leakage current distribution follows a lognormal distribution [2]. Fig. 3 shows the simulated lognormal distribution with different mean (μ) and standard deviation (σ) determined by statistical theory [3]. As aforementioned, very-high leakage current in extreme tail cell causes low DRAM tREF. In Fig. 4, the cell leakage distribution plot changes with standard variation. Small variation causes a steeper slope on the retention time–FBT plot. The slope is defined as the "time–FBT slope (TF slope)" herein, and σ is proportional to 1/TF slope in the low cumulative distribution function (CDF) region ($<10^{-7}$), which is proven in Fig. 5. The mismatch between GIDL in TEG and tREF (Fig. 2) can be explained by the TF slope. In Fig. 6, the slope of Processes B is much steeper than that of Process A. Thus, Process B has smaller variation, although their average leakage currents (i.e., GIDL from TEG) are larger than that of Process A. The DRAM tREF can increase with TF slope. Thus, TF slope plays an important role in estimating new processes in terms of distribution. The TF slope thus provides the first proper index for estimating processes in terms of distribution.

Conclusion

This study is the first to demonstrate that the TF slope is an outstanding index for evaluating the cell leakage distribution. The physical meaning of TF slope is the cell leakage distribution that is influenced by processes such as S/D iip or contact-etch. The TF slope represents a significant contribution to DRAM development and contributes DRAM lead time reduction

References

[1] Hurkx, G.A.M. et al., "A New Analytical Diode Model Including Tunneling and Avalanche Breakdown", IEEE TED Vol. 9, No. 9, p. 2090-2098, 1992

[2] Rao, R. et al. "Statistical estimation of leakage current considering inter- and intra-die process variation", ISLPED '03, p. 84-89, 2003.

[3] Limpert, E. et al, "Log-normal Distributions across the Sciences: Keys and Clues", BioScience, 51 (5), 2003

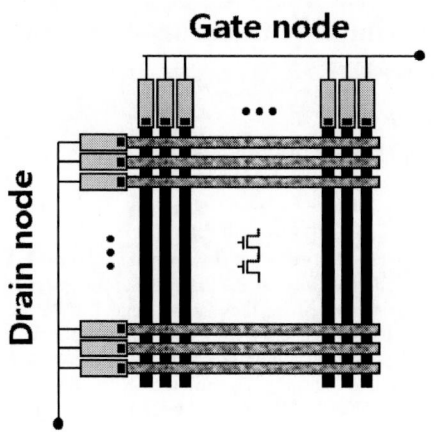

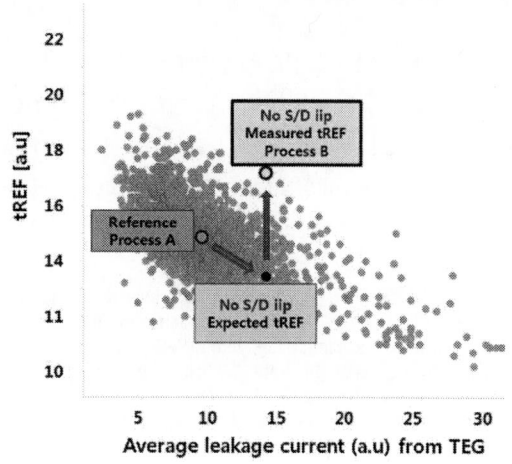

Fig. 1. Schematic of a TEG module. TEG is typically located in the scribe lane. Average GIDL and Lov can be measured with TEG.

Fig. 2. tREF *vs*. Average cell leakage current (*i.e.*, average GIDL) plot. The GIDL (X axis) is measured from TEG in this plot. Theoretically, tREF should be inversely proportional to GIDL. However, Process B does not follow this trend.

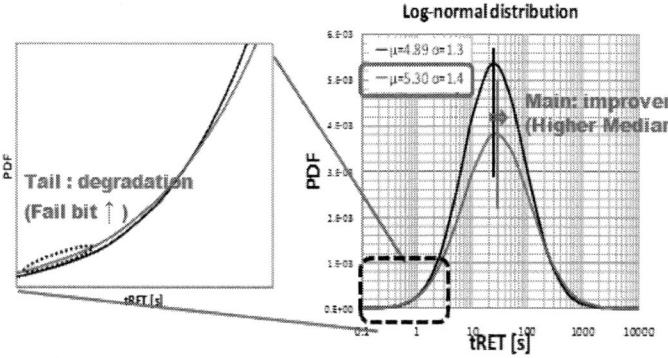

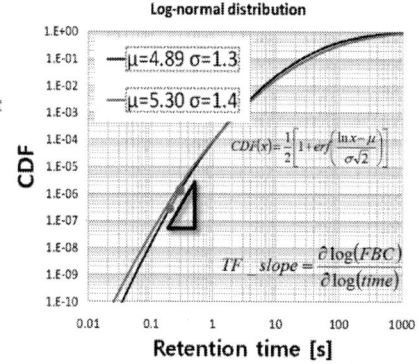

Fig. 3. Simulated lognormal distribution. The x-and y-axis are the log scale of time and probability distribution function (PDF), respectively. If cell leakage distribution is worse (*i.e.*, has a higher standard variation), the extreme tail cells have higher leakage current, even for larger μ.

Fig. 4. The typical form of the retention time *vs*. FBT plot; the y axis has been changed from PDF to CDF. Plot shown in Fig. 3 with the y-axis changed into CDF. A plot with a steeper slope has a lower standard variation.

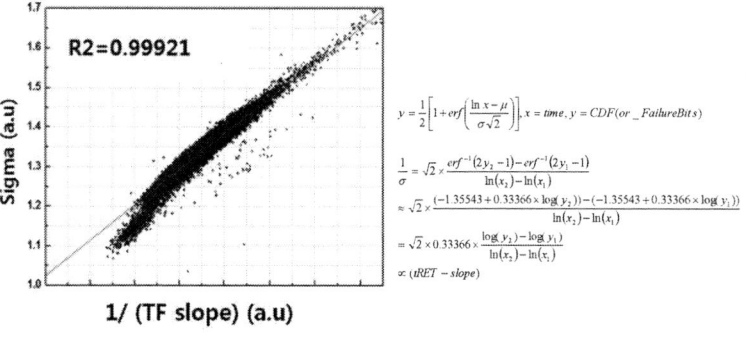

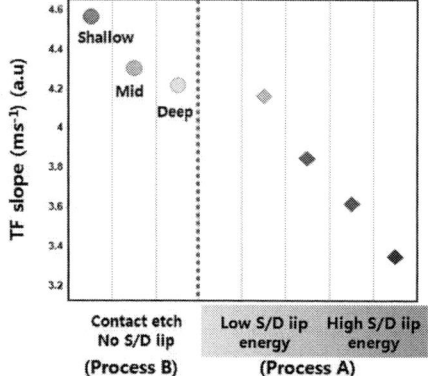

Fig. 5. The relationship between the 1/(TF slope) and standard variation. This relationship has been demonstrated statistically and experimentally. Leakage current standard deviation (sigma) is not easy to calculate but TF slope can be easily extracted from the retention time *vs*. FBT plot.

Fig.6. the TF slopes for each process are compared with various processes (S/D iip/ contact etch depth). As S/D iip energy or contact etch depth decreases, TF slope increases (*i.e.*, distribution is improved and longer tREF).

978-1-5090-4661-4/17 $31.00 © 2017 IEEE

7A-1

Analysis of Subthreshold Swing and Internal Voltage Amplification for Hysteresis-Free Negative Capacitance FinFETs

Pin-Chieh Chiu and Vita Pi-Ho Hu

Department of Electrical Engineering, National Central University, Taoyuan, Taiwan

Email:vitahu@ee.ncu.edu.tw

Abstract

We present the device design guideline for hysteresis-free negative capacitance FinFETs (NC-FinFETs) to enhance the internal voltage amplification (A_V) and reduce the subthreshold swing (SS). A_V can be increased by increasing fin width (W_{fin}), coercive field (E_c), and thickness of the ferroelectric layer (T_{FE}), and A_V can also be enhanced by reducing EOT, channel length (L_{ch}), buried oxide thickness (T_{box}), fin height (H_{fin}) and remnant polarization (P_0). The subthreshold swing improvements of NC-FinFETs over FinFETs become larger as A_V increases. With the same channel length, compared with the NC-FinFET without underlap design, NC-FinFET with underlap design exhibits better capacitance matching and larger A_V, hence showing larger subthreshold swing improvement and on-current improvement over FinFET. At shorter L_{ch} (= 12.5nm), NC-FinFETs with underlap design exhibit 73.6% improvements in intrinsic delay compared with the FinFETs due to its larger effective drive current.

(Keywords: Negative capacitance, NC-FinFET, subthreshold swing, amplification factor, hysteresis, ferroelectrics)

Introduction

Emerging IoT technologies demand extremely low power systems. Device with higher Ion/Ioff ratio is essential in order to achieve energy-efficient switching at given supply voltage. Negative capacitance FET (NCFET) [1] has been attracting interests because it reduces subthreshold swing below classical limit of ~60mV/dec, and therefore shows higher Ion/Ioff ratio. Negative capacitance FinFETs (NC-FinFETs) have been presented [2-4] for device and circuit analysis. However, the impact of FinFET device design on the subthreshold swing (SS) and internal voltage amplification (A_V) of NC-FinFET has rarely been examined. In this paper, based on the 14nm ITRS FinFETs [5], we analyze the improvements in hysteresis-free NC-FinFET over the baseline FinFET. The impacts of FinFET device parameters (including W_{fin}, H_{fin}, EOT, T_{box}, and L_{ch}) and ferroelectric parameters (including E_c, P_0, and T_{FE}) on the SS and A_V have been analyzed comprehensively for NC-FinFETs. The impact of gate-to-drain underlap design on the SS, A_V, Ion, and intrinsic gate delay of NC-FinFETs has also been examined.

Impacts of Device and Ferroelectric Parameters

Fig. 1(a) shows the Id-Vg characteristics of the baseline FinFET and NC-FinFET. The calibrated 3D TCAD simulations agree well with the experimental data for FinFETs [5], and Fig. 1(b) shows the FinFET structure. The extracted coercive electric field $E_c = 0.29$MV/cm and remnant polarization $P_0 = 39$ μC/cm² [4] are used for NC-FinFETs, and $T_{FE} = 69$nm is used for hysteresis-free design in this work. Fig. 2 shows the simulation flow of 3D TCAD coupled with Landau-Khalatnikov (LK) ferroelectric equation for modeling the NC-FinFET. Fig. 3 shows the simple capacitance model of NC-FinFET, and equations for internal voltage amplification (A_V), SS of NC-FinFET, and intrinsic gate delay. $|C_{FE}|$ should be closer to C_{MOS} to achieve higher A_V and smaller SS, but $|C_{FE}|$ cannot be smaller than C_{MOS} to avoid hysteresis phenomenon.

Fig. 4 shows the impacts of device parameters on the A_V and SS for FinFETs and NC-FinFETs at Vd=0.7V. In Fig. 4(a), A_V increases with W_{fin}, and therefore the SS improvements of NC-FinFET over FinFET increases with W_{fin}, i.e., compared with FinFET, NC-FinFET shows 33% SS improvements at $W_{fin} = 10$nm and 29% at $W_{fin} = 6$nm. SS improvement is defined as [($SS_{NC-FinFET} - SS_{FinFET})/SS_{FinFET}$]. Fig. 4(b) shows that reducing H_{fin} slightly increases A_V and SS improvements. Fig. 4(c) shows that reducing T_{box} increases A_V and shows larger SS improvements. Decreasing EOT shows larger Av and SS improvements as shown in Fig. 4(d). A_V increases with E_c (Fig. 5(a)) and T_{FE} (Fig. 5(c)), while A_V decreases with P_0 (Fig. 5(b)). Therefore, the SS improvements of NC-FinFET over FinFET increase

with E_c and T_{FE}, and decrease with P_0.

Impacts of Underlap Design for NC-FinFET

In this section, we analyze the impact of underlap design for NC-FinFET. For fair comparisons, the channel length (L_{ch}) is the same for both NC-FinFET with underlap (w/ Lund) and without underlap (w/o Lund) designs as shown in Fig. 6. The gate-to-drain underlap length (ΔLund) is fixed to 3nm. Fig. 7 shows the SS and SS improvements comparisons for NC-FinFETs w/o and w/ Lund. As can be seen, NC-FinFETs w/ Lund (Fig. 7(b)) exhibit larger SS improvements than NC-FinFETs w/o Lund (Fig. 7(a)) at various L_{ch}. The SS improvements increase as L_{ch} decreases. Fig. 8(a) shows the $|C_{FE}|$ and C_{MOS} comparisons for NC-FinFET w/o and w/ Lund. NC-FinFET w/ Lund exhibits better capacitance matching due to its larger drain-to-channel field coupling, hence showing larger A_V as shown in Fig. 8(b). Besides, A_V increases as L_{ch} decreases due to stronger drain field coupling. Fig. 9 shows the Ion comparisons for FinFET and NC-FinFET w/ and w/o Lund at Vd = 0.7V and fixed Ioff = 1E-8A/μm. As we can see in Fig. 9(a) and 9(b), as L_{ch} scales from 45nm to 16nm, Ion increases for NC-FinFETs and decreases for FinFET. In other words, as L_{ch} decreases, Ion/Ioff ratio increases for NC-FinFET and decreases for FinFET. Besides, NC-FinFET w/ Lund exhibits larger Ion improvements over FinFET than that w/o Lund. Fig. 10 shows the intrinsic gate delay comparisons for NC-FinFET and FinFET w/ Lund design at various L_{ch}. The intrinsic delay time of NC-FinFET w/ Lund becomes smaller as L_{ch} decreases. At L_{ch} > 20nm, NC-FinFET exhibits larger intrinsic delay than FinFET due to its larger gate capacitance. Note that the gate capacitance of NC-FinFET is larger than that of FinFET due to the ferroelectric layer negative capacitance. However, at L_{ch} < 20nm, NC-FinFET exhibits much smaller intrinsic delay than FinFET due to its larger effective drive current (Ieff). At L_{ch}=12.5nm, NC-FinFET shows significant improvements in intrinsic delay (73.6%) compared with FinFET. The intrinsic delay time increases as L_{ch} changes from 16nm to 12.5nm for FinFET, because the effective drive current decreases significantly as L_{ch} decreases.

Conclusion

Impacts of device and ferroelectric parameters, and underlap design on the internal voltage amplification (A_V) and subthreshold swing improvements are analyzed for hysteresis-free NC-FinFETs compared with the baseline FinFET. NC-FinFET w/ Lund shows larger SS improvements and Ion improvements than that w/o Lund. At shorter L_{ch} (< 20nm), NC-FinFET w/ Lund exhibits better SS, Ion, and intrinsic gate delay compared with the baseline FinFET. The performance of NC-FinFET can be further improved by tuning the device and ferroelectric parameters.

Acknowledgments

This work is supported by the Ministry of Science and Technology in Taiwan under Contract MOST 105-2628-E-008-006-MY3. The authors are grateful to the National Center for High-Performance Computing in Taiwan for computational facilities and software.

References

[1] S. Salahuddin et al., *Nano Lett.*, vol. 8, pp. 405, 2008.
[2] C. Hu et al., *Device Research Conference (DRC)*, pp. 39, 2015.
[3] A. I. Khan et al., IEEE Electron Dev. Lett., vol. 37, pp. 111, 2016.
[4] S. Khandelwal et al., *Symp. on VLSI Tech. Dig.*, pp. 230, 2016.
[5] S. Natarajan et al., *IEEE IEDM*, pp. 71, 2014.

978-1-5090-4661-4/17 $31.00 © 2017 IEEE

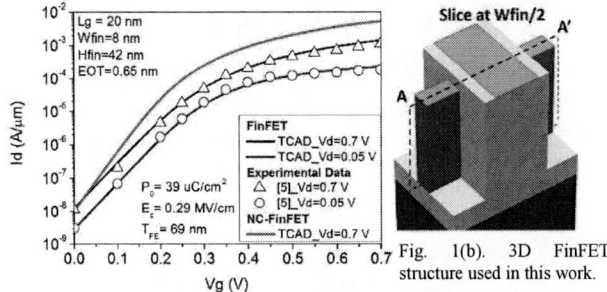

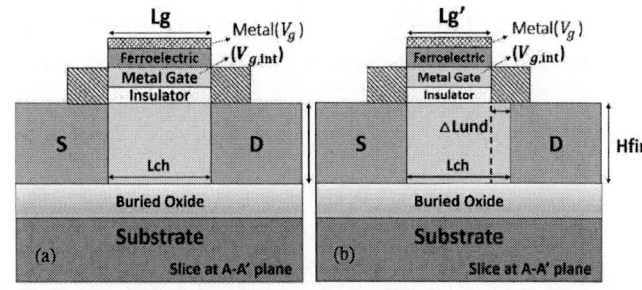

Fig. 1(b). 3D FinFET structure used in this work.

Fig. 1(a). The Id-Vg characteristics of the TCAD baseline FinFET show good agreement with the experimental data [5]. The NC-FinFET is modeled based on the baseline FinFET and LK model.

Fig. 6. Cross-sections (slice at plane AA' in Fig. 1(b)) of NC-FinFET (a) without underlap (w/o Lund, $L_{ch}=L_g$) and (b) with gate-to-drain underlap (w/ Lund, $L_{ch}=L_g'+\Delta Lund$). The channel length (L_{ch}) is the same for NC-FinFET w/o and w/ Lund. (underlap length $\Delta Lund$ is 3 nm.)

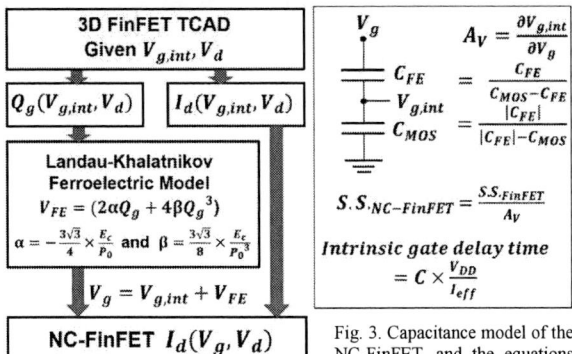

Fig. 2. 3D TCAD FinFET simulation coupled with LK equation is used to model NC-FinFET.

Fig. 3. Capacitance model of the NC-FinFET, and the equations for amplification factor (A_v), subthreshold swing (SS) and intrinsic delay time. I_{eff} is effective drive current.

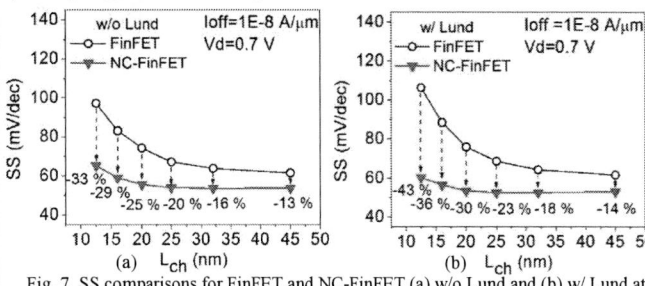

Fig. 7. SS comparisons for FinFET and NC-FinFET (a) w/o Lund and (b) w/ Lund at various L_{ch}. With the same L_{ch}, NC-FinFET w/ Lund exhibits larger improvements in SS than the NC-FinFET w/o Lund. $\Delta Lund$ is 3 nm for FinFET and NC-FinFET with underlap design (w/ Lund) at various L_{ch}.

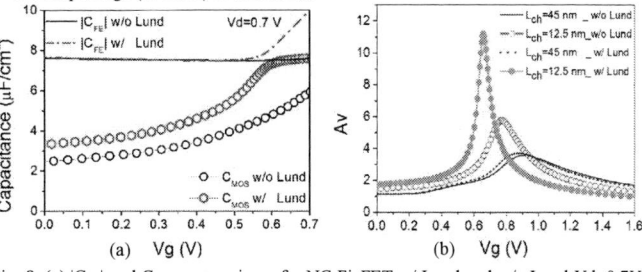

Fig. 8. (a) $|C_{FE}|$ and C_{MOS} comparisons for NC-FinFET w/ Lund and w/o Lund Vd=0.7V. (b) A_v comparisons for NC-FinFET w/ Lund and w/o Lund at Vd=0.7V. With the same L_{ch}, NC-FinFET w/ Lund exhibits better capacitance matching and larger A_v. NC-FinFET with shorter L_{ch} (12.5nm) shows larger A_v than that with longer L_{ch} (45nm).

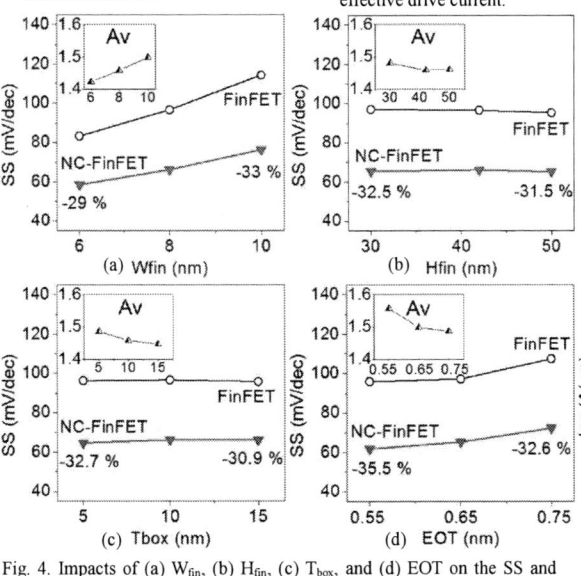

Fig. 4. Impacts of (a) W_{fin}, (b) H_{fin}, (c) T_{box}, and (d) EOT on the SS and amplification factor (A_v at Vg = 0V) at Vd=0.7V. NC-FinFET with higher A_v shows larger improvement in SS compared with the baseline FinFET.

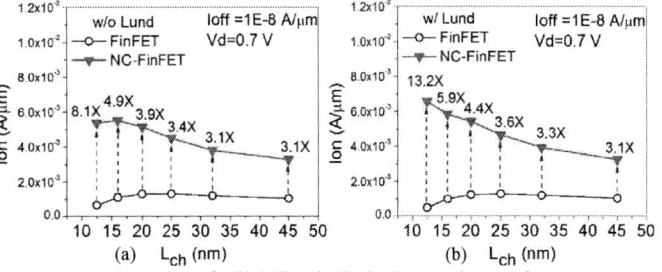

Fig. 9. Ion comparisons for FinFET and NC-FinFET at various L_{ch} for cases (a) w/o Lund and (b) w/ Lund. With the same L_{ch} and fixed Ioff, compared with FinFET, NC-FinFET w/ Lund shows larger improvements in Ion than that w/o Lund.

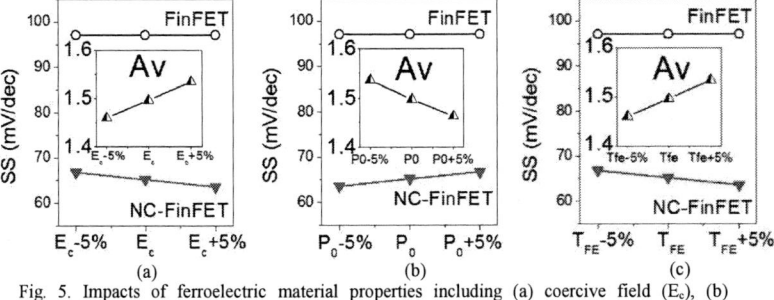

Fig. 5. Impacts of ferroelectric material properties including (a) coercive field (E_c), (b) polarization (P_0), and (c) thickness (T_{FE}) on the SS and A_v (at Vg = 0V) at Vd=0.7V. All cases for NC-FinFET are in hysteresis-free region.

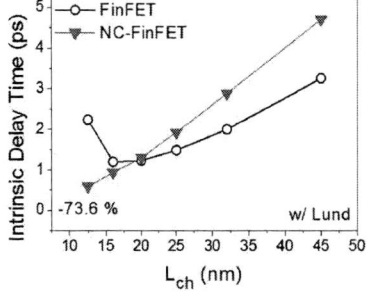

Fig. 10. Intrinsic delay time comparisons for FinFET and NC-FinFET w/ Lund. At $L_{ch} < 20$nm, NC-FinFET exhibits smaller delay than the FinFET due to its larger effective drive current.

978-1-5090-4661-4/17 $31.00 © 2017 IEEE   135

7A-2

Design Space Exploration Considering Back-Gate Biasing Effects for Negative-Capacitance Transition-Metal-Dichalcogenide (TMD) Field-Effect Transistors

Wei-Xiang You and Pin Su

Department of Electronics Engineering & Institute of Electronics, National Chiao Tung University, Taiwan
E-mail: wxyou.ee03g@nctu.edu.tw, pinsu@faculty.nctu.edu.tw

Abstract

In this work, with the aid of an analytical and scalable model, we explore the design space for negative-capacitance (NC) FETs with a 2D semiconducting transition-metal-dichalcogenide (TMD) channel with emphasis on the impact of back-gate biasing. Our study indicates that, to mitigate the conflict between subthreshold swing (SS) and hysteresis and to maximize the design space for the NC-TMDFET, a thin buried oxide (BOX) and an adequate reverse back-gate bias can be applied to achieve the optimum design.
(Keywords: Negative-capacitance FET, 2D material, transition-metal-dichalcogenide)

Introduction

Negative-capacitance field-effect transistor (NCFET) is one of the promising post-CMOS device candidates that may achieve a subthreshold swing (SS) smaller than 60 mV/dec at room temperature [1]. With their atomically-thin thicknesses, pristine surfaces and adequate bandgap energies, 2D transition-metal-dichalcogenides (TMDs) such as MoS_2 [2] are very attractive channel materials, and a NC-TMDFET (Fig. 1(a)) has been demonstrated recently [3]. How to optimize the device design for NC-TMDFET to achieve high voltage amplification (Av) and hysteresis-free operation is an important question and merits investigation.

In this work, with the aid of an analytical and parameter-free model, we explore the design space for NC-TMDFETs. The impact of back-gate biasing (V_{bs}) on the design space is addressed.

Methodology

We have derived an analytical NC-TMDFET model by which the design space can be constructed. The model is enabled by a scalable TMDFET model coupled with the 1-D Landau-Khalatnikov (L-K) equation [4] and has been verified with 2-D numerical simulation (Fig. 2(a)). Our model can also predict the hysteresis as demonstrated in Fig. 2(b).

Results and Discussion

Fig. 3(a) shows that the design space for NC-TMDFET with T_{box} = 10 nm is limited, and the minimum SS can only be designed to 50 mV/dec in the thicker equivalent oxide thickness (EOT) region to avoid hysteresis. Fig. 3(b) shows that the design space can be significantly enlarged by thinning T_{box} to 5 nm, and larger Av can be obtained before hysteresis taking place. Hence, the minimum SS can be designed to 10 mV/dec.

Fig. 4 shows that, for a given device design, the SS can be slightly improved by forward V_{bs}. This is because the decrease of the gate charges of underlying TMDFET ($Q_{g,mos}$) will reduce the ferroelectric capacitance (C_{FE}) and improve the Av and SS. However, it may encounter the hysteresis earlier.

To avoid the hysteresis, the design spaces of NC-TMDFET with reverse V_{bs} for T_{box} = 5 and 10 nm are shown in Fig. 5(a) and 5(b). It can be seen that the design spaces are enlarged compared with Fig. 3(a) and 3(b), respectively, by the reverse V_{bs}.

In Fig. 6, the red circles represent the point at which the hysteresis takes place. By applying reverse V_{bs} to the NC-TMDFET, the gate capacitance curve C_{mos} shifts to the right due to the increase of threshold voltage (V_T). The shift in C_{mos} extends the space to lower the C_{FE} curve and increase Av without hysteresis. Although, under reverse V_{bs}, the increasing $Q_{g,mos}$ enhances the C_{FE} and worsens the Av and SS under a given T_{FE}, the achievable minimum SS can be improved (from 50 to 40 mV/dec) by increasing the T_{FE} without inducing hysteresis. In other words, the design spaces can be enlarged by the reverse V_{bs}.

Fig. 7(a) and 7(b) show that, under reverse V_{bs}, the increase in the V_T of the underlying TMDFET may result in a hysteresis in the subthreshold region.

Fig. 8 shows that, to avoid the hysteresis near (Fig. 2(b)) or below (Fig. 7(a)) threshold, there exists an optimal back-gate bias (e.g. V_{bs} = -0.1V) to maximize the design space of the NC-TMDFET.

Acknowledgements

The authors acknowledge the support from Ministry of Science and Technology, Taiwan, under MOST 105-2221-E-009-147 and MOST 105-2911-I-009-301

References

[1] S. Salahuddin *et al.*, *Nano Lett.*, 8, 405, 2008.

[2] Y. Yoon *et al.*, *Nano Lett.*, 11, 3768, 2011.

[3] F. McGuire *et al.*, *APL*, 109, 093101, 2016.

[4] L. D. Landau *et al.*, *Dokl. Akad. Nauk*, 96, 469, 1954.

[5] M. Kobayashi *et al.*, *AIP*, 6, 025113, 2016.

978-1-5090-4661-4/17 $31.00 © 2017 IEEE

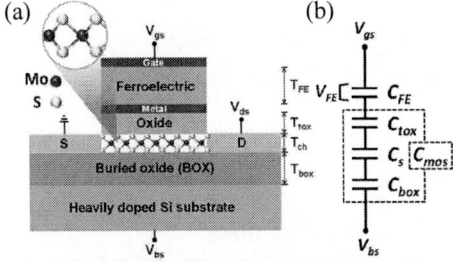

Fig. 1: (a) Schematic of an NC-TMDFET with a ferroelectric on top of a MOSFET with a 2D TMD channel ($T_{ch} \sim 0.65$ nm). (b) The equivalent capacitor network of the NC-TMDFET with C_{FE} the ferroelectric capacitance and C_{mos} the gate capacitance of the underlying TMDFET.

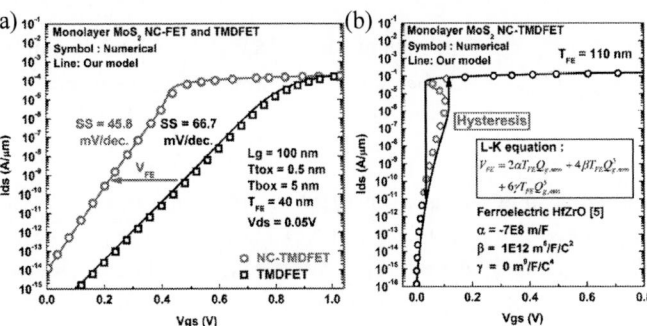

Fig. 2: (a) Comparison of the model and numerical simulation for a hysteresis-free NC-TMDFET and underlying TMDFET. V_{FE} is the voltage drop across the ferroelectric (b) The model can predict the hysteresis of the NC-TMDFET.

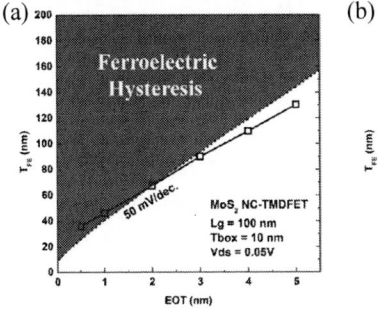

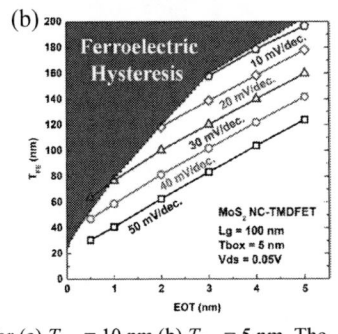

Fig. 3: Design spaces of NC-TMDFET for (a) $T_{box} = 10$ nm (b) $T_{box} = 5$ nm. The design space can be enlarged by thinning T_{box} to 5 nm, and thicker EOT is favorable to obtain steeper SS.

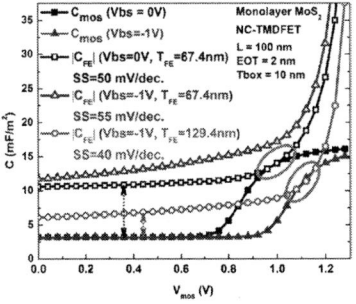

Fig. 4: The SS of NC-TMDFET changes with V_{bs}, and hysteresis may occur for NC-TMDFET with forward V_{bs}.

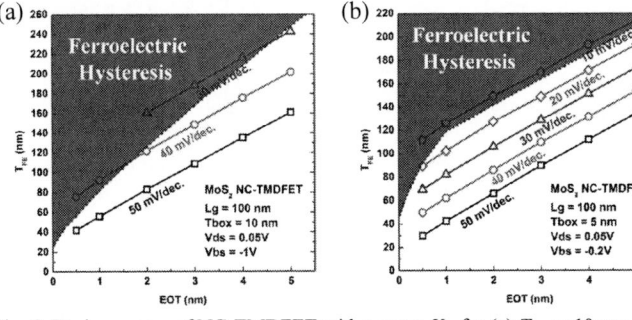
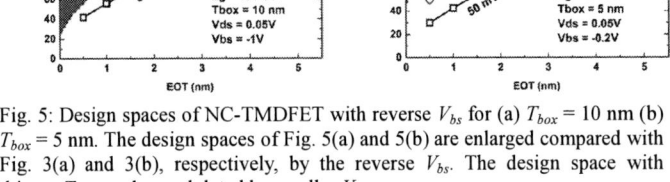

Fig. 5: Design spaces of NC-TMDFET with reverse V_{bs} for (a) $T_{box} = 10$ nm (b) $T_{box} = 5$ nm. The design spaces of Fig. 5(a) and 5(b) are enlarged compared with Fig. 3(a) and 3(b), respectively, by the reverse V_{bs}. The design space with thinner T_{box} can be modulated by smaller V_{bs}.

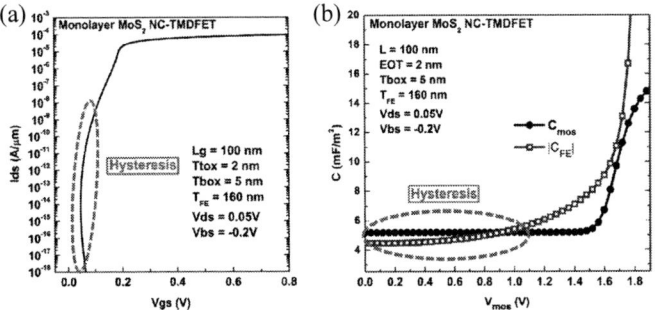

Fig. 6: By applying reverse V_{bs} to NC-TMDFET, the achievable minimum SS can be improved from 50 to 40 mV/dec by increasing the T_{FE} without inducing hysteresis.

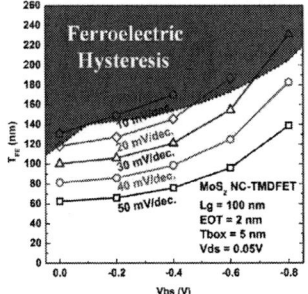

Fig. 7: (a) The hysteresis may occur in the subthreshold region under reverse V_{bs}. (b) The subthreshold hysteresis results from the intersecting of C_{FE} and C_{mos} at subthreshold.

Fig. 8: There exists an optimal back-gate bias (e.g. $V_{bs} = -0.1$V) to maximize the design space of NC-TMDFET with $T_{box} = 5$ nm.

978-1-5090-4661-4/17 $31.00 © 2017 IEEE   137

7A-3

A simple way to grow large-area single-layer MoS₂ film by chemical vapor deposition

Yan-Cong Qiao[1,2], Zhen Yang[1,2], Hai-Ming Zhao[1,2], Xue-Feng Wang[1,2], Lu-Qi Tao[1,2], Yi Yang[1,2], Tian-Ling Ren[1,2,*]

[1]Institute of Microelectronics, Tsinghua University, Beijing 100084, China,

[2]Tsinghua National Laboratory for Information Science and Technology (TNList), Tsinghua University, Beijing 100084, China.*corresponding author: RenTL@tsinghua.edu.cn

Abstract

We demonstrate a simple way to grow continuous and single-layer MoS₂ film by LPCVD (Low Pressure Chemical Vapor Deposition) in centimeter scale. The as-grown MoS₂ films are characterized by optical microscopy (OM), scanning electron microscopy (SEM), Raman spectroscopy, photoluminescence (PL) spectroscopy, atomic force microscopy (AFM) and X-ray photoelectron spectroscopy (XPS). The layer number of MoS₂ can be controlled by changing the amount of MoO_3 and S.

(Keywords: MoS₂, LPCVD and thickness control)

Introduction

Since the discovery of grapheme [1], two-dimensional materials have attracted much attention. However, logic devices fabricated with graphene have a very low on-off ratio [2]. Therefore, MoS₂ has been focused by more and more researchers due to its nonzero band gap [3]. At present, many ways to grow MoS₂ by chemical vapor deposition have been proposed [4,5,6,7], but most of these methods are expensive, complex and time-cost. Here, we propose a simple way to grow continuous and large-area single-layer MoS₂ film. The layer number of MoS₂ can be controlled by adjusting the amount of MoO_3 and S.

Experiment

The Si with 300 nm SiO_2 layer was chosen as the substrate. Before growing, the substrate was sonicated in acetone, isopropyl alcohol and deionized water for 5 minutes each by each, then treated by Piranha solution (1:3 mixture of H_2O_2 and H_2SO_4) for 15 min and O_2 plasma for 5 min, which is important to the growth. After plasma treatment, the substrate was placed face-up in a short quartz tube and put at the center of a horizontal tube furnace (MTI) immediately. 3 mg MoO_3 (>99.8%, Alfa Aesar) and 40 mg S (>99.5%, Alfa Aesar) were placed into the same short tubes as reactants, as shown in Fig.1. The CVD growth was performed at low pressure with ultrahigh-purity Argon as carrier gas. The furnace was heated to 300 ℃ and kept for 1h with Argon continuously streamed through the chamber under a flow rate of 500 sccm. After that, the temperature was ramped to 730 ℃ in 30 min and decreased to 700 ℃ under 100 sccm Argon in 30 min with the pressure of 100 Pa. Finally, the furnace was cooled down to 570 ℃ naturally and rapidly cooled to room temperature.

Result and discussion

Raman spectra shows the difference between two characteristic peaks E^1_{2g} and A_{1g} are 22 cm⁻¹ indicating the as-made MoS₂ are two or three layers. When the amount of MoO_3 and S is halved, the difference shifts to 18.6 cm⁻¹ revealing single-layer MoS₂ was achieved (Fig.2a). Fig.3 displays the OM and SEM graphs of samples where the size of single-layer MoS₂ film can be up to centimeter. The number of inset graphs correspond to area numbers on the substrate, respectively. Continuous film is located in the upstream direction and scattered triangular MoS₂ crystals are found in the downstream direction, where triangles become more and more small. It is confirmed by PL spectrum, where two pronounced emission peaks at 630.6 and 678.6 nm are observed (Fig.2b). Furthermore, XPS is used to probe the chemical composition of the as-grown simples, Mo 3d peaks at 232.6 eV and 229.5 eV correspond to the $3d_{5/2}$ and $3d_{3/2}$ doublet and S 2s peak is at 227.2 eV. Peaks at 164.0 eV and 162.8 eV correspond to the S $2p_{1/2}$ and S $2p_{3/2}$ orbitals (Fig 3c,d). The thickness of samples was confirmed by AFM shown in Fig.4. The height of single-layer MoS₂ is around 0.9nm.

To further optimize the optimum condition of the growth, parameters which influences the results of experiment such as temperature, substrate and mass ratio of MoO_3 to S were researched. When the temperature was adjusted to 770 ℃ without changing any other parameters, the difference between two characteristic peaks E^1_{2g} and A_{1g} became larger and emission peaks at 630.6 nm became weaker. Those all indicated that as the temperature increased, MoS₂ prepared by the method mentioned above tended to grow vertically rather than laterally. Triangular crystals couldn't be found on the Si/SiO₂ substrate if the substrate wasn't treated by Piranha solution and O_2 plasma previously. Therefore, active surface play an important role in the crystallization of MoS₂. Three experiments with different mass ratio of MoO_3 to S (1:8, 1:13 and 1:18) were also conducted. A large amount of contamination was found on samples with mass ratio of 1:8 and 1:18. However, samples with the mass ratio of 1:13 have the optimum result.

978-1-5090-4661-4/17 $31.00 © 2017 IEEE

Conclusion

In conclusion, we synthesized centimeter-scale single-layer MoS_2 thin films on Si/SiO_2 substrates by LPCVD using MoO_3 and S, whose characteristics were confirmed by OM, SEM, AFM, Raman and PL spectra and XPS. The optimum of temperature and mass ratio of MoO_3 to S are 730 ℃ and 1:13 respectively. These results will provide potentially important applications in electronic and photoelectronic devices.

Acknowledgments

This work was supported by the National Natural Science Foundation (61574083, 61434001), National Basic Research Program (2015CB352101), National Key Research and Development Program (2016YFA0200400), National Key Project of Science and Technology (2011ZX02403-002), and Special Fund for Agroscientific Research in the Public Interest (201303107) of China. The authors are also thankful for the support of the Independent Research Program (2014Z01006) of Tsinghua University, and Advanced Sensor and Integrated System Lab of Tsinghua University Graduate School at Shenzhen (ZDSYS20140509172959969).

References

[1] Novoselov K S, Geim A K, Morozov S V, et al. Electric field effect in atomically thin carbon films[J]. science, 2004, 306(5696): 666-669.

[2] Schwierz F. Graphene transistors[J]. Nature nanotechnology, 2010, 5(7): 487-496.

[3] Splendiani A, Sun L, Zhang Y, et al. Emerging photoluminescence in monolayer MoS2[J]. Nano letters, 2010, 10(4): 1271-1275.

[4] Kang K, Xie S, Huang L, et al. High-mobility three-atom-thick semiconducting films with wafer-scale homogeneity[J]. Nature, 2015, 520(7549): 656-660.

[5] Chen J, Tang W, Tian B, et al. Chemical Vapor Deposition of High‐Quality Large‐Sized MoS2 Crystals on Silicon Dioxide Substrates[J]. Advanced Science, 2016

[6] Van der Zande A M, Huang P Y, Chenet D A, et al. Grains and grain boundaries in highly crystalline monolayer molybdenum disulphide[J]. Nature materials, 2013, 12(6): 554-561.

[7] Wang S, Rong Y, Fan Y, et al. Shape evolution of monolayer MoS2 crystals grown by chemical vapor deposition[J]. Chemistry of Materials, 2014, 26(22): 6371-6379.

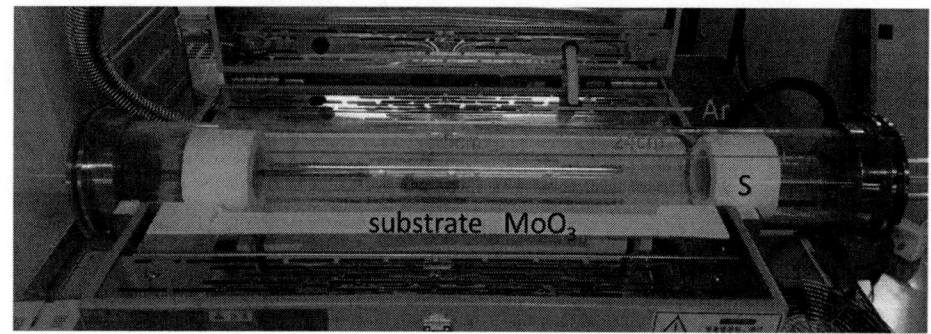

Fig.1: Configuration of the LPCVD furnace for growing single-layer.

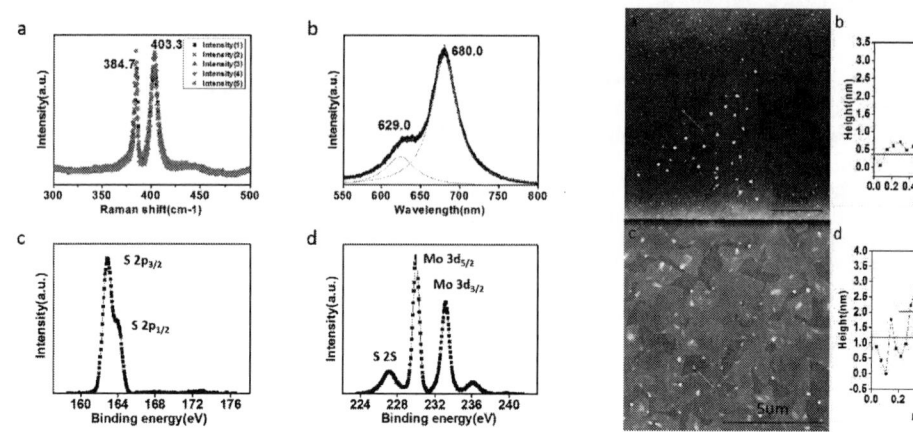

Fig.2: (a) Raman spectra and (b) Photoluminescence (PL) of as-grown single-layer MoS_2 film. (c)XPS spectra of single-layer MoS_2 films.

Fig.4: (a,c) AFM images and (b,d) cross-sectional profiles of the multi-layer and single-layer MoS_2, respectively.

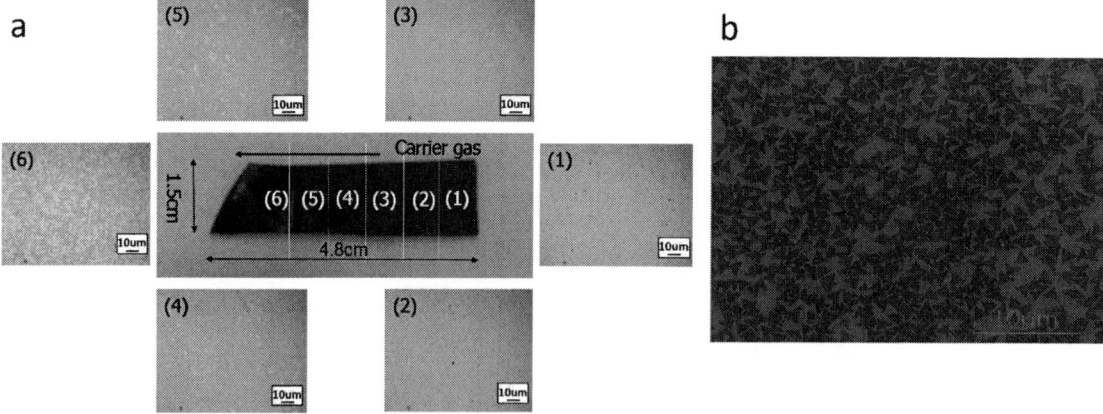

Fig.3: (a) Optical images of single-layer MoS_2 after LPCVD growth. (b) SEM images of the single-layer and multi-layer simples, respectively.

High-mobility and H$_2$-anneal Tolerant InGaSiO/InGaZnO/InGaSiO Double Hetero Channel Thin Film Transistor for Si-LSI Compatible Process

Nobuyoshi Saito, Kentaro Miura, Tomomasa Ueda, Tsutomu Tezuka, and Keiji Ikeda

Advanced LSI Technology Laboratory, Corporate Research & Development Center, Toshiba Corporation,

1, Komukai Toshiba-Cho, Saiwai-ku, Kawasaki, 212-8582, Japan

Email: nobuyoshi.saito@toshiba.co.jp

Abstract

We demonstrate a high-mobility and H$_2$-anneal tolerant InGaSiO/InGaZnO/InGaSiO double heterochannel (DH) TFT for 3D-LSI applications. A novel oxide semiconductor (OS) material, InGaSiO (Si/In ratio > 0.47) was found to exhibit semiconductor property even after H$_2$ annealing at 380°C, whereas a conventional InGaZnO layer changed into a metallic one. Moreover, the DH channel TFT was operated in an enhancement mode and achieved a high mobility of 30 cm^2/Vs after the H$_2$ annealing.

(Keywords: oxide semiconductor, thermal stability, mobility, and BEOL transistor, double heterochannel)

Introduction

Recently, wide-bandgap oxide semiconductor (OS) TFTs have attracted attention as BEOL transistors for 3D-LSI applications because of their unique characteristics i.e., extremely low off-state current (<10^{-22}A/μm), high breakdown voltage (V_{BD}>40V) and low-temperature process (<400°C). Thus, the various applications, such as high voltage I/Os and embedded memory, have been intensively investigated [1-5]. However, the thermal instability of OS TFTs caused by oxygen vacancy formation during H$_2$ annealing [6, 7] is an obstacle to integration with Si-CMOS LSIs. Moreover, typical OS channel InGaZnO shows low mobility (~10 cm^2/Vs), which limits the applications of OS TFTs. Although higher mobility values have been reported for other OS materials [8, 9], the stability against the H$_2$ annealing is not clear.

In this paper, for the first time, we demonstrate the improvement of mobility up to 30 cm^2/Vs and thermal stability against H$_2$ annealing, by sequentially stacking the novel InGaSiO bottom and top layers and an InGaZnO core layer to form DH channel structure. .

Optimization of Si composition for novel OS material; InGaSiO

Fig.1(a) shows a concept of improvement of thermal stability against H$_2$ annealing by replacing Zn in InGaZnO to Si, resulting in the novel material, InGaSiO. This idea is based on a comparison of the dissociation energies of In-O, Ga-O, Zn-O and Si-O bonds (Fig.1(b), [10, 11]). Since the Si-O bond exhibits much larger dissociation energy than the Zn-O bond, replacing Zn atoms to Si atoms are expected to suppress oxygen vacancy formation during H$_2$ annealing.

We fabricated InGaSiO films having various Si fractions by co-sputtering with a polycrystalline InGaSiO$_4$ target and a polycrystalline In$_2$O$_3$ target. As shown in Fig. 2, the Si/In ratio of the InGaSiO films was successfully controlled by changing the input RF power to the In$_2$O$_3$ target with a constant RF power to the InGaSiO$_4$ target.

The deposited OS films were annealed for 1 hour at various temperatures in N$_2$ and mixed atmosphere of N$_2$ and H$_2$ (2%), then the sheet resistance values were evaluated as shown in Fig.3. Compared to N$_2$ annealing, the sheet resistance of OS films decreased at lower temperature than in H$_2$ annealing. It is supposed that hydrogen atoms accelerate oxygen vacancy formation in OS film due to its reducing property. Although the sheet resistance of InGaZnO film decreased severely after annealing at 320°C, that of InGaSiO (Si/In ratio > 0.47) maintained higher resistivity of more than 5 orders of magnitudes after H$_2$ annealing at 320°C and higher temperatures.

Fig. 4 shows a TEM image of the fabricated OS channel bottom-gated TFT on a SiO$_2$ layer formed on a Si substrate. The thickness of the OS channel and the gate insulator was 25nm and 40nm, respectively.

Fig. 5 shows a comparison of I_d-V_g characteristics between (a) InGaZnO TFT and (b) InGaSiO TFT (Si/In ratio =0.47) after the H$_2$ annealing. Although the enhancement mode operation was kept for both InGaZnO and InGaSiO TFTs up to 360°C, only the InGaSiO TFT remained as enhancement mode while the InGaZnO TFT changed to depletion mode after the annealing at 380°C. The better stability against the H2 anneal of the InGaSiO TFT is consistent with the results in Fig.3.

Fig. 6 shows a clear trade-off between the V_{th} stability and high I_{on} as a function of Si/In ratio after the H$_2$ annealing at 380°C. It is shown that higher Si fraction is effective to improve the tolerance to the H$_2$ annealing. However, the current drivability is more degraded and the electron mobility for the InGaSiO TFT was estimated to be 0.14 cm^2/Vs for Si/In ratio =0.47, which is only about 1% of the mobility for the InGaZnO TFT.

Breaking the trade-off by InGaSiO/InGaZnO/InGaSiO DH TFTs

In order to improve both thermal stability and mobility, we examined a DH TFT composed of an InGaZnO core layer and thermally stable InGaSiO top/bottom barrier layers as shown in Fig. 7. Here, UPS and XPS analyses revealed that the band structure of the DH structure was type-I quantum well and the conduction band offset was 0.7eV. Since electrons are confined in the InGaZnO layer with higher mobility than the other, mobility improvement is expected by carrier separation from the SiO$_2$ interface via the InGaSiO barrier layers.

Fig. 8 shows I_d-V_g characteristics of the InGaSiO/InGaZnO/InGaSiO (5nm/10nm/2nm) DH TFT after H$_2$ annealing at 380°C. About two orders of magnitude higher current drive was observed as expected than that of the InGaSiO TFT. Furthermore, the enhancement mode operation was maintained, suggesting that the InGaSiO barrier layers suppress dissociation of Zn-O bonds in the InGaZnO core layer via blocking oxygen diffusion out form the core layer.

Fig.9 shows effective electron mobility in the InGaZnO TFT and the DH TFT extracted by split CV method (L/W=50μm/50μm). Almost 3 times higher mobility than that of the InGaZnO TFT, was realized at N_s=2x10^{12} cm^{-2} in the DH TFT. This result is consistent with the expectation of mobility improvement described above.

Fig.10 shows a relationship between mobility (at N_s=2x10^{12} cm^{-2}) and H$_2$-anneal tolerance (V_{th} after H$_2$ annealing at 380°C, L/W=2μm/2μm). It is shown that the proposed InGaSiO/InGaZnO DH TFT achieved both of better tolerance (stable enhancement mode operation) and higher mobility than those of InGaZnO and InGaSiO single-layer TFTs.

Conclusion

We have developed a high-performance OS TFT with excellent H$_2$-anneal tolerance and high mobility for the first time. A newly developed OS material, InGaSiO (Si/In ratio > 0.47) was found to show high resistivity (>1x10^{10} Ω/sq.) after H$_2$ annealing at 380°C. Enhancement mode operation was confirmed for the InGaSiO TFTs after the annealing. Novel DH TFTs composed of InGaSiO and InGaZnO layers were also examined and achieved both of enhancement mode operation and high mobility of 30 cm^2/Vs even after H$_2$ annealing. These results indicate that the proposed DH TFTs have great potential to serve high-performance BEOL transistors for 3D-LSI applications.

Acknowledgements

The authors thank S. Nakano, Y. Maeda, T. Morooka and T. Ohguro for sample preparation and helpful discussions.

References

[1] K. Nomura et al., Nature,432, 488(2004), [2] T. Aoki et al., ISSCC 2014, p502, [3] Y. Kobayashi et al., VLSI 2104, p170, [4] K. Kaneko et al., VLSI2011, p120, [5] K. Ota et al., VLSI 2105, p214, [6] Y. Hanyu et al.,APL, 103, 202114(2013), [7] M. Nakashima et al., JAP 116, 213703(2014), [8] J. C. Park et al., IEDM2009, p191, [9] C. W. Shih et al., IEDM2015, p145, [10] Y. R. Luo, Handbook of Chemistry and Physics, [11] N. Mitoma et al., APL, 104, 102103(2014).

978-1-5090-4661-4/17 $31.00 © 2017 IEEE

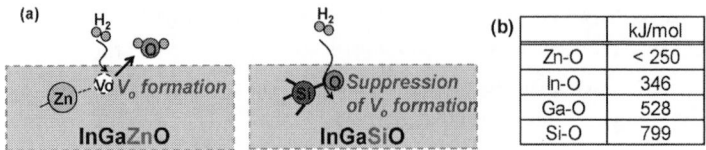

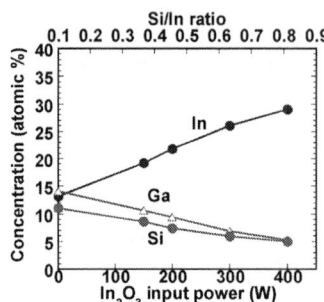

Fig.1 (a) Concept of the improvement of H₂-anneal tolerance and (b) list of metal-oxygen bond dissociation energy of InGaZnO related metals and Si [7]. Since the Si-O bond exhibits much larger dissociation energy than the Zn-O bond, replacing Zn atoms to Si atoms are expected to suppress oxygen vacancy formation during H₂ annealing.

Fig.2 In₂O₃ input power dependence of InGaSiO film composition. The input RF power to the InGaZnO₄ target was fixed at 400W. The Si/In ratio of the InGaSiO films was successfully controlled by changing the input RF power to the In₂O₃ target. O concentration was almost constant (61%).

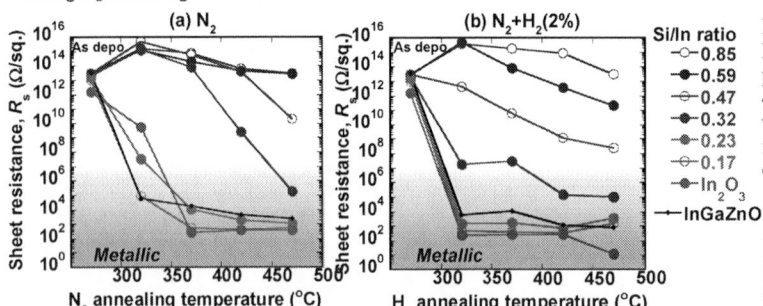

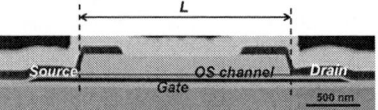

Fig.3 Annealing temperature dependence of sheet resistance of InGaSiO and InGaZnO film after annealing in (a) N₂ and (b) H₂ annealing for 1 hour. Although the sheet resistance of InGaZnO film decreases severely after annealing at 320°C, that of InGaSiO (Si/In ratio >0.47) keeps higher resistivity (>1×10¹⁰Ω/sq.) even after H₂ annealing at 380°C, which suggests that InGaSiO maintains semiconductor properties.

Fig.4 Cross-sectional TEM image of the fabricated OS channel bottom-gated TFT on a SiO₂ layer formed on a Si substrate. The thickness of the OS channel and the gate insulator was 25nm and 40nm, respectively.

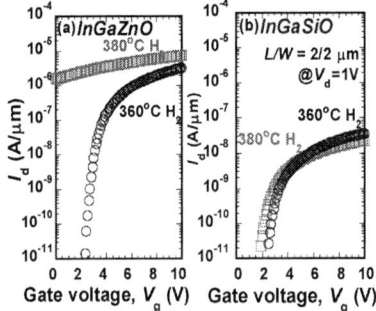

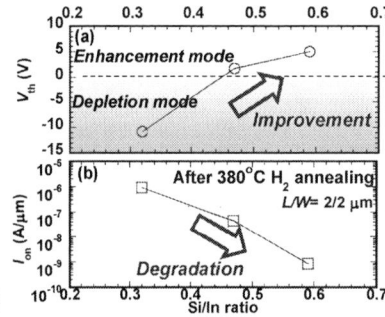

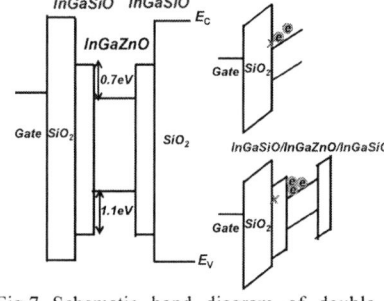

Fig.5 Comparison of I_d-V_g characteristics between (a)InGaZnO and (b)InGaSiO TFT. Although the enhancement mode operation was kept for both InGaZnO and InGaSiO TFTs up to 360°C, only the InGaSiO TFT remained as enhancement mode after the annealing at 380°C.

Fig.6 Si/In ratio dependence of (a) V_{th} and (b) I_{on} after H₂ annealing at 380°C. V_{th} is V_g at I_d=1nA/μm. I_{on} is I_d at V_g=20V and V_d=1V. Higher Si fraction is effective to improve the thermal stability of InGaSiO, however, there is a trade-off between current drivability and thermal stability.

Fig.7 Schematic band diagram of double heterochannel (DH) composed of InGaSiO and InGaZnO layers. The conduction band offset was 0.7 eV by UPS and XPS analyses. Mobility improvement is expected by carrier separation from the SiO₂ interface via the InGaSiO barrier layers.

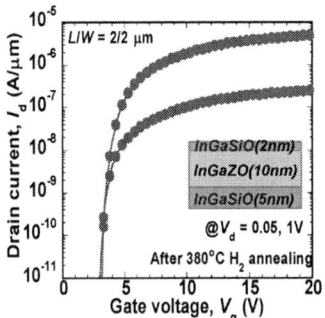

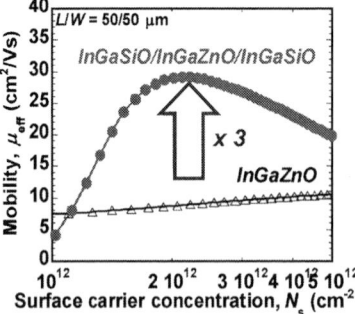

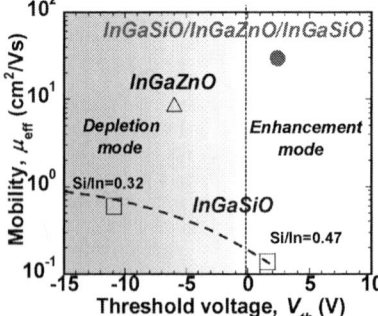

Fig.8 I_d-V_g characteristics of InGaSiO/InGaZnO/InGaSiO DH TFT after H₂ annealing at 380°C. Both of high current drivability and enhancement mode operation were achieved.

Fig.9 Comparison of mobility between characteristics between InGaZnO and InGaSiO/InGaZnO/InGaSiO DH TFT extracted by split CV method. Almost 3 times higher mobility was realized in the proposed DH TFT.

Fig.10 Relationship between mobility at Ns=2×10¹²cm⁻² and H₂-anneal tolerance (V_{th} after H₂ annealing at 380°C, L/W=2μm/2μm). The proposed InGaSiO/InGaZnO DH TFT achieved both of better tolerance (stable enhancement mode operation) and higher mobility than those of InGaZnO and InGaSiO single-layer TFTs.

978-1-5090-4661-4/17 $31.00 © 2017 IEEE

7B-1 (Invited)

Growing Market of MEMS and Technology Development in Process and Tools Specialized to MEMS

Hiroshi Yanazawa* and Kohji Homma

MEMS-CORE Co. Ltd., Sendai, Japan

Abstract

In this paper, we will review the MEMS device applications and market trend/forecast. And introduce technology development in MEMS industry.
(Keywords: MEMS (Micro Electro Mechanical System))

MEMS device application

A broad history of MEMS device application is shown in Fig.1. Lately, market seems to going into the second generation, by development of multi-functional combo and the idea of wearable sensors. Application of MEMS device will be expanded more and more into our daily life. This implies further market growth. Here the report of Yole Development [1] is cited in Fig.2. The global market in 2012 was 10,000M$, and it expanded to 13,000M$ in 2014 and be forecasted to be 24,000M$ in 2019, 13% annual growth.

Special technologies in MEMS fabrication

In MEMS device, different from LSI, not only surface but bulk silicon is processed.

Deep Si Etching

Highly anisotropic plasma is exposed to Si surface, and by repeating etching-deposition cycle deep trench with vertical wall is formed in Si substrate. In extreme case, trench reaches to the back surface which is used as feed through wiring.

Sacrificial layer etching

Usually SOI (silicon-on-insulator) wafer is adopted for this process. For example, after making a Si pattern on silicon dioxide, etching off the underneath silicon dioxide results empty room under silicon which can be used for movement.

Stealth dicing

In many of MEMS device, there are fine moving parts, so, such structure is fragile. The blade dicing easily destroy the structure. In stealth dicing, laser exposure into silicon single crystal makes small defects and expanding force makes chips separate without damage to the structure.

Technology development in MEMS-CORE

We have dedicated MEMS line for 6 and 4 inch Si.

Lithography; A contact type exposure system is used with double side alignment function. For resist coating, in addition to conventional spin-coating, we adopt spray-coating and dry film laminating depend on purpose. Resist stripping are carried out by wet chemical or plasma ashing.

Deposition; Multi-target load-lock sputtering machine is used for deposition Cr, Ti, Ni, Al, Pt, Au etc. Electro-plating of Cu, Au and Ni are available. Lift-off process for Al wiring is also available. For insulator deposition, atmospheric pressure plasma deposition is adopted, which deposits TEOS film with excellent step coverage. Conventional thermal oxidation of Si is available.

Etching; Deep Si etching and isotropic plasma etching is selected for device requirement. For most of metal etching, ion-milling technique is adopted. For sacrificial SiO_2 etching, HF vapor phase etching or HF gas phase etching is used. By gas phase etching system, hydrophobic surface treatment is available, which is effective to prevent sticking of moving part.

We are frequently using equipment in Tohoku Univ. Hands-on-access fab.

Fig.3 shows HF vapor phase etcher and Fig.4 shows an example of SiO_2 sacrificial etching. After Si active layer was etched by anisotropic deep etch, SiO_2 box layer was etched by HF vapor into 15 μ m depth.

Acknowledgments

The authors gratefully acknowledge to Professor Esashi, Professor Tanaka and Associate Professor Totsu of Tohoku University for their daily technical support and advice.

References

[1] Jean-ChristopheELOY "How to increase the value in MEMS devices" TST MEMS Sensor(2) Semicon Japan 2014.

*Contact person; e-mail yanazawa@mems-core.com

978-1-5090-4661-4/17 $31.00 © 2017 IEEE

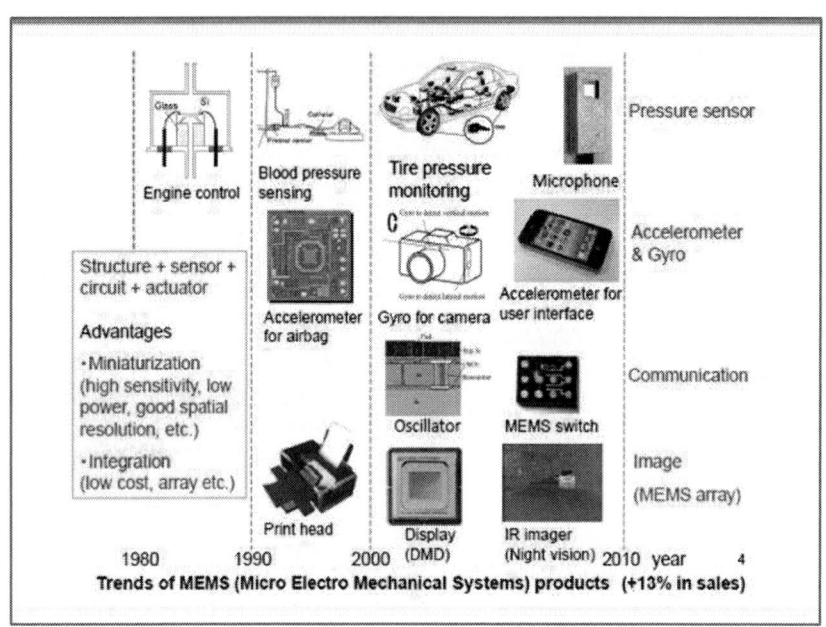

Fig.1 Various applications of MEMS (Micro Electro Mechanical System)

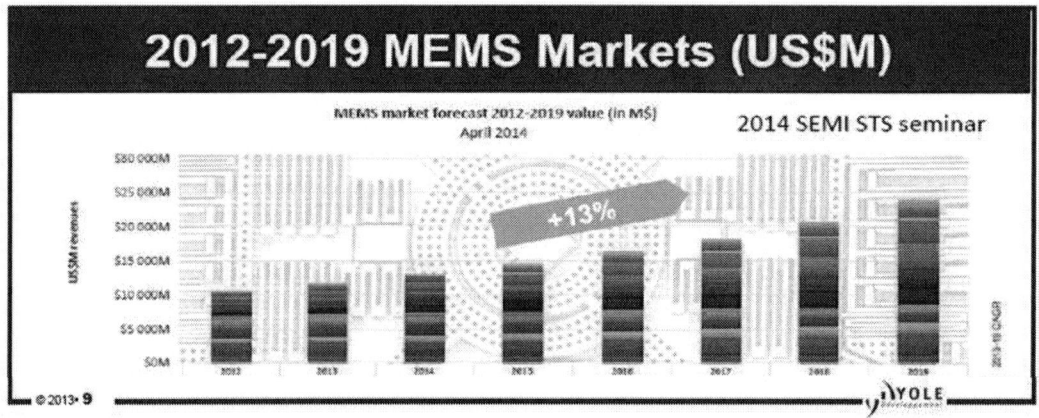

Fig.2 Market Trend Forecast, 13% annual increase up to 2019.

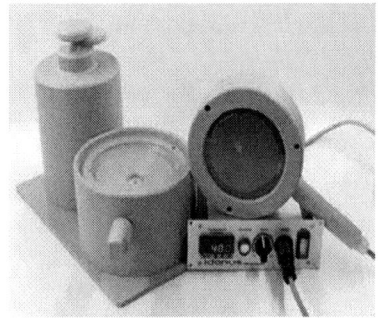

Fig.3 HF Vapor Phase Etcher
(product of idonus/Switzerland)

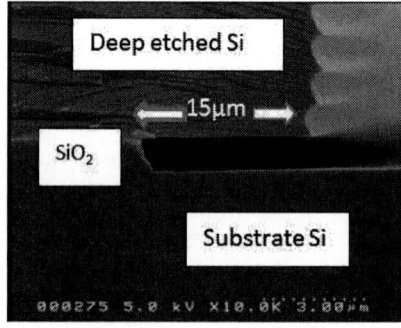

Fig.4 Example of SiO_2 sacrificial etch

7B-2

Development of MEMS Vibrating Sensor with Phase-Shifted Optical Pulse Interferometry

Yusaku Ohe[1], Hitoshi Kimura[1], Norio Inou[1],
Yoshiharu Hirayama[2] and Minoru Yoshida[2]

[1]Tokyo Institute of Technology, [2]Hakusan Corporation

Abstract

This study aims to develop a precision sensor to realize a real-time measurement system for ground motion or dynamic behaviors of large structures. The proposed sensor uses optical interferometry that requires no electric power, which has an advantage of long-term measurement. This paper describes an experimental study of a MEMS vibration sensor. The result shows that the sensor is able to measure vibration of the sensor mass part. In addition, we have confirmed the possibility of a real-time measurement of the acceleration.

(Keywords: Motion measurement system, Optical interferometry and MEMS sensor)

Introduction

A real-time measurement system with a large number of sensors under the sea is expected to be useful for research of oil and natural gas resource as well as prevention of earthquakes. It is desirable for reliable long-term measurement to realize a sensor which does not require electric power source [1] [2].

We proposed a measurement system with using phase-shifted optical pulse interferometry [3]. The proposed system measures the displacement of the sensor mass part by laser pulses and the sensor module does not require electric power. This paper describes a MEMS vibration sensor for reliable long-term measurement.

MEMS vibration sensor

The proposed optical sensor is composed of a spring -mass structure made with MEMS as shown in Fig.1. With reflecting laser on the mass part, displacement of the mass part is detected with optical interferometry.

Fig. 2 (right side) shows a prototype of the MEMS sensor. The vibrating part of the sensor which consists of sensor mass part and springs is manufactured of silicon single crystal with MEMS. The size is about 10 mm square, and the weight is about 170 mg.

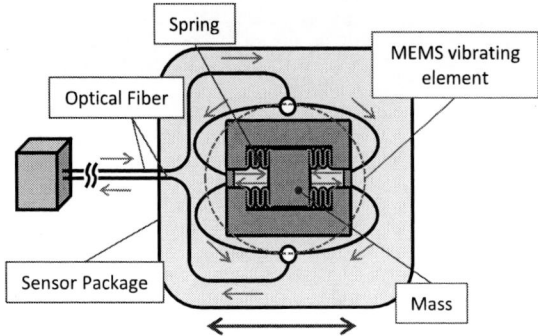

Fig. 1 Diagram of the optical interferometry sensor

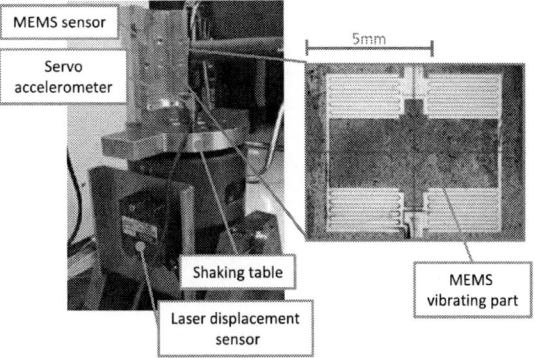

Fig. 2 Experimental system with MEMS sensor

Experiment

We made an excitation experiment to evaluate the performance of the sensor. Several patterns of acceleration are measured with a shaking table as shown in Fig. 2 (left side) under the following conditions.

- 10 Hz ~ 400 Hz sinusoidal input
- 0.1 Hz square wave input

With this experiment, the MEMS sensor and a servo accelerometer measure the vibration of the shaking table. The displacement of the table is measured by laser displacement sensor. The time stamps of the three sensors are synchronized by GPS timer and a vibration input of handy click by a hummer.

From the sinusoidal excitation test, natural frequency of the sensor was estimated as about 93.5 Hz. Fig. 3 shows the displacement of the shaking table measured with a laser displacement sensor, and

978-1-5090-4661-4/17 $31.00 © 2017 IEEE   145

Fig. 4 shows the acceleration measured with a servo accelerometer in the square excitation test. Fig. 5 shows the response from the optical sensor. The response curve shows that the system can detect the displacement of the sensor mass part with accuracy of nanometers.

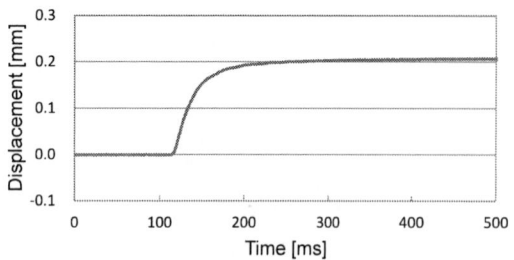

Fig. 3 Displacement of the shaking table

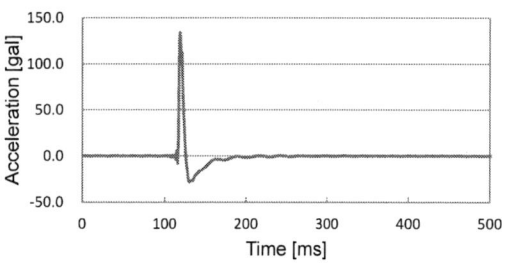

Fig. 4 Acceleration of the shaking table

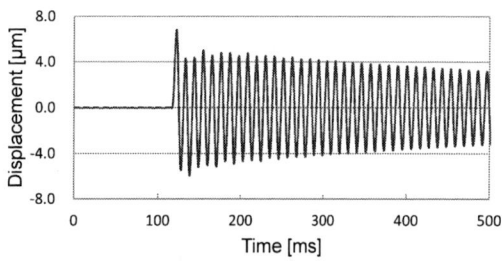

Fig. 5 Displacement of the sensor oscillator

Performance evaluation

We attempted to estimate the motion of the MEMS sensor package from the displacement of the sensor mass part by computer simulation. The sensor can be regarded as a spring-mass-damper model as shown in Fig. 6. The dynamic behavior is expressed by as the follows,

$$\ddot{y} = \ddot{e} + 2\zeta\omega\dot{e} + \omega^2 e \qquad (1)$$

where, y is displacement of the shaking table, and

e is the sensor output. Equation (1) shows that only acceleration of the shaking table affects the senor output.

For the numerical analysis, a finite element model is made as shown in Fig. 7 and the natural frequency was calculated as 101 Hz by eigenvalue analysis, which shows good agreement with the experimental result of 93.5 Hz. Next, we estimated the physical parameters of the MEMS sensor for the dynamic simulation as listed in Table 1 from FE Analysis and the experimental result.

Table.1 Characteristic of the sensor

Spring constant [N/m]	9.19
Mass [kg]	2.26×10^{-5}
Damping ratio	3.74×10^{-3}

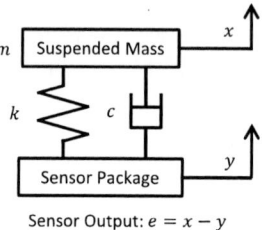

Sensor Output: $e = x - y$

Fig. 6 Spring-mass-damper model

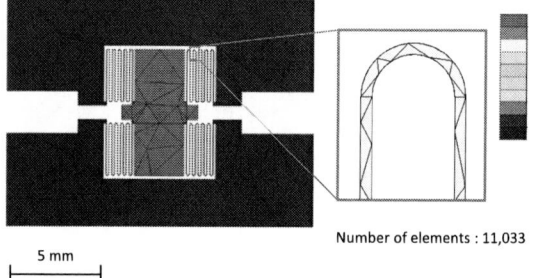

Fig. 7 FEM model of the sensor oscillator

To confirm the practical use of the MEMS sensor, we simulated the acceleration of the shaking table from the MEMS sensor data by using equation (1). In the simulation, high frequency components of the MEMS sensor data are cut to remove noises. Fig. 8 shows a Bode diagram of the low-pass filter. The simulated and experimental data are plotted in Fig. 9. Both profiles showed good agreement. The result supports the possibility of practical use of the

978-1-5090-4661-4/17 $31.00 © 2017 IEEE

proposed MEMS sensor as a real-time accelerometer by using spring-mass-damper model. However, according to FFT result of Fig. 9, the simulated acceleration includes a component of about 40 Hz. A possible reason of this 40 Hz vibration is fixation mechanism of MEMS sensor. Stiffer fixation will eliminate the 40 Hz vibration. Assuming the component is not derived from MEMS sensor, we subtracted the 40 Hz vibration from simulated acceleration data as shown in Fig. 10. The simulation result shows good agreement with experimental the data obtained by the servo accelerometer. For analyzing more detailed performance of the sensor, we will specify and eliminate the 40 Hz vibration.

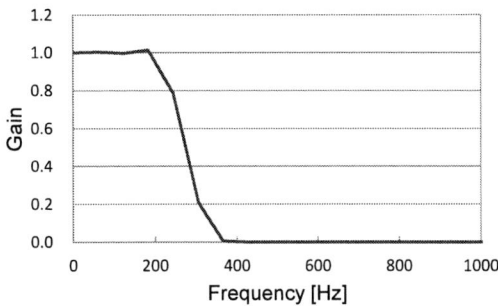

Fig. 8 Low-pass filter

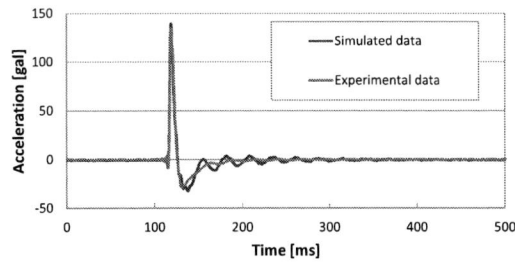

Fig. 9 Simulated acceleration of the shaking table

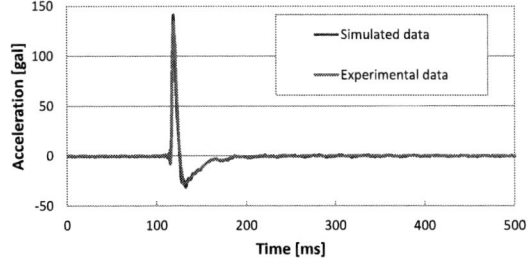

Fig. 10 Simulated acceleration without 40 Hz

Conclusions

We developed a new MEMS vibration sensor that does not require electric power, and proved the basic performance by the excitation experiment. We simulated acceleration of the MEMS sensor package from the sensor output. The simulated data showed almost good agreement with experimental data obtained by the servo accelerometer. The simulation method can realize a real-time measurement.

Acknowledgments

Part of this research was supported by the fund of the technical solution project of Japan Oil, Gas and Metals National Corporation (JOGMEC). We express our great thanks to JOGMEC for supporting this study. We express our great thanks to Prof. Kentaro Nakamura and Yosuke Mizuno, PhD of this institute of technology for their advice about optical interferometry. We also express our great thanks to Shuji Ikeda, PhD of tei solutions for his assistance in developing and manufacturing the MEMS sensor.

References

[1] Nakstad H. and Kringlebotn J. T. 2008 Proc. SPIE 7004 700436

[2] Kringlebotn J. T., Nakstad H. and Eriksrud M. 2009 Proc. SPIE 7503 75037U

[3] M. Yoshida, Y. Hirayama et al, Real-time displacement measurement system using phase-shifted optical pulse interferometry: Application to a seismic observation system Japanese Journal of Applied Physics, Vol.55, 022701(2016)

Microstructuring Polydimethylsiloxane Elastomer Film with 3D Printed Mold for Low Cost and High Sensitivity Flexible Capacitive Pressure Sensor

Bengang Zhuo, Sujie Chen, and Xiaojun Guo

Shanghai Jiao Tong University, Shanghai, China, x.guo@sjtu.edu.cn

Abstract

3D printing is used to fabricate molds for micro-structuring the polydimethylsiloxane (PDMS) film. The fabricated sensor device using the micro-structured PDMS film presents high sensitivity, excellent durability and fast response, and is also shown to perform reliable real-time wrist pulse monitoring.

(Keywords: PDMS, 3D Printing, Flexible Pressure Sensor)

Introduction

Demands for wearable healthcare and patient rehabilitation have brought wide research interests on developing flexible pressure sensors, capable of being directly wrapped on skin surfaces for real-time monitoring of diverse human body motions and physiological signals [1] [2]. The capacitor structure with an elastomeric dielectric layer sandwiched between two flexible electrodes is regarded as a promising flexible pressure sensor solution for its simple structure, ease of processing, and low voltage operation. Polydimethylsiloxane (PDMS) is a popular material of choice for the elastomer dielectric layer. However, the sensor device using a bulk PDMS layer presents a very low sensitivity, and is not capable of reliably sensing weak pressure force. To improve the sensitivity, different approaches have been developed to micro-structure the PDMS film, forming air voids inside the film to reduce elastic resistance and also induce additional change of the effective dielectric constant of the dielectric layer under compression [1] [2]. The most straightforward and reliable approach for forming the micro-structured PDMS film is using a pre-fabricated mold. Silicon wafer molds have been fabricated to obtain different well defined geometries for optimal sensor design. However, it requires expensive and complicated photolithography and chemical etching processes, and is thus challenging for low cost large area manufacturing.

In this work, 3D digital printing is used to fabricate the molds for micro-structuring the PDMS films with low cost. The fabricated sensor device can achieve sensitivity higher than the previously work based on micro-fabricated silicon wafer molds.

Results and Discussions

The fabrication of the mold, the micro-structured PDMS film and the pressure sensor are depicted in Fig. 1. Firstly, a mold with periodical micro-grooves on the surface was fabricated by a 3D printer using acrylonitrile butadiene styrene (ABS) (Fig. 1(a)). The pre-pared mixture of PDMS elastomer (Sylgard 184, Dow Corning) and cross-linker was casted on the mold followed by annealing process. Then, the PDMS layer was peeled off from the mold to obtain a free-standing film with micro-structured surface of periodical line geometries as shown in Fig.1(b). The film was then laminated onto an ITO coated PET film and cut into smaller pieces to be used for making the sensor device as illustrated in Fig. 1(c).

As shown in Fig. 2, in the low pressure regime (<100 Pa), the sensor achieves a sensitivity of 1.623 kPa-1, which is higher than the previous work using micro-fabricated silicon wafer mold of pyramid-structure (0.55 kPa^{-1}) [1]. With a very low detection limit (< 3 Pa), the sensor can reliably detect loading/unloading a rice on top (Fig. 3). The sensor also presents excellent durability (Fig.4) and fast response (34 ms) and recovery (54 ms) for real-time pressure monitoring applications. Finally, the sensor was integrated in a wearable system to demonstrate it capability for real-time monitoring wrist pulse, as shown in Fig. 5.

Conclusion

A low cost and "manufacturing-on-demand" approach based on 3D printed mold is developed for fabricate micro-structured PDMS film and flexible capacitive pressure sensors. The fabricated device presents high sensitivity, excellent durability and fast response, and is also shown to perform reliable real-time wrist pulse monitoring. With the exhibited attractive performance features and low cost processing, this technology is very promising to provide an economic and convenient way for real-time health and sport performance monitoring.

References

[1] S. C. Mannsfeld, et al., "Highly sensitive flexible pressure sensors with microstructured rubber dielectric layers," Nature Materials, vol. 9, no. 10, pp. 859-64, Oct 2010.

[2] S. Chen, B. Zhuo, and X. Guo, "Large area one-step facile processing of microstructured elastomeric dielectric film for high sensitivity and durable sensing over wide pressure range", ACS Appl. Mater. Interfaces, vol. 8, no. 31, pp. 20364-20370, Jul. 2016.

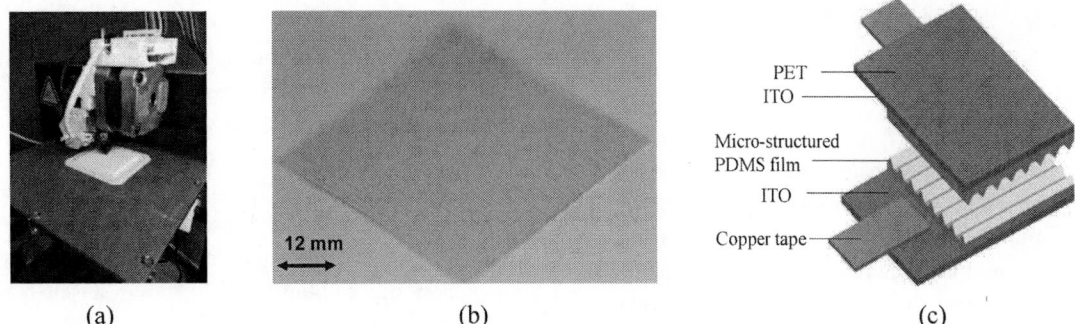

(a)   (b)   (c)

Fig. 1: (a) Photo of the 3D printer for fabricating the mold in ABS; (b) the fabricated free-standing PDMS film with micro-structured surface of periodical line geometries; (c) illustration of the capacitor structure pressure sensor fabricated using the micro-structured PDMS film.

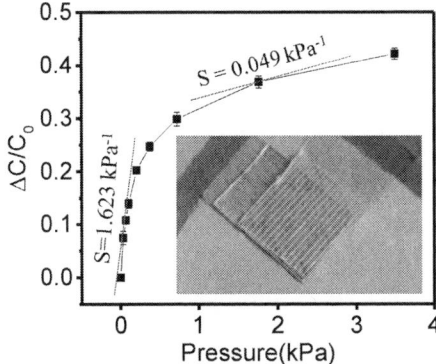

Fig. 2: The measure relative capacitance changes at different applied pressure for the fabricated sensor in the inset, showing a high sensitivity of 1.623 kPa⁻¹ in the low pressure regime.

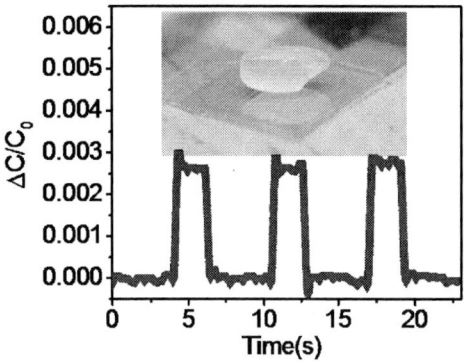

Fig. 3: The measured capacitance changes upon loading and unloading a rice with a corresponding pressure less than 3 Pa, showing the very low detection limit of the sensor.

Fig. 4: Durability test of the sensor by repeating loading/unloading a pressure of 1kPa for more than 1000 cycles.

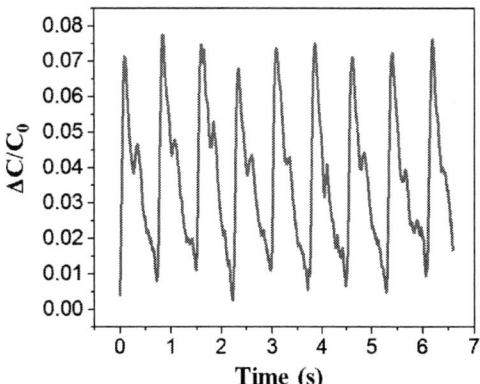

Fig. 5: Real-time wrist pulse measurement results using the fabricated flexible pressure sensor.

978-1-5090-4661-4/17 $31.00 © 2017 IEEE   149

7B-4

Effective Performance of a Tiny-chamber Plasma Etcher in Scallop Reduction

Sommawan Khumpuang[1,2], Hiroyuki Tanaka[1,2], and Shiro Hara[1,2]

[1]AIST, [2]Minimal Fab, Tsukuba, Japan, sommawan.khumpuang@aist.go.jp

Abstract

An inductively-coupled plasma-reactive ion etching system (ICP-RIE) for a half-inch wafer process is developed. The machine is in human-size and supporting clean-localized manufacturing system of minimal fab. A tiny chamber with a volume of $1/4\ell$ enhances performance of gas-replacement inside the etching chamber. The gas switching time between etching and polymeric passivation for a Bosch process at 1 second is applied where the gas resident time in the etching chamber is as short as 0.2 second. A high frequency of 100 MHz is employed for a high-density plasma formation with power consumption less than 50W. Etching result of a straight side-wall with scallop-less silicon structure is obtained without additional process applied.

(Keywords: Minimal fab, Bosch process and Plasma etching)

Introduction

In many applications of silicon etching, sidewall smoothness and taper angle are important especially in optical device fabrication. Bosch process [2] is well employed for etching a low-tapered silicon trench structure. A cycle of etching process and polymeric passivation is performed by switching gas between SF_6 and C_4F_8, respectively. The longer gas switching rate, the larger scallop structure generates at the sidewall and resulting in a strong roughness as shown in Fig.1. In conventional etching system, due to the larger chamber size, although the switching performance of gas valves is as fast as a millisecond, the gas evacuation performance in etching chamber is very low and finally minimum gas switching time is limited at 2s where the etched structure is still resulting in approximately 100 nm scallop height. A number of etching process engineers apply other side-wall treatment process after etching to remove scallop while maintaining the conventional gas switching performance of etching system. Due to the Minimal fab ICP-RIE is developed for proposing a straight side-wall and scallop-less silicon etching structure within one etching process.

Equipment

ICP plasma etching system is minimized from meter size to W30 cm, D45 cm, H144 cm, which is standardized (optimized) for half-inch wafer process. ICP plasma generally uses RF source of 13.56MHz. However, for a tiny chamber, ICP coil diameter size generates too wide wavelength for such a frequency,

Thus the plasma density is not sufficient for etching, therefore a very high frequency (VHF) is required and high density plasma is successfully formed. With the electric power efficiency of our tiny chamber to the plasma formation, 100MHz is generated with only 10W-40W power. Besides, 2MHz (2W) is applied to the substrate side to generate a sheath for an anisotropic etching. Fig. 2 shows a schematic diagram of the tiny chamber.

Experiment and Results

Experiment is performed on half-inch silicon (100) substrates. Photoresist are patterned using photolithography of minimal fab. Etching process parameters are as follows. SF_6 and C_4F_8 gases are applied with the time ratio of 1:1 for etching and passivation, respectively. Therefore twice of the gas switching time is a cycle time. Four silicon samples are etched with 1) a cycle of 2s for 20 cycles, 2) 4s for 10 cycles, 3) 6s for 7 cycles and 4) 13s for 3 cycles. Fixed parameters are chamber pressure: 10Pa, RF power: 40W and gas flow rate: 8sccm for both gases. The scallop heights are measured and shown in Fig. 3. At the 2 seconds' cycle (etching 1s, passivation 1s), smooth and scallop-less sidewall structure is realized. Fig. 4 illustrates a close-up straight side-wall image of etched structure where scallops could not detect obviously within sub 10-nm resolution where only slightly waving structure is seen instead. Fig. 5 shows that the etching rate decreases exponentially with a decrease in ICP power in the typical Bosch process. It is found that the etching rate of minimal Bosch is quite higher than the extrapolated rate at ~40W in the typical Bosch curve.

Conclusion

We successfully confirmed that a tiny chamber of ICP-RIE enhanced the performance of gas replacement and applicable to realize a straight side-wall with scallop-less silicon structure.

Acknowledgments

The authors gratefully thank SPP Technologies Co., Ltd. for developing ICP-RIE to comply with minimal fab standard.

References

[1] S. Khumpuang et al., IEEE Transactions on Semiconductor Manufacturing, 28(3), 393-398 (2015).

[2] F. Laermer et al., IEEE Proc. 12th Intl Conf. on MEMS 1999, 211-216 (1999)

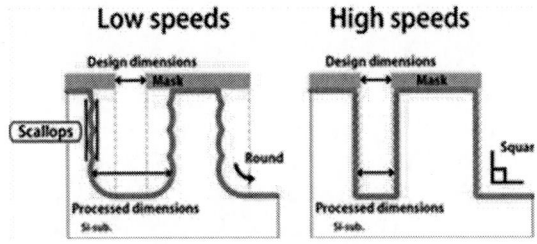

Fig.1. A schematic diagram of etched surface morphology for high and low speeds of gas switching.

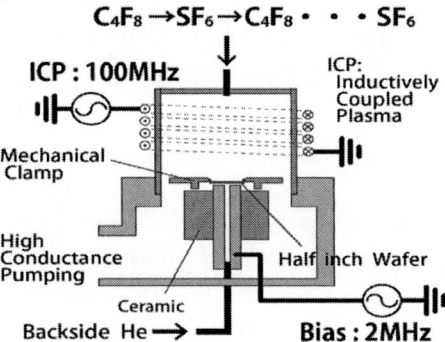

Fig.2. A schematic diagram of etched surface morphology for high and low speeds of gas switching.

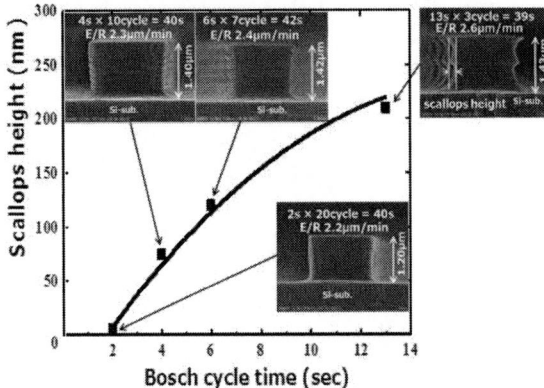

Fig.3 Plots of scallop heights corresponding with Bosch cycle times and illustrations of etching structures at each plot using minimal ICP etching tool.

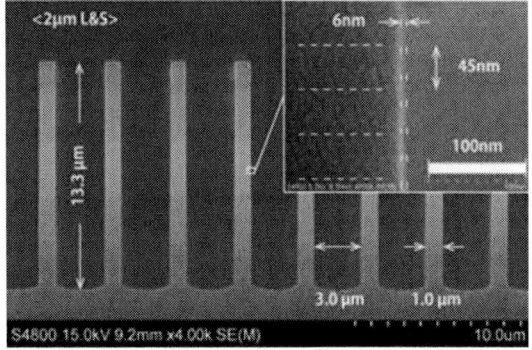

Fig.4.A cross-sectional image of smooth and straight 2μm line & space etched structure with total 2s-cycle for 300 cycles.

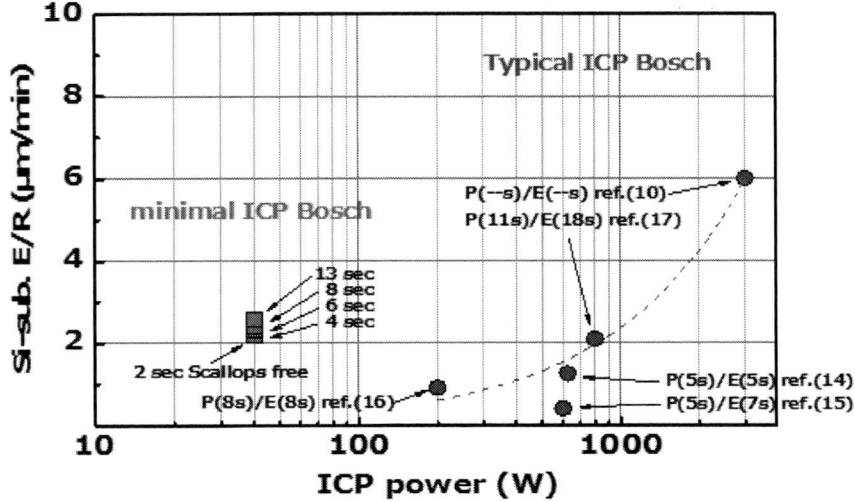

Fig.5 A plot of Si-etching rate dependent with ICP power at various Bosch cycles. The minimal ICP etching(E)/ passivation(P) ratio are all 1:1. Etching rates for typical ICP Bosch process are also plotted.

Exploiting NbO$_x$ Metal-Insulator-Transition Device as Oscillation Neuron for Neuro-Inspired Computing

Ligang Gao, Pai-Yu Chen, and Shimeng Yu

Arizona State University, Tempe, AZ 85281, USA   Email: shimengy@asu.edu

Abstract

In this work, we fabricated the Pt/NbO$_x$/Pt MIT device showing threshold switching. XPS results revealed that there are mixed NbO$_2$ and Nb$_2$O$_5$ phases in the NbO$_x$ thin film. The self-oscillation of NbO$_x$ device with a resistor has been demonstrated, showing its feasibility as an oscillation neuron.

(Keywords: Metal-insulator-transition, oscillation, neuron, NbO$_x$, neuro-inspired computing)

Introduction

Resistive memory (RRAM) has been proposed to emulate the synapses in the neural network [1]. Integration of the RRAM into a cross-point array architecture can efficiently implement the weighted sum (Fig. 1(a)), which is the most time/energy-consuming operation in the neuro-inspired learning algorithms. When an input vector is fed into the cross-point array, the weighted sum current (modulated by the weight or conductance of each RRAM synapse) will be sink to the neuron node at the end of the column. The neuron node integrates this analog current and convert to spikes or digital outputs. The conventional neuron node design generally employs the integrate-and-fire neuron model. Fig. 1(b) shows an example of the CMOS neuron design [2]. Fig. 2(a) shows the simulated waveform of the membrane voltage (V_{in}) and output spike (V_{spike}) for different weighted sum current (6 μA vs. 1 μA). The number of output spike is designed to be proportional to the amplitude of the input weight sum current. Apparently, such CMOS neuron node is complex and occupies much larger size than the column pitch of the crossbar array, thereby reducing the parallelism as the time-multiplexing is needed to sequentially read out all the weighted sum from the array. In this work, we propose to design a compact oscillation neuron node by using a NbO$_x$ based Metal-Insulator-Transition (MIT) device to replace the CMOS neuron (Fig. 2(b)). Due to the threshold switching I-V with hysteresis, the voltage on the MIT device will oscillate and emulate the V_{in} node in the CMOS neuron, thus oscillation frequency is expected to be proportional to the weighted sum current [3].

Results and Discussion

Pt/NbO$_x$/Pt devices were fabricated in the cross-point structure with an active area of 5 μm^2. A schematic of the Pt/NbO$_x$/Pt device is shown in the inset of Fig. 3. A forming voltage of ~3V is needed to trigger the subsequent threshold switching. The on/off ratio is about 100. A hysteresis exists: off-to-on switching's threshold voltage (V_{th}) is about 2.2 V and on-to-off switching's hold voltage (V_{hold}) is about 2 V (Fig. 3). The X-ray Photoelectron Spectroscopy (XPS) is employed to characterize the stoichiometry and the binding energy of the NbO$_x$ thin film. Fig. 4(a) shows the XPS spectrum of the Nb3d in the NbO$_x$ thin film. The two peaks of the green fitting curve located at 205.5 eV and 208.3 eV correspond to NbO$_2$ 3d$_{5/2}$ and 3d$_{3/2}$, respectively, while the two peaks of the blue fitting curve located at 207.3 eV and 210.1 eV correspond to Nb$_2$O$_5$ 3d$_{5/2}$ and 3d$_{3/2}$, respectively, which indicated that both NbO$_2$ and Nb$_2$O$_5$ phases were formed in the NbO$_x$ thin film. It is known that NbO$_2$ exhibits the MIT behavior while Nb$_2$O$_5$ may show the resistive switching [4]. The O1s peak shows that the Nb-O bonds are located at ~531 eV and other non-bridging oxygen bonds are located at ~532.4 eV (Fig. 4(b)). Next, we connected the NbO$_x$ device with an external resistor 3.5 kΩ to demonstrate the self-oscillation (Fig. 5(a)). We applied a pulse voltage on Channel 1 and monitored the node voltage on Channel 2 using the oscilloscope. Fig. 5(b) shows the measured oscillation waveform in this test set-up. The oscillation frequency is about 370 kHz. The oscillation frequency is much limited by the parasitic capacitance in the testing set-up as the resistor is externally connected to the pad of the NbO$_x$ device.

Conclusion

In summary, the NbO$_x$ MIT device has shown its potential as an oscillation neuron for the weighted sum operation in the resistive cross-point array.

Acknowledgments

This work is supported by NSF-CCF-1552687.

Reference

[1] D. Kuzum, et al., "Synaptic electronics: materials, devices and applications," Nanotechnology, vol. 24, p. 382001, 2013.

[2] D. Kadetotad, et al., "Parallel architecture with resistive cross-point array for dictionary learning acceleration," IEEE JETCAS, vol. 5, no. 2, pp. 194-204, 2015.

[3] P.-Y. Chen, et al., "Compact oscillation neuron exploiting metal-insulator-transition for neuromorphic computing," International Conference on Computer-Aided Design, 2016.

[4] S. Kim, et al., "Ultrathin (<10nm) Nb$_2$O$_5$/NbO$_2$ hybrid memory with both memory and selector characteristics for high density 3D vertically stackable RRAM applications," Symposium on VLSI Technology, 2012.

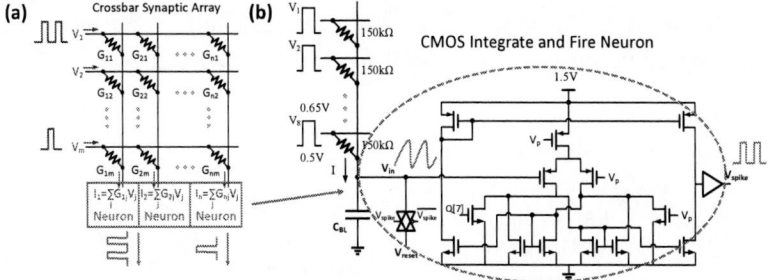

Fig. 1: (a) The resistive crossbar array implements the synaptic network and the neuron node at the end of the column integrates the weighted sum current from the array. (b) One example of the CMOS integrate-and-fire neuron design [3]. The membrane voltage integrates and discharges after triggering the output spike.

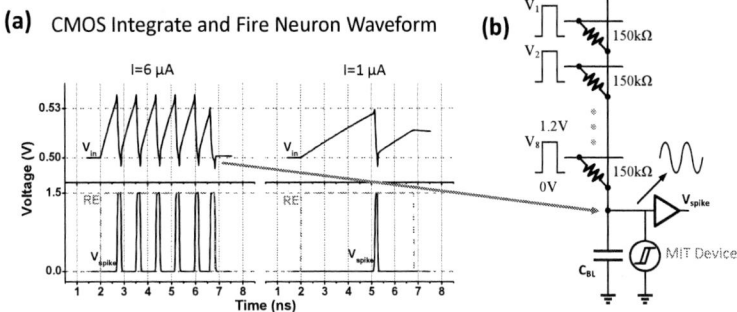

Fig. 2: (a) The waveform of the CMOS integrate-and-fire neuron for different weighted sum current (6 µA vs. 1 µA) [3]. (b) Proposed design of an oscillation neuron with a MIT device at the end of the column that emulates the V_{in} node.

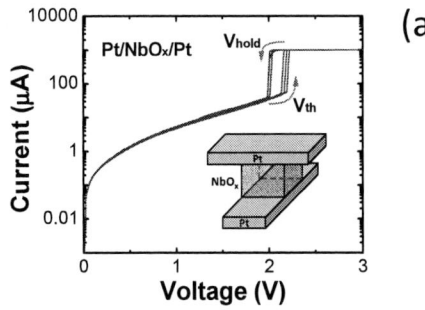

Fig. 3: Measured I-V threshold switching characteristics of Pt/NbOₓ/Pt device. The inset shows the schematic of the fabricated Pt/NbOₓ/Pt device.

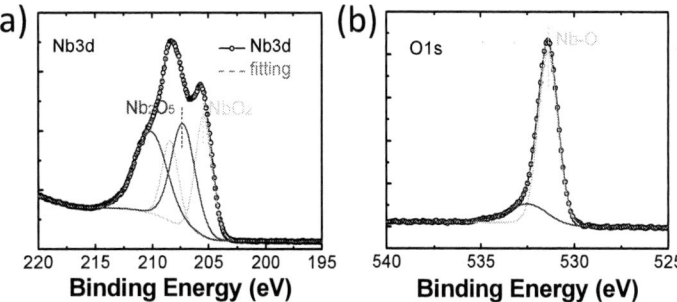

Fig. 4: The XPS spectra of (a) Nb3d and (b) O1s in the fabricated NbOₓ thin film, showing the co-existence of NbO₂ and Nb₂O₅ phases.

Fig. 5: (a) The set-up of an oscillation neuron. (b) Experimental measured oscillation waveform of the oscillation neuron in the set-up (by monitoring Channel 2 of the oscilloscope). The oscillation frequency is ~370 kHz.

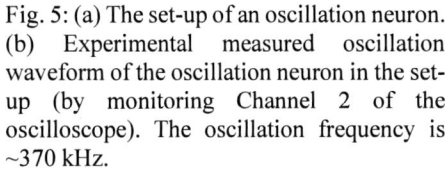

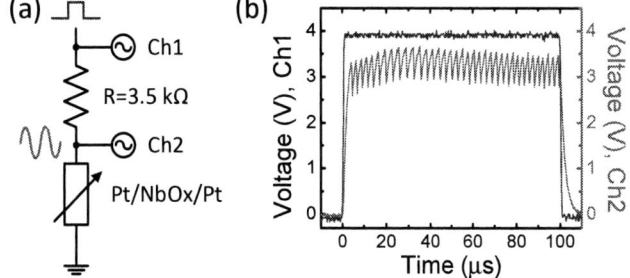

978-1-5090-4661-4/17 $31.00 © 2017 IEEE   153

Ohmic Contact Formation Between Ge₂Sb₂Te₅ Phase Change Material and Vertically Aligned Carbon Nanotubes

Panni Wang*, Suwen Li, Yihan Chen, Lining Zhang, Mansun Chan

Hong Kong University of Science and Technology, Hong Kong, pwangae@ust.hk

Abstract

The contact property between $Ge_2Sb_2Te_5$ (GST) with vertical carbon nanotubes (CNTs) is studied in this work. By careful catalyst design and process optimization, we have demonstrated the formation of ohmic contact between the CNT and the GST material. The developed process is CMOS compatible and can be used for form phase change memory over the vias in the interconnect layers.

(Keyword: Carbon Nanotube, Phase Change Memory, contact resistance)

Introduction

Resistive based non-volatile has many interest features for new applications in the memory hierarchy. In particular, Phase change memory (PCM) using $Ge_2Sb_2Te_5$ (GST) is the most well understood system [1]. As the thermal transition requires a relatively high temperature, it is desirable to have one electrode to be very small so that the thermal energy can be concentrated [2]. The concept can be realized using metallic CNT as one of the electrodes. It has been realized using horizontally aligned CNT as electrode and the programming can be performed at relatively low current [3][4]. To integrate the process into mainstream CMOS, GST material on vertically aligned metallic CNT synthesized with CMOS compatibility has to be used. And the most important aspect of the process is the formation of ohmic contact between the GST and vertically aligned metallic CNT. In this work, we have developed a CMOS compatible CNT via-filling process that forms ohmic with GST material.

Fabrication and characterization

The key steps to form the CNT-to-GST contact are shown in Fig.1. Silicide was used as the bottom electrode to mimic the condition with the GST material connected to an active device. After depositing the passivation oxide, a lithography step was used to open up the vias and expose the silicide. For CMOS compatibility reasons, Ni/Al/Ni catalyst composite as described in [5] was deposited over the entire wafer. The selected catalyst was CMOS compatible and can be used to achieve selective synthesis which the CNTs were only formed at the vias but not the passivation oxide. After the catalyst was formed, CNTs were synthesized using plasma enhanced chemical vapor deposition (PECVD). By adjusting the catalyst thickness, the diameter of the CNT can be controlled as shown in Fig. 2. After filling the via with Al_2O_3, the sample was polished by Copper Mechanical Polisher to expose the top of the CNTs. To make sure that the CNT tubes were exposed, the sample surface condition was checked by conducting atomic force microscope (C-AFM) as shown in the Fig.3 (a). Fig.3 (b) showed that the patterned via holes were conductive, so the CNT tube was opened. Then a layer of GST was sputtered and annealed at 300°C for 2 min in the N_2 atmosphere to crystallize the GST film. Afterward, TiW layer was deposited to serve as the top electrode. And the bottom electrode was opened by another photoresist pattern and dry etching.

The resistance of the structure was measured by Agilent 4156C precision semiconductor parameter analyzer. Fig.4 showed I-V characteristic of the structure between the top TiW electrode and the silicide. The GST and CNT can achieve ohmic contact. The experimental data showed that it is possible to form phase change memory between vertically aligned metallic CNT and GST material.

Conclusion

In this process, we have demonstrated the feasibility to form ohmic contact between vertically aligned CNT and GST material. While it is still in the stage of proof-of-concepts, it demonstrates the potential to use CNT filled via as a bottom electrode to achieve low power programming with the small cross-section of the CNT as the heating electrode.

Acknowledgments

This work is supported by a NSFC/RGC grant from the Research Grant Council of Hong Kong under project number N_HKUST_605/12.

References

[1] H.P.Wong, et al, Proc IEEE, vol. 98, pp. 2201-2227, 2010

[2] M. Boniardi, et al, IEDM, 2014, pp. 29.1.1-29.1.4.

[3] F. Xiong, et al, Science, vol. 332, pp. 568-570, Apr 29, 2011.

[4] J. Liang, et al, VLSIT, 2011, pp.100-101

[5] S.Li, et al, IEEE Electron Device Letter, vol.37, pp. 793-796, 2016.

978-1-5090-4661-4/17 $31.00 © 2017 IEEE

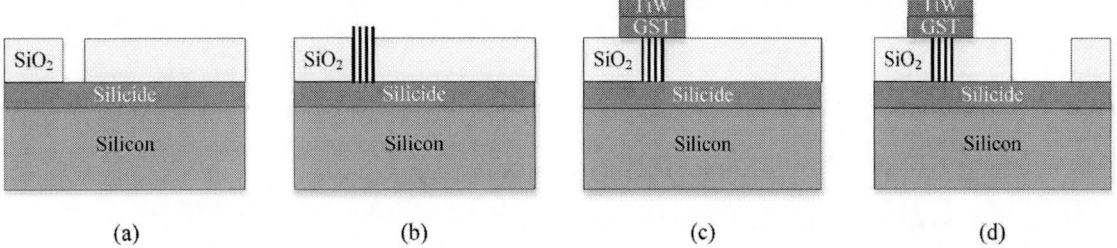

(a)   (b)   (c)   (d)

Fig. 1 Key steps to form the CNT-to-GST contact that include (a) the starting substrate with Ti silicide and LTO deposition followed by via holes opening; (b) catalyst deposition and CNT growth; dielectric filling; (c) CNT polishing and GST/TiW deposition and patterning; (d) bottom contact opening.

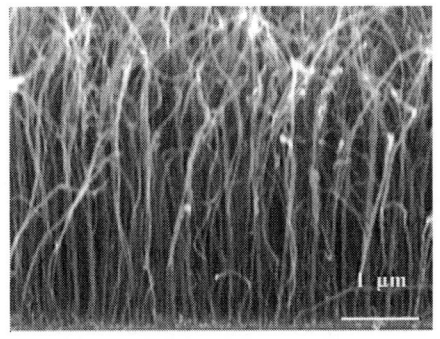 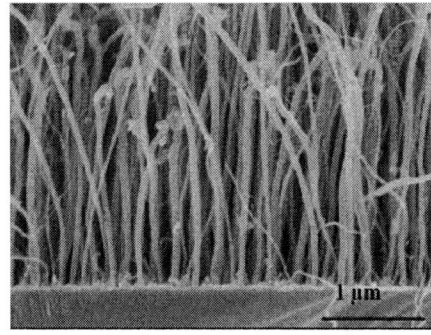

Fig. 2 SEM image of the synthesized CNT forest with Ni/Al/Ni catalyst composition of (a) 1nm/0.5nm/1nm; (b) 2nm/1nm/2nm.

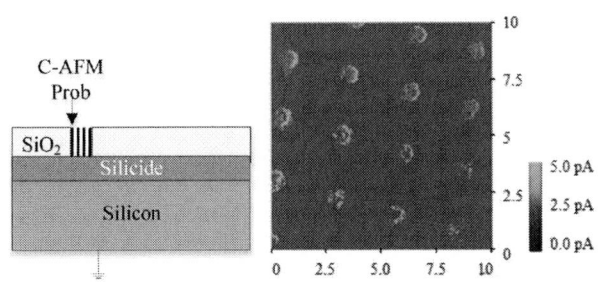

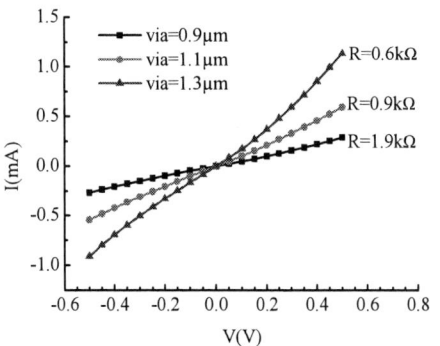

Fig. 3 Setting and result of the C-AFM testing. (a) 500mV was applied between scanning prob and the back of the wafer. The scanning prob scaned in the surface of the wafer, and measured the current flowing through the prob.(b) The current distribution on the surface of the wafer in the scanning process

Fig. 4 I-V characteristic of the structure between the TiW and silicide with via diamter of 0.9 μm, 1.1 μm and 1.3 μm.

978-1-5090-4661-4/17 $31.00 © 2017 IEEE   155

8M-3

Impact of Current Distribution on RRAM Array with High and Low I_{ON}/I_{OFF} Devices

Mohammed Zackriya V[1,2], Albert Chin[1], and Harish M Kittur[2]

[1]Dept. of Electronics Engineering, National Chiao-Tung Univ., Hsinchu, Taiwan, albert_achin@hotmail.com

[2]School of Electronics Engineering, VIT University, Vellore, India

Abstract

Using novel circuit design topology to include the sneak path current as a reference input, the performance of two RRAM devices with I_{ON}/I_{OFF} of 26 and 925 were compared for crosspoint array. The RRAM with better current distribution outperforms the RRAM with 36X higher I_{ON}/I_{OFF}, on crosspoint array. Thus, the RRAM devices should aim on tightening the current distribution, where high I_{ON}/I_{OFF} also consumes high power on circuit.

(Keywords: RRAM array, I_{ON}/I_{OFF} and distribution)

Introduction

RRAM crosspoint array is crucial for next generation 3D memory, but the array size is limited by the sneak current. Fig. 1 shows a simple RRAM 3×3 crosspoint array with read resistor to sense the current on BL3. The non-selected cells (NSEL) are biased at $V_{read}/2$, which features low voltage swing on wordlines (WLs) and bitlines (BLs) for low power operation. The resistive state at 1×3 (SEL) is being read, whereas the read current which is passing through R_{read} is disturbed by the current from half selected cells (HSEL). To suppress the sneak path current and increase the array size, RRAM devices including one-selector-one-resistor (1S1R) are presented with high resistance ratio [1], although it consumes high power due to high I_{ON}. In this work, we present a comparative study of RRAM devices with higher current ratio versus better current distribution.

Effect of current distribution

The *I-V* hysteresis curves of Ni/GeO$_x$/TiO$_y$/TaN and Ni/GeO$_x$/HfON/TaN RRAM devices [2]-[3] are presented in Figs. 2(a) and 2(b). Ni/GeO$_x$/TiO$_y$/TaN RRAM device has an I_{ON}/I_{OFF} of 26, much lower than the 925 I_{ON}/I_{OFF} of Ni/GeO$_x$/HfON/TaN RRAM device. Fig. 3 shows the current distribution of these two RRAM devices. The current coefficient of variation (CV) of Ni/GeO$_x$/TiO$_y$/TaN RRAM for HRS and LRS are 34% and 44%, which is better than its counterpart with HRS and LRS of 65% and 111%, respectively. The device parameters were summarized in Table 1 for RRAM array design.

Fig. 4 shows the ideal *I-V* switching curves of a RRAM device. The slope of I_{on} is higher compared to I_{off} slope. This strategy is used to determine the state of RRAM cell. The circuit architecture is presented in Fig. 5, which effectively predicts the sneak-path current on BL while reading a RRAM device in an array (in1 stores the sneak-path value when clock is at low-level). The V_{in1} is further compared with read voltage on in2 (V_{in2}) when SEL is biased at V_{read} on high-level of clock. Figs. 6-7 show the voltage difference (ΔV) between V_{in1} and V_{in2} of Ni/GeO$_x$/TiO$_y$/TaN and Ni/GeO$_x$/HfON/TaN RRAM array. Differential amplifier (DA) sensing ΔV as low as 10 mV [4] was used in the design. The ΔV while reading off-state (S_{off}) is <2 mV for both the RRAM devices as shown in Fig. 6(b) and 7(b). If $\Delta V > 10$ mV, the DA reads high to indicate the RRAM on-state (S_{on}). As shown in Fig. 6(a), if only nominal case is considered, the array size can go beyond 256×256, but over-estimated. While worst case (extreme distribution) is considered, array size is limited to 256×256 as shown in Fig. 8 for Ni/GeO$_x$/TiO$_y$/TaN device and power consumed by the array is as low as 6.83 μW. For Ni/GeO$_x$/HfON/TaN device, it is limited to only 64×64 even though it has high I_{ON}/I_{OFF} due to poor current distribution CV.

Conclusion

RRAM with lower I_{ON}/I_{OFF} and tight current distribution performs better than RRAM with high I_{ON}/I_{OFF}. Thus, emphasis should be given on attaining lower current variation during device design.

Acknowledgments

The authors thank the support to attend EDTM by Ministry of Science and Technology, Taiwan.

References

[1] L. Zhang, S. Cosemans, D. J. Wouters, G. Groeseneken, M. Jurczak and B. Govoreanu, "On the optimal on/off resistance ratio for resistive switching element in one-selector one-resistor crosspoint arrays," IEEE Electron Device Lett., vol. 36, pp. 570-572, June 2015.

[2] K. I. Chou, C. H. Cheng, Z. W. Zheng, M. Liu and A. Chin, "Ni/GeO$_x$/TiO$_y$/TaN RRAM on flexible substrate with excellent resistance distribution," IEEE Electron Device Lett., vol. 34, pp. 505-507, April 2013.

[3] C. H. Cheng, C. Y. Tsai, A. Chin, and F. S. Yeh, "High performance ultra-low energy RRAM with good retention and endurance," IEDM Tech. Dig., 2010, pp. 448-451

[4] D. Schinkel, E. Mensink, E. Klumperink, E. van Tuijl, and B. Nauta, "A double-tail latch-type voltage sense amplifier with 18ps setup+hold time," ISSCC Tech. Dig., 2007, pp. 314-605

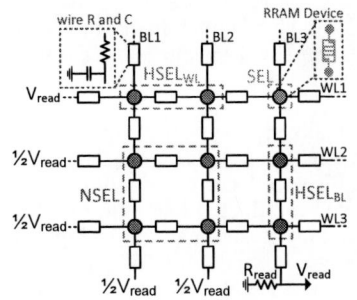

Fig. 1: RRAM crosspoint array structure with read resistor (R_{read}) and wire resistor and capacitor.

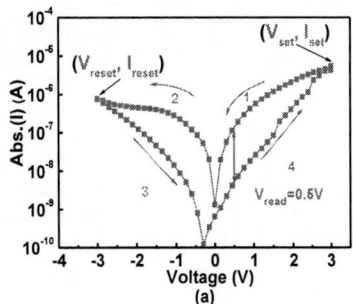

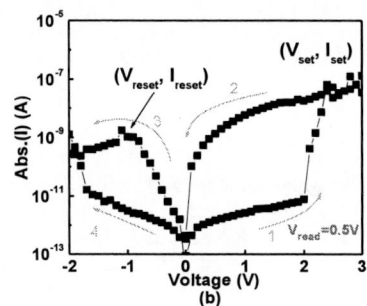

Fig. 2: Swept *I-V* hysteresis curves of (a) Ni/GeO$_x$/TiO$_y$/TaN and (b) Ni/GeO$_x$/HfON/TaN RRAM devices.

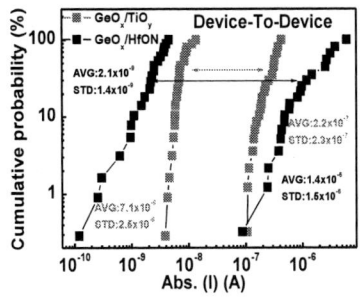

Fig. 3: Current distributions of Ni/GeO$_x$/TiO$_y$/TaN and Ni/GeO$_x$/HfON/TaN RRAM devices.

Table 1: Device parameters under different conditions with wire RC values.

Parameters	Ni/GeO$_x$/ TiO$_y$/TaN	Ni/GeO$_x$/ HfON/TaN
S_{off}-SEL (MΩ)	46.2	3.4×10^5
S_{on}-SEL (MΩ)	1.8	367.6
S_{off}-HSEL (MΩ)	103.7	3.6×10^5
S_{on}-HSEL (MΩ)	2.8	637.8
R_{read} (KΩ)	100	
R_{wire} (Ω)	2.82	
C_{wire} (fF)	0.046	
V_{read} (V)	0.5	

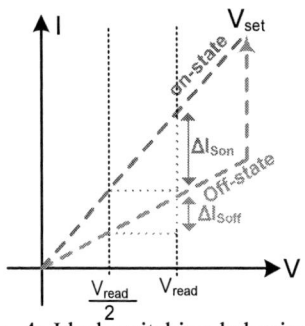

Fig. 4: Ideal switching behavior of RRAM devices.

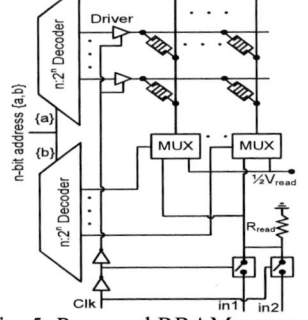

Fig. 5: Proposed RRAM crosspoint array structure to study the effect of current variation.

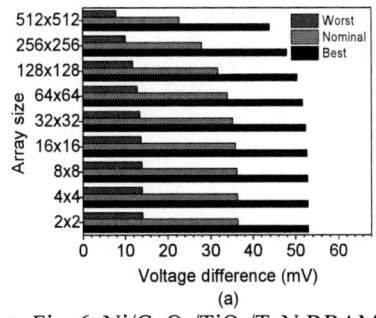

Fig. 6: Ni/GeO$_x$/TiO$_y$/TaN RRAM array: sense voltage of (a) S_{on} and (b) S_{off} for various array sizes under best, nominal and worst conditions.

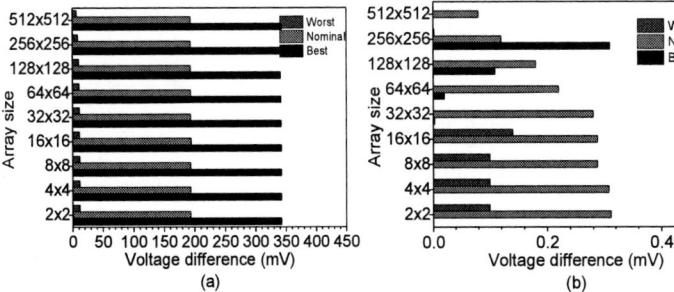

Fig. 7: Ni/GeO$_x$/HfON/TaN RRAM array: sense voltage of (a) S_{on} and (b) S_{off} for various array sizes under best, nominal and worst conditions.

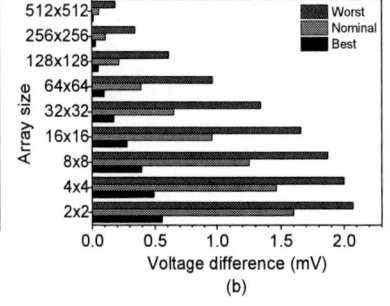

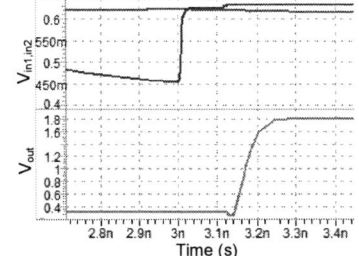

Fig. 8. Waveforms for Ni/GeO$_x$/TiO$_y$/TaN RRAM array while reading 256×256 array under worst case scenario.

978-1-5090-4661-4/17 $31.00 © 2017 IEEE

8M-4

3D Time-Contingent Physical Unclonable Function Array on 16nm FinFET Dielectric RRAM

Yi-Hung Chang[1], Po Shao Yeh[1], Yue-Der Chih[2], Jonathan Chang[2], Ya-Chin King[1], Chrong Jung Lin[1]

[1] Microelectronics Laboratory, Institute of Electronics Engineering, National Tsing Hua University, Hsinchu, Taiwan

[2] Design Technology Division, Taiwan Semiconductor Manufacturing Company, Hsinchu 300, Taiwan

Phone/Fax: +886-3-5162182, E-mail: cjlin@ee.nthu.edu.tw

Abstract

The randomness and unpredictability of Random Telegraph Noise (RTN) of 16nm FinFET Dielectric (FIND) RRAM is firstly implemented to Time-Contingent Physical Unclonable Function (PUF) application. A novel 3D Time-Contingent Physical Unclonable Function (TC-PUF) realized by 1Kbit 16nm FinFET Dielectric (FIND) RRAM has been newly proposed and demonstrated on a pure 16nm FinFET CMOS logic technology. The new TC-PUF RRAM shows wide operation ranges of voltages and temperatures, and Randomness test confirms the feasibility and stability of the TC-PUF RRAM in the frequency, unpredictability, and long-run continuity.

Introduction

With the rapid development of communication and internet of things (IoT) technologies, demand for information security attracts much attention recently. Due to this reason, the True Random Number Generator (TRNG) based on Physical Unclonable Function (PUF) has been proposed to serve the needs besides RTN is mostly adaptable to the advanced CMOS processes. In this paper, a novel 3D Time-Contingent Physical Unclonable Function (TC-PUF) realized by a 1Kbit array of 16nm FinFET Dielectric (FIND) RRAM has been firstly proposed and demonstrated in pure CMOS logic 16nm FinFET technology. The novel TC-PUF can generate 2D 1Kbit true random number data at each read point as shown in Fig.1, as a result the TC-PUF can efficiently generate a serial random and continuous Physical Unclonable Function (PUF) array along the third dimension of reading time. Besides, the generated data can be non-volatile stored in other mirror 1Kbit FIND RRAM array for PUF application and the data can be easily replaced by the next TC-PUF data coming if need to update. 1Kbit TC-PUF array data can be stably generated within 2ms as shown in Fig.1(c) and (d).

FIND RRAM Array for PUF Applications

The 1Kbit TC-PUF RRAM array was fabricated by a pure 16nm FinFET CMOS logic process. The FIND RRAM cell consists of an n-channel 16nm FinFET core transistor for cell selection as a wordline and an HfO_2-based resistive film to be a resistive switching node (1T1R). Due to charge trapping and detrapping in the current conducting path, the read current of resistive node appears a significant fluctuation at high or low resistance states [1]. The generated RTN data can be easily measured and converted to digital output as shown in Fig.1(d).

RTN Measurement and Analysis

Strong RTN of single FIND RRAM cell can be clearly measured by time with a proper voltage on the top electrode of resistive node as shown in Fig.2(a). Besides, Fig.2(b) exhibits a typical bi-state mode with a normalized fluctuation current of $\triangle I/I_{AVG}$. Under different temperatures, Fig.3 shows no much change in RTN cell current distribution [2]. The normalized current fluctuation ratio, $\triangle I/I_{AVG}$, of 1Kbit TC-PUF array is summarized in Fig.4. By the normalized noise current ratio of $\triangle I/I_{AVG}$, at each reading point, the 1Kbit sampling data can be grouped into: Quiet Bit with $\triangle I/I_{AVG}$ less than 1% and RTN Bit with $\triangle I/I_{AVG}$ much larger than 1%. In addition, the cumulative probability of $\triangle I/I_{AVG}$ of 1Kbit TC-PUF is shown in Fig.5 at some time point. The Power Spectrum Density (PSD) analysis for the two kinds of noise behaviors is shown in Fig.6. The PSD of RTN Bit shows a $1/f^2$ Lorentz PSD relation in frequency spectrum as a typical RTN behavior [3] [4]. Conversely, the Quiet Bit PSD shows a different frequency spectrum, which could be caused by thermal or environmental noises.

Randomness Evaluation of TC-PUF RRAM

The Hamming Distance for the 1Kbit 3D TC-PUF RRAM is summarized in Fig.7, the result exhibits the superior uniqueness of the 1Kbit Time-Contingent PUF RRAM array. Fig.8 shows the relatively uniform occurrence of 256 possible values of 8bit response among 100 times of 1Kbit RTN sampling, there is no cell correlation between the 100 read-out times of 1Kbit TC-PUF RRAM [5]. Fig.9 exhibits one million random data generated by the 1Kbit 3D TC-PUF RRAM, the converted digital random data is also exhibited in Fig.10. Finally, the key parameters of randomness examining result of the 1Kbit 3D TC-PUF RRAM is shown in Table 1, which shows the 3D TC-PUF can pass the items of randomness criteria provided by USA's NIST, the randomness capability of 1Kbit 3D TC-PUF RRAM is very superior and robust to be a reliable TRNG.

Conclusion

A new 3D TC-PUF realized by a 1Kbit 16nm FIND RRAM array has been successfully demonstrated by a pure CMOS logic 16nm FinFET process. The superior performance of the 3D TC-PUF includes: a stable RTN current fluctuation, a rapid RTN data generation rate, uniqueness and randomness in the 1Kbit FIND RRAM array, and unlimited 2D PUF 1Kbit data output per 2msec. The excellent results lead the 1Kbit 3D TC-PUF RRAM to be a very promising True Random Number Generator (TRNG) for the information security and identification applications in the future.

References

[1] Hsin Wei Pan, et al. IEDM, pp. 10.5.1-10.5.4, 2015.

[2] Yuan Heng Tseng, et al. IEDM, pp. 28.5.1-28.5.4, 2010.

[3] K. Uchida, et al. IEDM, pp. 177-180, 2002.

[4] N. Conrad, et al. IEDM, pp. 20.1.1-20.1.4, 2014.

[5] An Chen, IEDM, pp. 10.7.1-10.7.4, 2015.

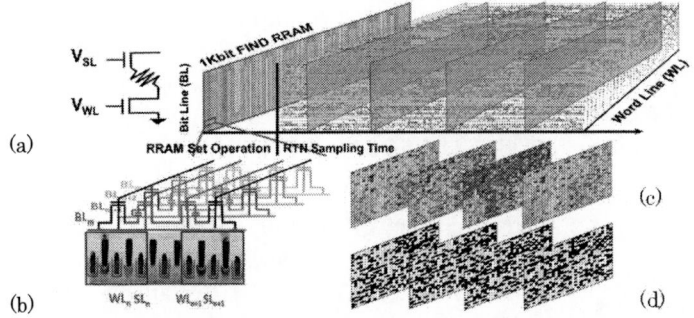

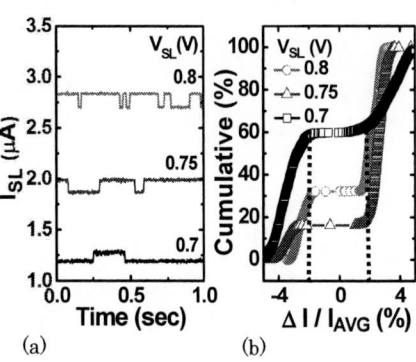

Fig.1 1Kbit 3D TC-PUF RRAM (a)1Kbit chip and RTN serial signal by time and (b)FIND RRAM single cell and array architecture (c)1Kbit current fluctuation of $\Delta I/I_{AVG}$ at t=0ms, 2ms, 4ms, and 6ms (d)converted to 1Kbit 2D digital PUF output

Fig.2 (a) Single cell RTN and (b) $\Delta I/I_{AVG}$ at different SL voltages reveals shift in the state occupancy probability

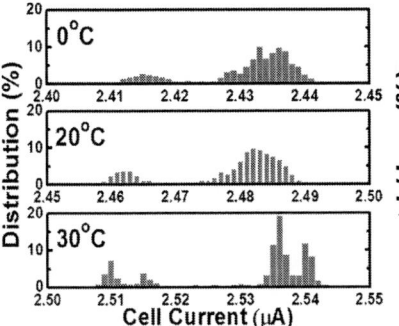

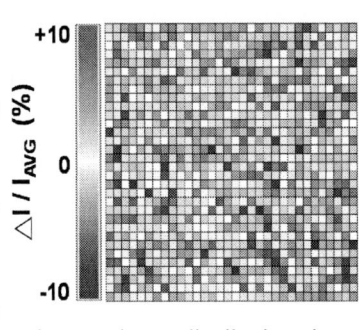

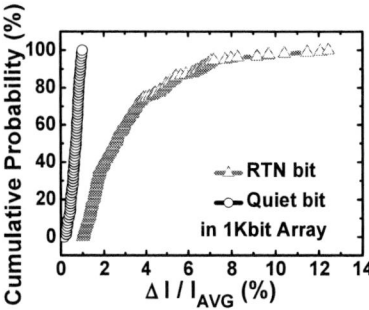

Fig.3 Single cell current distribution at 0^0C, 20^0C, and 30^0C, respectively

Fig.4 $\Delta I/I_{AVG}$ distribution in a 1Kbit FIND RRAM array with RTN % ranging from -10~10%

Fig.5 1Kbit RTN signal are grouped into: **Quiet Bit** and **RTN Bit** with $\Delta I/I_{AVG}$ larger than 1%

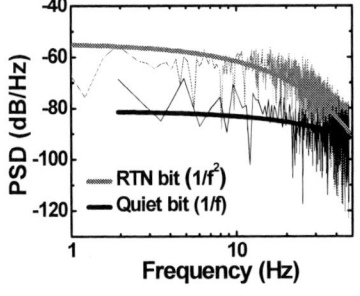

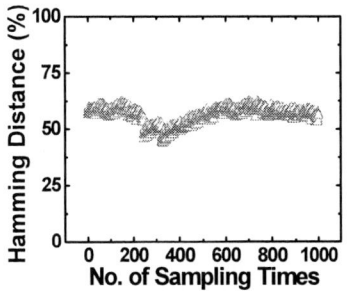

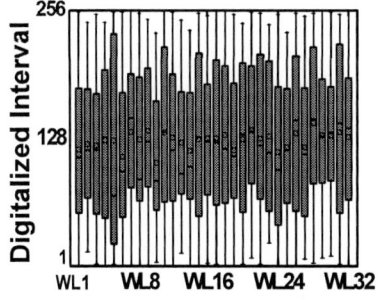

Fig.6 Power spectrum density of $1/f^2$ frequency function for RTN, indicating that the trapping/de-trapping is the dominant noise

Fig.7 Hamming Distance of 1Kbit steadily keeps around 50% with no correlation to RTN sampling times

Fig.8 No WL dependency of 1Kbit FIND RRAM array in digitized 2D PUF data output are shown

Fig.9 One million RTN data generated by the 3D TC-PUF RRAM with 1Kbits/2ms data rate with variation of ±10%

Fig.10 One million digitized RTN output data of the 3D TC-PUF RRAM, further test by NIST standard in Table.1

Test	Single Cell Pass / Fail	1Kbit TC-PUF Pass / Fail
Frequency	Pass	Pass
Block Frequency	Pass	Pass
Cumulative Sums	Fail / Fail	Pass / Pass
Runs	Pass	Pass
Longest Run	Pass	Pass
FFT	Fail	Pass
Non Overlapping Template	–	Pass
Serial	Fail / Fail	Pass / Pass

Table.1 NIST randomness tests of single FIND RRAM cell and 1Kbit 3D TC-PUF RRAM after Von Neumann correction

978-1-5090-4661-4/17 $31.00 © 2017 IEEE

Surface Preparation and Wet Cleaning for Germanium Surface

Masayuki Otsuji[1], Yukifumi Yoshida[1], Hiroaki Takahashi[1],

Farid Sebaai[2], Kurt Wostyn[2], Frank Holsteyns[2],

Masanobu Sato[1] and Hajime Shirakawa[1]

[1]Screen Semiconductor Solutions Co., Ltd, Japan, [2]imec vzw, Belgium

Abstract

The CMOS devices with Ge, considered as one of new materials for the later 5-nm generations, has been investigated since Ge should be mandatory material to enhance the electron mobility as replacement for Si. In this paper, we will propose new two techniques for surface preparation and wet cleaning, one of two techniques is in terms of PRE (Particle Removal Efficiency) on Ge surface, and the other is Ge corrosion caused by the dissolved oxygen effect. High PRE without Ge loss was achieved for surface preparation by using O3/NH4OH mixture. And to add to it, excellent selective Ni removal along germanidation (NiGe generation) without Ge corrosion was realized by reducing dissolved oxygen in chemistry.

Introduction

In recent year, requirement of low power consumption and of electron devices improving high-performance, as typified smartphone, has become enlarged. The improvement includes integration of the structure change such as FinFET and Nanowire, and also changes the materials for CMOS using Ge and III-Velements. However the use of Ge is high impact to apply into wet process because there are some challenges to be solved such as Ge loss derived from oxidation by oxidant in the chemical solution and dissolved into DIW (De-Ionized Water). Hence, in this study, we focused on the wet cleaning of Ge surfaces and established a new chemistry and a technique achieving a high PRE without Ge loss, and also selective removal process of excess Ni for contact cleaning.

As for surface preparation, wet cleaning is necessary for Ge device fabrication, while Ge is a sensitive material that easily can be oxidized and its oxide dissolves even into DIW. APM (Ammonia Peroxide Mixture) is one of the conventional chemistry for surface preparation on Si surface however it could not be applied on Ge surface in terms of Ge loss caused by a high etching rate due to reaction mechanism of oxidant agent and reductant agent in APM as shown in Fig. 1. Therefore less oxidizing chemistries have to be applied. Particulate contaminations can be removed by the lift-off phenomena using ozonated water (1ppm), but a certain amount of Ge loss (>3nm) is required to obtain sufficient PRE. New studies show that by applying diluted APM (200:1:20,000) for high pH oxidative chemistry, Ge loss can be minimized (<1nm) while having effective particle removal (>90%) as shown in Fig. 2 [1]. However since the evaluated diluted APM still etches Ge, we studied its mechanism of particle removal and finally suggested to using O3/NH4OH mixture which has a high PRE performance minimizing Ge loss.

For the selective Ni removal during germanidation with a conventional chemistry condition on NiGe as a contact electrode, Ge void formation shown in Fig. 3, was caused by the occurrence of a galvanic corrosion reaction between Ge and NiGe interface. In this reaction, dissolved oxygen is considered to play a role in the Ge corrosion when exposed, as explained by the Eq. (1), (2), (3) and Fig. 4.

Anode:
$$Ge + 1/2\ O_2 \rightarrow GeO + e^- \qquad (1)$$

Cathode:
$$H^+ + e^- \rightarrow H_2 \qquad (2)$$
$$O^2 + 2H_2O + 4e^- \rightarrow 4OH^- \qquad (3)$$

Therefore we evaluated the dissolved oxygen effect on Ge void occurrence in different conditions as shown in Fig. 5 [2], and demonstrated that the Ge void formation can be suppressed by reducing the dissolved oxygen concentration in chemistry as shown in Fig. 6.

Conclusion

We demonstrated the benefits of using a new chemistry and technique, and suggested that the controlled O3/NH4OH mixture in preferable concentration for Ge surface preparation which can effectively remove surface contamination on Ge without Ge loss. In addition, the chemistry with a reduced dissolved oxygen concentration for the other applications, e.g. selective excess Ni removal from NiGe surface, could suppress Ge corrosion.

References

[1] H. Takahashi, et al., ECS Transactions, 2011 41 (5) 163-170

[2] Y. Yoshida et al., *Solid State Phenomena* 219 (2014) 85-88

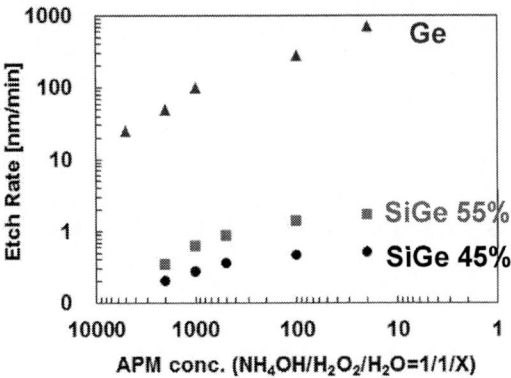

Fig. 1: SiGe/Ge loss related with APM concentration.

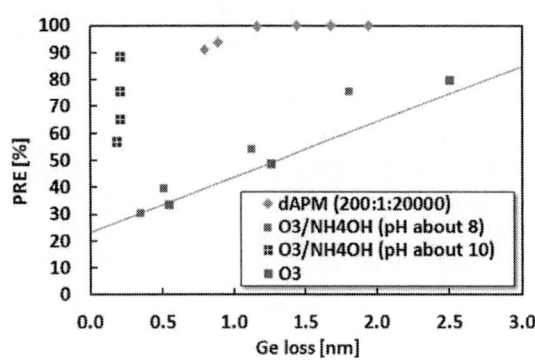

Fig. 2: Removal efficiency of a 30-nm SiO_2 slurry particles on a Ge as a function of Ge loss with diluted APM and O3/NH4OH mixture.

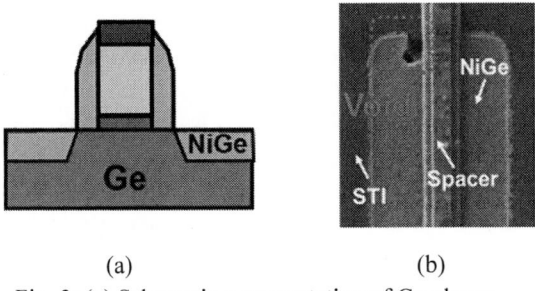

(a)   (b)

Fig. 3: (a) Schematic representation of Ge planar device, (b) SEM Top-View image of a typical void.

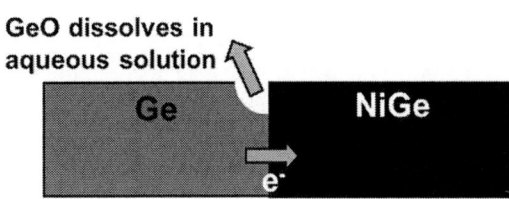

Fig. 4: Schematic representation of Ge void occurrence during chemical process.

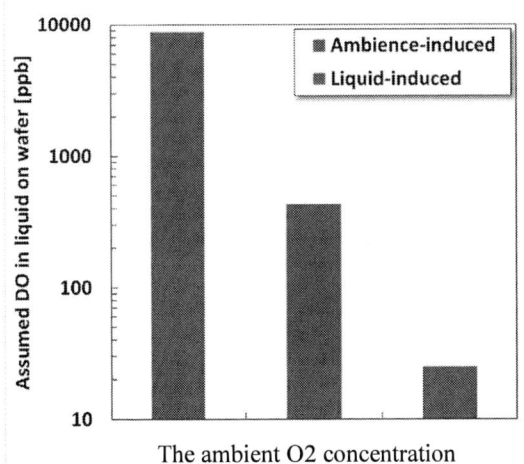

Fig. 5: Dissolved oxygen control in the chemical solution.

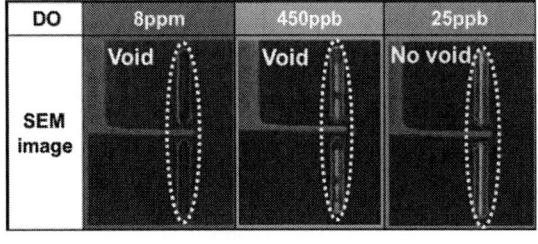

Fig. 6: Void occurrence dependency on the dissolved oxygen concentration in the chemical solution.

978-1-5090-4661-4/17 $31.00 © 2017 IEEE   161

Oxidation Mechanism and Surface Passivation of Germanium by Ozone

Xiaolei Wang[1], Jinjuan Xiang[1], Chao Zhao[1,2], Tianchun Ye[1,2], and Wenwu Wang[1,2]

[1]Key Laboratory of Microelectronics Devices & Integrated Technology, Institute of Microelectronics, Chinese Academy of Sciences, Beijing 100029, China, wangwenwu@ime.ac.cn
[2]School of Microelectronics, University of Chinese Academy of Sciences, Beijing 100049, China

Abstract

Oxidation mechanism and passivation of Ge surface by ozone is experimentally investigated. The GeO_x oxidation process contains two regions: initial linear growth region and following parabolic growth region. The linear growth region contains reaction of oxygen atoms with surface bond and back bonds of outmost Ge layer. The parabolic growth region starts when the oxygen atoms diffuse into back bonds of second outmost Ge layers. Furthermore, in the ozone oxidation it is not O_3 molecules but O radicals that go through the GeO_x film. The interface state density (D_{it}) is found to decrease with increasing the GeO_x thickness (0.26-1.06 nm). X-ray photoelectron spectroscopy (XPS) results show that Ge^{3+} oxide component is responsible to the decrease of D_{it}.

(Keywords: Germanium, Ozone oxidation, interface state density, passivation)

Introduction

Ge is of great interest as a channel material for future technology nodes [1, 2]. Passivation of Ge surface is critical for excellent electrical performance. GeO_x interlayer by ozone oxidation shows favorable GeO_x/Ge interface [3, 4]. In order to further improve GeO_x/Ge interface by ozone oxidation, the oxidation and passivation mechanisms are essential to investigate. In this paper, we address these issues.

Experimental

After HF clean, the Ge surface was subjected to a flow of ozone gas to form GeO_x. The GeO_x thickness was determined by XPS. In addition, MOS capacitor with Ge/GeO_x/Al_2O_3/TiN/W structure ware prepared.

Results and Discussion

Fig. 1 shows the GeO_x thickness vs. ozone oxidation time in temperature range from 80 °C to 400 °C. An initially linear growth of GeO_x thickness is observed below ~10 s, and then it becomes parabolic as the oxidation time increases. In linear growth region, the oxidation is determined by chemical reaction occurring at the GeO_x/Ge interface. While in parabolic growth region, the oxidation is limited by diffusion process of oxygen atoms through GeO_x.

In order to accurately understand the reaction mechanisms of Ge oxidation by ozone, the Arrhenius temperature dependence of oxidation process is investigated for each oxidation growth region. **Fig. 2** shows the Arrhenius plot of linear rate constant (B/A) in the initially linear growth region. The activation energy is calculated to be 0.06 eV. This rather small activation energy indicates that the initially linear growth is nearly barrier-less. And this activation energy is nearly equal to that of Si surface oxidation by ozone. Also shown in **Fig. 2** is Arrhenius plot of parabolic rate constant (B). The activation energy is found to be 0.54 eV, which is dramatically reduced compared to the reported value of thermal oxidation in O_2 [5]. Thus it can be concluded that in the ozone oxidation it is not O_3 molecular but O radicals that go through the GeO_x film. In order to further investigate the oxidation kinetics, the XPS spectra of Ge 3d core level were recorded to obtain the oxidation states and atomic structures of grown GeO_x/Ge samples. **Fig. 3** shows show Ge $3d$ spectra at 300 °C for 1 s and 60 s. **Fig. 4** shows the change in areal intensity of Ge sub-oxide component vs. GeO_x thickness.

Then based on a layer-by-layer calculation of the Ge oxide components (Ge^{1+}, Ge^{2+}, Ge^{3+} and Ge^{4+}) to Ge^0 ratios in **Table I**, atomic structures are obtained for GeO_x/Ge with GeO_x of 4.3 Å (1 s oxidation, linear growth region) and 5.7 Å (60 s oxidation, parabolic growth region), not shown here. Topmost two layers of Ge atoms are oxidized for 1s oxidation (linear region), and three layer for 60s oxidation (parabolic region). In other words, the linear growth region includes the oxidation of the topmost and second layers of Ge (layers 1 and 2), and the parabolic growth region contains oxidation of third layer (layer 3) and following layers. **Fig. 5** schematically shows the linear and parabolic growth region.

Fig. 6 shows D_{it} vs. GeO_x thickness. The D_{it} decreases with thicker GeO_x. From results in Fig. 4, it can be concluded that the D_{it} decreases with larger Ge^{3+} amounts, indicating Ge^{3+} is key for passivation.

Conclusion

The oxidation and passivation mechanisms of Ge surface by ozone is investigated. The oxidation process is composed of initially linear growth region and following parabolic growth region. The Ge^{3+} state is key for surface passivation. Our finding is helpful to further improving device performances.

Acknowledgments

This work was financially supported by National Natural Science of China (No. 61574168 and No. 61504163).

References

[1] C. H. Lee et al., TED, 58 (2011) 1295. [2] R. Zhang et al., TED, 60 (2013) 927. [3] D. Kuzum et al., EDL, 29 (2008) 328. [4] X. Wang et al., ASS, 357 (2015) 1857. [5] M. Kobayashi et al., JAP, 106 (2009) 104117.

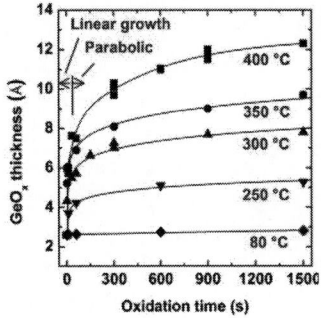

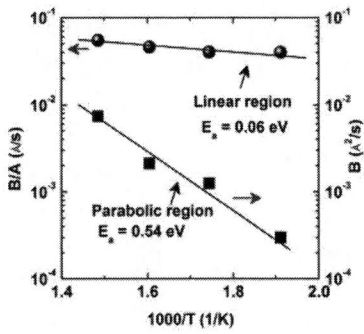

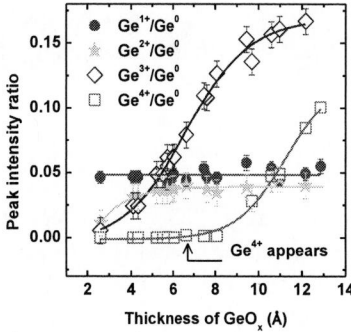

Fig. 1. GeO$_x$ thickness vs. ozone oxidation time in temperature range from 80 °C to 400 °C. An initially linear growth is observed below ~10 s, and then it becomes parabolic growth. This indicates two different chemical/physical oxidation mechanisms.

Fig. 2. Arrhenius plots of linear rate constant (B/A) and parabolic rate constant (B). The activation energy in parabolic region is 0.54 eV, which is dramatically reduced compared to thermal oxidation in O$_2$. Thus in the ozone oxidation it is not O$_3$ molecular but O radicals that go through the GeO$_x$ film.

Fig. 4. Areal intensity of Ge oxide component vs. GeO$_x$ thickness. Ge^{1+} is nearly unchanged. The Ge^{2+} state increases initially, and then becomes saturated for thicker GeO$_x$. The Ge^{3+} increases with GeO$_x$ thickness. The Ge^{4+} cannot be detected for GeO$_x$ less than ~6.6 Å, then it appears and increases for GeO$_x$ more than ~6.6 Å.

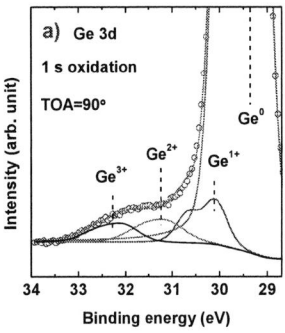

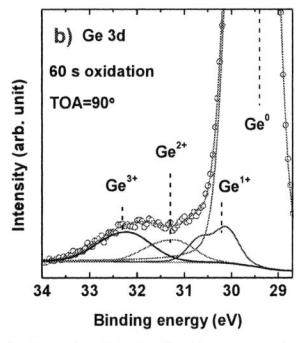

Fig. 3. Ge $3d$ spectra at 300 °C oxidation for (a) 1 s in linear region and (b) 60 s in parabolic region.

Fig. 5 (right figure) Schematic of Ge oxidation by ozone for the linear and parabolic regions. The linear growth region includes the oxidation of the topmost and second layers of Ge (layers 1 and 2), and the parabolic growth region contains oxidation of third layer (layer 3) and following layers.

Table I Summary of equations in the layer-by-layer calculation. The area ratios of Ge oxidation states to Ge^0 are calculated. θ_{xy} means the proportions of Ge^{x+} atoms in layer y. λ is the photoelectron mean free path. Then θ_{xy} can be obtained using XPS results in Fig. 4, i.e., atomic structure can be obtained.

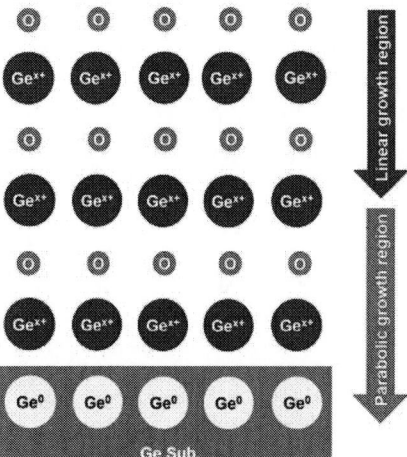

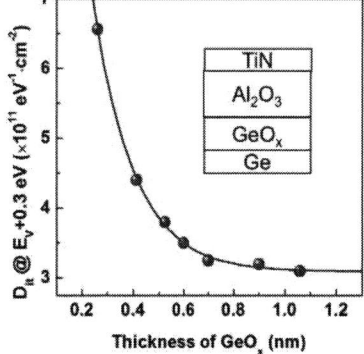

Fig. 6 D$_{it}$ vs. GeO$_x$ thickness. The D$_{it}$ decreases with thicker GeO$_x$. Combined with results in Fig. 4, the D$_{it}$ decreases with larger Ge^{3+} amounts. This shows that Ge^{3+} is key to achieve excellent passivation of Ge surface.

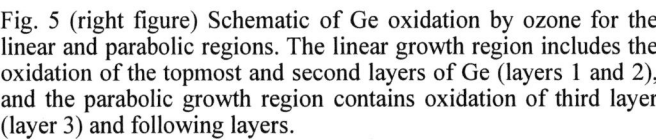

$$\frac{Ge^{1+}}{Ge^0} = [\theta_{12} + \theta_{13}\exp(-\frac{1.41}{\lambda_{Ge}})] / [(1-\theta_{13})\exp(-\frac{1.41}{\lambda_{Ge}}) + \sum_{n=2}^{\infty}\exp(-\frac{1.41n}{\lambda_{Ge}})]$$

$$\frac{Ge^{2+}}{Ge^0} = \sum_{m=1}^{2}\theta_{2m}\exp(-\frac{3.15(m-1)}{\lambda_{GeOx}}) / [(1-\theta_{13})\exp(-\frac{1.41}{\lambda_{Ge}}) + \sum_{n=2}^{\infty}\exp(-\frac{1.41n}{\lambda_{Ge}})]\exp(-\frac{3.15}{\lambda_{GeOx}})$$

$$\frac{Ge^{3+}}{Ge^0} = \sum_{m=1}^{2}\theta_{3m}\exp(-\frac{3.15(m-1)}{\lambda_{GeOx}}) / [(1-\theta_{13})\exp(-\frac{1.41}{\lambda_{Ge}}) + \sum_{n=2}^{\infty}\exp(-\frac{1.41n}{\lambda_{Ge}})]\exp(-\frac{3.15}{\lambda_{GeOx}})$$

8A-3

The impact of atomic layer depositions on high quality Ge/GeO₂ interfaces fabricated by rapid thermal annealing in O₂ ambient

Laura Žurauskaitė[1], Per-Erik Hellström[1] and Mikael Östling[1]

[1]KTH Royal Institute of Technology, Stockholm, Sweden, lauraz@kth.se

Abstract

This work demonstrates high quality Ge/GeO₂ interfaces fabricated by O₂ RTA that are degraded by a good quality SiO₂ layer deposited by ALD. However, neither O₃ and H₂O precursors commonly used during subsequent high-k ALDs nor Si precursor AP-LTO-330 do not degrade the interface. Thus D_{it} increase after SiO₂ deposition is likely due to intermixing. Therefore, the effect of subsequent ALDs on the interface quality has to be considered while designing Ge-based gate stacks.
(Keywords: Germanium, GeO₂, high-k, ALD, D_{it})

Introduction

One of the most crucial challenges in realizing high-performance Ge devices is the gate stack formation. GeO₂ provides the lowest interface state density D_{it} values [1] and can be combined with high-k dielectrics to achieve low equivalent oxide thickness (EOT) [2]. However, the gate process conditions are important, e. g. O₃-based atomic layer deposition (ALD) on a thin GeO₂ layer has been shown to cause oxidation of Ge surface [3]. In this work we show that subsequent ALD on thermally grown GeO₂ layers can influence the Ge/GeO₂ interface quality.

Experimental

The fabrication process of Ge MOS capacitors is displayed in Fig. 1. After the cleaning of the n-Ge substrate native germanium oxide was removed with aqueous HF and HCl solutions. The samples were immediately loaded into the rapid thermal anneal (RTA) chamber and oxidation was carried out at 550°C for 5 s to 5 min. Then either SiO₂ layer was deposited by ALD at 350°C employing O₃ and AP-LTO-330 precursors or vacuum, H₂O, O₃ or AP-LTO-330 anneals were performed in ALD reactor at 350°C. Reference samples without SiO₂ deposition and annealing were also fabricated. Al gate metal was deposited by PVD and patterned.

Results and discussion

GeO₂ layer thermally grown by RTA displays an increasing growth characteristic with time (Fig. 2). Well-behaved capacitance-voltage (CV) curves with almost no frequency dispersion in accumulation obtained from n-Ge/GeO₂ gate stacks are shown in Fig. 3. D_{it} at the midgap was estimated from the CV curves in the following way. Theoretical CV curves [4] that assume a parabolic D_{it} distribution within

the bandgap were fitted to the measured curves. The parabolic distribution with a minimum at the midgap was employed due to the close resemblance to the symmetrical D_{it} distribution for GeO₂ observed in literature [1]. A good agreement between the measured and the theoretical curves from midgap to accumulation can be judged from Fig. 4. The influence of the D_{it} value at the midgap on low frequency CV characteristics is displayed in Fig. 5 where theoretical curves are plotted for a 5 nm EOT MOS capacitor showing an increase of the capacitance minimum with a rising D_{it} value. The lowest interface state density at the midgap that can be estimated using this method is around $1 \cdot 10^{11}$ cm^{-2}eV^{-1}. Measured CV characteristics of the n-Ge/GeO₂ MOS capacitors were used to extract low D_{it} at the midgap values in the range of $2.5 \text{-} 5 \cdot 10^{11}$ cm^{-2}eV^{-1} which are shown in Fig. 6 at different GeO₂ CETs. However, a subsequent ALD of a high quality SiO₂ layer can damage the Ge/GeO₂ interface and to a greater extent for thinner GeO₂ as shown in Fig. 6 where a D_{it} increase is displayed after SiO₂ deposition. The following experiments were performed to investigate a possible source of the interface degradation in Ge/GeO₂/SiO₂ gate stacks as well as the effects other ALDs could have on good quality interfaces.

For scaled Ge channel MOSFETs GeO₂ is commonly integrated with high-k dielectrics that are deposited by ALD at elevated temperature. To investigate the effect of the main precursors employed during the high-k ALDs, H₂O and O₃ anneals were performed at 350°C on n-Ge/GeO₂ gate stacks that were fabricated by O₂ RTA at 550°C for 2 min. A vacuum anneal was also performed at 350°C to examine a possible effect of GeO desorption that has been reported to occur at elevated temperature in vacuum [5] and might take place during the deposition. Finally, the influence of the Si precursor AP-LTO-330 employed in SiO₂ ALD was investigated by performing an AP-LTO-330 anneal at 350°C. Almost no change in D_{it} was observed regardless of the anneal ambient (see Fig. 7).

Another factor that can cause the defects in the gate stack is metal deposition, i.e., if the metal reacts or intermixes with the oxide, or the deposition itself damages the underlying layers. The latter can be ruled out because Ge/GeO₂ gate stacks with good interface quality and employing a rather thin oxide of ~3 nm CET were achieved while $Ge/GeO_2/SiO_2$ gate stacks are much thicker and thus would be affected less. To examine if the interface degradation in $Ge/GeO_2/SiO_2$ gate stacks was caused by the employment of Al as a gate metal, TiW/Al stack was used instead. The results are summarized in Fig. 8 showing that employing TiW instead of Al as a gate metal does not reduce D_{it}. Therefore, the choice of metal is not likely to be the factor that increases the interface state density in $Ge/GeO_2/SiO_2$ MOS capacitors.

A cross sectional scanning transmission electron microscope (STEM) bright field image of a $Ge/GeO_2/SiO_2/Al$ MOS capacitor is displayed in Fig. 9a. Separate GeO_2 and SiO_2 layers (thicknesses of 6-7 nm and 7-7.5 nm respectively) can be distinguished as indicated in the figure. Another important feature of the image is an interfacial layer of Ge/GeO_2 that can be more distinctively seen as a bright line in the high resolution transmission electron microscope (HR TEM) image in Fig. 9b. This 3-4 nm thick layer is both crystalline and amorphous, and can appear this way due to the surface roughness since the information is gathered from a ~100 nm thick lamella. We have observed a RMS surface roughness of ~1.8 nm measured by AFM after HF and HCl dips. Moreover, GeO_2 and SiO_2 layers displayed in the HR TEM image are less distinguishable than in STEM image, and of slightly different thicknesses (6.5-7.5 nm and 8.5-9.5 nm respectively). This can be due to the surface roughness or an indication of an intermixture between layers.

Intermixing between GeO_2 and SiO_2 might be responsible for D_{it} degradation in $Ge/GeO_2/SiO_2$ gate stacks. If Si atoms are diffusing to Ge/GeO_2 interface and causing defects, the decreasing D_{it} with increasing GeO_2 thickness trend could be explained since less Si atoms would be able to reach Ge/GeO_2 interface if GeO_2 thickness is higher. Furthermore, Ge atoms could be diffusing into SiO_2 layer and leaving broken bonds at Ge/GeO_2 interface. This effect has been reported in $Ge/GeO_2/HfO_2$ gate stacks [6] where Ge diffused through HfO_2 layer and gathered at the surface. Ge diffusion would also explain D_{it} vs. GeO_2 thickness trend because more diffusion would occur from parts of GeO_2 layer that are closer to SiO_2 layer and thus further from the interface for thicker GeO_2 layers. Therefore, it is likely that either Si or Ge diffusion is responsible for the D_{it} degradation in $Ge/GeO_2/SiO_2$ gate stacks.

Conclusion

Ge/GeO_2 gate stacks fabricated by O_2 RTA exhibit low D_{it} at the midgap in the range of $2.5\text{-}5 \cdot 10^{11}$ $cm^{-2}eV^{-1}$. The interface is not degraded by further vacuum, O_3, H_2O or AP-LTO-330 anneals in ALD chamber but a D_{it} increase is observed after a SiO_2 deposition that employs O_3 and AP-LTO-330 precursors. This degradation is likely to be caused by either Si diffusion to the interface or Ge diffusion towards SiO_2 layer. Therefore, the effect of subsequent ALDs on the interface quality has to be considered while designing Ge-based gate stacks.

Acknowledgments

This work was supported by the Swedish Foundation for Strategic research. The authors would like to thank Konstantinos Garidis for the AFM measurement.

References

[1] H. Matsubara, T. Sasada, M. Takenaka, and S. Takagi, "Evidence of low interface trap density in GeO_2 / Ge metal-oxide- semiconductor structures fabricated by thermal oxidation," Appl. Phys. Lett., vol. 93, no. 3, p. 32104, 2008.

[2] R. Zhang, P. C. Huang, J. C. Lin, N. Taoka, M. Takenaka, and S. Takagi, "High-mobility Ge p- and n-MOSFETs with 0.7-nm EOT using $HfO_2/Al_2O_3/GeO_x/Ge$ gate stacks fabricated by plasma postoxidation," IEEE Trans. Electron Devices, vol. 60, no. 3, pp. 927–934, 2013.

[3] A. Delabie, A. Alian, F. Bellenger, M. Caymax, T. Conard, A. Franquet, S. Sioncke, S. Van Elshocht, M. M. Heyns, and M. Meuris, "H_2O- and O_3-Based Atomic Layer Deposition of High-κ Dielectric Films on GeO_2 Passivation Layers," J. Electrochem. Soc., vol. 156, no. 10, pp. G163–G167, 2009.

[4] Y. Tsividis, Operation and Modeling of the MOS Transistor. 1999.

[5] K. Kita, S. Suzuki, H. Nomura, T. Takahashi, T. Nishimura, and A. Toriumi, "Dramatic Improvement of GeO_2/Ge MIS Characteristics by Suppression of GeO Volatilization," ECS Trans., vol. 11, no. 4, pp. 461–469, 2007.

[6] S. Ogawa, R. Asahara, Y. Minoura, H. Sako, N. Kawasaki, I. Yamada, T. Miyamoto, T. Hosoi, T. Shimura, and H. Watanabe, "Insights into thermal diffusion of germanium and oxygen atoms in $HfO_2/GeO_2/Ge$ gate stacks and their suppressed reaction with atomically thin AlO_x interlayers," J. Appl. Phys., vol. 118, no. 23, pp. 1–6, 2015.

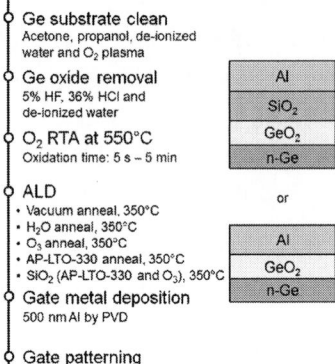

Fig. 1: Process flow of Ge MOS capacitor fabrication. Schematic view of the final structures is displayed on the right.

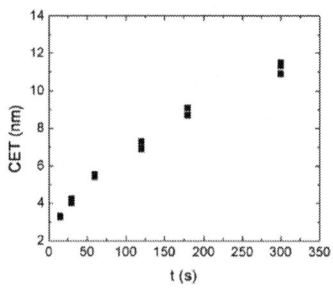

Fig. 2: GeO_2 CET versus O_2 oxidation time for n-Ge/GeO_2 MOS capacitors.

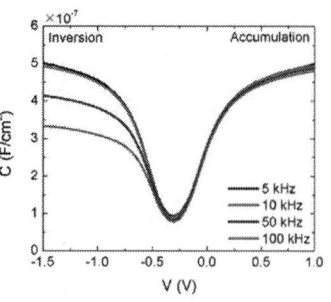

Fig. 3: Well-behaved CV characteristics of n-Ge/GeO_2 MOS capacitors oxidized by O_2 RTA. Inversion response is observed at room temperature.

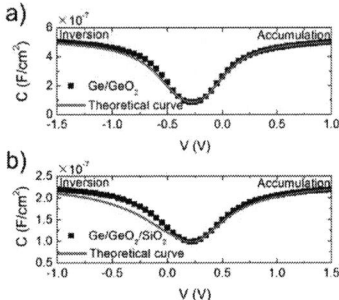

Fig. 4: Comparison between a measurement and a theoretical CV curve that assumes a parabolic D_{it} distribution for: a) Ge/GeO_2 and b) Ge/GeO_2/SiO_2. Good agreement is observed around the midgap.

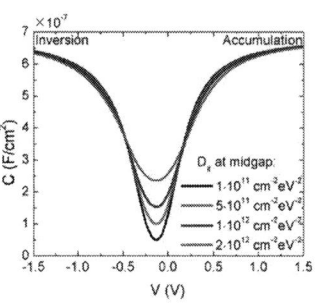

Fig. 5: Simulated CV characteristics of 5 nm EOT MOS capacitor with varying D_{it} in the midgap displaying an increase in midgap capacitance with increasing D_{it}.

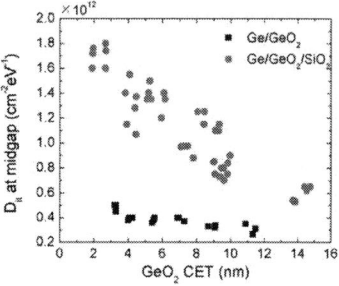

Fig. 6: High D_{it} at midgap values for n-Ge/GeO_2/SiO_2 MOS capacitors decreasing with GeO_2 thickness in reference to n-Ge/GeO_2 MOS capacitors with good quality interface.

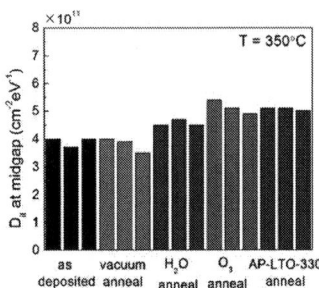

Fig. 7: D_{it} at midgap for as deposited, vacuum, H_2O, O_3 and AP-LTO-330 annealed n-Ge/GeO_2 gate stacks.

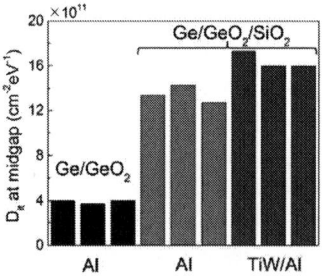

Fig. 8: D_{it} at midgap for Ge/GeO_2/Al, Ge/GeO_2/SiO_2/Al and Ge/GeO_2/SiO_2/TiW/Al stacks showing higher D_{it} for stacks with SiO_2 regardless of the gate metal employed.

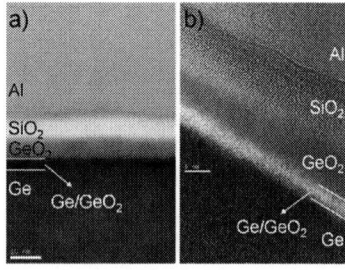

Fig. 9: a) STEM Bright Field and b) HR TEM cross sectional images of Ge/GeO_2/SiO_2/Al MOS capacitor with comprising layers indicated.

978-1-5090-4661-4/17 $31.00 © 2017 IEEE

Experimental Investigation on Growth Mechanism of GeO$_x$ Layer Formed by Plasma Post Oxidation Based on Angle Resolved X-ray Photoelectron Spectroscopy

Zhiqian Zhao[1], Xiaolei Wang[2], Jing Zhang[1,a)], Chao Zhao[2,3], Tianchun Ye[2,3], and Wenwu Wang[2,3,b)]

[1] Microelectronics Department, North China University of Technology, Beijing 100041, China
[2] Key Laboratory of Microelectronics Devices & Integrated Technology, Institute of Microelectronics, Chinese Academy of Sciences, Beijing 100029, China, E-mail: a) zhangj@ncut.edu.cn; wangwenwu@ime.ac.cn
[3] School of Microelectronics, University of Chinese Academy of Sciences, Beijing 100049, China

Abstract

The growth mechanism of GeO$_x$ layer formed by plasma post oxidation at room temperature (RT PPO) is investigated based on angle resolved X-ray photoelectron spectroscopy (AR-XPS). The experimental results show that the GeO$_x$ grown by RT PPO does not obey layer-by-layer growth mode. And the distribution of Ge oxidation states is random during RT PPO. These findings are helpful to the optimization of the Ge based gate stacks for future CMOSFET devices.

(Keywords: Ge, plasma post oxidation, MOS, XPS)

Introduction

Ge is a potential channel material [1-3]. However, several issues still need to be solved. One of the critical issues is to provide a high-quality interfacial layer [2]. Zhang et al. [4] demonstrated that a high-quality interfacial layer can be achieved by using the electron cyclotron resonance oxygen plasma post-oxidation method. In order to further improve interface property using PPO, the oxidation mechanism is necessary to understand. In this paper, the growth kinetics of GeO$_x$ layer formed by PPO technology is systematically investigated using angle resolved X-ray photoelectron spectroscopy (AR-XPS). We find that the GeO$_x$ grown by PPO dose not obey layer-by-layer growth mode.

Experimental

After HF clean, the wafers were immediately capped with 3-nm-thick Al$_2$O$_3$ layers by atomic layer deposition (ALD) at substrate temperature of 300 °C, with trimethylaluminum (TMA) and H$_2$O as the precursors. At last, the Al$_2$O$_3$/GeO$_x$/Ge structure was fabricated at room temperature (~20 °C) by PPO.

Results and discussion

Fig. 1 shows the XPS spectra of Ge $3d$ for different GeO$_x$ thicknesses. Based on the spectrum deconvolution in Fig.1, the area intensity ratios of Ge^{4+} and Ge^{1+} for takeoff angle of 35° and 90° are shown in Fig. 2. Considering that the spectrum obtained with smaller take off angle is more sensitive to the surface, this experimental phenomenon indicates that some Ge^{1+} states appear above Ge^{4+} components. This is in conflict with layer-by-layer growth mode, in which Ge^{4+} is localized above Ge^{1+}. Therefore, we consider that the layer-by-layer growth

mode cannot be applied to the growth mechanism of GeO$_x$ layer formed by RT PPO. Moreover, Fig. 3 shows the area intensity ratio of Ge^{x+} (x=1, 2, 3, 4) and Ge^{1+} vs. GeO$_x$ thickness. Both co-direction and reverse oscillation of Ge^{1+} and Ge^{2+} can be observed, which is inconsistent with the phenomenon of reverse oscillation of Ge^{1+} and Ge^{2+} in layer-by-layer growth mode [4]. Fig. 4 shows the area intensity ratios among various states of Ge for 0.63 nm GeO$_x$. From Fig. 4(a), it can be observed that the ratios decrease with increasing takeoff angel, indicating that all the oxidation states are located above the Ge^0. The similar phenomenon arises in Fig. 4(d), suggesting that Ge^{4+} states appear above Ge^{3+}. On the contrary, the ratios in Fig. 4(b) and (c) increase with increasing takeoff angel, suggesting that some Ge^{1+} are located above the Ge^{x+} (x=2,3,4) and some Ge^{2+} appear above the Ge^{x+} (x=3,4), respectively. Based on above discussion, a schematic diagram about distribution of various states of Ge is given in Fig. 5(a). It can be seen that Ge^{1+}, Ge^{2+}, Ge^{4+} and Ge^{3+} are localized from GeO$_x$ surface to GeO$_x$/Ge interface, but not Ge^{4+}, Ge^{3+}, Ge^{2+}, and Ge^{1+}. This is explained as follows. In the plasma oxidation, the oxygen plasmas are distributed with peak at a projected depth below the Ge surface as shown in Fig. 5(b), similar to "implantation process" [5]. Consequently, higher Ge oxidation state appears (Ge^{4+}) below the surface. Thus, we consider that the deeper atomic layers of Ge may be preferentially oxidized by Oxygen plasma instead of the outermost ones.

Conclusion

The growth mechanism of the GeO$_x$ layer formed by RT PPO is investigated based on the technology of AR-XPS. The layer-by-layer growth mode cannot be applicable to PPO.

Acknowledgments

This work was financially supported by National Natural Science of China (No. 61574168 and No. 61504163).

References

[1] S. Gupta et al., MRS Bull. 39 (2014) 678. [2] Q. Xie et al. Semicond. Sci. Technol. 27 (2012) 074012. [3] K. C. Saraswat et al., Microelectron. Eng., 80 (2005) 15. [4] R. Zhang et al., APL, 102 (2013) 081603. [5] J. D. Plummer et al., Silicon VLSI Technology: Fundamentals, Practice and Modeling, Prentice Hall, 2000.

978-1-5090-4661-4/17 $31.00 © 2017 IEEE

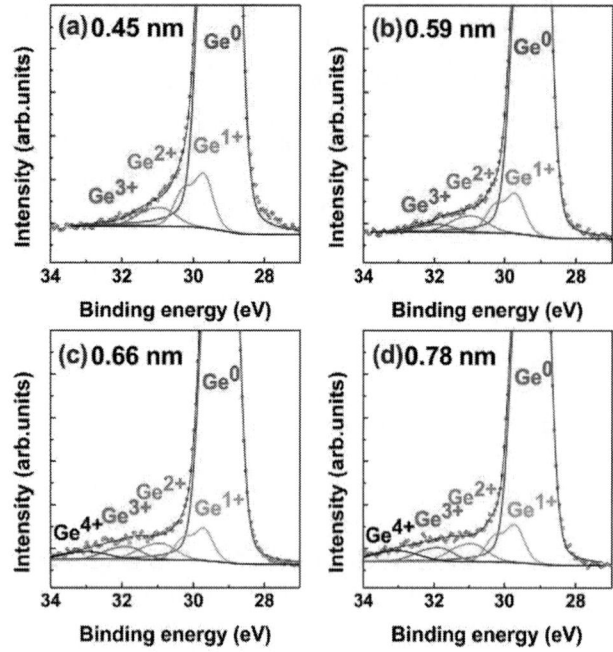

Fig. 1: XPS spectra of Ge 3d for $Al_2O_3/GeO_x/Ge$ structure with different GeO_x thicknesses at the takeoff angle of 90°.

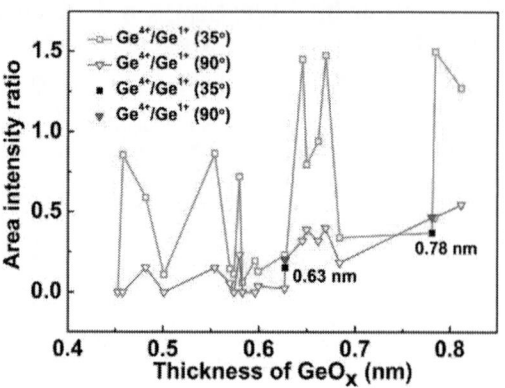

Fig. 2: Area intensity ratio of Ge^{4+} and Ge^{1+} vs. GeO_x thickness for the takeoff angle of 35° and 90°. For GeO_x of 0.63 nm and 0.78 nm, the ratios measured at takeoff angle of 35° is smaller than that at 90° at 0.63 nm and 0.78 nm. This indicates that Ge^{1+} is localized above Ge^{4+}, which is inconsistent with the layer-by-layer growth mode.

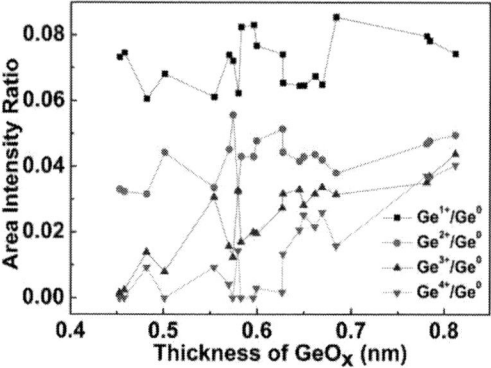

Fig. 3: Area intensity ratio of Ge^{x+} (x=1, 2, 3, 4) and Ge^{1+} vs. GeO_x thickness. Both co-direction and reverse oscillation of Ge^{1+} and Ge^{2+} can be observed, which is inconsistent with the phenomenon of reverse oscillation of Ge^{1+} and Ge^{2+} in layer-by-layer growth mode.

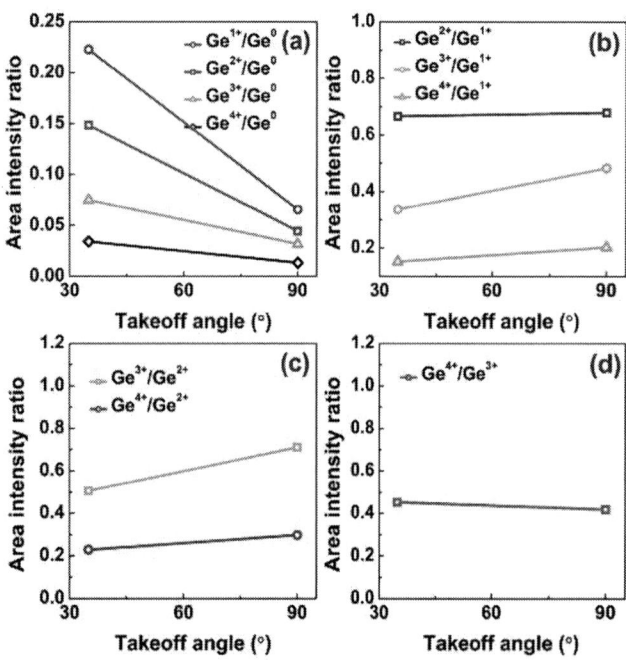

Fig. 4: Area intensity ratios of various states of Ge component as a function of the takeoff angle for the GeO_x thickness of 0.63 nm. According to these results, the distribution of various states of Ge in GeO_x layer is obtained, as shown in Fig. 5(a).

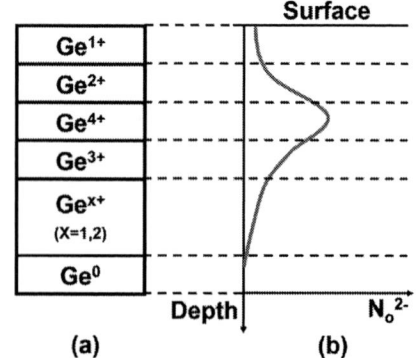

Fig. 5: (a) Distribution of various states of Ge components in GeO_x layer at 0.63 nm; (b) Schematic of oxygen concentration distribution as a function of the depth of GeO_x. In the plasma oxidation, the oxygen plasmas are distributed with peak at a projected depth below the Ge surface as shown in Fig. 5(b), similar to "implantation process" that the more the oxygen atoms are, the higher the state of germanium is.

978-1-5090-4661-4/17 $31.00 © 2017 IEEE

P-1

UV-Annealing-Enhanced Stability in High-Performance Printed InO$_x$ Transistors

William J. Scheideler[1] and Vivek Subramanian[1]

[1]University of California Berkeley, Berkeley, CA, USA

Abstract

We report on low-temperature additive processing methods for fabrication of high-performance InO$_x$ thin film transistors (TFTs) based on UV-annealed, printed high-k AlOx gate dielectrics. The impact of UV annealing on dielectric properties, TFT performance, and bias-stress stability is studied.
(Keywords: Printed Transistors, UV Annealing, Additive Manufacturing)

Introduction

Metal oxide thin-film transistors (TFTs) have good performance ($\mu_{eff} \sim$ 10-100 cm^2/Vs) and high visible range transparency (>90%) that make them well suited for IoE (Internet of Everything) applications involving sensors and information display. Solution-based fabrication of metal oxide TFTs can utilize printing to boost manufacturing throughput and enhance utilization of scarce elements like Indium. While various printed metal oxide semiconductors (IGZO, etc) have been well-studied, two major barriers to wide-scale adoption remain: the use of high processing temperatures (> 400 °C) and a lack of robust printed gate dielectrics [1]. Here we address these challenges by developing high-performance UV-annealed InO$_x$ TFTs with printed AlOx dielectrics, processed at 250 °C in air, allowing compatibility with a wide range of flexible substrates for high-throughput roll-to-roll compatible fabrication.

Experimental Studies

We demonstrate bottom-gated thin film transistors using AlOx dielectrics, InO$_x$ as a semiconductor, and CdO:Al as conductors. AlOx dielectrics (t$_{ox}$=15nm–200nm) were inkjet printed (DMP 2800) on UV ozone treated (2 min, Jelight 42) n++ Si substrates from a 400mM solution of Al(NO$_3$)$_3$ (Sigma 229415, 99.997 %) in 2-methoxyethanol, dried at 150 °C for 5 min, and UV annealed (1 W/cm^2, Loctite Zeta7401) for 5-20 minutes in air at a temperature of approximately 100 °C (Fig. 1a) before a 1 hour thermal anneal at 250 °C. The InO$_x$ films (15nm) were printed from aqueous In(NO$_3$)$_3$ (200mM) (Sigma 326127, 99.99 %) and air annealed at 250 °C (2 hours), before printing CdO:Al top contacts from 200mM aqueous Cd(NO$_3$)$_3$ (Sigma 229520, 99.997 %) with 3% Al(NO$_3$)$_3$ and annealing at 250 °C (1 hour) to form the substrate-gated, top contact TFTs illustrated in Fig. 1b.

The influence of UV annealing was first studied in MIM capacitors with inkjet printed metal contacts (Ag nanoparticle ink, ANP DGP 40LT-15C) annealed at 150 °C in air for 10 minutes. The dielectric constant of printed MIMS (Fig. 2a) without UV annealing exhibited strong low frequency dispersion in the range of 20Hz - 1kHz, which may indicate the presence of residual impurities. The UV-annealed films in contrast, showed progressively flatter frequency response, with the dielectric constant approaching the expected $\varepsilon_r \sim$ 6- 7 of AlOx. This could suggest that residual organic ligands are decomposed by UV[2] and can then be volatilized to produce denser, higher quality AlOx dielectric films. Indeed, alternating 2 min cycles of thermal (250 °C) and UV produced the flattest and most ideal frequency response (series '4 x 2 min', Fig. 2a).

Thickess measurements of these printed films (Fig 2b), show that UV annealing exposures as short as 5 min cause significant thickness reduction, consistent with the decomposition and densification of the precursor film. The effect is particularly noticeable for the thicker precursor films, which exhibit a > 30% volume reduction following a 10 min exposure. After longer UV exposures (20 min), the film thickness closely approaches the film thickness otherwise achieved following the 1 hour thermal annealing step at 250 °C. Interestingly, while the UV dose seems to have a large impact on the dielectric response and average film thickness, the leakage current density and breakdown fields of the printed MIM capacitors were not sensitive to the duration of UV exposure. AlO$_x$ MIMs with varying UV exposures showed consistently low leakage current density of approximately 10^{-7} A/cm^2 at an electric field of 1 MV/cm and had a median breakdown field of 3.5 MV/cm (Fig. 2c), making these gate dielectrics appropriate for use in printed transistors.

The principal benefits of utilizing the thicker AlO$_x$ gate dielectrics afforded by UV annealing are the improved operational stability and control over TFT operating voltages. Without the use of UV annealing, t$_{ox}$ > 100nm leads to extensive film cracking and inactive TFTs, but with UV annealing, TFTs with AlOx gate dielectrics from 15nm to 200nm thick were fabricated, allowing stable operation at voltages from ± 2 V to ± 30 V for a

978-1-5090-4661-4/17 $31.00 © 2017 IEEE

variety of potential IoE applications. As illustrated in the transfer characteristics (Fig. 3a-c), these TFTs showed excellent performance with minimal hysteresis ($\Delta V_t < 100$mV), Subthreshold Swing < 150mV/dec, and an average μ_{lin} of 12 ± 1.6 cm^2/Vs. These benchmarks show that the printed devices have characteristics comparable to InO$_x$ TFTs fabricated with vacuum-based deposition methods such as ALD[3] and DC sputtering[4]. Indeed, the low subthreshold swing and small hysteresis significantly outperform previously reported oxide TFTs with printed dielectrics [1].

The bias-stress stability of the printed TFTs was also investigated to understand their viability for long-term sensing and display applications. As shown in Fig. 4, the normalized V_t shifts caused by positive and negative bias-stress were evaluated for the TFTs with different gate dielectrics. TFTs with the thinnest AlOx (15nm) dielectrics exhibit V_t shifts of 25-30% of their operating voltage (V_{DD}) after a 1000s period of stress, whereas thicker dielectrics (200nm AlOx) show reduced V_t shifts (Fig. 4a,b) of under 15% of V_{DD}. As a result, the devices with thicker dielectrics have a considerably more stable I_{on}, that drops less than 10% over the course of the measurements (Fig. 4c), compared to a 50-80% loss in I_{on} for the TFTs with thinner AlOx dielectrics.

Conclusion

Low-temperature UV annealing is demonstrated for InO$_x$ transistors based on printed high-k AlOx dielectrics. The high-performance and improved bias-stress characteristics of these devices emphasize the utility of UV annealing for enabling low-temperature fabrication of stable high-k gate dielectrics and thin film transistors suitable for ubiquitous IoE applications involving sensing and information display.

Acknowledgments

William Scheideler was supported by the National Science Foundation Graduate Fellowship Program. We acknowledge helpful discussions with Jeremy Smith and Rajan Kumar.

References

[1] J. Jang, H. Kang, H. C. N. Chakravarthula, and V. Subramanian, "Fully Inkjet-Printed Transparent Oxide Thin Film Transistors Using a Fugitive Wettability Switch," *Adv. Electron. Mater.*, vol. 1, no. 7, p. n/a-n/a, Jul. 2015.

[2] S. Park *et al.*, "In-Depth Studies on Rapid Photochemical Activation of Various Sol–Gel Metal Oxide Films for Flexible Transparent Electronics," *Adv. Funct. Mater.*, vol. 25, no. 19, pp. 2807–2815, May 2015.

[3] H.-I. Yeom, J. B. Ko, G. Mun, and S.-H. K. Park, "High mobility polycrystalline indium oxide thin-film transistors by means of plasma-enhanced atomic layer deposition," *J Mater Chem C*, vol. 4, no. 28, pp. 6873–6880, 2016.

[4] M.-H. Lee *et al.*, "15.4: Excellent Performance of Indium-Oxide-Based Thin-Film Transistors by DC Sputtering," *SID Symp. Dig. Tech. Pap.*, vol. 40, no. 1, pp. 191–193, Jun. 2009.

Figure 1. UV photoannealing diagram (a). Printed TFT structure (b).

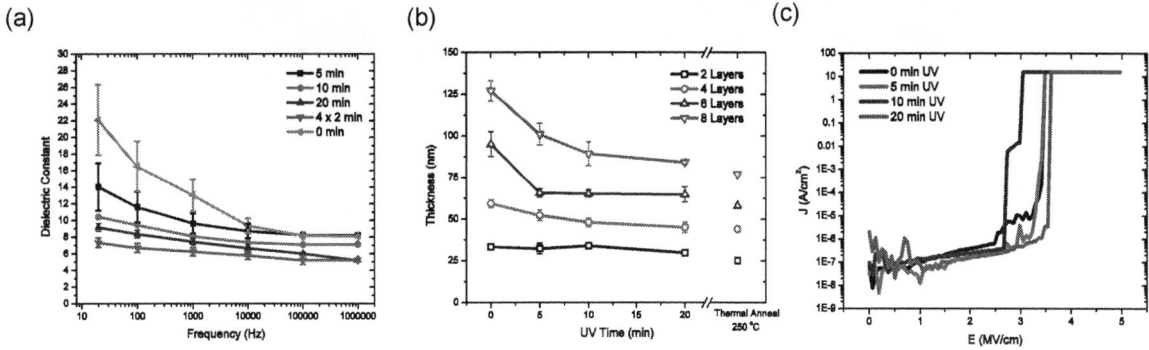

Figure 2. AlOx dielectric constant vs frequency and UV exposure for 60nm AlOx films (a). AlOx thickness vs UV exposure (b). AlOx leakage current vs electric field (c).

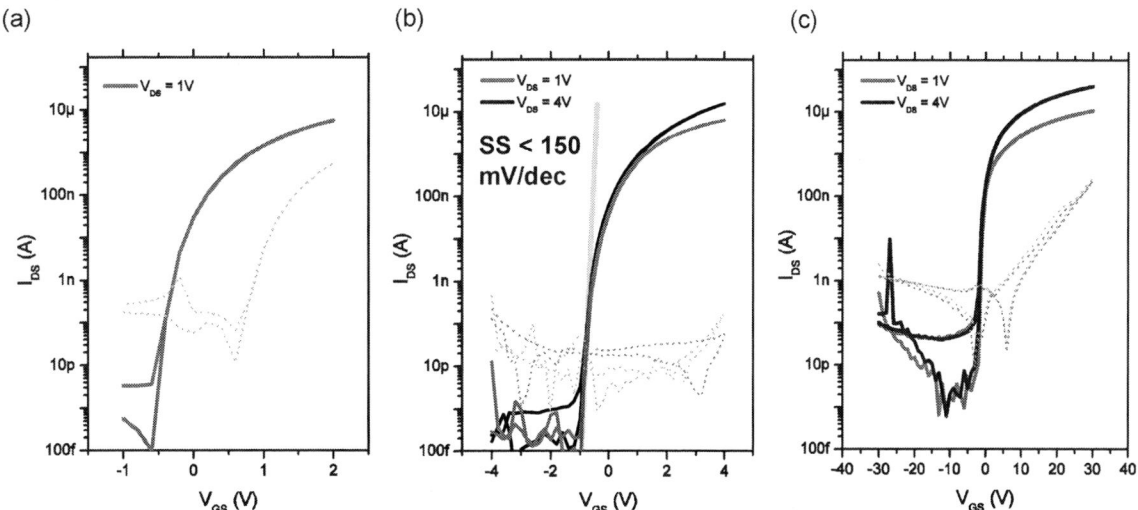

Figure 3. Transfer curves of InO_x TFTs with AlOx dielectrics of t_{ox} = 15nm (a), 30nm (b), or 200nm (c).

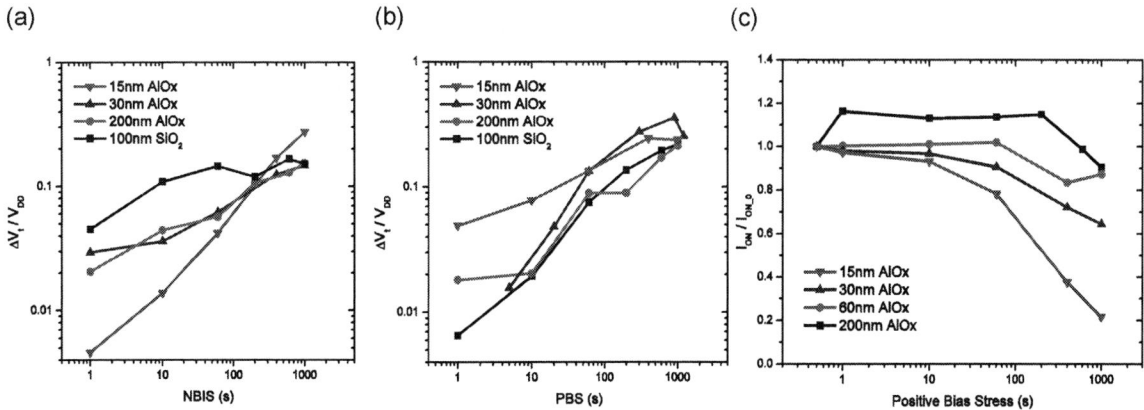

Figure 4. Normalized V_t shifts for printed TFTs with various dielectrics under negative illumination bias stress (a) and positive bias stress (b). Normalized on current vs positive bias (c).

978-1-5090-4661-4/17 $31.00 © 2017 IEEE

P-2

Random-Telegraph-Noise by Resonant Tunnelling at Low Temperatures

Z. Li[1], M. Sotto[1], F. Liu[1], M. K. Husain[1], I. Zeimpekis[1], H. Yoshimoto[2], K. Tani[2], Y. Sasago[2], D. Hisamoto[2],
J. D. Fletcher[3], M. Kataoka[3], Y. Tsuchiya[1], and S. Saito[1]

[1]Nano Research Group, ZI, ECS, FPSE, Univ. of Southampton, UK. email: S.Saito@soton.ac.uk
[2]Research & Development Group, Hitachi, Ltd., 1-280 Higashikoigakubo, Kokubunji, Tokyo 185-8601, Japan.
[3]National Physical Laboratory (NPL), Hampton Road, Teddington, Middlesex TW11 0LW, UK.

Abstract

We have found a systematic way to identify the bias conditions to observe the Random-Telegraph-Noise (RTN) in advanced Metal-Oxide-Semiconductor Field-Effect-Transistors (MOSFETs). We measured a p-type MOSFET at 2K, and found narrow bias conditions to observe the RTN presumably caused by charge trapping and de-trapping, which were only observed at low temperatures. It will pave the way to address the nature of a trap, which will be useful to understand the mechanism of RTN to secure the reliability. (Keywords: Random-telegraph-noise, charge trap, low temperatures)

Introduction

The RTN is becoming one of the major concerns to secure reliabilities in MOSFETs, when the technology node is scaled down to sub-20nm [1-2]. The RTN would be caused by the threshold voltage fluctuations due to the trapping/de-trapping process of carriers at charge traps near the Si/SiO2 interface [3]. However, it is difficult to identify the nature of these processes, since it is difficult to find a typical MOSFET statistically showing the RTN at room temperatures. The low temperature measurement of the RTN was reported previously [4], but the precise conditions under which bias conditions are required for the RTN, were missing. We measured MOSFET devices at low temperatures and identified the narrow bias conditions to observe the RTN.

Experimental Data and Discussion

The measured device was a standard p-type MOSFET with the width of 10μm and the length of 75nm. The device with a large width was chosen to increase the chance to find the RTN. The gate electrode was made of doped poly-Si, and the gate oxide was 2.4nm-thick SiON. Subthreshold characteristics are shown in Fig. 1. The subthreshold slope was 80.8mV/decade at 300K, 48.4mV/decade at 150K, and 9.8mV/decade at 2K,

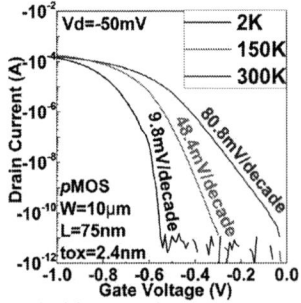

Fig. 1. Subthreshold properties at 2, 150, and 300K.

respectively. The background noise was less than 10pA.

We measured drain current (I_d) by changing the gate voltage (V_g) and drain voltage (V_d) to observe single-hole

charging events. We found Coulomb diamonds [5], as shown in Fig. 2.

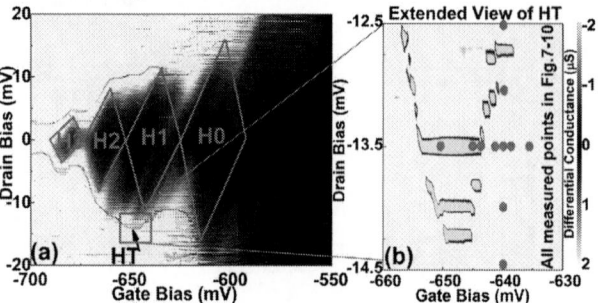

Fig. 2. (a) Current stability diagram of the device at 2K. An expanded diagram is shown in (b).

We extracted the capacitance of the quantum dot related to single-hole transistor characteristics (Table 1). The estimated diameter of the quantum dot is about 25nm, which might be coming from the poly-Si grains of the gate electrode or the surface roughness. The charging energies of the quantum dot were 14meV, 11meV, 8.5meV, and 4.1meV for H0, H1, H2, and H3 states, depending on the number of holes trapped in the quantum dot (Fig. 2(a)), respectively.

Tab. 1. Extracted capacitance of different hole states in the quantum dot.

	C_g (aF)	C_d (aF)	C_s (aF)
H0	4.8	2.6	4.0
H1	6.5	2.8	4.6
H2	7.6	3.8	6.5
H3	8.5	17.2	13.6

Near the edge of Coulomb diamonds, we observed sharp current peaks in the stability diagram. We measured the time domain characteristics of I_d under the bias conditions for the sharp peaks, and observed RTN, as shown in Fig. 4 (a). We identified the switching with large and small amplitudes, respectively. We investigated on the temperature dependence of the RTN, as shown in Fig. 3. The RTN was observed at 2K, while it was not observed

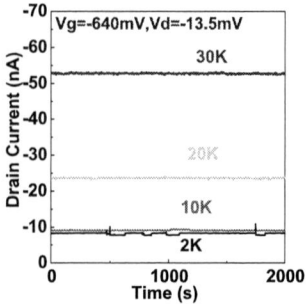

Fig. 3. RTN behaviour at 2, 10, 20, and 30K.

978-1-5090-4661-4/17 $31.00 © 2017 IEEE   172

above 20K. This implies that the thermal assisted transport hinder the observation of the RTN.

We have analysed the frequency to observe the current within a certain range of a fixed step of 4pA, which is actually the quantum mechanical probability, $P(I_d)$, to find the system under a certain current state, which should contain some information about the wave function of the traps. We found 4 different current levels, which are shown in Fig. 4 (b). From 4 levels, we can clearly identify 2 amplitudes; the small one was 71pA, and the large one was 528pA. The RTN with 2 different amplitudes implies that 2 traps were responsible. The schematic potential diagram across the channel is shown in Fig. 5.

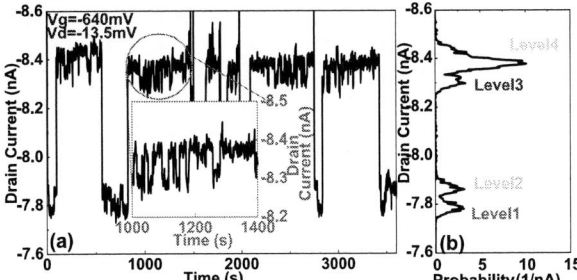

Fig. 4. RTN behaviour of I_d at HT. (a) shows the time domain measurement result of I_d.(b) shows the probability of current obtained from frequency counting.

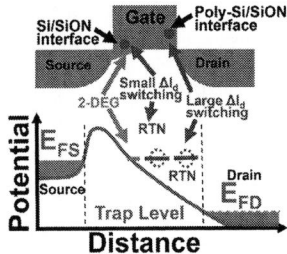

Fig. 5. Schematic potential diagram across the channel.

The current shows different correlation behaviours in time domain with different time lag [6]. The autocorrelation of I_d was strong if the time lag is 1s (Fig. 6(a)). This means that the switching was not frequently observed in this time scale. At the lag of 10s, the autocorrelation was weaker, and we started to observe switching with the small amplitude. At the lag of 100s, the autocorrelation was almost random (Fig. 6(c)), and we can recognize both small and large amplitudes.

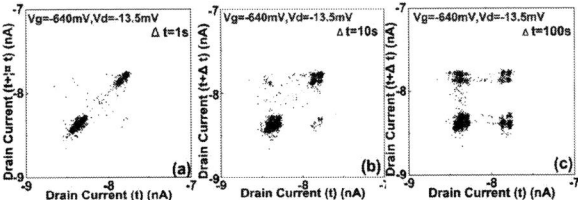

Fig. 6. lag plot of RTN under certain bias conditions. The shape of lag plot shows the correlation behaviour of current in time domain with different time lag.

We studied the V_g dependence of the RTN at V_d of -13.5mV (Figs. 7 and 8). We can explain these by assuming that the traps responsible for the large amplitude was located at the poly-Si/SiON interface, while the trap

responsible for the small amplitude was located at the SiON/Si interface (Fig. 5). We observed switching with both small and large amplitudes at V_g between -645mV and -640mV, since the trap level at the top interface is aligned with the Fermi level of the poly-Si, while the trap level at the bottom interface was aligned to the 2D hole-inversion layer. At V_g of -635mV, only the bottom trap was trapped/de-trapped through resonant tunnelling and the top trap was occupied, so that the switching of the small amplitude was observed.

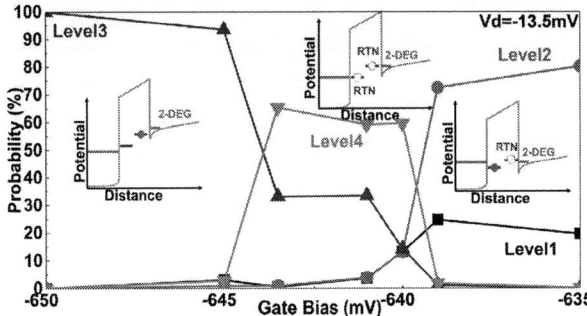

Fig. 7. Gate modulation of probability to find I_d. Schematic potential diagram under each bias condition is shown.

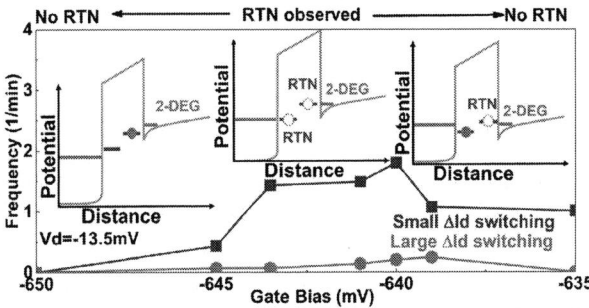

Fig. 8. Gate modulation of the RTN frequency. Schematic potential diagram under each bias condition is shown.

Finally, we examined the V_d dependence of the RTN at V_g of -640mV (Figs. 9 and 10). The switching of the small amplitude was observed in the entire V_d range, which implies that the bottom trap did not change significantly upon the change of V_d. This means that the bottom trap was presumably located at the source edge. On the other hand, the switching of the large amplitude was observed only at V_d between -13.0mV and -14.0mV, which means that the top trap level was resonated to the gate in the very narrow bias window. Considering the sensitivity of the top trap on V_d, we think that the top trap is located near the drain edge. The narrow bias condition

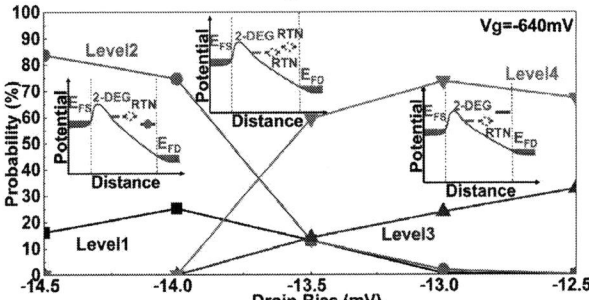

Fig. 9. Drain modulation of probability to find I_d. Schematic potential diagram under each bias condition is shown.

to observe the RTN implies resonant tunnelling was responsible for the trapping/de-trapping process.

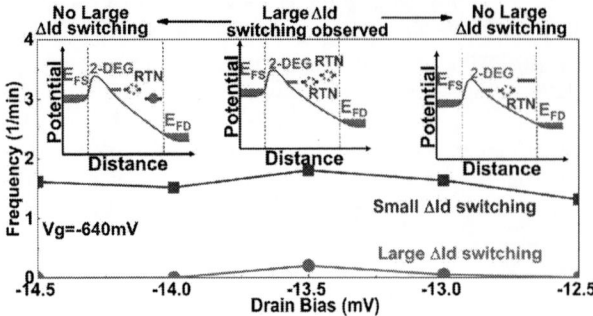

Fig. 10. Drain modulation of the RTN frequency. Schematic potential diagram under each bias condition is shown.

Conclusion

We have found resonant peaks in I_d under narrow bias conditions and observed the RTN at 2K. We addressed the 2 traps are responsible for the switching of large and small amplitudes. By measuring the current stability diagram, we can systematically identify bias conditions to find shallow traps through resonant tunnelling. We can apply this technique to investigate the reliability of MOSFETs in more detail for understanding the mechanism of the RTN for the future.

Acknowledgments

This work is supported by EPSRC Manufacturing Fellowship (EP/M008975/1), EU FP7 Marie-Curie Carrier-Integration-Grant (PCIG13-GA-2013-61811),

and the EMPIR programme co-financed by the Participating States and from the European Union's Horizon 2020 research and innovation programme, and University of Southampton. The data from the paper can be obtained from the University of Southampton ePrint research repository: http://dx.doi.org/ 10.5258/SOTON/399158.

References

[1] K. K. Hung, K. K Ping, C. Hu, and C. C. Yiu. "Random telegraph noise of deep-submicrometer MOSFETs." *IEEE electron device letters* **11**, no. 2 (1990): 90-92.

[2] K. S. Ralls, W. J. Skocpol, L. D. Jackel, R. E. Howard, L. A. Fetter, R. W. Epworth, and D. M. Tennant. "Discrete Resistance Switching in Submicrometer Silicon Inversion Layers: Individual Interface Traps and Low-Frequency (1/f) Noise." *Physical review letters* **52**, no. 3 (1984): 228.

[3] C. M. Compagnoni, R. Gusmeroli, A. S. Spinelli, A. L. Lacaita, M. Bonanomi, and A. Visconti. "Statistical model for random telegraph noise in Flash memories." *IEEE Transactions on Electron Devices* 55, no. 1 (2008): 388-395.

[4] E. Prati, M. Fanciulli, G. Ferrari, and M. Sampietro. "Giant random telegraph signal generated by single charge trapping in sub-micron n-metal-oxide-semiconductor field effect transistors." *Journal of Applied Physics* 103, no. 12 (2008): 3707.

[5] M. A. Kastner "The single-electron transistor." *Reviews of Modern Physics* 64, no. 3 (1992): 849.

[6] H. Miki, N. Tega, M. Yamaoka, D. J. Frank, A. Bansal, M. Kobayashi, K. Cheng et al. "Statistical measurement of random telegraph noise and its impact in scaled-down high-κ/metal-gate MOSFETs." In *Electron Devices Meeting (IEDM), 2012 IEEE International*, pp. 19-1. IEEE, 2012.

Fabrication of E-mode InGaN/AlGaN/GaN HEMT using FIB based Lithography

Shubhankar Majumdar[1,3], Chitrakant Sahu[2], and Dhrubes Biswas[3]

[1]National Institute of Technology Raipur, India, shubuit@gmail.com, shubhankar.majumdar@atdc.iitkgp.ernet.in,
[2]Malaviya National Institute of Technology Jaipur, India [3]Indian Institute of Technology Kharagpur, India

Abstract

In this paper, growth and fabrication of enhancement mode InGaN (5nm)/AlGaN (35nm)/ GaN (1.8 μm) HEMT is presented through focused ion beam (FIB) lithography technique for different alloy material as a source/drain contact. The X-TEM, AES, AFM has been done of the metal deposited surfaces and DC characterization reveals threshold voltage of 1.2V.

(Keywords: Fabrication, Epitaxial Growth, MBE, and characterization)

Introduction

To achieve E-mode operation of the AlGaN/GaN HEMTs, various techniques have been reported till now which are based on the two basic innovations one is physics based innovation [1-6] and process based etching [7]–[10]. A detailed view can be seen in the Fig.1. Among these, normally-off AlGaN/GaN HEMT with InGaN as cap layer is an epitaxial layer parameter based device. Hence, an attempt has been done to growth and fabricate the E-mode InGaN (5nm)/AlGaN (35nm)/ GaN (1.8μm) HEMT. The structure concept is taken from the the recently published article [11].

Experimental Details

The InGaN/AlGaN/GaN HEMT structure has been grown epitaxially on silicon (111) substrate by Molecular Beam Epitaxy (MBE) with nitrogen RF Plasma source. Thin AlN nucleation layer has been grown first to minimize the effect of lattice mismatch (16.9%), thermal expansion coefficient (TEC) mismatch (56%) between GaN buffer and silicon substrate. First, the SVT four chamber cluster tool with plasma assisted molecular-beam epitaxy (PAMBE) system for epitaxial growth is utilized. It has a modular configuration, i.e. growth chamber, sample exchange load-lock and loading chamber. These three chambers in the MBE system are connected by a UHV transfer tube. A base pressure of 10^{-10} Torr is maintained by using an ion pump and a cryogenic helium closed cycle pump as well as a liquid Nitrogen-cooled shroud. The growth chamber has five ports, with Ga, In, Al, Mg K-cells sources and Nitrogen RF plasma source. The RF-Nitrogen source offers better incorporation of Nitrogen due to a low ion count and a high atomic dissociation yield [12]. The substrate is placed in a special holder, which faces the K-cells. There is a heater behind the

holder to control the temperature for epitaxial growth. The growth temperature is monitored by a thermocouple. The MBE system is equipped with a reflection high energy electron diffraction (RHEED) apparatus, In-Situ 4000 for in-situ thickness and growth rate measurement, and a movable ion gauge works as density monitor.

When the Si substrate is transferred into the growth chamber after in-situ cleaning, the chamber is pumped down to 10^{-9} Torr and the substrate are heated. This desorbs the oxide and makes the surface atomically clean. Substrate quality is measured by seeing 7×7 RHEED pattern shown in Fig. 2. An electron beam focused at a low angle to the substrate surface is used in the RHEED. This apparatus is used for in situ monitoring of epitaxial film quality, to know the surface cleanliness and surface reconstruction. It is also used to measure the growth rate of the film (RHEED intensity oscillation with time can be used to determine the growth rate). Surface quality can be determined from RHEED pattern; a clean surface shows a streaky profile whereas surface with roughness shows a spotty pattern. RHEED uses high energy electron beam in the range of 5-40 KeV with incidence angle at 1-2 degree to the substrate surface, diffraction pattern forming on the screen opposite to source side [12]. This lower incidence angle makes it surface sensitive. The RHEED intensity pattern completes one cycle with the completion of every monolayer (ML) due to the change in surface coverage. On the layer with increasing surface coverage, the intensity drops from maximum to minimum till half surface coverage; then the intensities again increases to the maximum until full surface coverage. As the growth is done in the Si substrate so, AlN layer is utilized as a nucleation layer for the formation of the GaN surface. The 1×1 shows the growth of AlN. Then after the 1×1 streakier and brighter RHEED pattern signifies the proper GaN growth. The growth rate of GaN is about 0.3 μm/hr. When the ALGaN is opened then 1×1 streaky RHEED confirms the smooth surface of AlGaN. At last when InGaN is deposited the streaky line confirms a smooth deposition of the InGaN layers.

MBE grown sample is processed with 10 μm gate width and 1 μm gate length with Al as a schottky contact in gate and Au/Cr alloy is used as ohmic contacts for source and drain. The temperature profile

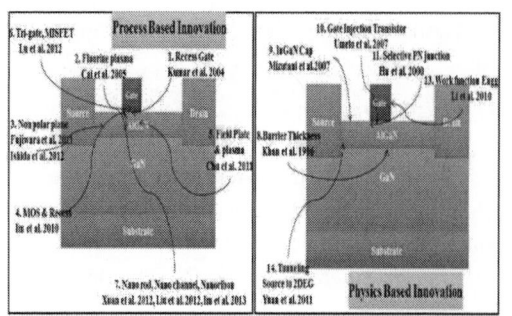

Fig.1: Literature review of E-mode Nitride HEMT

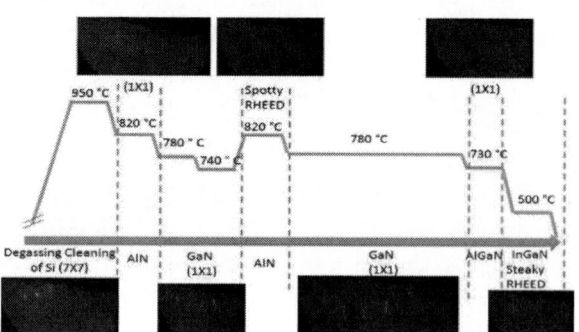

Fig.2: Temperature profile of substrate during the entire growth

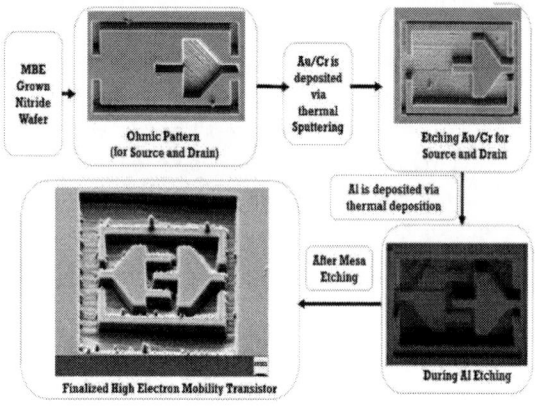

Fig.3: Focused Beam Lithography (FIB) based fabrication of E-Mode InGaN/AlGaN/GaN HEMT

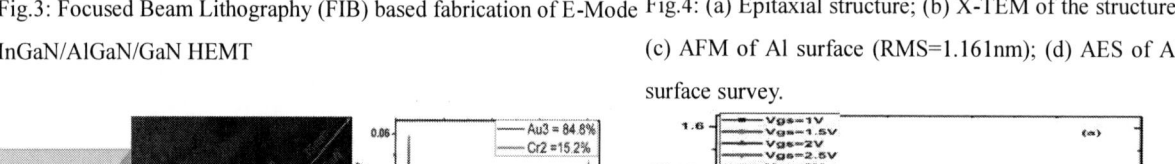

Fig.4: (a) Epitaxial structure; (b) X-TEM of the structure; (c) AFM of Al surface (RMS=1.161nm); (d) AES of Al surface survey.

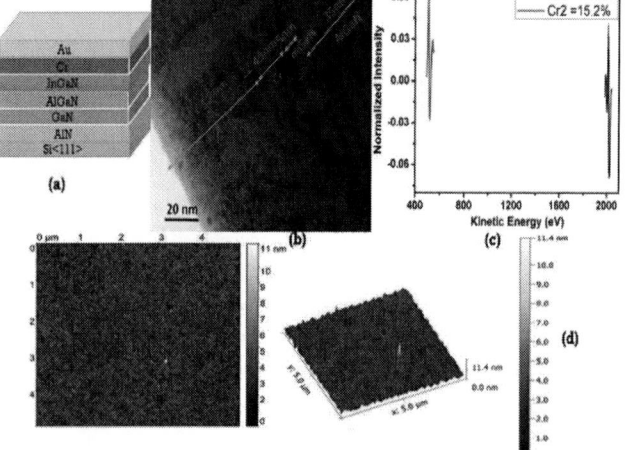

Fig. 5: (a) Epitaxial structure; (b) X-TEM of the structure; (c) AFM of Al surface (RMS=0.8220nm); (d) AES of Au and Cr elemental composition (Au=84.8%; Cr=15.2%)

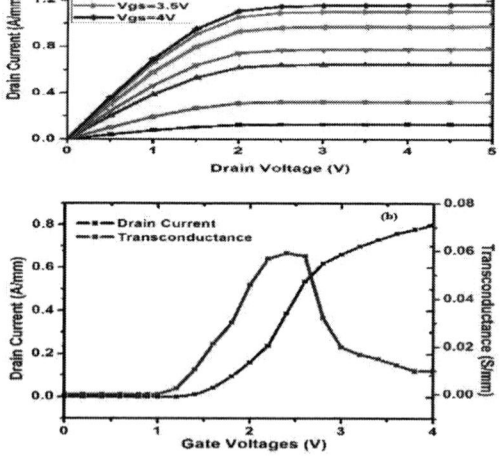

Fig. 6: (a) Id-Vds of E-mode GaN HEMT; (b) Id-Vgs and transconductance on Vds=4 V of E-mode GaN HEMT.

978-1-5090-4661-4/17 $31.00 © 2017 IEEE   176

during growth of the structure has been shown in Fig.2. Step by step FIB based fabrication flow (contact formation and the mesa isolation) of the HEMT structure has been shown in the Fig. 3. First the ohmic patterns are created by using 500 pA current for 5min then, for etching Au/Cr a 2 nA current for 2 min is utilized, whereas for the etching Al 5 nA for 3 min is used, finally for mesa isolation the current 10 nA for 10 min is utilized. Ga ions are utilized in the FIB for the etching purpose.

Results and Discussion

The Al and Au/Cr deposited on the grown epitaxial wafer has been characterized through AFM, X-TEM and AES that are shown is the Fig. 4 and 5. After examining the deposited Au/Cr layer through AES, it has been found that Au=84. 8% and Cr=15.2, and the thickness of the layers are Au= 137.32 nm and Cr=22.45 nm, whereas the roughness (rms) of the layers 0.822 nm. On the other hand, the deposited Al layer examination shows that the thickness of Al is 250 nm and roughness rms is about 1.161nm. The DC characterization of fabricated device is performed using Keithley 4200A-SCS parameter analyzer. Fig. 6 shows output and transfer characteristics of fabricated HEMT device. The extracted value of threshold voltage is found to be around 1.2 V, which ensure E-mode of operation of InGaN/AlGaN/GaN HEMT.

Conclusion

In this paper, fabrication of E-mode HEMT has been done by utilizing different alloy materials for ohmic contact (Au/Cr). The whole process is done through FIB technique and DC characterization of proposed device is also performed.

Acknowledgments

The authors thank Ministry of Human Resource and Development (MHRD), and Department of Electronics and Information Technology (DeitY), Government of India for financial support and sanctioning the ENS project, respectively.

References

[1] X. Hu, G. Simin, J. Yang, M. A. Khan, R. Gaska, and M.S.Shur, "Enhancement mode AlGaN/GaN HFET with selectively grown pn junction gate," Electron. Lett. vol. 36, no. 8, pp. 753–754, 2000.

[2] M. A. Khan, Q. Chen, C. J. Sun, J. W. Yang, M. Blasingame, M. S. Shur, and H. Park, "Enhancement and depletion mode GaN/AlGaN heterostructure field effect transistors," Appl. Phys. Lett., vol. 68, no. 4, pp. 514–516, 1996.

[3] J. S. Moon, D. Wong, T. Hussain, M. Micovic, P. Deelman, M. Antcliffe, C. Ngo, P. Hashimoto, and L. McCray, "Submicron enhancement-mode AlGaN/GaN HEMTs," 60th DRC. Conf. Dig. Device Res. Conf., pp. 23–24, 2002.

[4] Y. Cai, Y. Zhou, K. M. Lau, and K. J. Chen, "Control of Threshold Voltage of AlGaN/GaN HEMTs by Fluoride-Based Plasma Treatment: From Depletion Mode to Enhancement Mode," IEEE Trans. Electron Devices, vol. 53, no. 9, pp. 2207–2215, Sep. 2006.

[5] T. Mizutani, H. Yamada, S. Kishimoto, and F. Nakamura, "Normally off AlGaN/GaN high electron mobility transistors with p-InGaN cap layer," J. Appl. Phys., vol. 113, no. 3, p. 34502, 2013.

[6] C. S. Suha, A. Chinia, C. Poblenzb, and U. K. Mishraa, "p-GaN/AlGaN/GaN," vol. 41, no. 805, pp. 2005–2006, 2006.

[7] C. Tang, G. Xie, and K. Sheng, "Enhancement-mode GaN-on-Silicon MOS-HEMT using pure wet etch technique," in Power Semiconductor Devices IC's (ISPSD), 2015 IEEE 27th International Symposium on, 2015, pp. 233–236.

[8] H. With, C. Density, R. Wang, P. Saunier, S. Member, X. Xing, S. Member, C. Lian, X. Gao, S. Guo, G. Snider, P. Fay, D. Jena, and H. Xing, "Gate-Recessed Enhancement-Mode InAlN / AlN / GaN," vol. 31, no. 12, pp. 1383–1385, 2010.

[9] B. Lu, E. Matioli, and T. Palacios, "Tri-gate normally-off GaN power MISFET," IEEE Electron Device Lett., vol. 33, no. 3, pp. 360–362, 2012.

[10] J. J. Freedsman, T. Egawa, Y. Yamaoka, Y. Yano, A. Ubukata, T. Tabuchi, and K. Matsumoto, "Normally OFF Al2O3/AlGaN/GaN Metal–Oxide–Semiconductor High-Electron-Mobility Transistor on 8in. Si with Low Leakage Current and High Breakdown Voltage (825V)," Appl. Phys. Express, vol. 41003, pp. 7–10, 2014.

[11] S. Majumdar and D. Biswas, "Performance Variability Projection of InGaN/AlGaN/GaN E-mode HEMT for RF Switch Application," ECS Solid State Lett., vol. 4, no. 10, pp. P72–P74, 2015.

[12] S. Majumdar, A. Bag and D. Biswas, "Implementation of veriloga GaN HEMT model to design RF switch" Microwave and Optical Technology Lett., vol. 57, no. 7, pp. 1765–1768, 2015.

Endurance characterization of the Cu-dope HfO₂ based selection device with One Transistor-One Selector structure

Qing Luo, Xiaoxin Xu, Hangbing Lv*, Tiancheng Gong, Shibing Long,
Qi Liu, Ling Li, and Ming Liu

Key Laboratory of Microelectronics Devices and Integrated Technology, Institute of
Microelectronics, Chinese Academy of Sciences, Beijing, China; Email: lvhangbing@ime.ac.cn

Abstract

We investigated the endurance characteristics of a Cu-doped HfO_2 selector device in one transistor-one selector (1T1S) structure, which is fully compatible with standard BEOL process. The device exhibits high endurance of 10^{10} under 10 μA compliance current. However, reduced endurance (10^5) was observed as increasing the compliance up to 100 μA. Under the condition of high operation, intrinsic defect in the HfO_2 layer was possibly generated and resulted in endurance degradation.

(Keywords: selector, endurance and 1T1S)

Introduction

Crossbar structure not only offers a sensible architecture for high density integration of two-terminal memory deivces [1], but also provides a possibility to configure non-volatile logic and neuromorphic computation architecture [2]. However, the sneaking current issue prohibits these various applications from reality. To suppress the unexpected sneaking current, a high performance selector was much required. In the previous work, we have demonstrated a Cu-doped HfO_2 based selector, which show outstanding performance [3-4]. However, the endurance failure mode of this selector was not well characterized. Here, we systematically investigated the endurance failure mechanism of Cu-doped HfO_2 selector device in a 1T1S structure under different pulse conditions. The transistor acts as the current limiter to prevent the selector from hard breakdown.

Results and Discussion

Fig.1 shows the schematic of 1T1S structure. The selector cell is composed by Cu/Cu-doped HfO_2/HfO_2/Pt structure and Cu plug acts as the bottom electrode. The HfO_2 layer was deposited on top of Cu plug by magnetron sputtering at room temperature, followed by annealing process to drive the Cu dopant into the HfO_2 layer. Additional tunneling layer and top electrode were deposited by sputtering in success. **Fig.2** is the TEM image of the transistor and the selector cell (inset). The in-depth XPS data shows quite few Cu element was detected in the HfO_2 layer, whereas the gradient Cu concentration was observed in Cu-doped HfO_2 layer (**Fig.3**). **Fig.4** shows the output characteristic curves of the CMOS used in this experiment at the gate voltage of 0.8 V to 2 V, which is used for current limit in pulse test process. **Fig5**.shows the endurance test of bilayer selector device with pulse width and compliance current fixed at 50 ns and 10 uA. High endurance over 10^{10} cycles was successfully achieved under this condition. **Fig.6** shows the I-V curve of

selector device under higher compliance up to 800 μA, demonstrating the on current density as high as 10^6 A/cm². As increasing the compliance current and pulse width to 100 μA and 1 μs, the endurance was reduced to 2.5×10^5 (**Fig. 7**). **Fig. 8** shows the dependence of pulse duration on the device endurance with fixed compliance current of 10 μA. A clear degradation trend was observed as the the pulse width increased from 50 ns to 1 ms. Similar tendency could also be observed under the test condition of varying compliance current when fixing the pulse duration (**Fig. 9**). **Fig.10** illustrated a possible failure model of the selector device, which was thought to be related with the generation of defects in the HfO_2 layer in the case of continuous loss of Cu ions in the doped HfO_2 layer.

Conclusion

We systematically investigated the endurance failure behavior of Cu-doped HfO_2 selector device in 1T1S structure under different pulse conditions. A failure mechanism was proposed for this selector. The endurance characteristics was highly dependent on the compliance current and pulse duration. A failure mechanism related with loss of Cu ions and defect generation was proposed to account for the failure mode.

Acknowledgments

This work was supported by the National Natural Science Foundation of China under grant 61221004, 61422407, 61322408, 61522408, 61574169, 61334007, 61474135 and 61274091, the MOST of China under Grant No 2016YFA0203800

References

[1] J. Y. Seok, S. J. Song, J. H. Yoon, K. J. Yoon, T. H. Park, D. E. Kwon, H. Lim, G. H. Kim, D. S.Jeok and C. S. Hwang," A Review of Three-Dimensional Resistive Switching Cross-Bar Array Memories from the Integration and Materials Property Points of View," Adv. Mater., Jul. 2014, pp. 5316-5339.

[2] A. Siemon, S. Menzel, A. Chattopadhyay, R. Waser and E.Linn, "In-memory adder functionality in 1S1R arrays," in Proc. IEEE ISCAS, May. 2015, pp. 1338-1341

[3] Q. Luo, X. Xu, H. Liu, H. Lv, T.Gong, S. Long, Q. Liu, H. Sun, W. Banerjee, L. Li, N. Lu and M. Liu, " Cu BEOL Compatible Selector with High Selectivity (>10^7), Extremely Low Off-current (~pA) and High Endurance (>10^{10})," in Proc. IEEE IEDM, Dec. 2015, pp. 253-256.

[4] Q. Luo, X. Xu, T. Gong, H. Lv, S. Long, Q. Liu and M. Liu,"Self-compliance Cu-doped HfO2-based selector with low leakage and high reliability," in Proc. IEEE SNW, Jun. 2016, pp. 94-95.

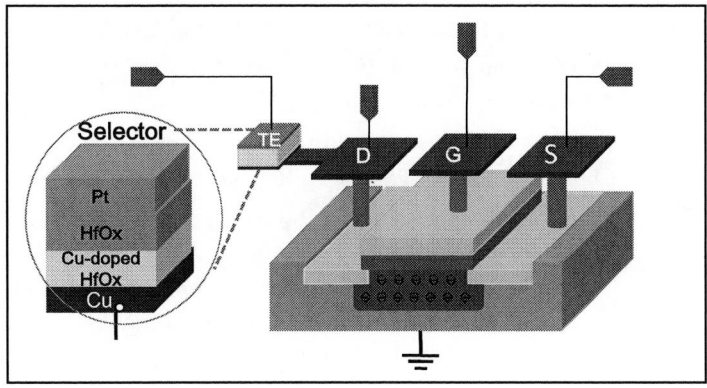

Fig.1 Schematic of the 1T1S structure.

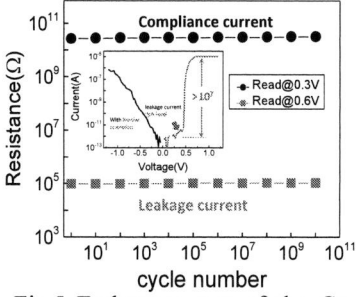

Fig.2 The cross-sectional TEM image of the selector device.

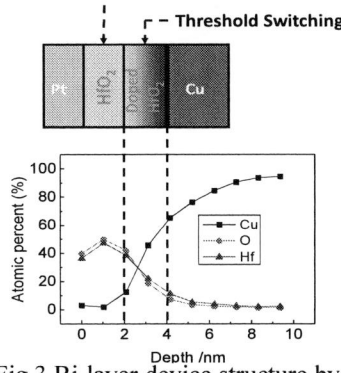

Fig.3 Bi-layer device structure by adding barrier layer on the TS layer and XPS depth profile of HfO₂/doped-HfO₂/Cu stack.

Fig.4 I_{DS}-V_{DS} curves of the 0.13-um channel length NMOS in the 1T1S configuration. (V_G: 0.8V~2V).

Fig.5 Endurance test of the Cu-doped HfO₂ selection device with pulse width and amplitude fixed at 50 ns and 1V. Typical IV curves of the device.

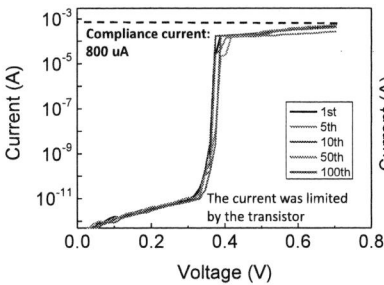

Fig.6 DC I-V sweeps (V_G=2V) from the 1T1S cell. This device can provide a current close to 1mA.

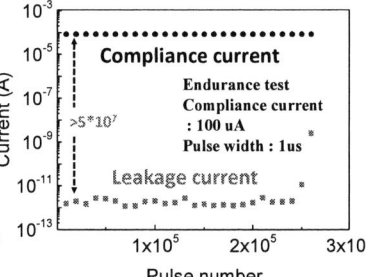

Fig.7 The leakage change from low to high after cycles under high current and wide pulse.

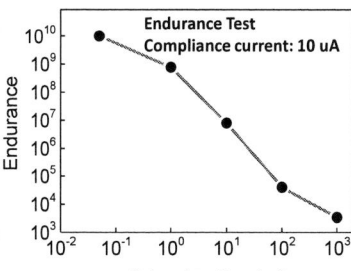

Fig.8 Endurance with different pulse duration under compliance current of 10 uA.

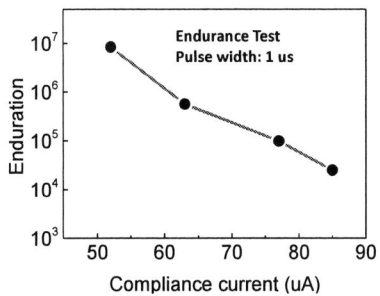

Fig.9 Endurance with different compliance current with the pulse duration of 1 us.

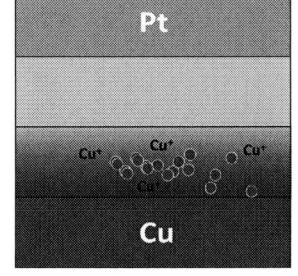
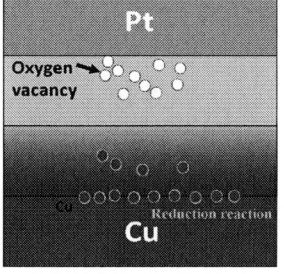

Fig.10 The proposed failure mode of the device. The defects generated after pulse cycles in the harsher test condition.

978-1-5090-4661-4/17 $31.00 © 2017 IEEE   179

Investigation on Direct-Gap GeSn Alloys for High-Performance Tunneling Field-Effect Transistor Applications

Lei Liu[1], Renrong Liang[1], Guilei Wang[2], Henry H. Radamson[2], Jing Wang[1], and Jun Xu[1]

[1]Tsinghua National Laboratory for Information Science and Technology, Institute of Microelectronics, Tsinghua University, Beijing 100084, China, liangrr@tsinghua.edu.cn

[2]Key Laboratory of Microelectronics Devices & Integrated Technology, Institute of Microelectronics, Chinese Academy of Sciences, University of Chinese Academy of Sciences, Beijing, 100029, China

Abstract

GeSn alloys are investigated for high-performance tunneling device applications. Samples with relatively high Sn compositions are characterized. GeSn electronic band structures are calculated and basic material parameters are extracted. Based on the established GeSn parameter sets, direct-gap GeSn tunneling field-effect transistors are simulated and analyzed. A higher Sn composition enhances device performance, but subthreshold swing is affected by the increased leakage level. For ultra small supply voltages, device structure should be optimized to improve device characteristics.

(Keywords: GeSn alloys, tunneling FETs and electronic band structures)

Introduction

GeSn alloys are promising materials owing to their tunable band gaps [1]. Direct-gap GeSn alloys can be achieved by increasing Sn composition (x), which is beneficial for optoelectronic applications [2]. In addition, direct-gap properties of GeSn alloys are also favorable for tunneling field-effect transistors (TFETs) which enable the power reduction for ultra-scaled supply voltages (V_{dd}) [3,4]. GeSn alloys for TFET applications are analyzed in this work.

Material Characterization

GeSn samples A and B (grown on (001) Ge/Si substrates) are characterized and discussed. Figure 1 presents the HRTEM image for sample A. Thickness of the top GeSn layer is ~37 nm, and that of sample B is ~20 nm. To further examine the crystal quality and determine the Sn composition, HRXRD ω-2θ scan is performed, and results for (004) reflection are plotted in Fig. 2. In addition, reciprocal space map (RSM) for (224) reflection is also measured, and result of sample A is presented in Fig. 3 (result of sample B is similar). It can be seen that the GeSn layers of samples A and B are nearly fully strained. Based on this assumption, Sn composition and the induced compressive strain can be determined from the diffraction peaks. By interpolated calculation, the calculated x of sample A is 10%, and the in-plane strain $e_\parallel$ is -1.7%; for sample B, the determined x is 14.8%, and $e_\parallel$ is -2.48%. According to the determined x and $e_\parallel$, electronic band structures for samples A and B are developed by a non-local empirical pseudopotential method [5], as shown in Fig. 4. Note that the induced compressive strain makes sample A indirect-gap, while sample B keeps direct-gap owing to its high enough x.

Device Simulation

GeSn material parameters are extracted from the calculated electronic band diagrams [6]. Based on the obtained parameter sets, direct-gap GeSn-source TFETs are simulated using *Sentaurus* device simulator. The simulated GeSn alloys (x=0.12, 0.15) are unstrained, and the device schematic is presented in Fig. 5 (double-gate line-tunneling pTFET). Ge is used as the channel and drain materials. Transfer characteristics of $Ge_{0.88}Sn_{0.12}$ and $Ge_{0.85}Sn_{0.15}$ devices are plotted in Fig. 6. Off-state current (I_{off}) is set to be 10^{-10} A/μm. For V_{dd}=0.4 V, the $Ge_{0.85}Sn_{0.15}$ device achieves a higher on-state current (I_{on}) than that of the $Ge_{0.88}Sn_{0.12}$ device (1.5-fold enhancement) owing to its enhanced tunneling capability, although its subthreshold swing (SS) and leakage level are worse. To improve SS, channel thickness T_B is reduced for the $Ge_{0.85}Sn_{0.15}$ device, and the corresponding simulation results are compared in Fig. 6. Reducing T_B benefits the leakage and SS reduction. However, the transfer curve deteriorates for too small T_B. Figure 7 presents the tunneling rates for the device with T_B=15 nm. Reduced T_B affects the electric field, leading to the weakened line-tunneling rate (at large $|V_{gs}|$). Instead, the undesired point-tunneling occurs first (at small $|V_{gs}|$) and worsens SS. For further scaled V_{dd}, $Ge_{0.85}Sn_{0.15}$ devices with T_B=25 and 20 nm are examined (Fig. 8). Device with reduced T_B still achieves better SS at V_{dd}=0.2 V. For the ultra small V_{dd}, better SS directly corresponds to a higher I_{on}. Hence, the device with small T_B achieves better device performance at V_{dd}=0.2 V.

Conclusion

Direct-gap GeSn alloys are very promising for high-performance TFET applications. High Sn composition may achieve better I_{on}. Device structure should be optimized for ultra small V_{dd}.

Acknowledgments

This work was supported in part by the National Natural Science Foundation of China (No. 61306105).

References

[1] X. Gong, G. Han, S. Su, R. Cheng, P. Guo, F. Bai, Y. Yang, Q. Zhou, B. Liu, K. H. Goh, G. Zhang, C. Xue, B. Cheng and Y.-C. Yeo, Symposium on VLSI Technology, T34, 2013.

[2] S. Gupta, R. Chen, B. Magyari-Köpe, H. Lin, B. Yang, A. Nainani, Y. Nishi, J. S. Harris and K. C. Saraswat, IEEE IEDM Technical Digest, 398, 2011.

[3] A. M. Ionescu and H. Riel, Nature, Vol. 479, 329, 2011.

[4] Y. Qiu, R. Wang, Q. Huang and R. Huang, J. Appl. Phys., Vol. 115, 234505, 2014.

[5] L. Liu, R. Liang, J. Wang and J. Xu, J. Appl. Phys., Vol. 117, 184501, 2015.

[6] Y. Yang, K. L. Low, W. Wang, P. Guo, L. Wang, G. Han and Y.-C. Yeo, J. Appl. Phys., Vol. 113, 194507, 2013.

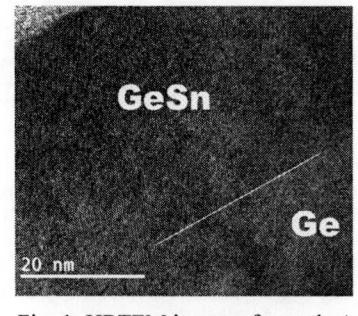

Fig. 1: HRTEM image of sample A.

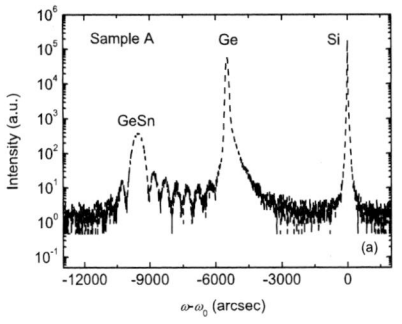

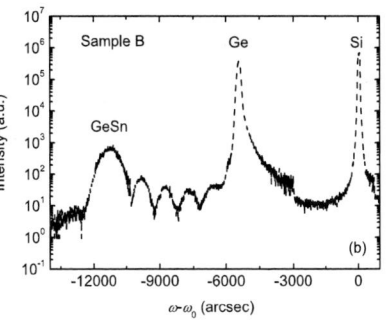

Fig. 2: HRXRD (004) ω-2θ scan curves of samples (a) A and (b) B.

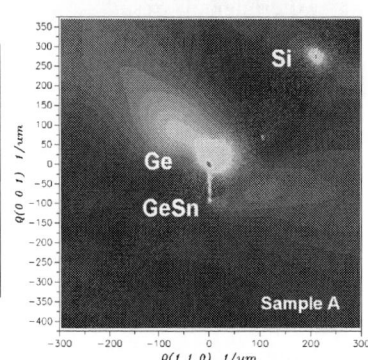

Fig. 3: (224) RSM of sample A.

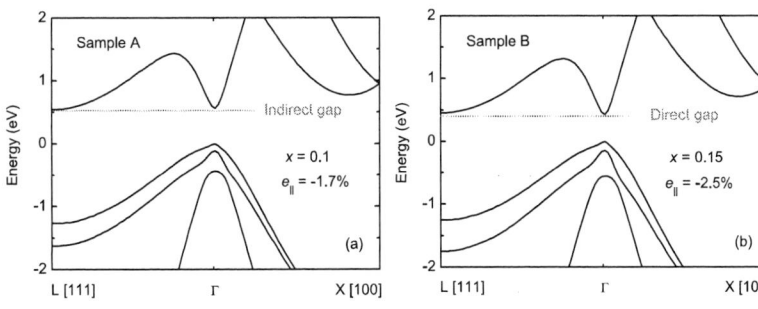

Fig. 4: Calculated electronic band structures of samples (a) A and (b) B.

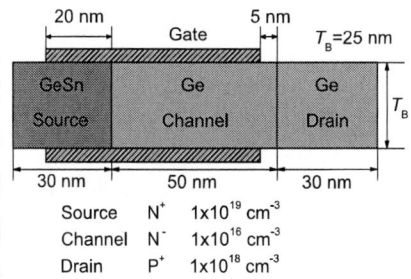

Fig. 5: Schematic of the simulated double-gate GeSn-source line-pTFETs.

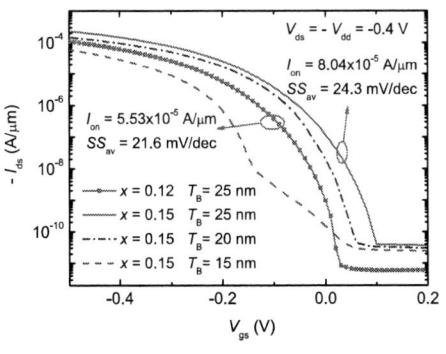

Fig. 6: Simulated transfer curves of $Ge_{0.88}Sn_{0.12}$ and $Ge_{0.85}Sn_{0.15}$ pTFETs. Average SS (SS_{av}) is evaluated from I_{sd} range of 0.1~10 nA/μm.

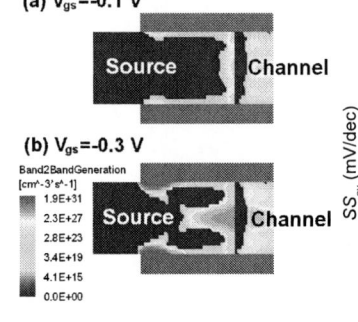

Fig. 7: Band-to-band generation rates of the $Ge_{0.85}Sn_{0.15}$ TFET device with T_B=15 nm.

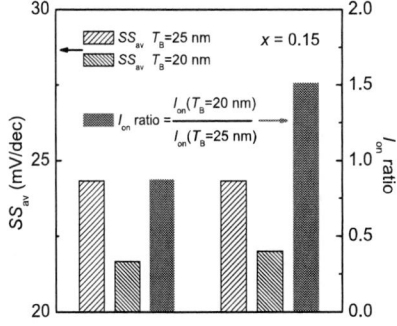

Fig. 8: SS_{av} and I_{on} ratio of the $Ge_{0.85}Sn_{0.15}$ TFET devices with T_B=25 and 20 nm at varied V_{dd}.

978-1-5090-4661-4/17 $31.00 © 2017 IEEE

Online Training on RRAM based Neuromorphic Network: Experimental Demonstration and Operation Scheme Optimization

Peng Yao[1], Huaqiang Wu[1*], Bin Gao[1], Ning Deng[1], Shimeng Yu[2], and He Qian[1]

[1]Tsinghua University, China, [2]Arizona State University, USA Email: wuhq@tsinghua.edu.cn

Abstract

In this work, online training is experimentally demonstrated on a neuromorphic network built with 1k-bit 1T1R RRAM array. The 1T1R RRAM cells in the array exhibit excellent synaptic behavior. Patterns with up to 11.83% noises can be correctly classified by the network after training. Based on the analysis of experimental results, we find the device characteristics and operation schemes significantly affect the system performance. The impact of voltage dependence and asymmetry effects are evaluated, and optimization guidelines are provided.

(Keywords: network, RRAM, synaptic behavior)

Introduction

Brain-inspired computing is a computing paradigm that is efficient in processing cognitive tasks (e.g. image recognition) with fast speed and low energy consumption [1]. Resistive random access memory (RRAM) is a promising candidate to implement a neuromorphic system due to its high density, multilevel capability and CMOS compatibility. In particular, the multilevel capability enables RRAM to implement the analog weights on chip for a neural network that can be trained online. So far, only few work has experimentally demonstrated the online training on a neuromorphic system built with a relatively large-scale RRAM array [2,3]. It is difficult to verify the results due to the lack of real characteristics measured on a large-scale RRAM array [4]. Furthermore, the resistive synaptic behavior is highly correlated with the operation scheme. A simplified model cannot be used for neuromorphic network device-system-algorithms co-design considering variations in real RRAM devices. To address these challenges, in this work, we experimentally demonstrate online training on a 1k-bit 1-transistor-1-resistor (1T1R) RRAM array based neuromorphic network. Based on the RRAM characteristics measured from the array, the system performance can be evaluated under different operation conditions. The impact of initial conductance distributions, programming voltage, asymmetry and nonlinearity effects are discussed.

Experiments

The RRAM stack TiN/TaOx/HfAlyOx/TiN was integrated on the drain of the transistor. The fabricated 1k-bit 1T1R RRAM array and RRAM cell stack structure are presented by Fig.1 and Fig.2

respectively. Typical I-V curve of the resistive switching in a 1T1R cell is shown in Fig. 3. Gradual switching is observed during both SET and RESET processes. The multilevel switching of the 1T1R cell under programming pulses with increased amplitudes was measured and shown in Fig. 4. The continuous conductance modulations under identical pulse train with different amplitudes during SET and RESET processes are presented in Fig. 5.

A perceptron neural network was demonstrated successfully by experiment to identify a grey scale image which contains 169 pixels organized by 13×13 size. There are 27 patterns in the training set and 5,400 patterns which are constructed by introducing random noise pixels to random location of the training patterns in the test set. Fig. 6 presents the inference operation and update operation applied on the array during the online training process.

Results and Discussion

To further investigate the impact of operation conditions on system performance, simulations were performed based on the compact model (Fig. 7) extracted from the experimental results. Besides, the asymmetry of conductance tuning performance has been modeled by Fig. 8. Fig. 9 shows how varied programming pulses and initial conductance distribution states effect iteration number, energy consumption during training process and the recognition rate on the test. Fig. 10 presents how the asymmetry influences the system performance. As for this specific neuromorphic network, the latency and energy consumption during the training process decreases and the recognition rate on the test set increases when the asymmetry is larger.

Conclusion

Online training on RRAM based neuromorphic network was demonstrated experimentally using the fabricated 1k-bit 1T1R RRAM array. This system reaches a low energy consumption (2.23 nJ) per iteration and a high 89.1% recognition rate on the test set with noises. A relatively low pulse amplitude as well as the initial conductance distribution range could result in improved system performance.

References

[1] S. B. Eryilmaz, IEDM 2015, p.64. [2] M. Prezioso, Nature 521.7550 (2015): 61-64. [3] S. Park, Scientific reports 5 (2015). [4] G. W. Burr, IEEE Transactions on Electron Devices 62.11 (2015): 3498-3507.

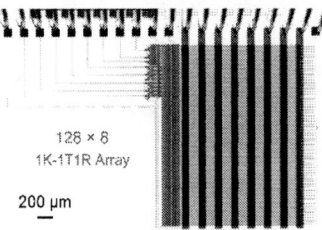

Fig. 1: Photo image of the fabricated 1k-bit 1T1R RRAM array using CMOS compatible fabrication process.

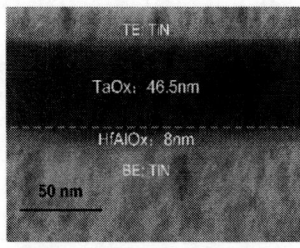

Fig. 2: TEM image of the fabricated TiN/TaOx/HfAlyOx/TiN RRAM stack in the array.

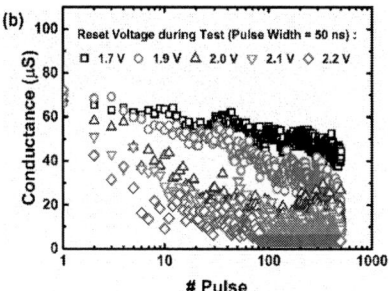

Fig. 3: Typical I-V curve of the 1T1R cell. Inset is the schematic of the 1T1R structure.

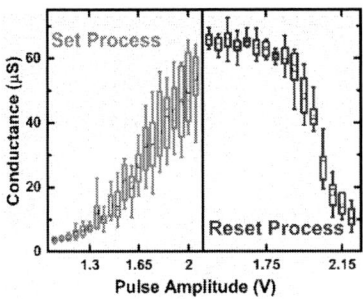

Fig. 4: Continuous conductance tuning characteristics using identical SET pulses with different amplitudes. Conductance increases faster under larger voltage.

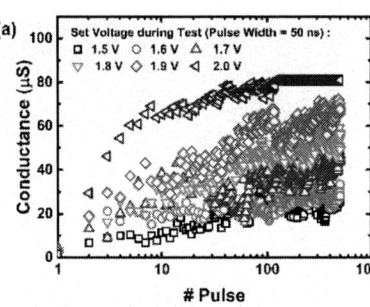

Fig. 5: Continuous conductance tuning characteristics using (a) identical SET pulses and (b) identical RESET pulses with different voltages. Conductance increases / decreases faster under larger voltage.

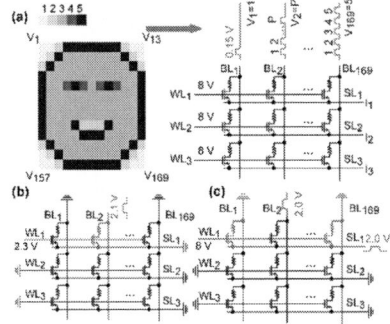

Fig. 6: Schematic of array operations: (a) parallel reading, (b) SET row by row and (c) RESET row by row to update the weights at the same time.

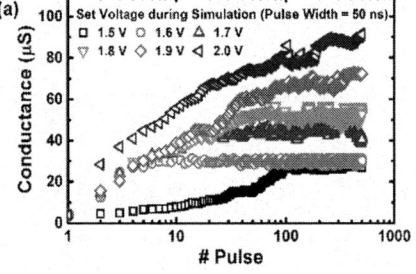

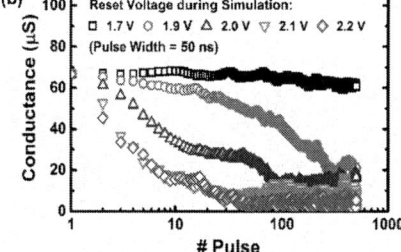

Fig. 7: Simulated continuous conductance tuning behaviors during (a) SET process and (b) RESET process. The data are fitted from the measured results (Fig. 5). A larger pulse amplitude implies a higher strenth of SET condition, ranging from 1 to 6 in (a) and a higher strenth of RESET condition, ranging from 1 to 5 in (b).

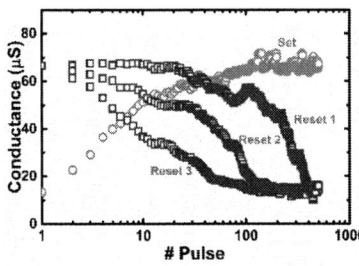

Fig. 8: Simulated asymmetric conductance modulation of RRAM device. The nonlinearity of "Reset 3" is larger than "Reset 2", and much larger than "Reset 1".

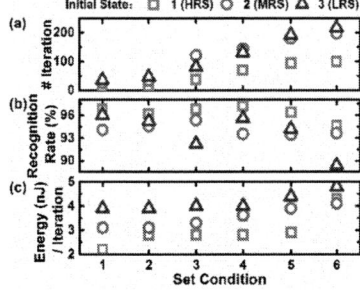

Fig. 9: Simulated programming pulse condition dependence of (a) training iteration required for convergence, (b) energy consumption per iteration during the training process and (c) recognition rate after convergence. Different initial states (HRS, MRS and LRS) of RRAM devices in the array are simulated.

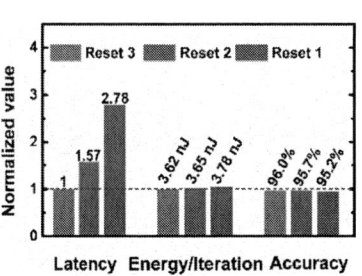

Fig. 10: System performance as a function of the asymmetry of conductance modulation behavior. Latency, energy consumption per iteration, and recognition accuracy on the test set are simulated.

978-1-5090-4661-4/17 $31.00 © 2017 IEEE   183

Uniformity improvements of low current 1T1R RRAM arrays through optimized verification strategy

Shan Wang, Xinyi Li, Huaqiang Wu[*], Bin Gao, Ning Deng, Dong Wu, and He Qian

Institute of Microelectronics, Tsinghua University, Beijing, China, wuhq@tsinghua.edu.cn

Abstract

This paper systematically analyzed and optimized the operation parameters of low current 1T1R RRAM arrays. Considering both thermal and electrical field driven effects, a current and voltage joint verification strategy has been proposed. Highly uniform multilevel resistive switching performances with LRS resistance higher than 100kΩ and HRS resistance higher than 10MΩ were obtained on 130nm CMOS process fabricated 1T1R RRAM arrays using the optimized verification strategy. (Keywords: RRAM, low current, uniformity)

Introduction

Resistive random access memory (RRAM) has attracted great interests due to its outstanding performances and promising widespread applications [1]. It can be used for the next generation integrated non-volatile memory chips, and also be adopted to neuromorphic computing and information security field like physical unclonable function(PUF) circuit. However, most of the RRAM devices suffer the large operation current issue. Even though the RESET current can be reduced to below 50 μA with certain methods, the variation increases significantly due to the formation of unstable oxygen vacancy filaments and the reduced local temperature [2]. Therefore, the uniformity improvement mechanism and methodology might be different between RRAM with large current and small current. On the other hand, performance variation is another major obstacle for massive storage applications. The traditional incremental step pulse programming (ISPP) only adjusts the SET/RESET voltages without considering the adjustment of thermal factor. Thereby, new verification strategies considering both operation current and voltages should be developed.

In this work, we successfully demonstrated 1T1R RRAM arrays with stable switching under 20 μA. A novel verification strategy considering both thermal and electrical field driven effects during RRAM write operations for low power applications was developed. At least seven resistance levels with 100kΩ as the lowest LRS level were obtained on the HfO$_x$/TaO$_x$ bilayer based 1T1R RRAM arrays fabricated using 130nm CMOS process.

Experiments and results

Fig. 1 is the optical image of the fabricated 1T1R RRAM based 1k-bit 1T1R RRAM arrays used to verify the developed RRAM operation methods. In the 1T1R architecture, the transistor is used to select a RRAM cell in the array and control the current through the RRAM cell by applying various voltage on the gate terminal. The HfOx/TaOx bilayer RRAM shows bipolar switching characteristics, which means that the switching direction depends on the polarity of the applied voltage. During set process, voltage pulse is applied on the top electrode of RRAM. And during reset process, voltage pulse is applied on the source of transistor. Fig. 2 is the cross-sectional TEM image of the HfOx/TaOx based RRAM stack. For the fabrication of 1k-bit 1T1R RRAM arrays, the MOSFETs were fabricated using 130nm CMOS process in foundry first. Then RRAM cells were fabricated using BEOL process in the lab. HfOx/TaOx bilayer structure is used as the resistive switching layers and TiN layers are used for both bottom and top electrode. The device exhibits excellent resistance switching performance in the uniformity aspect by using the newly developed verification strategy. The detailed device fabrication process can be found in our previous publications [3]. Fig. 3 shows the DC sweep I-V curves of the 1T1R RRAM device. Inset of Fig. 3 is the I-V curves of the integrated transistor. Gradual switching during RESET processes and sharp switching during SET processes are observed. RESET current can be controlled below 20 μA owning to the optimized device structure and the effective current limit from transistor, which can decrease power consumption significantly.

Fig. 4 shows the proposed current compliance (CC) and SET/RESET pulses incremental programming joint verification scheme. During SET, CC was used to avoid hard breakdown. During RESET, CC was used to avoid overshoot. Traditional ISPP scheme was used to adjust the SET/RESET pulses. Simultaneously, CC was gradually increased by transistor gate controlling considering both the surge current generated by the parasitic capacitance and the V_{th} sift of transistor caused by voltage drop on the RRAM. Fig. 5 shows the measured cycle to cycle LRS resistance distribution of the RRAM devices. The LRS resistance is set below 500kΩ. The lines with various colors express different operation schemes. The black line is LRS resistance distribution measured by traditional ISPP scheme, namely, constant CC scheme. The LRS resistance distribution measured by CC increased before the operation voltage increasing scheme (C prior V) and CC increased after the operation voltage increasing scheme (C post V) are shown in the red and blue line, respectively. For low current operation, the resistance measured by constant CC scheme shows much wider distribution, which has the worst uniformity with the maximum σ / μ 2.5502. Under this circumstance, the LRS resistance is larger than 500kΩ in some cycles. Between the other two operation schemes, CC increased after the operation voltage increasing scheme shows the best uniformity with the minimum σ / μ 0.2468. Meanwhile, cycle to cycle HRS resistance distribution of the RRAM devices was also obtained using three kinds of various operation schemes, as shown in Fig. 6. The HRS resistance is set above 10MΩ, using verification strategy to keep the on-off window larger than 20 times. The black line is the HRS resistance distribution measured by constant CC scheme. The HRS resistance distribution measured by CC increased before the operation voltage increasing scheme and CC increased after the operation voltage increasing scheme are shown in the red and blue line, respectively. Similarly, among the three kinds of operation schemes, the resistance measured by constant CC scheme shows the worst uniformity with the maximum σ / μ 1.0817, and CC increased after the operation voltage increasing scheme shows the best uniformity with the minimum σ / μ 0.2468. In the case that taking the verification algorithm by using constant CC scheme, the HRS resistance may be smaller than 10MΩ. Based on the above analysis, it is not difficult to find that the newly developed verification scheme is better than the traditional ISPP scheme for the low current RRAM resistance uniformity to a large extent. In Fig. 7, the device to device resistance distributions of the 1T1R RRAM arrays can be tightly controlled using the newly developed verification operation scheme. The LRS resistance is set below 100kΩ and the HRS resistance is set above 2MΩ, with the on-off window larger than 20 times. Whatever for CC increased before the operation voltage increasing scheme or CC increased after the operation voltage increasing scheme, both the HRS and LRS resistance can reach the goal level which is set before. Form CC increased before the operation voltage increasing scheme to CC increased after the operation voltage increasing scheme, the σ / μ of the LRS resistance reduces from 0.1311 to 0.0927, while the σ / μ of the HRS

resistance reduces from 1.5208 to 1.1756, which shows that CC increased after the operation voltage increasing scheme is still better than CC increased before the operation voltage increasing scheme for the improved resistance uniformity under low current operation. At least seven resistance levels with the LRS resistance higher than 100kΩ and HRS resistance higher than 10MΩ were obtained under low current operation using the proposed CC increased after the operation voltage increasing scheme, as shown in the Fig. 8.

The related physical mechanisms behind the current and voltage joint verification strategy to improve uniformity can be expressed as follow: besides controlling the O^{2-} migration, the redox reactions induced by thermal effect during resistive switching process was also controlled, which was schematically shown in Fig. 9.

Conclusion

Considering both thermal and electrical field driven effects, a new RRAM verification strategy was developed. By using the compliance current and voltage joint verification strategy, resistance distributions of the fabricated RRAM arrays under low current operation were tightly controlled. This work provides valuable guidelines for designing RRAM arrays with large scale, low power and non-volatile multilevel storage technology.

Acknowledgements

The authors would like to thank ICAC@IMECAS for the device fabrication. This work was supported in part by: National Hi-tech (R&D) project of China (2014AA032901), China key research and development program (2016YFA0201801), National Natural Science Foundation of China (61076115, 61674089, 61674087), and 863 program (2015AA016501).

References

[1] Byung Joon Choi, et al., "High-speed and low-energy Nitride memristors," Adv. Funct. Mater., may 2016.

[2] Sungho Kim, et al., "Comprehensive physical model of dynamic resistive switching in an oxide memristor," ACS Nano, vol 8, 2369, Feb. 2014.

[3] Xueyao Huang, et al., "HfO$_2$/Al$_2$O$_3$ multilayer for RRAM arrays: a technology technique to improve tail-bit retention," Nanotechnology, vol. 27, 395201, June 2016.

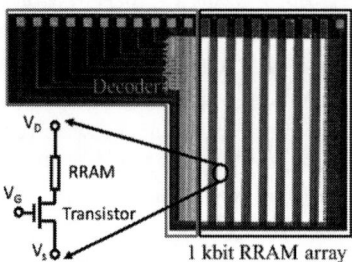

Fig. 1: Optical image of the 1k-bit 1T1R RRAM array.

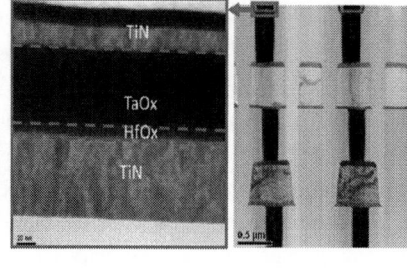

Fig. 2: Cross-sectional TEM image of the RRAM device.

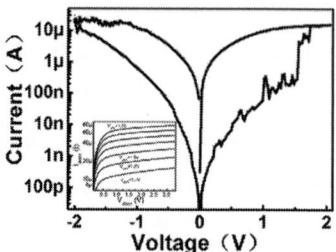

Fig. 3: DC I-V of the 1T1R RRAM, inset is the transistor I-V curves.

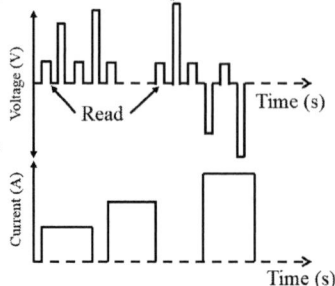

Fig. 4: Operation scheme of the proposed compliance current and SET/RESET pulse joint adjustment strategy, include current prior voltage change and current post voltage change.

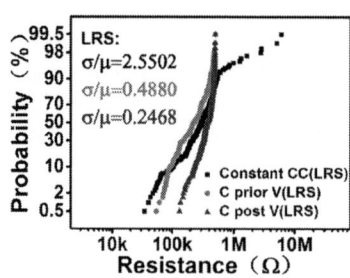

Fig. 5: LRS resistance distribution measured by traditional ISPP scheme and the newly developed current joint voltage verification scheme. Current post voltage change scheme shows the best uniformity.

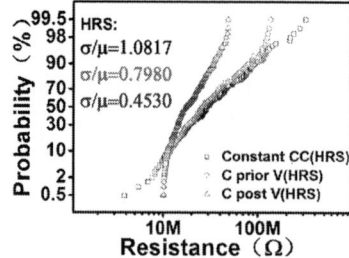

Fig. 6: HRS resistance distribution measured by traditional ISPP scheme and the developed current joint voltage verification scheme. Current post voltage change scheme shows the best uniformity.

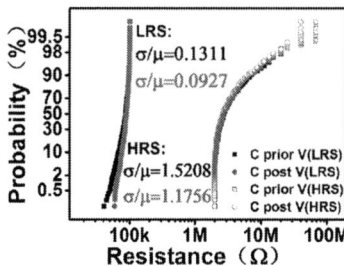

Fig. 7: The resistance distribution of 1k-bit 1T1R RRAM arrays measured by the developed current and voltage joint verification strategy. Current post voltage change scheme also shows the best improved uniformity.

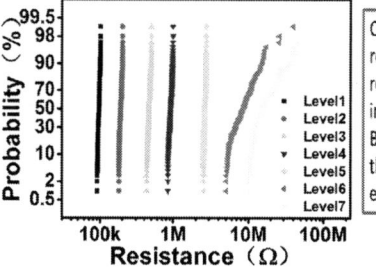

Fig. 8: using the developed current post voltage change verification strategy, seven resistance levels were measured with lowest level higher than 100kΩ.

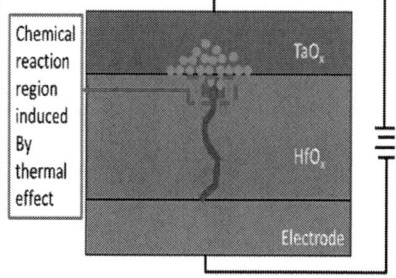

Fig. 9: Schematic image of the physical mechanisms to explain the newly developed verification schemes.

978-1-5090-4661-4/17 $31.00 © 2017 IEEE

Electroluminescence Characteristics of Rare Earth Doped Silicon Based Light Emitting Device

Fumihiro Hattori[1], Hideyuki Iwata[1], Toshihiro Matsuda[1], and Takashi Ohzone[2]

[1]Toyama Prefectural University, Kurokawa, Imizu, Toyama 939-0398, Japan, matsuda@pu-toyama.ac.jp

[2]Dawn Enterprise Co., Ltd, Nagoya 467-0808, Japan

Abstract

Electroluminescence (EL) of MOS devices with rare earth related oxide layer, which were fabricated with the mixtures of organic liquid sources of Tb and (Tb+Eu) spin-coated on the Si substrate and annealed, are reported. Visible green and red EL were observed. The spectral peaks of the EL correspond to radiative transitions of Tb^{3+} and Eu^{3+} ions. Effects of compounding ratios of (Tb+Eu) devices on EL characteristics are analyzed.
(Keywords: EL, MOS, and Rare earth)

Introduction

Although Si is not suitable for light-emitting devices due to the smaller and indirect band gap, Si-based light-emitting devices are attractive because of its process compatibility to Si metal oxide semiconductor (MOS) LSIs for applications such as mobile systems with micro displays. A number of literatures on photoemission from Si-based devices have been published, but sufficient performance has not been achieved yet. Recently, visible electroluminescence (EL) from MOS devices with rare-earth doped oxide has been reported.

In this study, we demonstrate green (Tb) and red (Tb+Eu) EL from MOS devices with spin-coated organic rare earth compounds on silicon. Effects of atomic ratios of Tb/Eu on EL are also discussed.

Experimental Procedure

Fig. 1(a) shows a schematic cross section of a MOS device with a rare-earth related oxide layer of [Tb/(Tb+Eu)–Si–O]. As for the (Tb+Eu) devices, eight kinds of atomic ratios (Tb : Eu = 10 : 0.01 ~ 10 : 3) of the liquid source mixture were examined. Fig. 1(b) shows a fabrication process of the devices. Organic liquid source of Tb and Eu were spin-coated on the Si substrates at 2000 rpm for 20 s. After drying, the chips were furnace-annealed at 850 °C for 30 min in air to form insulator layers on the Si substrate. The ITO electrodes of 1.0 mm diameter were formed by a screen printing technique using ITO ink. Finally, the contact electrode on the Si substrate was formed on the surface using a conductive silver composition after the selective removal of the insulator layer at the chip edge.

The EL spectra were measured under a constant current supply. The emitted light from the EL device was directly guided to a monochromator, and finally received by a CCD. The spectral data were corrected with the total wavelength response curves of the measurement system.

Results and Discussion

Fig. 2 shows (a) current (I_G) vs. voltage (V_G) characteristics and Fowler-Nordheim (FN) plot [i.e., $\log(I/V^2)$ vs $1/V$ plot], which supports that the steep increase in I_G is FN tunnel current.

Fig. 3 shows EL microphotographs of the green and red EL color images of devices (a) Tb and (b) Tb+Eu. Fig. 4 shows EL spectra measured at a gate current I_G of 100 μA for devices (a) Tb and (b) Tb+Eu, respectively. The EL spectra are normalized to the maximum intensity for each device. The Tb device have peaks at the wavelengths of 494, 550, 592, and 627 nm, which correspond to the intrashell transitions ($^5D_4$ - ^7F_J) of Tb ion. The peaks for the device Tb+Eu are observed at 594, 620, 655 and 702 nm, which can be attributed to the energy transitions ($^5D_0$ - ^7F_J) in Eu ion. Although the device of the lowest Eu ratio (Tb : Eu = 10 : 0.01) shows the peaks corresponding to both Tb and Eu, the peaks originated from Tb ion disappear for the highest ratio device (Tb : Eu = 10 : 3).

Fig. 5 shows Eu ratio dependences of the peak intensities at 550 and 594 nm normalized by the intensity of 620 nm peak. As Eu ratio increases, the peak intensity at 550 nm decreases down to about 1/30. Since both Tb and Eu ions have the transition energy around 2.1 eV (590 nm), Eu ratio scarcely affects the normalized intensity of 594 nm peak.

Figure 6 shows Eu ratio dependence of the integrated EL intensities calculated from the measured spectra. The EL intensities increase with the Eu ratio, and saturate around 1.5. The intensity of the highest Eu ratio device is larger than that of the lowest Eu ratio device by about 10 times.

The results suggest that the direct impact excitation by the energetic hot electrons in the conduction band of the oxide layer and subsequent energy transfer to the rare-earth ions. As for the Tb+Eu devices, the energy transitions in Eu ion become dominant with the increase of Eu ratio.

Conclusion

MOS devices with rare earth related oxide were fabricated by the spin-coating of Tb and (Tb+Eu) on Si. Green and red EL were observed, and the spectral peaks correspond to radiative transitions of Tb^{3+} and Eu^{3+} ions. The increase of Eu ratios in (Tb+Eu) devices enhances Eu related EL spectral peaks and intensity.

References

[1] T. Matsuda, et al., Ext. Abst. Int. Conf. on Solid State Devices and Materials, pp.328-329, 2005.

[2] T. Ohzone, T. Matsuda, S. Hase, S. Nohara and H. Iwata, Japanese Journal of Applied Physics, vol. 50, pp. 064102-1-7, 2011.

[3] T. Ohzone, T. Matsuda, R. Fukuoka, F. Hattori, and H. Iwata, Japanese Journal of Applied Physics, vol. 55, pp. 082102-1-13, 2016.

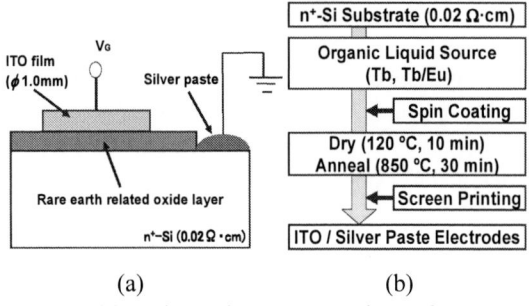

Fig. 1. (a) Schematic cross section of a MOS device with a rare-earth related oxide layer and (b) a fabrication process of the devices.

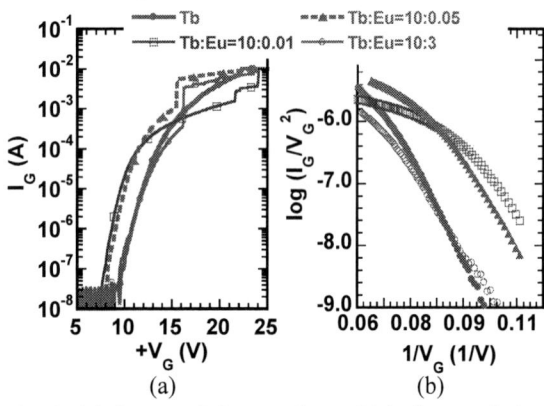

Fig. 2. (a) Current (I_G) vs. voltage (V_G) characteristics and Fowler-Nordheim (FN) plot.

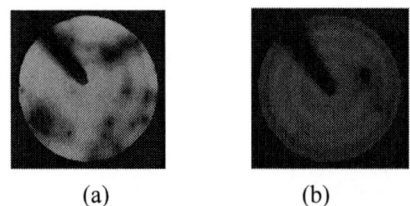

Fig. 3. EL microphotographs of the green and red EL color images of devices (a) Tb and (b) Tb+Eu.

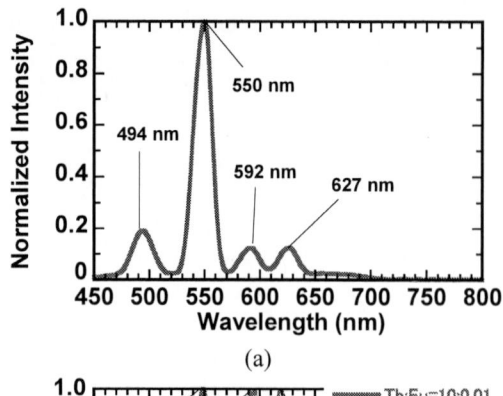

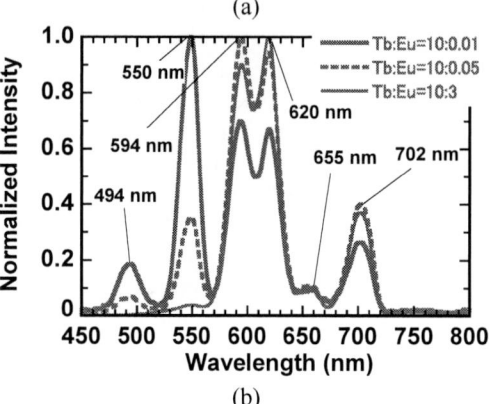

Fig. 4. EL spectra measured at a gate current I_G of 100 µA for devices (a) Tb and (b) Tb+Eu, respectively.

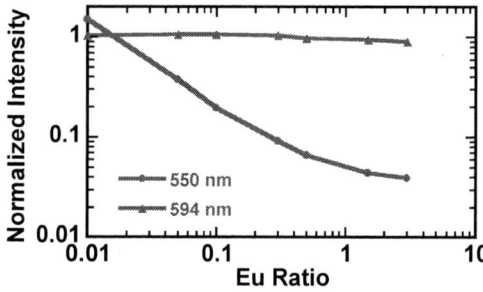

Fig.5. Eu ratio dependences of the peak intensities at 550 and 594 nm normalized by the intensity of 620 nm peak.

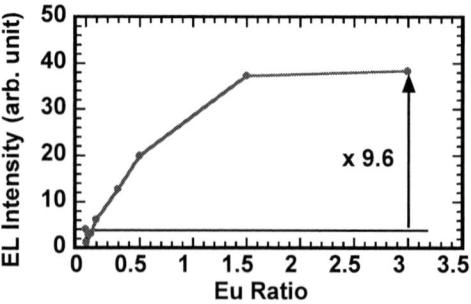

Fig. 6. Eu ratio dependence of the integrated EL intensities calculated from the measured spectra.

P-9

High Photoresponsivity Germanium Nanodot PhotoMOSFETs for Monolithically-Integrated Si Optical Interconnects

Ming-Hao Kuo[1], Morris M. Lee[2], and Pei-Wen Li[1,2]

[1]National Central University, ChungLi, Taiwan, pwli@nctu.edu.tw

[2]National Chiao Tung University, Hsinchu, Taiwan

Abstract

We presented a self-organized, MOS gate-stacking structure of SiO_2/Ge-dot/SiO_2/$Si_{1-x}Ge_x$-shell, using thermal oxidation of poly-$Si_{0.85}Ge_{0.15}$ nanopillars over buffer Si_3N_4 on the Si substrate, for the fabrication of high performance Ge-dot photoMOSFETs. Low dark current of $3\mu A/\mu m^2$, Superior high photoresponsivity of 1400–710A/W, and short response time of <0.8ns are measured on 90nm Ge-dot photoMOSFETs with $W/L = 70\mu m/3\mu m$ under 850nm–1550nm illumination, providing a practically-achievable, core building block for monolithically-integrated Si optical interconnects. (Keywords: Ge-dot, phototransistor, and optical interconnects)

Introduction

Si optical interconnects is receiving intense attention because of its promise to provide greater bandwidth, lower power consumption, decreased latency, resistance to EM interference, and reduced signal crosstalk in comparison to their counterpart metal wiring. The major challenge for Si optical interconnects lies in seamless, monolithic integration of photonics and electronics in standard CMOS processes. This is particularly vital for the receiver end that requires to include high-responsivity, high-speed and ultra-low capacitance photodetectors integrated in close proximity of TIA and limiting amplifiers.

Despite the fact that Ge BiCMOS HBTs [1] and Ge-gate MOSFETs [2] have been reported with good performance using epitaxy and rapid melting growth approaches. However, the required technology is not available for mass production. Recently, we proposed a self-organized, MOS structure of SiO_2/Ge dots/SiO_2/$Si_{1-x}Ge_x$-shell over the Si substrate [3]. The key novelty of our Ge-dot MOS structure is its simplicity and elegancy of being simultaneously produced in a single oxidation step. Another feature for our Ge-MOS structure is its size-tunability on Ge-dot's diameter (5–100nm), gate-oxide thickness (3–40nm), and $Si_{1-x}Ge_x$-channel's thickness and composition (2–22nm, $x = 0.5$–0.8) [3], enabling practically-achievable fabrication of Ge phototransistors in a standard MOS configuration. In this paper, we advance our exquisite MOS structure for fabricating Ge-dot photoMOSFETs (PTs) to boost photoresponsivity, quantum efficiency, and response speed by tailoring the Ge-dot diameter and the gate oxide thickness.

Experimental and Results

Cross-sectional transmission electron microscopy (CTEM) micrographs and schematic diagram of the Ge-dot PTs are illustrated in Figure 1. We firstly grew a 30 nm-thick Si_3N_4 layer over the Si substrate followed by the deposition of 60-nm-thick poly-$Si_{0.85}Ge_{0.15}$ and 5-nm-thick SiO_2 layers. 100–210nm- wide $Si_{0.85}Ge_{0.15}$ nanopillars were generated by using a combination of electron-beam lithography and SF_6/C_4F_8 plasma etching. Thermal oxidation at 900 °C in an H_2O ambient successfully converts SiGe nanopillars to array of 35nm-thick SiO_2/50–90nm Ge dot/3.5nm-thick SiO_2/$Si_{0.4}Ge_{0.6}$-shell. Lastly, a 150 nm-thick transparent ITO top-gate, source/drain implantation and metallization processes were conducted to complete the fabrication of Ge-dot PTs.

Figures 2 and 3 plot I_D-V_G and I_D-V_D characteristics of Ge-dot n- and p-MOSFETs ($W_G/L_G = 70\mu m/3\mu m$), respectively, with $t_{ox} = 38.5nm$ and Ge-dot size of 90nm measured under variable-power illumination at 850nm–1550nm. Illumination indeed increases drain current for Ge-dot PTs across the entire experimental gate voltage range. For example, optical pumping at $P_{IN} = 87\mu W$ increases I_{ON} and I_{OFF} for Ge-dot n-PTs by a factor of 40 and 4×10^7 in magnitude, respectively, in comparison to that measured in the darkness.

Very high photoresponsivity of $\Re = 1400$–710A/W is measured under $P_{IN} = 6nW$ at 850–1550nm illumination with external quantum efficiency of EQE = $\Re[h\nu/e] = 1.5\times10^6$ % and specific detectivity of $D^* = \Re(A/2eI_{dark})^{0.5} \approx 4.2\times10^{12} cm/(W \cdot s^{1/2})$. Our Ge-dot PTs exhibits a power dependent photoresponse with the largest responsivity obtained in the nW range due to light affecting the gate potential. A response time of 0.8ns with a 3dB frequency of 1.15GHz is measured on 70nm Ge-dot n-PTs. We expect to further improve the response time and 3dB frequency by increasing the spatial areal density of Ge-dots within the gate stack, since considerable parasitic capacitance is induced by buffer Si_3N_4 and growing SiO_2 surrounding Ge dots.

Conclusion

We demonstrated Ge-dot n- and p-photoMOSFETs with gate stacking of Ge dot/SiO_2/SiGe-channel in standard CMOS fabrication processes. Very large photoresponsivity of 1400–710A/W under 850nm–1550nm illumination and fast response time of 0.85ns are achieved, offering a great promise for future Si-based optical interconnection applications.

978-1-5090-4661-4/17 $31.00 © 2017 IEEE

Acknowledgments

This work was supported by the Ministry of Science and Technology of Republic of China (MOST 105-2221-E-009-134-MY3) and by the Asian Office of Aero-space Research and Development under contract no. FA 2386-15-1-4025.

References

[1] K. Ang, M. Yu, G. Lo, and D. Kwonh, "Low-voltage and high-responsivity germanium bipolar phototransistor for optical detections in the near infrared regime," IEEE Trans. Electron. Dev. Vol. 29, No. 10, pp. 1124–1127 (2008).

[2] R. W. Going, J. Loo, T. K. Liu, and M. C. Wu, "Germanium gate photo MOSFET integrated to Silicon photonics," IEEE J. Sel. Top. Quantum Electron. Vol. 20, No. 4, 8201607 (2014).

[3] W. T. Lai, K. C. Yang, T. C. Hsu, P. H. Liao, T. George, and P. W. Li, "A unique approach to generate self-Aligned SiO_2/Ge/SiO_2/SiGe gate-stacking heterostructures in a single fabrication step," Nanoscale Research Letters, vol. 10, 224 (2015).

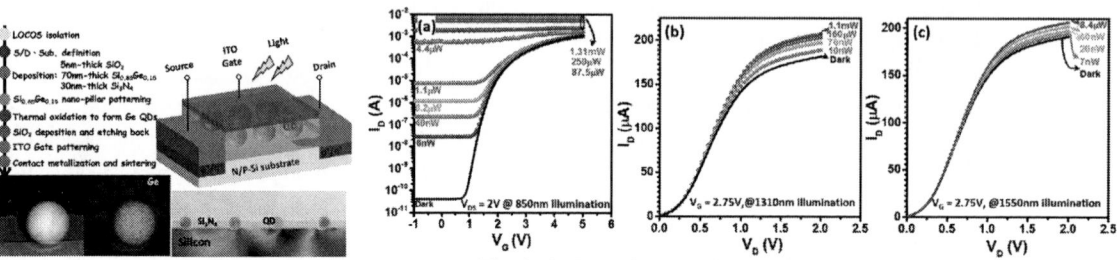

Fig. 1: Experimental procedure for the fabrication of Ge-dot photoMOSFETs with corresponding CTEM/EDX micrographs.

Fig. 2: I_D-V_G and I_D-V_D characteristics of 90mn Ge-dot n-PTs under variable-power 850–1550nm illumination.

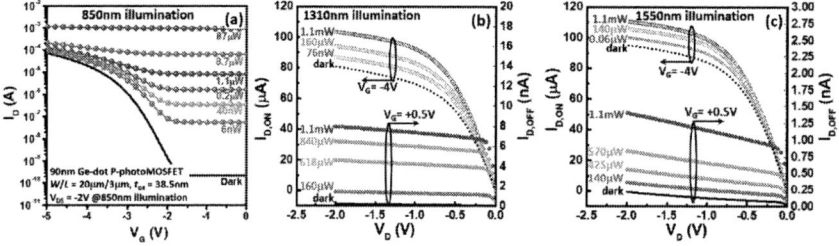

Fig. 3: I_D-V_G and I_D-V_D characteristics of 90mn Ge-dot p-PTs under variable-power 850–1550nm illumination.

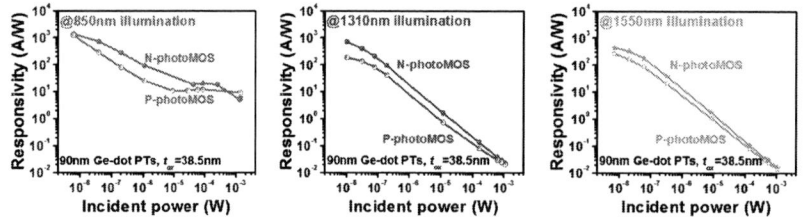

Fig. 4: Photoresponsivity of 90mn Ge-dot n- and p-PTs under variable-power 850–1550nm illumination.

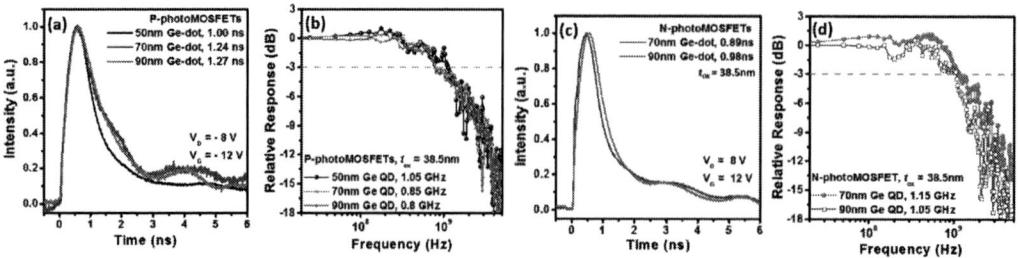

Fig. 5: Normalized temporal response time and relative response to optical excitation as a function of frequency.

978-1-5090-4661-4/17 $31.00 © 2017 IEEE

The Impact of Oxygen Insertion Technology on SRAM Yield Performance

A Marshall[1], S Nimmalapudi[1], WK Loh[2], J Krick[3], L Hutter[4],

H Takeuchi[5], R Stephenson [5], RJ Mears [5] and S Ikeda[6]

[1]UT Dallas, USA, [2]Consultant, [3]Logix Consulting, Inc., [4]Lou Hutter Consulting, [5]Atomera Inc., [6]tei solutions
Andrew.Marshall@utdallas.edu

Abstract

Oxygen Insertion (OI) to enhance MOSFET performance has been demonstrated previously. We expand the analysis of OI to determine the impact of Negative Bias Temperature Instability (NBTI), and examine its effect on an optimized SRAM cell. (Keywords: Memory; NBTI; OI; SRAM)

Introduction

Insertion of partial monolayers of oxygen (OI) into the channel of a MOSFET has been shown to provide an increase in mobility, and improvement to circuit and memory performance [1-5]. We evaluate SRAM performance of OI-enhanced CMOS technology in terms of static noise margin, write margin and yield performance at low V_{DD} operation.

Method

The analysis used an industry standard 28nm low power CMOS technology. OI electrical parameter shifts were then used to create OI-enhanced SPICE models for the 28nm CMOS node. 28nm 6T SRAM cells ($0.12\mu m^2$) (Fig. 1) were optimized for OI and non-OI flows, using SNM and WM results over V_{DD} and temperature. The optimized 28nm 6T SRAM cells were then simulated across a full characterization range of V_{DD} (0.75V and 1.4V) and temperatures (-40°C to 125°C) using Monte Carlo techniques. SRAM yield performance vs. V_{DD} for various memory arrays and repair schemes were compared for OI and non-OI material. Finally, NBTI effects were incorporated for OI and non-OI material and new SRAM yield vs. V_{DD} assessments were done. The transistor-level benefits of the OI film on the NMOS and PMOS devices are summarized in Tables 1 & 2. These electrical parameter shifts in Fig. 3 were determined by measuring OI-enhanced and non-OI NMOS devices on a similar technology. The baseline SPICE model for the given technology was then adjusted to an OI-enhanced SPICE model using the shifts summarized in the tables 1 and 2 for NMOS and PMOS, respectively. The retargeted characteristics are shown in Fig. 3.

Results

SRAM level analysis of NBTI shows that with OI we observe significant improvement in NBTI. This is because there is less PMOS VTP shift over time, so less guard-banding is needed during screening (Fig. 4). The OI film enhanced PMOS has a Vt shift of only 19.7mV (Fig. 5) compared to control Vt shift of 50mV. 1000 Monte-Carlo runs were made at each condition of V_{DD} and temperature, and Iread, SNM, and WM parameters were determined for mean, global standard deviation, and mismatch standard deviation conditions (Table 3). No circuit assist techniques were used. Fig. 6 highlights the worst-case 0.25Mb with natural yield conditions (without bit correction). At V_{DD} = 1.0V, the OI devices allow 10% higher yield. At the 80% yield point, OI allows 150mV lower V_{DD}. Fig. 7 displays the worst-case 64Mb Yield with 1/64kb repair over the range -40°C to 125°C. The baseline yield is around ~80% at V_{DD} = 1.0V, as expected. However, we observe a 20% yield gain with OI at V_{DD} = 1.0V following repairs. The OI devices extend V_{DD} down to 0.825V with 100% yield. At 60% yield point, OI allows V_{DD} = 0.8V, 160mV lower than the baseline.

Conclusion

Approximately 50% of the SRAM V_{DD} reduction capability is due solely to NBTI shift reduction for the OI devices. The remaining 50% is due to the inherent mobility and matching improvements of the OI technology. In conclusion, OI technology offers a significant improvement for low-power SoC applications with embedded SRAM. Coupled with already demonstrated logic benefits, this represents more than a full node of performance improvement without circuit assist.

References

[1] R.J. Mears, et.al., IEEE International SOI Conference, October 1-4, 2007, pp23-4

[2] R.J. Mears, et.al., IEEE Silicon Nanoelectronics Workshop, June 10-11, 2012

[3] N. Xu, et.al., IEEE IEDM, December 12, 2012, pp6.4.2-6.4.4

[4] N. Damrongplasit, et.al., IEEE Trans. Electron Devices, vol. 60, no. 5, pp 1790-1793, May 5, 2013

[5] Xi Zhang, et.al., Trans. Electron Devices, vol. 63, no. 4, pp 1502-7 April 2016

978-1-5090-4661-4/17 $31.00 © 2017 IEEE

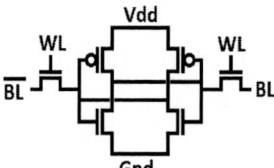

Figure 1: 6T SRAM Circuit used for this analysis

NMOS Performance Table

Optimized MST Film:	NMOS	Bias Condition	Comments
Ion	15%	Vds=1.0V, Vgs=Vtsat+0.667V	Vtsat=Vgs at Idrain=1e-07*w/l, Vds=1.0V
Ieff	24%	Ieff=(Ilow+Ihigh)/2	
Ilow	50%	Vds=1.0V, Vgs=Vtsat+0.167V	
Ihigh	22%	Vds=0.5V, Vgs=Vtsat+0.667V	
Ilin	21%	Vds=0.05V	
DIBL	-19%	Vds=0.05, 1.0V	
Subthreshold Slope	-4mV/dec	Vds=1.0V	
Rseries	0%		No change
Cgd	-10%		
C,junction	0%		No change
Vtsat sigma total	-18.8%		
Vtsat matching	-19.0%		
Idsat matching	-20.2%		
Leff	Longer		Cgate,total at LMIN unchanged
Igate (inversion) - L=LMIN	-50.0%	Vgs=1.0V, Vds=0V	

PMOS Performance Table

Optimized MST Film:	PMOS	Bias Condition	Comments
Ion	5%	Vds=-1.0V, Vgs=Vtsat-0.7V	Vtsat=Vgs at Idrain=0.25e-07*w/l, Vds=-1.0V
Ieff	8%	Ieff=(Ilow+Ihigh)/2	
Ilow	16%	Vds=-1.0V, Vgs=Vtsat-0.2V	
Ihigh	7%	Vds=-0.5V, Vgs=Vtsat-0.7V	
Ilin	6%	Vds=-0.05V	
DIBL	-5%	Vds=-0.05, -1.0V	
Subthreshold Slope	-2mV/dec	Vds=-1.0V	
Rseries	0%		No change
Cgd	0%		No change
C,junction	0%		No change
Vtsat sigma total	-9.2%		
Vtsat matching	-15.3%		
Idsat matching	-14.0%		
Leff	No Change		
Igate (inversion) - L=LMIN	-50.0%	Vgs=-1.0V, Vds=0V	

Tables 1 & 2: Transistor-level benefits of the OI film on the NMOS and PMOS device.

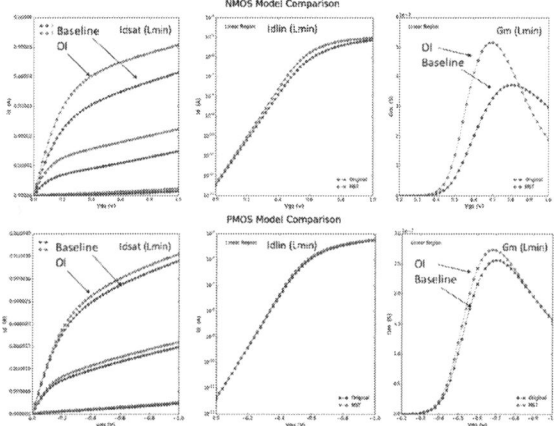

Figure 3: Simulated LMIN NMOS and PMOS characteristics after retargeting due to OI film

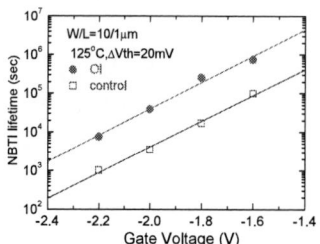

Figure 4: PMOS NBTI lifetime as a function of applied voltage

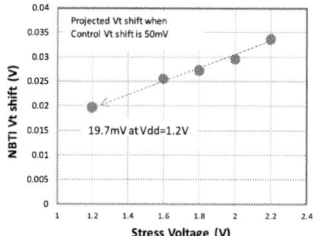

Figure 5: Projected NBTI-induced Vt shift when using the OI film.

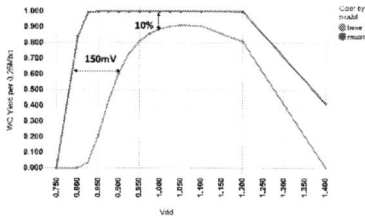

Figure 6: Worst-case 0.25Mb (under natural yield condition).

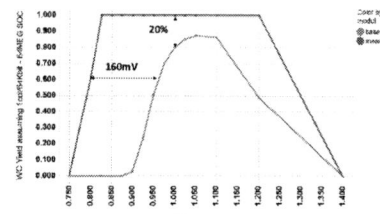

Figure 7: Worst-case 64Mb Yield with 1/64kb repair over the range -40°C to 125°C.

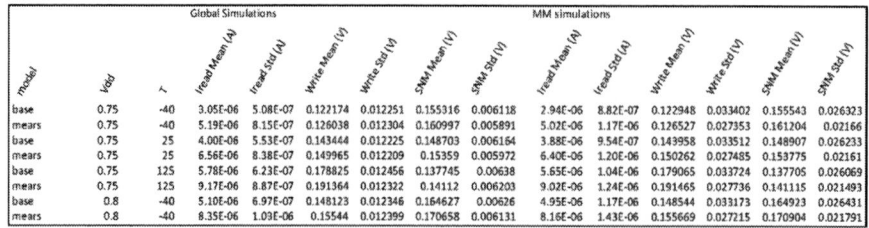

model	Vdd	T	Global Simulations						MM simulations					
			Iread_Mean (A)	Iread_Std (A)	Write_Mean (V)	Write_Std (V)	SNM_Mean (V)	SNM_Std (V)	Iread_Mean (A)	Iread_Std (A)	Write_Mean (V)	Write_Std (V)	SNM_Mean (V)	SNM_Std (V)
base	0.75	-40	3.05E-06	5.08E-07	0.122174	0.012251	0.155316	0.006118	2.94E-06	8.82E-07	0.122948	0.033402	0.155543	0.026323
mears	0.75	-40	5.19E-06	8.15E-07	0.126038	0.012304	0.160997	0.005891	5.02E-06	1.17E-06	0.126527	0.027353	0.161204	0.02166
base	0.75	25	4.00E-06	5.53E-07	0.143444	0.012225	0.148703	0.006164	3.88E-06	9.54E-07	0.143958	0.033512	0.148907	0.026233
mears	0.75	25	6.56E-06	8.38E-07	0.149965	0.012209	0.15359	0.005972	6.40E-06	1.20E-06	0.150262	0.027485	0.153775	0.02161
base	0.75	125	5.78E-06	6.23E-07	0.178825	0.012456	0.137745	0.00638	5.65E-06	1.04E-06	0.179065	0.033724	0.137705	0.026069
mears	0.75	125	9.17E-06	8.87E-07	0.191364	0.012322	0.14112	0.006203	9.02E-06	1.24E-06	0.191465	0.027736	0.141115	0.021493
base	0.8	-40	5.10E-06	6.97E-07	0.148123	0.012346	0.164627	0.00626	4.95E-06	1.17E-06	0.148544	0.033173	0.164923	0.026431
mears	0.8	-40	8.35E-06	1.03E-06	0.15544	0.012399	0.170658	0.006131	8.16E-06	1.43E-06	0.155669	0.027215	0.170904	0.021791

Table 3: Global and mismatch standard deviations for both baseline and OI film at various voltage and temperature conditions.

P-11

Transport properties in silicon nanowire transistors with atomically flat interfaces

F. Liu[1], M. K. Husain[1], Z. Li[1], M. S. H. Sotto[1], D. Burt[1], J. D. Fletcher[2], M. Kataoka[2],

Y. Tsuchiya[1], and S. Saito[1]

[1]Nanoelectronics & Nanotechnology Research Group, Univ. of Southampton, UK. Email: S.Saito@soton.ac.uk

[2]National Physical Laboratory (NPL), Hampton Road, Teddington, Middlesex TW11 0LW, UK.

Abstract

We have fabricated ultra-narrow (sub-10 nm) short channel (100 nm) silicon (Si) nanowire transistors with atomically flat interfaces based on Si-on-Insulator (SOI) substrates. The raised source and drain electrodes were patterned together with the gate electrode. The smaller threshold voltage in the narrower nanowire suggests self-limiting oxidation during the gate oxide formation.

(Keywords: narrow channel effect, silicon nanowire, SOI, TMAH, self-limiting oxidation)

Introduction

Due to the stronger immunities against short channel effects compared with FinFETs [1], Si nanowire transistors have attracted significant interests for various applications [2-5] in nanoelectronics beyond conventional Complementary Metal Oxide Semiconductor Field Effect Transistors (CMOSFETs). Si Nanowire FETs are promising for sensors [2], single electron devices [3], and quantum technologies [4]. However, fabrication of highly uniform sub-10 nm Si nanowires is still a big challenge even with the state-of-the-art large-scale manufacturing technologies.

In order to avoid line-edge roughness, we used

tetramethylammonium hydroxide (TMAH) etching [5] to define Si nanowires with the atomically flat interfaces. The raised source and drain method is implemented to pattern source/drain electrodes and gate electrode at the same time.

Device Fabrication

The device design and structure are shown in Fig. 1. We used SOI wafers with the standard (100) SOI layer (100 nm) on the 145 nm-thick buried-oxide (BOX). The SOI layer was thinned to 24 nm before making the nanowire. The 21 nm-thick SiO_2 layer was formed by oxidation for the hard-mask to pattern Si nanowires. The initial patterning was made by electron-beam lithography with a Hydrogen-Silsesquioxane (HSQ) layer as a resist. Then dry etching was implemented to etch the SiO2 layer down to 3.6 nm. The remaining SiO2 layer was removed by hydro-fluoric (HF) etching. The combination of dry etching and wet etching prevents the damage of the Si layer caused by dry etching. Subsequent TMAH etching was carried out to form triangle or trapezoid nanowires. The planar view of a nanowire was taken by Scanning Electron Microscope (SEM, Fig. 2). The narrowest Si nanowire after TMAH etching was 12.5 nm in width before gate oxidation. We found the

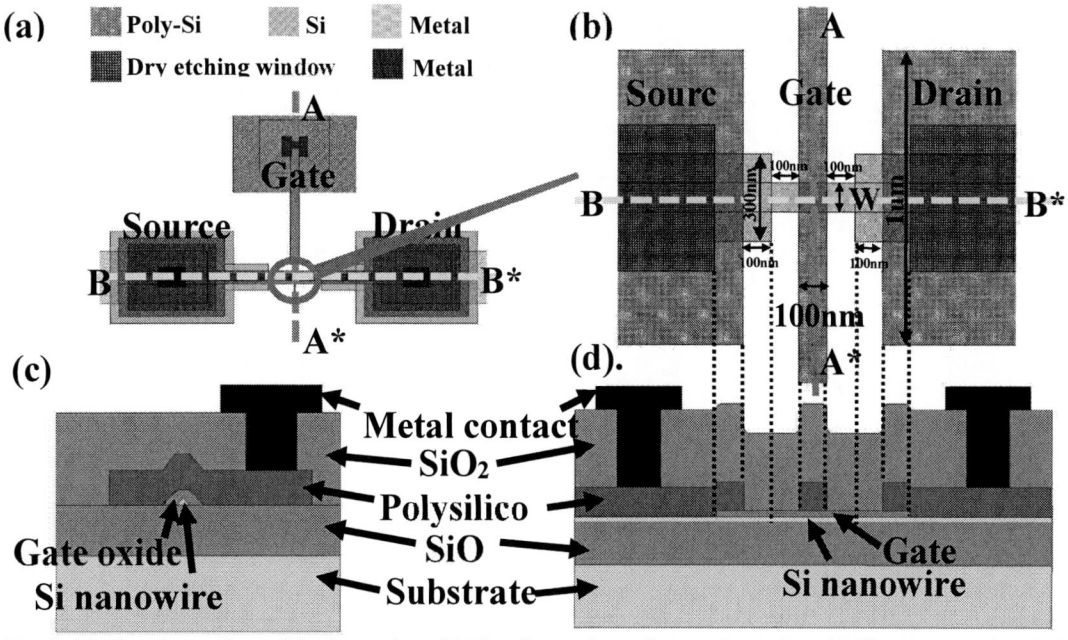

Fig. 1: (a) The planar view of device region. (b) The planar view of nanowire region. (c) The cross sectional view in AA* direction (blue dash line). (d) The cross sectional view in BB* direction (yellow dash line).

978-1-5090-4661-4/17 $31.00 © 2017 IEEE   193

suspended SiO_2 hard mask at the corners of the source and drain (Fig. 2), which show the quasi-stable planes of (113) and (114) in addition to (111). We then oxidized the surface of Si nanowire to form a gate oxide layer. The nominal widths of the Si nanowires after oxidation were 1 nm, 21 nm and 71 nm respectively. Then the contact windows for source and drain were opened by dry etching. Subsequently, the polycrystalline-Si (Poly-Si) layer was deposited by low-pressure chemical vapour deposition (LPCVD). After doping the Poly-Si layer with phosphorus, we used electron-beam lithography and

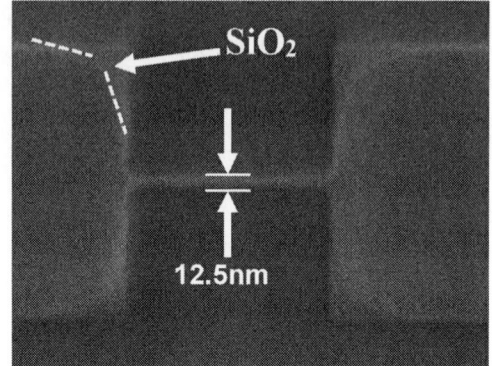

Fig. 2: SEM view of nanowire after TMAH etching (12.5 nm). At the corners of the source and drain, the SiO_2 hard mask was suspended.

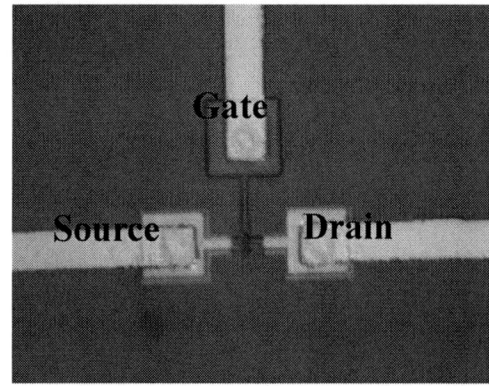

Fig. 3: Optical micrograph of Si nanowire transistor. The gate electrode is made of LPCVD Poly-Si, while the same Poly-Si layer was used for raised source and drain.

inductively coupled plasma (ICP) dry etching to make the source, drain, and gate electrodes simultaneously. The gate length was 100 nm. After deposition of the SiO_2 layer by plasma-enhanced CVD (PECVD), metal contact windows were opened by wet etching, followed by Al/Ti layers deposition and patterning. The optical microscope image of the final device is shown in Fig. 3.

Electrical Characteristics

We characterized the transport properties by using a semiconductor parameter analyser (B1500A, Agilent), and a probe station at room temperature. The background noise level was below 100 fA. We confirmed excellent sub-threshold characteristics (Figs. 4 and 5). The narrowest device (nominal 1 nm width) showed an ON-current (I_{on}) of 342 nA (0.342 mA/μm) at 1.0 V. The variation of threshold voltages

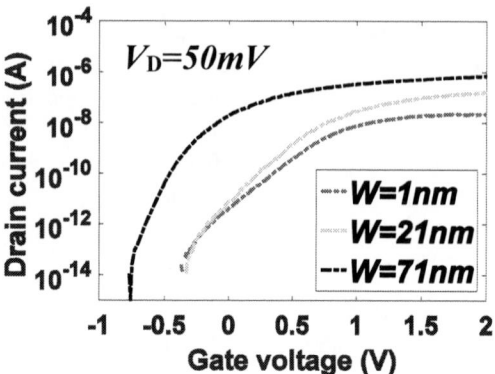

Fig. 4: Drain current vs Gate voltage curve at 50mV Drain voltage

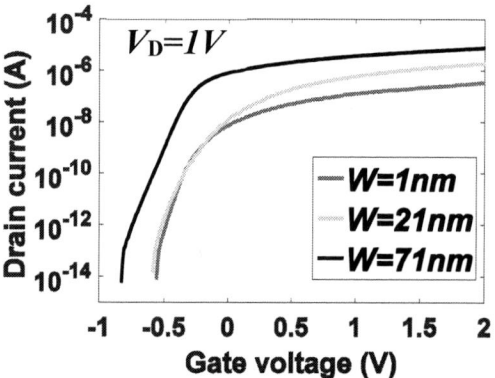

Fig. 5: Drain current vs Gate voltage curve at 1V Drain voltage

and sub-threshold slopes are shown in Figs. 6 and 7 against nominal Si nanowire width. The sub-threshold slope in the narrowest device (nominal 1nm) was 60 mV/decade at the drain voltage of 1.0 V, which shown strong controllability [6], since our device is nearly a gate-all around nanowire FET. We also found that the threshold voltage of the device with the nominal width of 21 nm was higher than the device with the width of 71 nm. This could be attributed to the quantum mechanical confinement and subsequent band-gap increase in the narrower transistor [7]. However, the threshold voltage of the

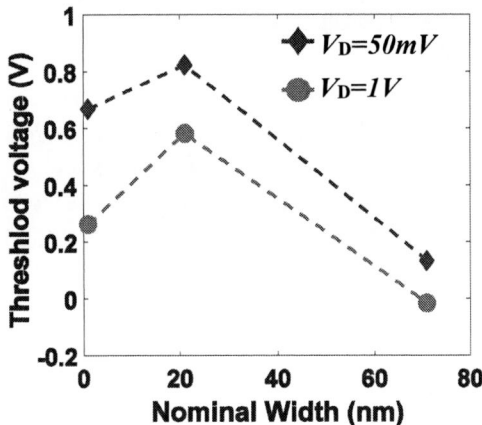

Fig.6 Threshold voltage with silicon nanowire width size dependent.

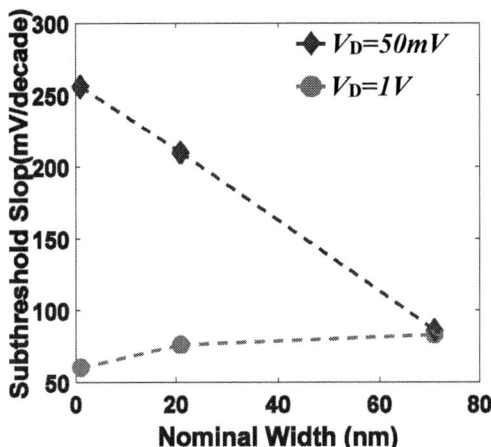

Fig.7 Subthreshold slope with silicon nanowire width size dependent.

transistor with the nominal width of 1 nm was smaller than the device with the width of 21 nm. This can be explained by self-limiting oxidation due to the strain in the nanowire [8]. We expect the actual width of the nanowire should be much larger than the nominal value.

Conclusion

We fabricated nanowire transistors with TMAH etching of Si to achieve atomically flat interfaces. The raised source/drain and gate electrodes were patterned simultaneously by dry etching. The threshold voltage shift due to quantum confinement was found in sub-100 nm MOSFETs. We found that self-limiting oxidation phenomenon occurred during gate oxidation.

Acknowledgments

This work is supported by EPSRC Manufacturing Fellowship (EP/M008975/1), EU FP7 Marie-Curie Carrier-Integration-Grant (PCIG13-GA-2013-61811), and the EMPIR program co-financed by the Participating States and from the European Union's Horizon 2020 research and innovation program, and University of Southampton. The data from the paper can be obtained from the University of Southampton e-Print research repository: http://doi.org/10.5258/SOTON/402332.

References

[1] D. Hisamoto, W. Lee, J. Kedzierski, H. Takeuchi, K. Asano, C. Kuo, E. Anderson, T. King, J. Bokor and C. Hu, "FinFET—A Self-Aligned Double-Gate MOSFET Scalable to 20 nm," IEEE Trans. Electron Dev. Vol. 47, No. 12, 2320, December 2000, doi: 10.1109/16.887014.

[2] F. Patolsky, G. Zheng and C.M. Lieber, "Fabrication of silicon nanowire devices for ultrasensitive, label-free, real-time detection of biological and chemical species," Nature Protocols, Vol. 1, 1711-1724, November 2006, doi:10.1038/nprot.2006.227.

[3] C. Wang, M.E. Jones and Z.A.K. Durrani, "Single-electron and quantum confinement limits in length-scaled silicon nanowires," Nanotechnology, Vol. 26, No. 30, July 2015, doi:10.1088/0957-4484/26/30/305203 C. Wang, *et al.*, *Nanotechnology* 26, 305203 (2015).

[4] K.S. Yi, K. Trivedi, H.C. Floresca, H. Yuk, W. Hu and M.J. Kim, "Room-Temperature Quantum Confinement Effects in Transport Properties of Ultrathin Si Nanowire Field-Effect Transistors," Nano Lett., Vol. 11, No. 12, pp 5465–5470, November 2011, doi: 10.1021/nl203238e.

[5] S. Ramadan, K. Kwa, P. King and A. O'Neill, "Reliable fabrication of sub-10 nm silicon nanowires by optical lithography," Nanotechnology, Vol. 27, No. 42, 425302, September 2016, doi:10.1088/0957-4484/27/42/425302.

[6] J.P. Colinge, M.H. Gao, A. Romano-Rodriguez, H. Maes and C. Claeys, "Silicon-on-insulator 'gate-all-around device'," IEDM'90, 90-595, doi:10.1109/IEDM.1990.237128.

[7] H. Majima, H. Ishikuro and T. Hiramoto, "Threshold Voltage Increase by Quantum Mechanical Narrow Channel Effect in Ultra-Narrow MOSFETs," IEDM'99, 99-381, doi: 10.1109/IEDM.1999.824174.

[8] H. I. Liu, D. K. Biegelsen, F. A. Ponce, N. M. Johnson, and R. F. W. Pease, "Self-limiting oxidation for fabricating sub-5 nm silicon nanowires," Appl. Phys. Lett., 64, 1383, 1994, doi: 10.1063/1.1119.

InAs MOS-HEMT power detector for 1.0 THz on quartz glass

Eiji Kume[1], Hiroyuki Ishii[2], Hiroyuki Hattori[2], Wen-Hsin Chang[2],

Mutsuo Ogura[1], Haruichi Kanaya[3], Tanemasa Asano[3] and Tatsuro Maeda[2]

[1]IRspec Corporation, Tsukuba, Ibaraki, Japan, kume@irspec.com

[2]AIST, [3]Kyushu University

Abstract

Terahertz wave was detected in 1.0 THz using InAs MOS-HEMT as a non-biased (cold) FET power detector at room temperature. Terahertz power detector was fabricated on quartz glass substrate by direct-wafer-bonding technique. 1.0 THz signal power was directly input to the gate terminal with drain coupling in MOS-HEMT detector through the GSG THz probe. The high responsivity of around 60 V/W was achieved at the bias voltage of -0.4 V.

(Keywords: Terahertz detector, and InGaAs/InAs MOSHEMT)

Introduction

In recent years, there has been impressive progress in Terahertz technologies. Recently, the feasible sub-THz detection using Si MOSFETs have been reported [1][2]. Therefore, MOS-HEMTs can be suitable for high sensitive and low noise THz detectors due to the high electron mobility with high quality heterointerface. In this work, InAs MOS-HEMT detector was fabricated on quartz glass substrate which is suitable for THz optics, and demonstrated 1.0 THz detection at room temperature.

Experimental and Results

$In_{0.53}Ga_{0.47}As$ / $In_{0.71}Ga_{0.29}As$ / InAs channel / $In_{0.71}Ga_{0.29}As$ / $In_{0.53}Ga_{0.47}As$ / InP etch stop layer / InGaAs etch stop layer was epitaxially grown on 2-inch p-InP wafer as a donor wafer (Fig. 1(a)). Then, the hetero-epitaxial wafer was directly bonded to a quartz glass as shown in Fig. 1(b). After bonding, InP wafer, the first InGaAs etch stop layer and the second InP etch stop layer were wet-etched subsequently, resulting HEMT structure on glass (Fig. 1(c)) [3]. The recess channel structure with Al_2O_3 / TaN gate stack was fabricated and the source/drain terminals were formed. Fig. 1(d) shows the final schematic structure of InAs MOS-HEMT on quartz glass.

Fig. 2 shows a photograph of the MOS-HEMT detector. The detector has two-fingers gates with a gate length of 0.4 µm and a total gate width of 80 µm. Terahertz signal power is input to the gate terminal with drain coupling through the ground-signal-ground (GSG) THz probe. THz wave is generated by signal generator and frequency extension modules. The input signal power is -32dBm in 1.0 THz. Fig. 3 shows the equivalent circuit in conjunction with by cold FET power detector. Self-mixing which is provided with the gate-drain coupling capacitor ($C_{gd,couple}$), leads to a DC current response following the square-law.

Fig. 4 shows DC characteristics at the drain bias (V_d) of 50 mV. Threshold voltage (V_{th}) of the device is -0.37 V. The subthreshold slope and the maximum transconductance of the device were 170 mV and 31.6 mS/mm, respectively. We observed the output drain DC voltage with or without 1.0 THz signal power. The typical V_{out} shift was detected about -15 µV at Vg = -0.2 V as shown in Fig. 5. The voltage response in 1 THz increased with gate bias (V_g-V_{th}) decreasing and achieved around 60 V/W at V_g-V_{th} = -0.4 V, as shown in Fig. 6.

Conclusion

In summarize, we fabricated InAs MOS-HEMT THz detector on quartz glass, and demonstrated to detect Terahertz wave in 1.0 THz at room temperature. The responsivity was around 60 V/W at bias voltage was -0.4 V.

Acknowledgments

This work was supported by Core Research for Evolutional Science and Technology (CREST) Program from Japan Science and Technology Agency (JST). A part of the device fabrication was carried out at AIST-NPF.

References

[1] E. Öjefors, U. R. Pfeiffer, A. Lisauskas, and H. G. Roskos, "A 0.65 THz focal-plane array in a quarter-micron CMOS process technology," IEEE Journal of Solid-State Circuits 44.7 (2009) 1968-1976.

[2] W. Knap, M. Dyakonov, D. Coquillat, F. Teppe, N. Dyakonova, J. Łusakowski, K. Karpierz, M. Sakowicz, G. Valusis, D. Seliuta, I. Kasalynas, A. El Fatimy, Y.Meziani, T. Otsuji, "Field effect transistors for terahertz detection: physics and first imaging applications," Journal of Infrared, Millimeter, and Terahertz Waves 30.12 (2009) 1319-1337.

[3] E. Kume, H. Ishii, H. Hattori, W. H. Chang, M. Ogura, T. Maeda. "Demonstration of InGaAs FETs on quartz glass toward terahertz applications," CSW Includes 28th IPRM & 43rd ISCS (2016).

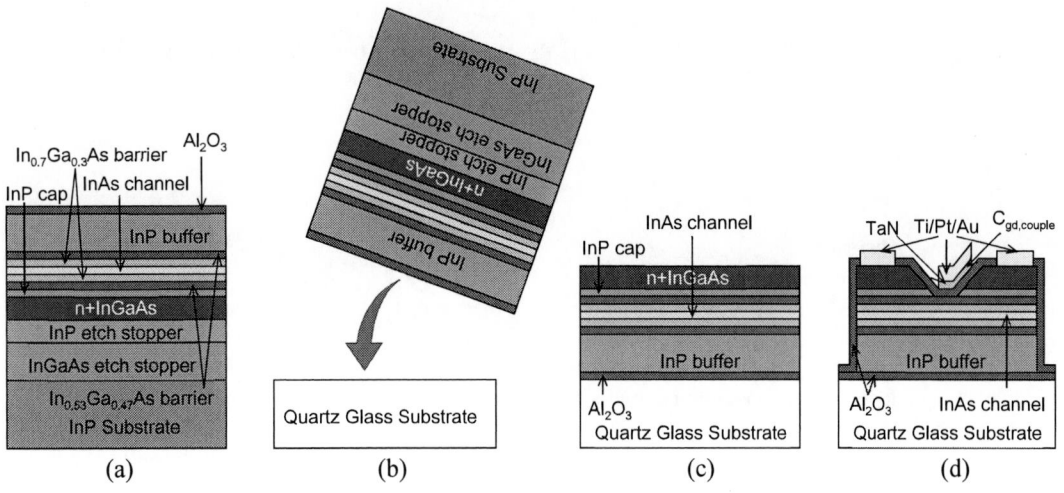

Fig. 1: Schematic of fabrication process of InGaAs/InAs MOSHEMT on quartz glass.

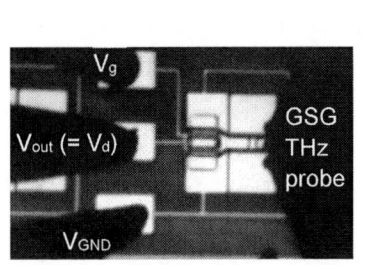

Fig. 2: Photograph of MOSHEMT detector.

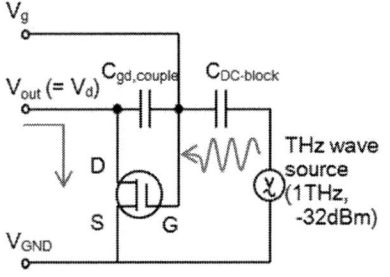

Fig. 3: Equivalent circuit of MOSHEMT detector.

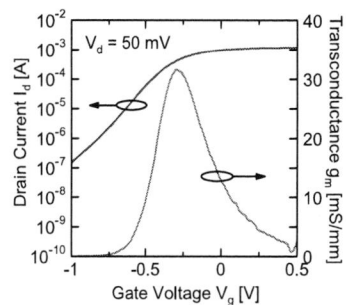

Fig. 4: Transfer characteristics of MOSHEMT at V_d = 50 mV.

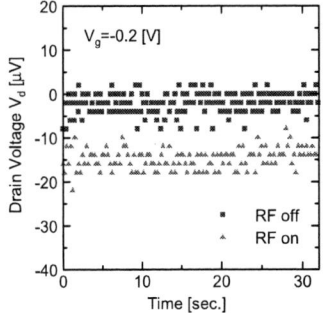

Fig. 5: Drain output voltage when 1THz wave was input (red triangle) and when not (blue rectangle).

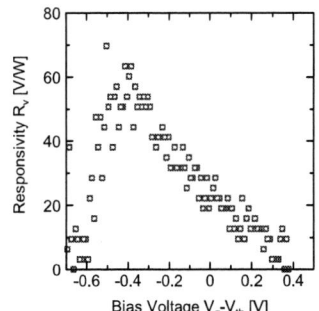

Fig. 6: Gate bias voltage dependence of the responsivity in 1THz.

978-1-5090-4661-4/17 $31.00 © 2017 IEEE

P-13

Analysis and Modeling of Capacitances in Halo-Implanted MOSFETs

Chetan Gupta*, Harshit Agarwal‡, Sagnik Dey [†], Chenming Hu‡ and Yogesh S. Chauhan*

*IIT Kanpur, India, [†]SPICE Modeling Lab, TI, Dallas, USA, ‡UC Berkeley, Berkeley, USA

chetang@iitk.ac.in

Abstract

In this paper, we report the anomalous behavior of capacitances in halo channel MOSFET for the linear and saturation regions. Unlike MOSFETs these devices have different threshold voltage (V_{TH}) for the DC and CV operations, and therefore cannot be modeled by conventional methods. We have investigated various cases of doping non-uniformity: Source side halo (SH), Drain side halo (DH), both side halos (Halo) and uniformly doped (UD) transistors using TCAD simulations under various bias conditions. A computationally efficient SPICE model is used to model these trends which shows excellent matching with the measured and TCAD data.
Keywords—BSIM6, Capacitances, Halo Doping

Introduction

Halo implants are necessary to suppress short channel effects and enable low leakage for digital circuit blocks. However, they have been shown to have detrimental impact on device DC and AC characteristics, important for analog circuits. The impact is more significant for longer channel lengths [1]–[4]. Fig. 1(a) shows the TCAD simulations [5] of threshold voltage values (V_{TH}), extracted from $C - V$ and $I - V$ characteristics with Halo doping (N_{HALO}) for different T_{ox}. Fig. 1(b) and 1(c) compare the CV and DC behavior of the Halo and UD devices. C_{GG} is almost equal for both cases, this indicates nearly the same V_{TH} values for both the cases, since in case of Halo device, some of the gate charge is balanced with the channel charge, when channel inverts while the Halo regions (SH, DH, or Halo) are still in weak inversion i.e. $V_{TH,ch} < V_{GS} < V_{TH,h}$ (where V_{GS} is the effective gate voltage at source side and $V_{TH,ch}$, $V_{TH,sh}$ represents the threshold voltage of channel [6] and Halo regions respectively). Due to the above reason, if we extract the V_{TH} from the $C - V$ characteristics for the Halo device, it comes quite less as compare to what we extract from Halo device $I - V$ characteristics [7]. In Fig. 1(c), although the threshold voltage for UD device is much smaller than that of the other non-uniformly doped devices, but they all carry nearly the same on current (due to which a peak in g_m is observed), since effective channel resistance is same in all the cases [8]. Fig. 1(d) shows C_{GD} - V_{GS} comparison for UD and non-uniform devices in the linear region. As expected, the C_{GD} at $V_{DS} = 0$ for UD MOSFET remains at very low values for $V_{GS} < V_{TH,ch}$. As V_{GS} starts to increase beyond $V_{TH,ch}$, the C_{GD} increases rapidly and saturates to $0.5.C_{ox}.W.L$ (where W and L are the channel Width and Length and C_{ox} is the oxide capacitance.) [9]. The situation is very different for source Halo (SH) devices. Due to Halo doping at source end, the C_{GD} starts increasing as

soon as effective gate voltage at drain side (V_{GD}) is more than the $V_{TH,dh}$ (drain side Halo region threshold voltage). It means that the inversion at the source side starts at higher V_{GS} values as compare to drain side. As gate voltage keeps on increasing, the inversion in the channel reaches from the drain towards source and C_{GD} keeps on rising. Once V_{GS} is greater than $V_{TH,sh}$ the C_{GD} starts to fall and again saturates to $0.5.C_{ox}.W.L$. Also it is easily observed that the C_{GD} for the DH rises at higher V_{GS} than the C_{GD} for the Halo device, this is because in the DH case, the gate charge is balanced by the source charge, so the impact of drain has postponed to high V_{GS}. Fig. 2(a) - Fig. 2(d) show the TCAD simulations to explain the impact of Halo doping and oxide thickness on the C_{GD} for the SH and Halo transistors. Fig. 2(a) shows, as we increase the halo doping at the source side, the source-body barrier or $V_{TH,sh}$ increases. To balance the gate charge, drain contributes the more inversion charge due to which C_{GD} peak shifts towards the right. And finally when $V_{GS} > V_{TH,sh}$, source side Halo also starts contributing the inversion charge and C_{GD} asymptotically reaches to $0.5.C_{ox}.W.L$. Fig. 2(b) shows the impact of oxide thickness on C_{GD} of SH device. As we increase the T_{ox}, since it decreases the gate control so it further increases the drain side barrier and V_{TH} shift can be observed in C_{GD}. But at higher V_{GS} when $V_{GS} > V_{TH,dh}$, drain starts to respond, now C_{GD} starts increasing. As T_{ox} increases, the C_{ox} decreases. So, for each T_{ox} value, the corresponding C_{GD} saturates at different values. Similarly Fig. 2(c) shows the impact of halo doping on C_{GD} of Halo device. Doping increases the barrier at both source and drain sides and due to which V_{TH} shift is definitely be noticeable. And finally Fig. 2(d) shows the impact of T_{ox} on the C_{GD} of Halo. Increasing of T_{ox} reduces the overall gate control, the source/drain barrier increases due to which C_{GD} starts to increase at higher V_{GS}. And for each value of Halo doping, it obtains the different values of peak, since change in T_{ox} leads to the change in C_{ox}.

Model Description and Results

The peculiar characteristics of the non-uniformly doped devices need a dedicated compact model which can emulate threshold voltage properties across DC and CV behavior. The proposed model uses a sub-circuit approach as shown in Fig. 3(a) [8]. The channel transistor is implemented using the BSIM6 model [10]–[12], while the source and drain side regions have a new model to capture the sub-threshold behavior of source and drain side halo barrier [8] [13], and is given by the equations from (1) to (4). Since source/drain Halo regions properties are different than that of the channel region, and therefore the parameters like mobility **U0(U0H)**, channel

978-1-5090-4661-4/17 $31.00 © 2017 IEEE

length **L(LH)**, threshold voltage $V_{TH}(V_{TH},$**halo**$)$, Subthreshold slope **NFACTOR(NH)** are separately needed for both regions. If we run the AC simulations with our model corresponding to a Halo device, when $V_{TH,ch} < V_{GS} < V_{TH,h}$ the channel part is fully inverted but the halo regions are not. So some of the gate charge is balanced by the channel inversion charge and for the same reason the V_{TH} extracted from the $C-V$ characteristics (getting from our model) is quite low, as compare to the extracted value from the DC simulations (Fig. 1(a)). Model-data (measured as well as TCAD data) overlay for long channel NMOS transistor are shown in Fig. 3(b)-3(d). Fig. 3(b) shows the validation of our model with the measured data for $I_{DS} - V_{GS}$ and $C_{GD} - V_{GS}$ characteristics for a long channel length Halo device in the linear region. Fig. 3(c) shows the TCAD validation for $I_{DS} - V_{GS}$, $C_{GD} - V_{GS}$ characteristics of Halo device and $C_{GD} - V_{GS}$ of SH device in the linear region. Finally, Fig. 3(d) shows the TCAD validation of a new model for $I_{DS} - V_{GS}$, $C_{GD} - V_{GS}$ characteristics for Halo device in the saturation region. Incorporating the new model for the halo region allows the proposed model to accurately capture the trend of DC as well for AC data for SH, Halo transistors.

$$I_{R_{S(D)}} = U0H.C_{ox}\frac{W}{L_H}.V_{s_i s(d_i d),eff}.V_{gsteff(gdteff)} \quad (1)$$

where

$$V_{gsteff(gdteff)}$$
$$= \frac{2N_H V_t.ln\left[1 + exp(\frac{V_{GS_i(GD_i)} - VTH,halo}{2N_H V_t})\right]}{1 + 2N_H.exp(-\frac{V_{GS_i(GD_i)} - VTH,halo}{2N_H V_t})}$$

$$V_{gsteff(gdteff)}$$
$$= \begin{cases} V_{GS_i(GD_i)} - V_{TH,halo} & V_{GS_i(GD_i)} > V_{TH,halo} \\ exp(\frac{V_{GS_i(GD_i)} - V_{TH,halo}}{n_h.V_t}) & V_{GS_i(GD_i)} < V_{TH,halo} \end{cases} \quad (2)$$

$$V_{s_i s(d_i d),eff} = \frac{V_{s_i s(d_i d)}}{\left[1 + \left(\frac{V_{s_i s(d_i d)}}{V_{dsat,H}}\right)^{\frac{1}{DELTAH}}\right]^{DELTAH}} \quad (3)$$

$$V_{dsat,H} = V_{gsteff(gdteff)} + 2V_t \quad (4)$$

Here **DELTAH** is a smoothing parameter for the effective drain to source voltage function in the halo region. [12].

Conclusion

Halo doped devices show anomalous behavior of of current and AC capacitances, which cannot be modeled by conventional compact models. The physical mechanism leading to such capacitance behavior is explained using TCAD, and a compact model is used that can capture all the characteristics of halo devices in the linear and saturation region for different biases.

ACKNOWLEDGMENT

This work was partially supported by Semiconductor Research Corporation, Science and Engineering Research Board, Council of Scientific and Industrial Research and Ramanujan fellowship research grant.

REFERENCES

[1] A. S. Roy, S. P. Mudanai, and M. Stettler, "Mechanism of Long-Channel Drain-Induced Barrier Lowering in Halo MOSFETs," *IEEE Trans. Electron Devices*, vol. 58, no. 4, pp. 979–984, April 2011.

[2] A. Cathignol, S. Bordez, A. Cros, K. Rochereau, and G. Ghibaudo, "Abnormally high local electrical fluctuations in heavily pocket-implanted bulk long mosfet," *Solid-State Electronics*, vol. 53, no. 2, pp. 127–133, 2009.

[3] C. McAndrew and P. G. Drennan, "Analysis of halo implanted mosfets," in *Technical Proceedings of the 2007 NSTI Nanotechnology Conference and Trade Show,(Nano Science and Technology Institute, 2007)*, vol. 3, 2007, pp. 594–598.

[4] K. M. Cao *et al.*, "Modeling of pocket implanted MOSFETs for anomalous analog behavior," in *Proc. IEEE IEDM*, 1999, pp. 171–174.

[5] U. Manual and J. Version, "Synopsys tcad sentaurus," *Synopsys, San Jose, CA, USA*, 2014.

[6] H. Agarwal, C. Gupta, P. Kushwaha, C. Yadav, J. P. Duarte, S. Khandelwal, C. Hu, and Y. S. Chauhan, "Analytical modeling and experimental validation of threshold voltage in bsim6 mosfet model," *IEEE Journal of the Electron Devices Society*, vol. 3, no. 3, pp. 240–243, 2015.

[7] R. Rios, W.-K. Shih, A. Shah, S. Mudanai, P. Packan, T. Sandford, and K. Mistry, "A three-transistor threshold voltage model for halo processes," in *Electron Devices Meeting, 2002. IEDM'02. International.* IEEE, 2002, pp. 113–116.

[8] H. Agarwal, C. Gupta, S. Dey, S. Khandelwal, C. Hu, and Y. S. Chauhan, "Anomalous transconductance in long channel halo implanted mosfets: Analysis and modeling," *IEEE Transactions on Electron Devices*, vol. PP, no. 99, pp. 1–8, 2017.

[9] Y. S. Chauhan, R. Gillon, M. Declercq, and A. M. Ionescu, "Impact of lateral non-uniform doping and hot carrier degradation on capacitance behavior of high voltage mosfets," in *ESSDERC 2007-37th European Solid State Device Research Conference.* IEEE, 2007, pp. 426–429.

[10] Y. S. Chauhan *et al.*, "BSIM6: Analog and RF Compact Model for Bulk MOSFET," *IEEE Trans. Electron Devices*, vol. 61, no. 2, pp. 234–244, Feb 2014.

[11] H. Agarwal, S. Venugopalan, M.-A. Chalkiadaki, N. Paydavosi, J. P. Duarte, S. Agnihotri, C. Yadav, P. Kushwaha, Y. S. Chauhan, C. C. Enz, A. Niknejad, and C. Hu, "Recent enhancements in BSIM6 bulk MOSFET model," in *Simulation of Semiconductor Processes and Devices (SISPAD), International Conference on*, Sept 2013.

[12] "BSIM6 Technical Manual," 2014. [Online]. Available: http://www-device.eecs.berkeley.edu/bsim/?page=BSIM6_LR

[13] "BSIM4 Technical Manual." [Online]. Available: http://www-device.eecs.berkeley.edu/bsim/?page=BSIM4

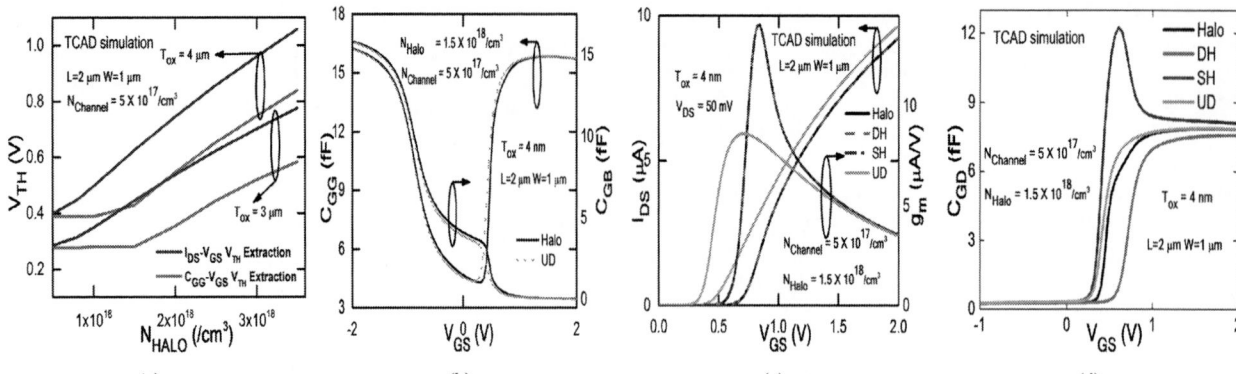

(a)   (b)   (c)   (d)

Fig. 1. (a) It shows the comparison of extracted V_{TH} from $C-V$ and DC characteristics with Halo doping. V_{TH} extracted from $C-V$ is lower because the gate charge is balanced by the already inverted channel charge. In case of DC operation, current can not flow unless both Halo sides get inverted, so extracted V_{TH} value from $I-V$ is high. (b) Comparison of C_{GG} and C_{GB} of a halo implanted and uniformly doped MOSFET, C_{GG} is almost same that indicates nearly the same V_{TH} values for both the cases, since in case of Halo some of the gate charge is balanced with the channel charge, this is also when the source side or drain side or both side halo are not inverted yet i.e. $V_{TH,ch} < V_{GS} < V_{TH,h}$ (c) Comparison of I_{DS} - V_{GS} and g_m for UD, SH, DH and Halo in Linear region. We can easily see from the plot that there is a V_{TH} shift in the I_{DS} - V_{GS} characteristics of UD and Halo. Although the threshold voltage for UD device is much smaller than the other Halo devices, but in strong inversion they all carry nearly same on current since effective channel resistance is same in all the cases.(d) C_{GD} comparison for UD, SH, DH and Halo in Linear region. For the SH case C_{GD} rises above $0.5.C_{ox}.W.L$, since the $V_{TH,sh}$ is very high, due to which most of the gate charge is balanced by the drain charge, and when $V_{GS} > V_{TH,sh}$, source also starts to contribute the charge and C_{GD} has to fall and saturates to $0.5.C_{ox}.W.L$. For DH case, C_{GD} rises at higher V_{GS} as compared to the Halo case. This is due to the fact- the source barrier is much lower than the drain barrier, so source contributes the more inversion charge, so drain shows its impact at higher V_{GS} and correspondingly C_{GD} too.

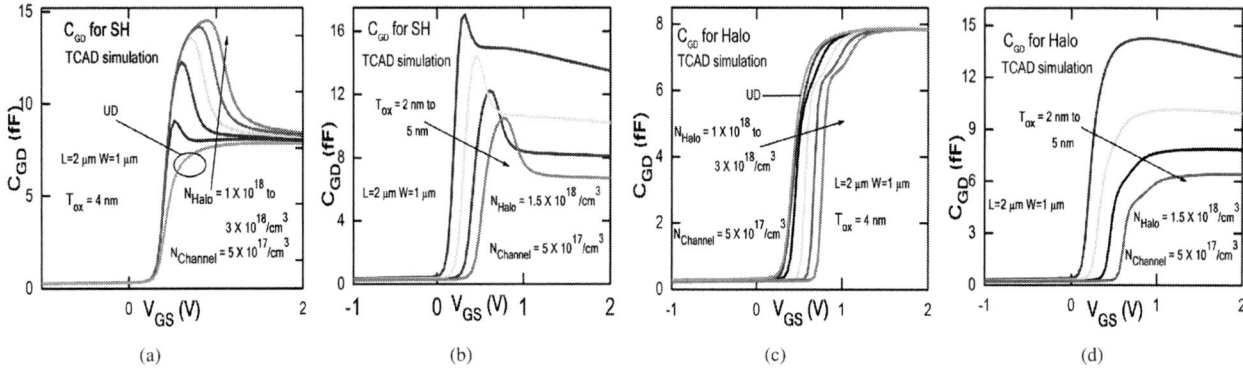

(a)   (b)   (c)   (d)

Fig. 2. (a) Impact of halo doping on C_{GD} of SH and UD devices. As we increase the halo doping at the source side, the source-body barrier or $V_{TH,sh}$ increases. So to balance the gate charge, drain contributes the inversion charge due to which C_{GD} peak shifts towards the right. And finally when $V_{GS} > V_{TH,sh}$, source side Halo also starts contributing the inversion charge, C_{GD} asymptotically reaches to $0.5.C_{ox}.W.L$ (b) Impact of oxide thickness on C_{GD} of SH device. As we increase the T_{ox}, since it decreases the gate control so it further increases the drain side barrier and V_{TH} shift can be observed in C_{GD}. But at higher V_{GS} when $V_{GS} > V_{TH,dh}$, drain starts to respond, now C_{GD} starts increasing. As T_{ox} increases, the C_{ox} decreases. For each T_{ox} value, the corresponding C_{GD} saturates at some different value. Similarly (c) Shows the impact of halo doping on C_{GD} of Halo device. Doping increases the barrier both at source and drain sides and due to which V_{TH} shift is definitely be noticeable. (d) It shows the impact of T_{ox} on the C_{GD} of Halo. Increasing of T_{ox} reduces the overall gate control, the source/drain barrier increases and C_{GD} starts to increase at higher V_{GS} and for each value of Halo doping, it obtains the different values of peak, since change in T_{ox} leads to the change in C_{ox}.

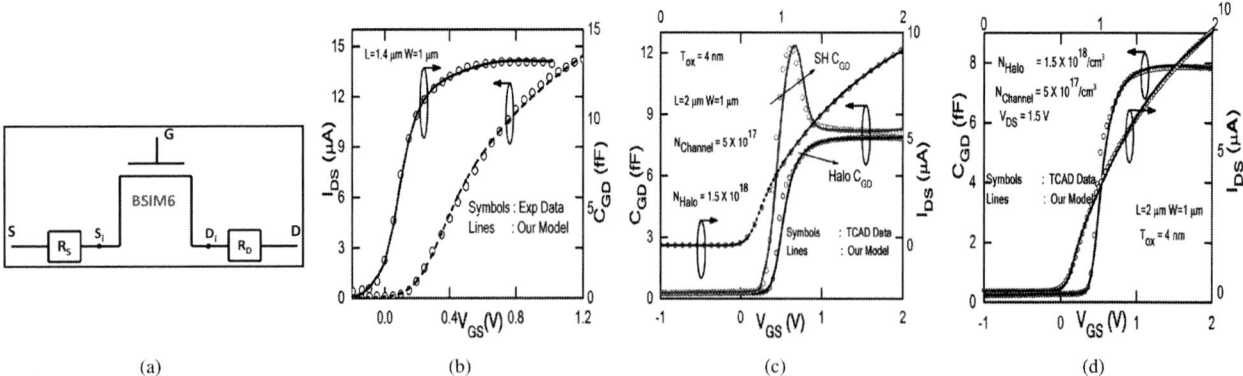

(a)   (b)   (c)   (d)

Fig. 3. (a) Sub-circuit based new model of halo implanted device. The channel region is modeled using BSIM6 MOS model and source, drain side halo regions by bias dependent Verilog-A based element. NMOS - Model validation with experimental data in linear region (b) I_{DS} - V_{GS} and C_{GD} - V_{GS} for a long channel length Halo device in linear region [7]. NMOS - Model validation with TCAD (c) I_{DS} - V_{GS}, C_{GD} - V_{GS} for Halo and SH devices in linear region. (d) I_{DS} - V_{GS}, C_{GD} - V_{GS} for Halo devices in saturation region. The proposed model accurately capture the DC and AC device trends. Measured data from LP $130nm$ CMOS SOC technology node [7].

P-14

Design and performance of thin-film µTEG modules for wearable device applications

Tsuyoshi Kondo, Nana Chiwaki, and Satoshi Sugahara

Tokyo Institute of Technology, Yokohama, Japan, tsuyoshi.k@isl.titech.ac.jp

Abstract

A thin-film micro thermoelectric generator (µTEG) applicable to wearable devices is investigated. This µTEG module has a transverse configuration of the Seebeck elements and thus can adapt to microfabrication process with a thin-film thermoelectric material. A design method to maximize the output of the µTEG is developed, and its performance is analyzed. A high output power of 10mW with an output voltage of 1V could be achieved for a wrist-band-style module-mounting. (Keywords: µTEG, Thin-film thermoelectric material, Module design)

Introduction

Micro thermoelectric generators (µTEGs) [1,2] are promising as an energy harvester for wearable devices, such as self-powered smart-watches, smart-glasses, and healthcare devices. The functionality and/or performance of these devices strongly correlate with their power consumption, i.e., higher functionality/performance requires higher power consumption. In addition, to drive CMOS microcontrollers or microprocessors in wearable devices, specific voltages are needed, e.g., ~1V for high performance operations and ~0.5V for high energy efficiency operations. A thin-film thermoelectric material suitable for µTEGs would also be requested for size/weight saving and cost reduction. Although, in general, thin-film thermoelectric materials degrade the output characteristics of µTEGs, these materials enable them to introduce micro- or nano-fabrication process based on lithography. In addition, these fabrication techniques highly enhance the degree of freedom of design for µTEGs, which is inaccessible to ordinary µTEG fabrication process based on MEMS technique.

In this paper, we investigate a transverse-type µTEG module adaptable to microfabrication process with a thin-film thermoelectric material. A design method to maximize the output of the µTEG is developed, and its performance is analyzed.

Module structures

Figs. 1 (a) and (b) show schematics of π- and transverse-type thin-film µTEG modules, respectively. Here, we assumed the thermoelectric-film thickness of a few hundreds of nanometers for these modules, which is for adaptability to micro- or nano-fabrication process. For the π-type thin-film module, the heat flow through the thermoelectric (Seebeck) elements is parallel to the direction of the temperature difference in the module. Since the thermal resistance of this module is considerably lowered owing to the thin-film thermoelectric material, the thermoelectric elements need to be scaled down to several tens or hundreds of nanometers to achieve sufficient thermal resistance. Therefore, this type of module can be referred to as nanoTEG. In contrast to the π-type module, the heat flow through the thermoelectric elements is perpendicular to the module-temperature difference for the transverse-type thin-film module shown in Fig. 1(b). This type of module can achieve sufficiently high thermal resistance even for thin-film thermoelectric elements, and the thermal resistance can be controlled by adjusting the element width along to the heat flow. As discussed in this paper, the order of micrometers or more is applicable to the thermoelectric element width for this module. Therefore, the name of µTEG is still appropriate for this transverse-type module. Here, it is slightly modified to µTTEG to accentuate the structural feature of the transverse-type module. The electrical resistance of the µTTEG, that is also an important factor to determine the output power, can be sufficiently lowered by setting the thermoelectric elements to a long length shape (see Fig. 1(b)). Although there exits the trade-off relation between the thermal and electrical resistances, a maximized output power with an appropriate output voltage of the µTTEG can be designed by optimizing the trade-off condition. In this paper, the µTTEG is investigated, compared with a π-type µTEG (µPTEG). Here, our discussions are mainly limited in micron or more ranges for the thermoelectric-element feature size, which would be beneficial for the production cost.

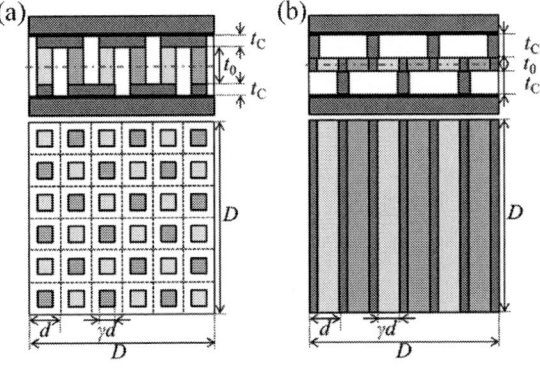

Fig. 1: Schematics of (a) π-type and (b) transverse-type thin-film µTEG (or nanoTEG) modules.

Table 1: Physical constants used for simulations.

Module size $D \times D$	1 cm × 1 cm
Temperature difference between human body and ambient air ΔT_s	10 K
Thermal resistance for ambient air K_{air}	212.5 K/W
Thermoelectric material	
Seebeck coefficient $S\ (=S_p - S_n)$	434 µV/K
Thermal conductivity $\lambda\ (=\lambda_p + \lambda_n)/2$	1.43 Wm^{-1}K^{-1}
Electrical resistivity $\rho\ (=\rho_p + \rho_n)/2$	8.11 µΩm
Interconnection/contact material (Cu)	
Thickness t_C	$t_{C_{opt}}$ (≤10 µm) (π-type)
	10 µm (transverse-type)
Thermal conductivity λ_{Cu}	386 Wm^{-1}K^{-1}
Electrical resistivity ρ_{Cu}	1.69 × 10^{-8} Ωm

978-1-5090-4661-4/17 $31.00 © 2017 IEEE 201

Design method

Table 1 shows physical constants used for the following simulations. In this study, n- and p-type Bi-Te alloys were selected as thermoelectric materials of the Seebeck elements. The thickness t_C of the contacts for the µTTEG is fixed to 10µm, while it is optimized within a range of $t_C \leq 10$µm for the µPTEG. The thickness t_0 of the thermoelectric elements is fixed to 100nm, unless otherwise noted.

To achieve a requested output voltage with the maximized output power, the thermal and electrical resistances that are in the trade-off relation need to be optimized. For this purpose, we introduced the trade-off parameter γ ($0<\gamma<1$) that is the area occupancy ratio of the thermoelectric elements to the module area. The number and size of thermoelectric elements and the resulting output voltage and power can be completely determined by γ instead of the thermal and electrical resistances.

Figs. 2 (a) and (b) show models of the system consisting of a human body and wearable device. In these models, the temperature at outside located sufficiently far from the module is the same as the ambient temperature. In ordinary µTEG applications, the constant heat flow model (Fig. 2(a)) is used, i.e., the surface temperature of the heat source is determined by the thermal resistance of the module. However, the system including homeothermic human cannot be adapted by this model owing to the isothermal body temperature. Therefore, we approximated this system as the ideal homeothermic model shown in Fig. 2(b).

Figs. 3 (a) and (b) show the thermal-resistance circuit models of the π- and transverse-type modules, respectively. For wearable device applications, the effects of Joule heating and Peltier cooling (shown in Fig. 3) can be neglected. In this paper, the vacuum isolation (that is indicated by the white regions of the upper figures in Figs. 2(a) and (b)) is adapted for both the models.

Design and performance

All the parameters shown in the following discussion are defined in Figs. 1-3, and V_S, P_{out} and m_0 represent the electromotive force, the output power, and the number of pairs of thermoelectric elements, respectively. The chip size is set to 1cm×1cm for both the modules. The target P_{out} and V_S are 100µW and 100mV, respectively, which would achieve $P_{out} = 10$mW and $V_{out} = 1$V for wrist-band-style mounting of a module.

Fig. 4 (a) shows γd, $(1-\gamma)d$, ΔT, V_S and P_{out} as a function of γ for µTTEG, in which t_0 and m_0 are set to t_0=100 nm and $m_0=10^2$. Note that m_0 is selected so as to extract a possibly high performance. For the µTTEG module, P_{out} has a peak at an optimized γ value. This γ value results in $\gamma d = 40$µm and $(1-\gamma)d = 8$µm, i.e. the sizes satisfy the feature-size range of micron. Solid curves in Fig. 4(b) shows these characteristics of µPTEG for $m_0=10^5$ (that is selected so as to extract a possibly high performance), in which the minimum feature size is limited to 1µm for γd or $(1-\gamma)d$. In this case, P_{out} has no peak position in the feature size range above 1µm. Therefore, the nanoTEG structure is required for the π-type generator. In the following discussions, the minimum size of γd and $(1-\gamma)d$ is limited to 1µm. The detail of the nanoTEG analysis is discussed later.

Fig. 4 (c) and solid curves in Fig. 4(d) show γd, $(1-\gamma)d$, ΔT, V_S, and P_{out} as a function of m_0 for t_0=100 nm for µTTEG and µPTEG, respectively, in which these parameters are optimized by γ at each m_0 and P_{out} is optimized within a condition of $V_S \geq 100$mV for the µTTEG module. Although P_{out} shows a peak for each module, it is restricted by the limited minimum feature size for µPTEG.

Fig. 4(e) and solid curves in Fig. 4(f) show γd, $(1-\gamma)d$, m_0, ΔT, V_S, P_{out} as a function of t_0 for µTTEG and µPTEG, respectively, in which these parameters are optimized by γ at each t_0. P_{out} for the µPTEG module severely decreases with decreasing t_0, while P_{out} for the

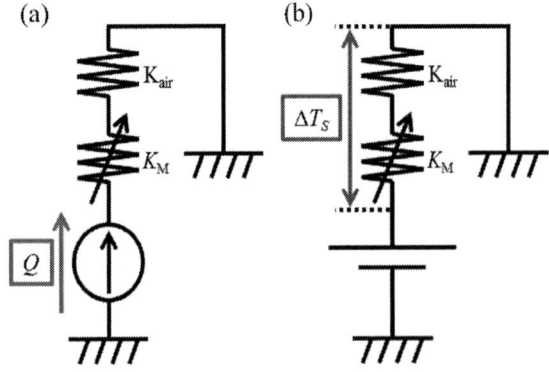

Fig. 2: Models of the systems consisting of a human body and wearable device. (a) Constant heat flow and (b) constant temperature models.

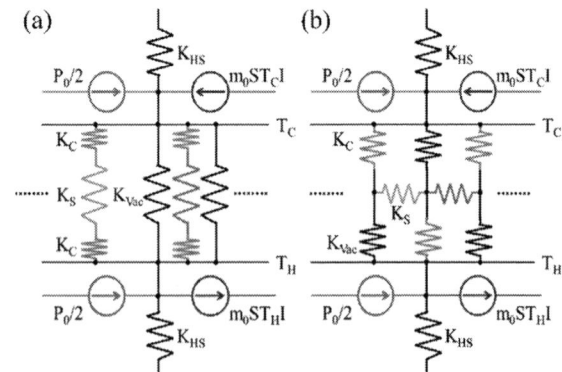

Fig. 3: Thermal-resistance circuit models for (a) µPTEG and (b) µTTEG.

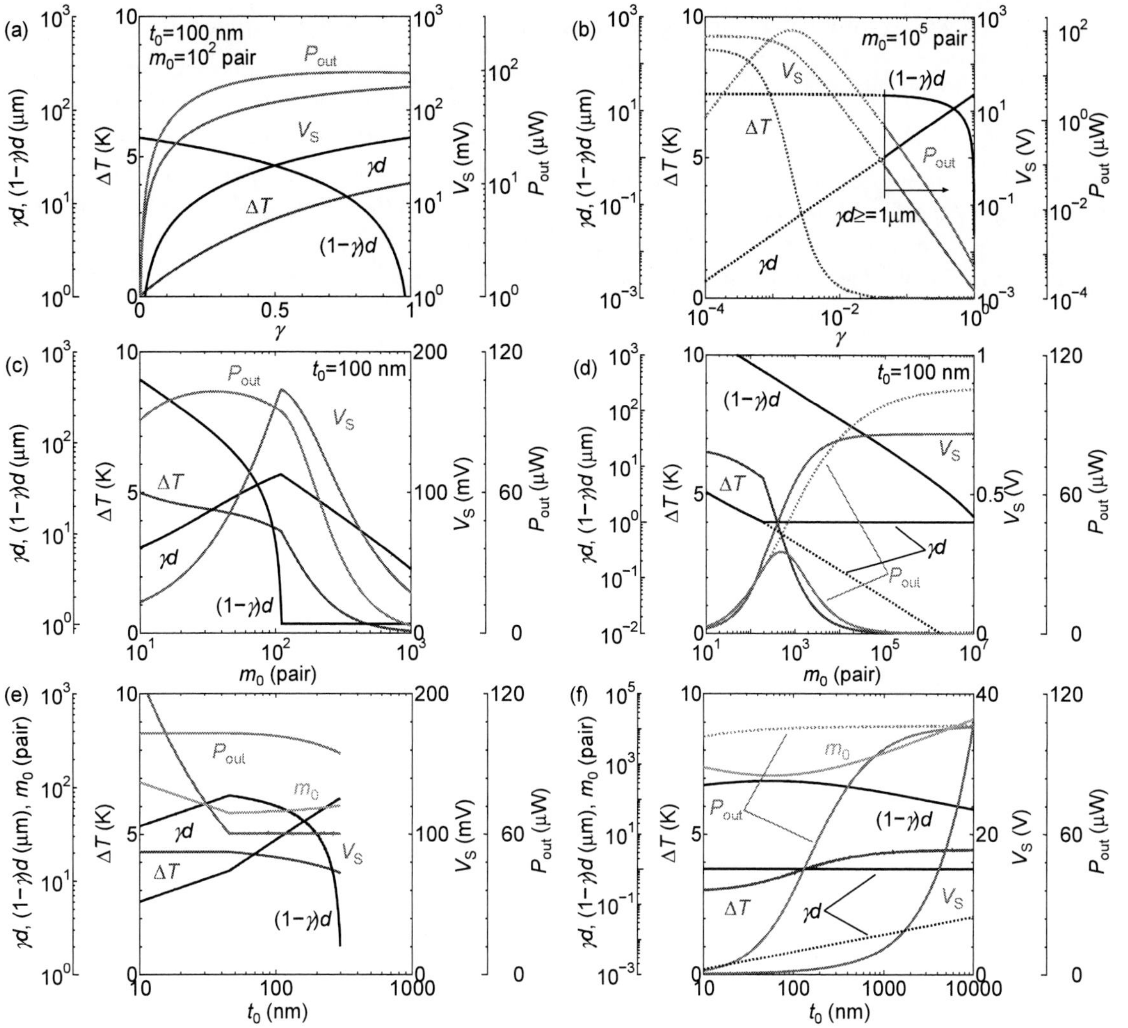

Fig. 4: γd, $(1-\gamma)d$, ΔT, V_S, and P_{out} as a function of γ, m_0, and t_0 for μTTEG ((a), (c), (e)) and μPTEG ((b), (d), (f)). Dashed curves represent γd and P_{out} for π-type nanoTEG without the feature size restriction.

μTTEG module remains at an almost constant with a high value for t_0 less than a few hundreds of nanometers. Therefore, a thin film of the thermoelectric elements is applicable to the μTTEG module even for the micron-scale minimum feature size (see γd and $(1-\gamma)d$ in Fig. 4(e)), which is the remarkable advantage of this type of module. Note that π-type generators with the feature size of 10-100nm, that is, nanoTEG, show a high performance comparable to μTTEG, as shown by dotted curves in Figs. 4(b), (d), and (f). Although nano-scale lithography is required, nanoTEG could be another candidate for wearable device applications.

Finally, performance for mounted μTTEG modules is examined. Fig. 5 shows the output characteristics of μTTEG modules, in which the module area is varied from 20 to 120 cm². When the module area is 100cm² or more, $P_{out} = 10$mW with $V_{out} = 1$V is achieved.

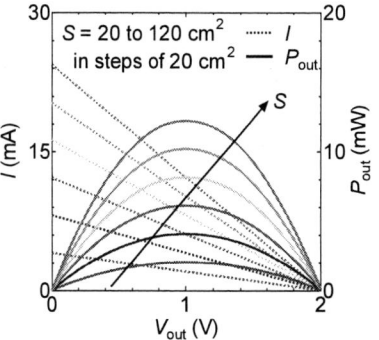

Fig. 5: Output characteristics of μTTEG modules, in which the module area is varied from 20 to 120 cm².

References

[1] M. Strasser, *et al.*, Sensors and Actuators A, **97-98**, 2002, 535.

[2] M. Kishi, et al., 18th Int. Conf. of Thermoelectrics, 1999, 301.

P-15

Analysis of spin accumulation in a Si channel using CoFe/MgO/Si spin injectors

Taiju Akushichi, Daiki Kitagata, Yusuke Shuto, and Satoshi Sugahara

Tokyo Institute of Technology, Yokohama, Japan, taiju.aku7@isl.titech.ac.jp

Abstract

Spin accumulation in a Si channel using CoFe/MgO/Si spin injectors is investigated. Hanle-effect signals from spin-polarized electrons accumulated in the Si channel are observed using three-terminal spin-accumulation (3T-SA) devices with the spin injectors. The Hanle-effect signals are decomposed into two components, i.e., channel-spin and trap-spin components. The proportion of the channel-spin component depends on the process condition of the spin injectors. The spin-injector fabricated by an optimized process condition exhibits a single channel-spin component, although this phenomenon strongly depends on a bias applied to the spin-injector. The energy-dependent trap-density distribution could affect this bias-dependent spin-injection phenomenon.
(Keywords: spintronics, spin injection, Si channel)

Introduction

Efficient spin injection and controllable spin transport for Si channels are required to realize Si-based spin devices [1]. The Hanle effect observed in 3T-SA devices has been widely used to investigate spin injection phenomena. However, Hanle-effect signals obtained from the 3T-SA technique need to be carefully verified [2]. Recently developed analysis technique of 3T-SA Hanle-effect signals [3] is quite useful, which is based on a decomposition procedure employing analytically obtained signal formulae of channel-spin and trap components. This technique can distinguish the origin of observed signals. i.e., only the intrinsic channel spin component can be extracted.

Recently, we investigated spin injection/ extraction for a Si channel using CoFe/MgO/n$^+$-Si and CoFe/AlO$_x$/n$^+$-Si spin injectors [4]. Hanle-effect

signals originated from spin accumulation in the channel were successfully observed. It was found that the spin injection ratio of the channel spin component to the trap spin component strongly depended on the quality of the tunnel barriers of the injectors.

In this paper, we optimize the fabrication process of the MgO tunnel barrier of CoFe/MgO/n$^+$-Si spin injectors. Spin injection behavior of a high quality CoFe/MgO/n$^+$-Si spin injector is demonstrated.

Experimental procedure

An Al/CoFe/MgO tunnel-contact stack was formed on a highly phosphorous-doped n-type Si substrate (N_D=4×10^{19}cm^{-3}) using a multi-chamber (metal-sputter / radical-oxidation / molecular beam deposition (MBD)) system without breaking ultrahigh vacuum. The ultrathin MgO barrier layer was formed by radical oxidation of an Mg thin film (0.5nm) deposited on the Si substrate at room temperature and then it was annealed under radical-oxygen exposure to improve the film quality. The CoFe electrode (30nm) was deposited on the MgO barrier layer using MBD technique. Spin injection behavior was evaluated employing the 3T-SA method shown in Fig. 1(a).

Results and Discussion

Hanle-effect signals due to the trap-spin and channel-spin components are simultaneously observed by the 3T-SA technique as shown in Fig. 1(b). These signals can be reproduced by a superposition of a Lorentz function S_{tr} (that represents the trap-spin component) and a much sharper non-Lorentz function S_{ch} (that represents the channel-spin component) [3]:

$$\Delta V_{det} = C_1 S_{ch} + C_2 S_{tr}, \qquad (1)$$

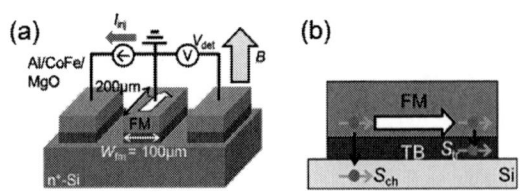

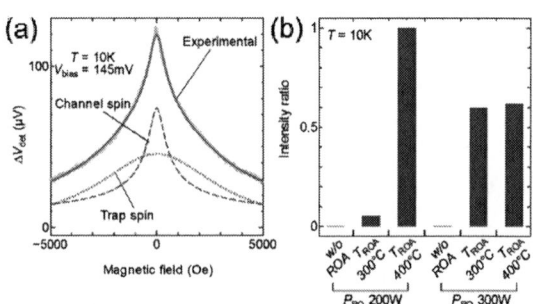

Fig. 1: (a) Schematic of a 3T-SA device. Spin polarized current is injected into the Si channel from the center ferromagnetic contact and detected by the same contact. (b) Spin injection mechanism using tunnel-contact-type spin injector. Channel-spin component (S_{ch}) and trap-spin component (S_{tr}) are simultaneously observed by the 3T-SA technique.

Fig. 2: (a) A typical Hanle-effect signal at 10K, measured using a CoFe/MgO/n$^+$-Si spin injector fabricated with the process condition of P_{RO}=300W and T_{ROA}=400 °C. The fitting curves are also shown in the figure. (b) Intensity ratio of the channel-spin component for various conditions of P_{RO} and T_{ROA}.

978-1-5090-4661-4/17 $31.00 © 2017 IEEE   204

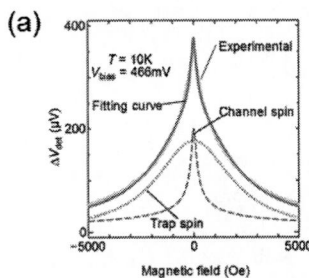 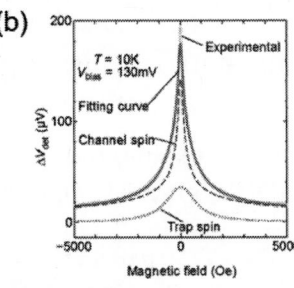 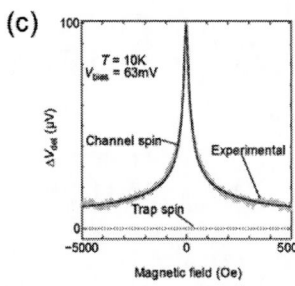

Fig. 3: Hanle-effect signals observed using the 3T-SA device with the high-quality CoFe/MgO/n^+-Si spin injector, measured at 10K with (a) V_{bias} = 466 mV, (b) V_{bias} = 130 mV, and (c) V_{bias} = 63 mV.

in which C_1 and C_2 are constants related to the magnitude of each component. Fig. 2 (a) shows a typical Hanle-effect signal observed using the 3T-SA device. The observed signal was closely fitted by a superposition of S_{ch} and S_{tr}. To evaluate the effect of the radical-oxygen annealing (ROA) treatment on the spin injection mechanism, we defined the intensity ratio of $C_1 / C_1 + C_2$. The channel-spin component was dominated for the condition of P_{RO}=200W and T_{ROA}=400°C, as shown in Fig. 2 (b).

Figs. 3(a)-(c) show bias dependence of the Hanle-effect signals for the CoFe/MgO/n^+-Si spin injector fabricated with the condition of P_{RO}=200W and T_{ROA}=400°C. The trap-spin component in the observed signals decreased with increasing bias voltage V_{bias}. Furthermore, the trap-spin component completely disappeared for V_{bias}≤63mV. The signal was closely fitted by a single channel-spin component, i.e., there was no observation of a trap-spin component for lower bias conditions. This bias-dependent behavior could imply energy-dependent trap density in the barrier. However, the MgO film was too thin to estimate its trap density by conventional capacitance or related method. Therefore, we tried estimating the trap density using the tunnel current characteristics of the spin injector (Fig. 4 (a)). Trap density N_{it} was derived from the difference between ideal and measured tunnel currents, which is caused by trapped charges in the tunnel barrier. Fig. 4 (b) shows bias dependence of estimated N_{it} of the CoFe/MgO/n^+-Si spin injector fabricated with the condition of P_{RO}=200W, T_{ROA}=400°C. The trap density decreased with decreasing bias voltage. This could be the origin of the bias-dependent spin injection phenomenon.

Conclusion

We investigated spin accumulation in a Si channel using 3T-SA devices with a high-quality CoFe/MgO/n^+-Si spin injector fabricated with an optimized condition of radical oxygen annealing. Hanle-effect signals caused by a single channel-spin component were successfully observed near zero biases. The trap density estimated by the tunnel current of this spin injector rapidly decreased with decreasing bias. This result suggests that the energy-dependent trap density could be the origin of the bias-dependent spin injection behavior.

Acknowledgments

The authors acknowledge Research Hub for Advanced Nano Characterization, The University of Tokyo. This study was partly supported by the Ministry of Education, Culture, Sports, Science and Technology (MEXT), Japan.

References

[1] S. Sugahara, IEE Proc. Circuits Devices Syst. **152**, 355, 2005.

[2] M. Tran, et al., Phys. Rev. Lett. **102**, 036601, 2009.

[3] Y. Takamura, et al., J. Appl. Phys. **115**, 17C307, 2014.

[4] T. Akushichi, et al., J. Appl. Phys. **117**, 17B531, 2015.

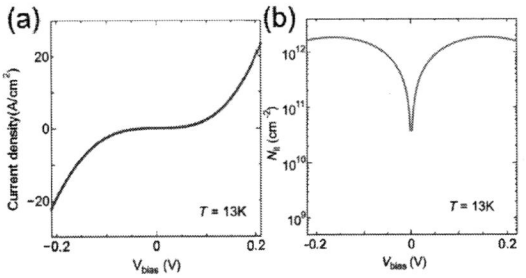

Fig. 4: (a) Current density characteristics of the CoFe/MgO/n^+-Si spin injector. (b) Bias-dependence of the estimated trap density N_{it} at 13K.

SER Scaling and Trends in Planar Submicron Technology Nodes

Krishna Mohan Chavali, Quality Reliability Engineering
GLOBALFOUNDRIES Inc. Malta, New York, USA

Abstract: **It is important to analyze and understand the soft error rate (SER) scaling and trends when selecting a technology node and process for a product that needs to meet robust SER requirements. In this paper the SER scaling and trends normalized to FITs/Mb in planar sub-micron nodes from 90nm to 20nm are analyzed and discussed. This paper discusses and presents the SER scaling based on normalized FITs/Mb. This improves comparability versus published SER trends and scaling based on fail counts.**

Index Terms—Soft Error Rate (SER), Scaling and Trends

I. INTRODUCTION

The evolution of planar sub-micron manufacturing technologies has led to an ever increasing density of SRAM by decreasing size of the individual SRAM cells. In parallel the total size of SRAM test vehicles in terms of Mbit has increased significantly. The size has doubled between each of the shown nodes. This is because of the requirement to cover a meaningful physical area of the tested wafer for technology qualification and because SRAM size in customer products has increased as well. So, this implies that SER trends that compare only fail counts are problematic if the size of the test vehicle is not unambiguously identified. This paper shows scaling trends based on normalized FITs/Mb. This allows a direct comparison for customer and designers to choose which process and node are more suitable for a particular product with given density to meet robust SER requirements.

II. TEST VEHICLES & METHODOLOGY

SER testing follows JESD-89 in terms of methodology for sample size, accelerated testing, and data assessment for both alpha particles and cosmic rays (that is to say high energy neutrons). The results of measurements on SRAM TQVs are compared that have been manufactured with technology nodes from 65nm to 20nm for bulk silicon and from 90nm to 22nm for SOI (silicon on isolator). Die-facing-up package types with de-capped encapsulation on top only are used for alpha particle SER. This exposes the active area of the sample to the particles of the alpha source. Flip chip packages are used for cosmic SER. The plots show the used test voltage on x-axis that

represents nominal voltage for each technology node, respectively. The y-axis shows FITs/Mb with arbitrary scales. All SRAM TQVs were manufactured with at least 6 metal layers. The size of the TQVs ranges from 8Mb to 128Mb.

III. SER TEST RESULTS

First alpha particle SER data on bulk process is plotted across technology nodes from 65nm to 20nm (Fig-1). The scaling across these nodes follows logarithmic trend.

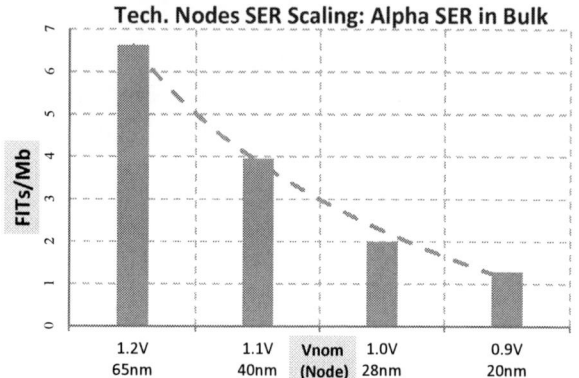

Fig-1: Alpha particle SER trend on bulk processes

Then the Cosmic SER data from 65nm to 20nm from bulk processes is plotted as shown in Fig-2 and the trend observed.

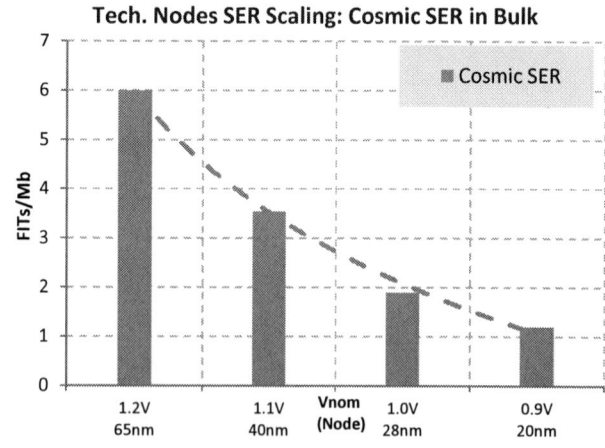

Fig-2 Cosmic SER trend on Bulk Processes

Alpha particle and Cosmic SER data on Bulk processes across planar technology nodes from 65nm to 20nm are shown in Fig-3 for side by side comparison. It can be observed that across the planar technology nodes from 65nm to 20nm the Alpha particle SER is pre-dominantly higher compared to Cosmic SER with in each node.

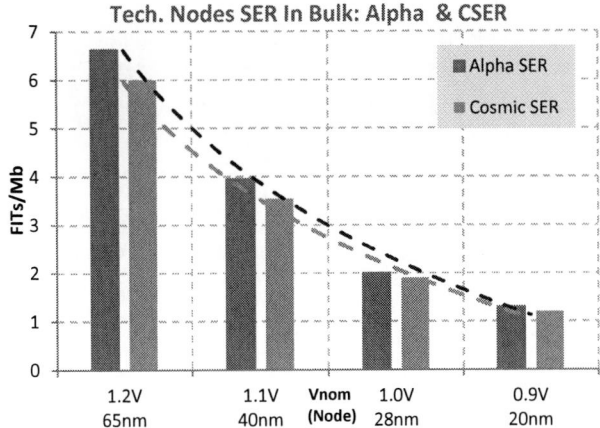

Fig-3 Alpha & Cosmic SER on Bulk side-by-side

The total SER values on bulk process are plotted to observe the trend from same 65nm to 20nm technology nodes as shown in Fig-4, which also follows the logarithmic trend.

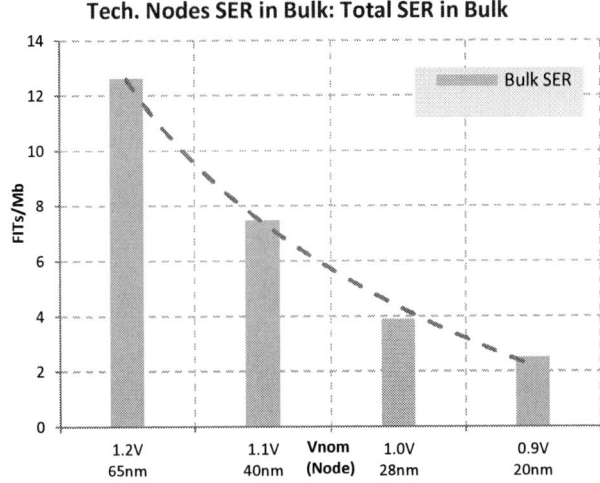

Fig-4 Total SER trend in bulk processes.

It is observed on the SER scaling in bulk that the FITs/Mb are reducing by around 40% from node to node in bulk planar submicron technology nodes. This represents a combination of the influence of several factors: 20% node to node bit-cell scaling, reduction in overall charge collection, and improved process integration techniques that enabled a reduced scaling of operation voltages which in turn reduced the scaling of critical charge for soft errors.

The alpha particle SER data on SOI processes for technology nodes from 90nm to 22nm are plotted for trend and to observe the scaling as depicted in Fig-5.

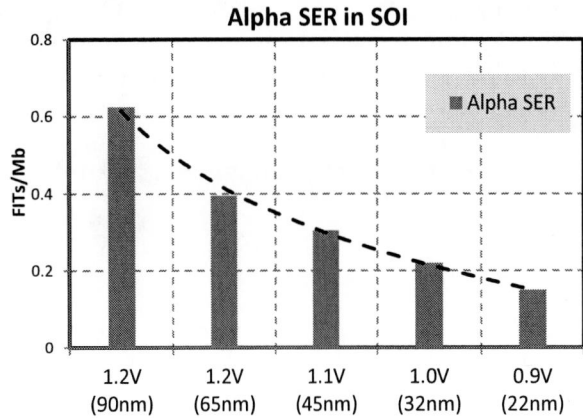

Fig-5: Alpha SER trend on SOI Processes

Then the Cosmic SER data on SOI processes for technology nodes from 90nm to 22nm are plotted for trend and to observe the scaling as furnished in Fig-6.

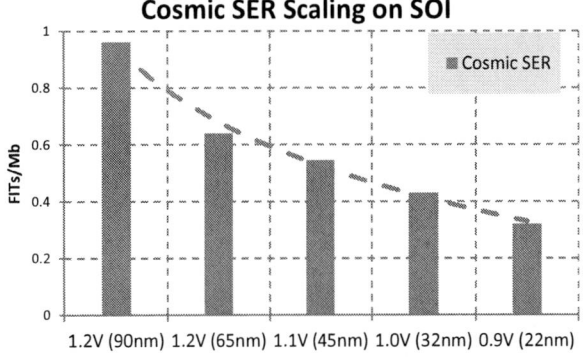

Fig-6: Cosmic SER trend on SOI Processes

Both alpha particle - and Cosmic SER data are plotted together to see which one is pre-dominant in SOI nodes, as shown in Fig-7. The Cosmic SER is the higher contribution factor compared to alpha particle SER for SOI technologies. That is the opposite relation as for bulk technologies.

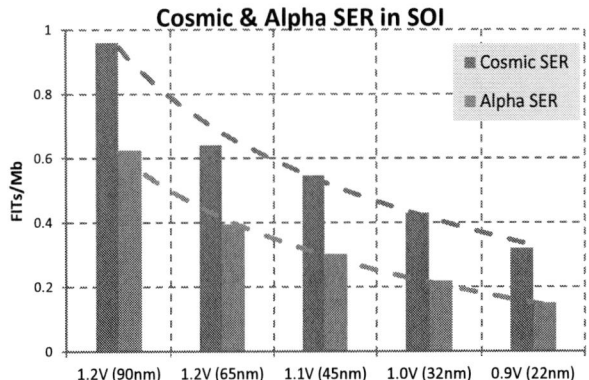

Fig-7: Cosmic and Alpha SER on SOI Processes

Fig-8 provides a side-by-side comparison of SER scaling on Bulk- and SOI-technologies. There is stable ratio of 1/5 to 1/6 between Bulk vs SOI SER across the range of compared technology nodes.

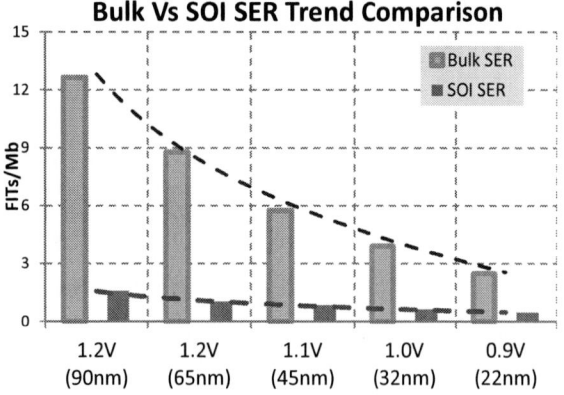

Fig-8: Bulk Vs SOI SER Scaling comparison

Fig-9 shows a different view on the data in Fig-8. The relative change of soft error rate is plotted for Bulk and SOI. The reduction of soft error rate follows a similar trend for Bulk and SOI manufacturing technologies, respectively.

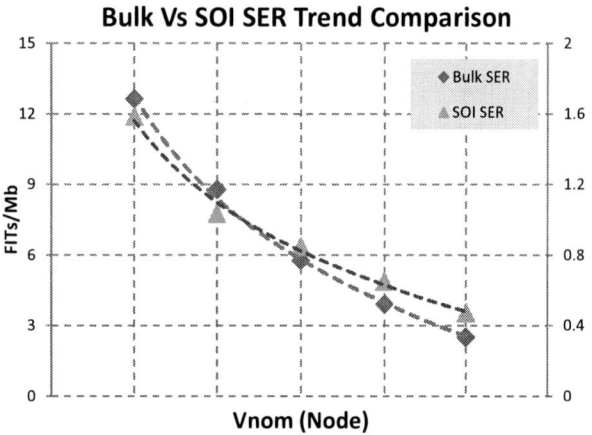

Fig-9: Bulk Vs SOI SER trend comparison

IV. DISCUSSION

The SER scaling on both Bulk and SOI processes across several technology nodes follows a Logarithmic downward trend with decreasing technology feature size. It is the expected behavior from one node to the next, as the charge collection and diffusion areas reduces by @ 20%. This will directly impact the charge collection. The reduced area and the increased accumulation time are making the cell less vulnerable (in terms of Qcrit) compared to earlier node. The opposing factor is the reduction of nominal voltage from a node to its next. That will make the Qcrit smaller and thus more sensitive. But this is only a 10% increase versus the bigger reduction in the charge collection area effect on Qcrit.

The SER is also influenced by the composition of the metallization layers. Alpha particles lose energy, get deflected or get even stopped while they penetrate to the active cells. Thus the number of metal layers, the density of autofill patterns, different materials for inter-layer and inter-metallic materials, and the thickness of passivation layer all will contribute to the effective soft error sensitivity. This is naturally of much reduced significance in the case of high energy cosmic irradiation.

SER on SOI in each technology node still shows much lower FITs/Mb to its corresponding Bulk process, mainly due to the well-known fact that in SOI the incident radiation funnel will be truncated by the insulation layer underneath.

V. CONCLUSION

Compared to existing trends based on fail counts vs test voltages across technology nodes which are not normalized to FITs/Mb, the trends based on normalized FITs/Mb also shows same expected Logarithmic Trends both on Bulk and SOI process.

VI. ACKNOWLEDGEMENT

The author would like to thank GLOBALFOUNDRIES Product Reliability team members at both Dresden and Singapore sites for their support in relentlessly collect the data across technology nodes spanning over past decade. Author would also to thank the paper reviewers for their thoughtful questions and recommendations to improve the quality of the paper and also acknowledge the full and endless encouragement from GLOBALFOUNDRIES management.

REFERENCES

[1] Alpha Particle induced Single-Event Error Rates and Scaling Trends in Commercial SRAM Cells I. Chatterjee, B. L. Bhuva, S.-J. Wen, R. Wong

[2] Modeling the Effect of Technology Trends on the Soft Error Rate of Combinational Logic Premkishore Shivakumar et. all, Proceedings of the 2002 International Conference on Dependable Systems and Networks

[3] JESD89A, Measurement and Reporting of Alpha Particle and Terrestrial Cosmic Ray-Induced Soft Errors in Semiconductor Devices.

Gap in pagination due to unavailable paper.

Pages 209-210

A Metal Micro-Casting Method for Through-Silicon Via(TSV) Fabrication

Jiebin Gu[1], Bingjie Liu[1], Heng Yang[1], Xinxin Li[1],

[1]Shanghai Institute of Microsystem and Information Technology, Shanghai, China, j.gu@mail.sim.ac.cn

Abstract

TSV is an important interconnection for advanced packaging. Via-filling by molten alloy is considered as an alternative to electroplating process. In this paper, a micro-casting method for TSV fabrication is proposed and demonstrated. This TSV fabrication method owns advantages of fast-filling and lost-cost. (Keywords: Packaging, TSV)

Introduction

On many TSV applications, relative large TSVs are needed, e.g. in MEMS Wafer level packaging or 2.5D Silicon/Glass interposer. Currently, electroplating is still the only solution for via-filling of these relatively large TSVs. However electroplating suffers from low deposition rate. Via-filling by molten alloy is continually studied as a fast and uncostly alternative method [1]. Previous we've developed an alloy via-filling method that utilizes a combinative effect of surface tension[2]. This TSV technology has also been successfully demonstrated for MEMS packaging [3]. Because of volume shrinkage during solidification, this method however is either limited to specific alloy or results dents on via surface [2]. To solve this problem, here we present a metal micro-casting method for via-filling, which is a further development of this method.

Mechanism

A. Process

The process of the micro-casting via-filling method is illustrated in Fig.1. The TSV wafer to fill is sandwiched by a cap wafer and a nozzle wafer, which is the most critical part of the method. As shown in Fig.2a, in the center of nozzle wafer is a through-wafer hole, which is used to feed molten alloy underneath the wafer. The surface of the nozzle wafer is etched with a cavity, which is used to compensate the volume shrinkage during molten alloy solidification. In addition, the most important structure is the grooves between the cavity and the through-wafer hole. The grooves sever as capillaries to form capillary bridges. The wafer sandwich is placed on top of a molten alloy surface(Fig.1a). When the pressure the liquid molten alloy is increased high enough, molten alloy can flow into the micro-mold formed by the compensation cavity and the via-holes(Fig.1b). In the groves, molten alloy forms capillary bridges. The pressure of molten alloy is then slowly decreased, when it reaches to the rupture pressure of the capillary bridge, molten alloy in the

grooves rupture(Fig.1c). After rupture, molten alloy is solidified by naturally cooling(Fig.1d), then the nozzle wafer and the cap wafer is released from the TSV wafer(Fig.1e). The residual on the compensation side of the TSV wafer is finally removed by CMP process(Fig.1f).

B. Theory

As we can see, capillary bridge rupture in groove is the most critical step. In order to make breakage happen, the rupture pressure of capillary bridge($P_{rupture}$) should meet the following condition[2]:

$$P_{rupture} > P_{via} \& P_{rupture} > P_{cavity}$$

Where P_{via} and P_{cavity} is the capillary pressure of vias and the cavity, respectively, as shown in Fig.2b.

Experiments & Results

A. Experiments

The experiment is carried out on die-level. The nozzle die is fabricated by KOH etching and laser drilling. The experiment is carried out on a self-made specific equipment, as shown in Fig.3a.

B. Results

Fig.3b-c shows the TSV of blind holes. As we can see, all the vias are fully filled without any void.

Conclusion

Here we proposed and demonstrated a micro-casting method for TSV fabrication. As an alternative to electroplating, the method has the potential for industrial applications.

Acknowledgments

The research is supported by National Natural Science Foundation of China (NSFC 61504158).

References

[1] Y.K. JEE, J. Y., K. W. PARK, T.S. OH (2009). "Zinc and Tin-Zinc Via-Filling for the Formation of Through-Silicon Vias in a System-in-Package." Journal of Electronic Materials 38(5).

[2] Jiebin Gu, Bingjie Liu, Gaoli Chen and Xinxin Li "Study of a through-silicon/substrate via filling method based on the combinative effect of capillary action and liquid bridge rupture." JMM 26(7): 075009, 2016.

[3] Jiebin Gu., Bingjie Liu, Heng Yang and Xinxin Li (2016). A fast and CMP-free TSV process based on wafer-level liquid-metal injection for MEMS packaging. 2016 IEEE 29th International Conference on Micro Electro Mechanical Systems (MEMS).

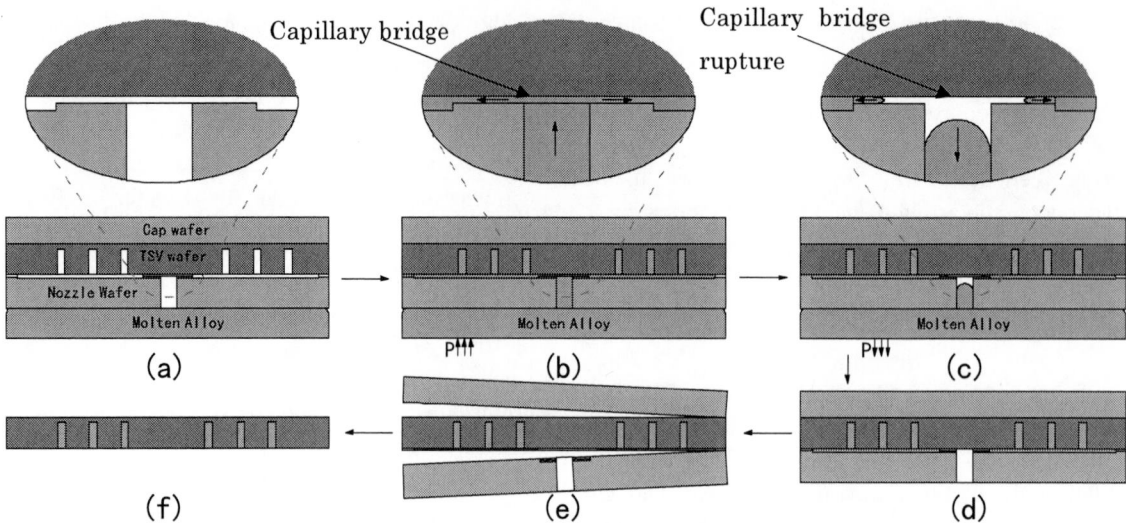

Fig. 1: The process flow of the micro-casting via-filling method. (a)a cap wafer, via wafer and nozzle wafer is sandwiched to formed a micro-mold. (b)injection the micro-mold by increasing pressure of molten alloy. (c) cut off the molten alloy in groove by decreasing pressure of molten alloy. (d)molten alloy is solidified by cooling. (e)release the cap and the nozzle wafer. (f)backside CMP of filled wafer.

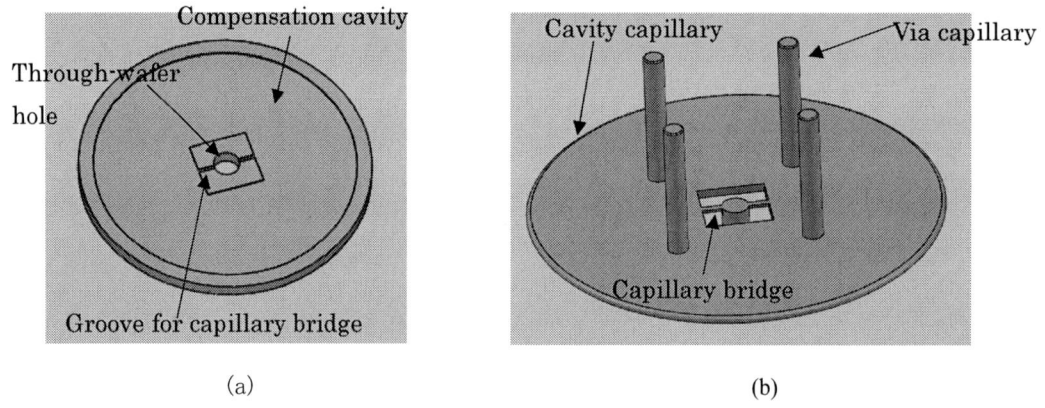

(a)   (b)

Fig. 2: (a)Schematic of nozzle wafer (not in scale). (b) The form of molten alloy in the micro-mold.

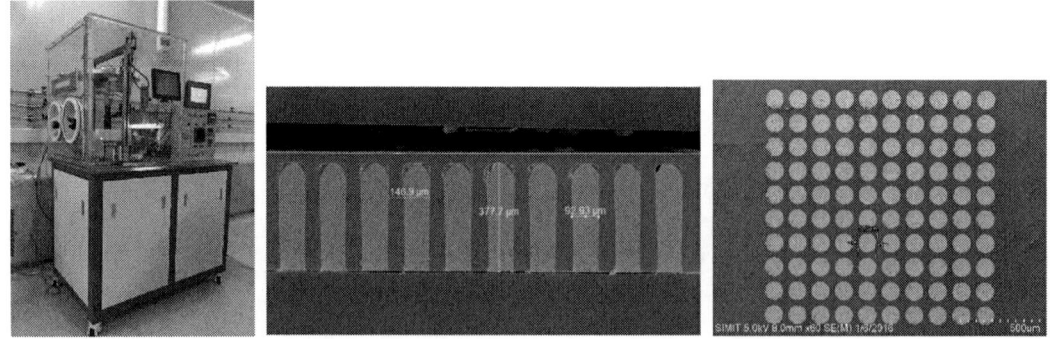

Fig. 3: (a) The specific equipment made for alloy TSV filling. (b) Blind TSV array (Φ100μm) after filling. (c)Top view of the blind TSV array.

978-1-5090-4661-4/17 $31.00 © 2017 IEEE   212

Acoustic Emission Wave Sensor with Thermally Controllable Force-Enhancement Mechanism for Acoustic Emission Source Detection and Biomedical Application

Guo-Hua Feng [1,2] and Wei-Ming Chen [1]

[1] Department of Mechanical Engineering, National Chung Cheng University, Taiwan, imeghf@ccu.edu.tw

[2] Advanced Institute of Manufacturing with High-Tech Innovations, National Chung Cheng University, Taiwan

Abstract

This paper presents a card-type acoustic emission wave (AE) sensor with symmetrically four-beam structures and thermal actuators providing normal enhanced forces to increase signal-to-noise ratio (SNR). The enhanced force applied to the sensing beam can be observable by thermal imager. The location of AE source can be detected by this developed single sensor, greatly improving the conventional method using multiple AE sensors and large installation area. Distinguishing the health and injured elbow joints is also demonstrated.

(Keywords: Sensor, acoustic emission, piezoelectric)

Introduction

Acoustic emissions can be considered as pressure waves produced by transient release of energy while the material is subjected to mechanical, thermal effects causing deformations or changes in atomic arrangement [1]. Acoustic emission testing has become a well-known method applied to detect and locate faults in mechanically loaded structures and components because of being able to provide information on the origination or development of a flaw [2]. Commercial AE sensors are commonly manufactured with rigid sensing surfaces and their contact forces to the target are determined once installed. In this work, we create the advanced fabrication process of piezoelectric film and thermal actuator for the AE sensor (Fig. 1). The double hydrothermal PZT film growth causes a dense film besides thickness increase. The film possesses higher piezoelectric constant and better lattice arrangement. A new composition of the polymer-based thermal actuator is discovered with faster response time for electrical actuation. The crucial fabrication process and testing results of the developed sensor is reported.

Device Fabrication

Four-beam and cross structures were formed by micromachining a 100 µm-thick titanium sheet (Fig. 2). A double hydrothermal process to grow PZT film was followed. The precursor solution for hydrothermal growth was prepared with TiO_2, $Pb(C_2H_3O_2)_2$, $ZrOCl_2$, KOH and DI water [3, 4]. Then the processed titanium chip was placed into an autoclave filled with precursor solution. The first PZT growth was processed 36 hr at 180 °C. After cooling to room temperature, the second PZT growth was performed 36 hr again at 190 °C with new precursor solution. The electrode was formed by sputtering 1µm-thick silver with a patterned shadow mask. After that, fabrication of SU-8 micropillars using photolithography was followed. The 1-mm-diameter micropillars were glued to the front ends of four piezoelectric beams to complete the AE sensing part.

The polymer-based thermal actuator started with Poly(N-isopropylacrylamide) (PNIPA) gel fabrication. PNIPA powder was mixed with cross-linking agent EGDMA ($C_{10}H_{14}O_4$), alcohol, DI water and cooled down for 15 hr at 4 °C to form the gel. Different from the previous report using a constant ratio of EGDMA to PNIPA [5], we fixed the amount of PNIPA as 10 g and vary the amount of EGDMA by 0.05, 0.1 and 0.2 g. The mixed alcohol and DI water were set at 30 and 10 mg, respectively. Then the catalyst TEMED 50 mg and initiator APS (10%) 30 mg were added to harden PNIPA gel for each case. A PDMS-micromold was made to shape the gel as a designed thermal actuator. It is followed by sputtering a silver thin-film as an electrode on the top surface of processed device. After two wires were glued to the electrode pads, the thermal actuator was complete for testing. Finally, the thermal actuator and AE sensing element were integrated with epoxy and packaged with an adhesive silicone pad (Fig. 2).

Experimental Results

The SEM and XRD results show the high quality of PZT film by a double hydrothermal process (Fig. 3). About 15 µm thick PZT films were grown on both sides of the patterned titanium substrate.

To measure the generated force of the thermal activated sensing beam at the micropillar, a constant electrical power setting was provided with the dc power supply. During characterization, the micropillar directly pushed a copper cantilever beam of 20mm x 2mm x 0.1mm. A laser displacement meter was used to measure the deformation of the copper beam (Fig. 4). According to Newton's law of motion, the micropillar applied force could be calculated using the measured deformation of copper beam based on the beam deflection formula. Results show the force can reach 160 μN with a 1.5 W heating process after 60 s. In addition, the force releasing time is approximately 60 s due to cooling effect at room temperature environment.

The PNIPA gels with three different amount of EDGMA mixing with PNIPA powders were tested. Experimental results revealed that the maximum displacement ratios are 1:1:1.7 for the cases with EDGMA 0.05, 0.1 and 0.2 g, respectively, under the same actuation time and power. This indicates the thermal actuator performance can be significantly improved by properly increasing the content of cross-linking agent.

Figure 5 shows the temperature response of thermal actuator acquired from thermal imager. The temperature curves from the recorded images displayed a standard transient response of a linear first-order system. This trend is highly consistent with the applied force, allowing us to monitor applied force by observing thermal images after appropriately calibration.

The experiment for locating AE source was performed (Fig. 6). The experimental setup was as follows. The fabricated AE sensor was anchored on a 10 mm thick acrylic plate, which has a sound velocity of 2730 m/s. A tested acoustic emission wave was produced about 1 m away from the sensor by a standard pencil-lead breakage method. Three detection points (at micropillars) of the AE sensor touched the acrylic plate and picked the transmitted AE signals. The high-speed data acquisition card was used to simultaneously obtain the digital signals for analysis. By examining the starting points of the received waves, the time differences between each other could be derived. Setting the origin of AE wave was x away from the sensing point 1, then the distance between the origin and point 2, 3 was approximately x+2.24 mm, x+4.30 mm, respectively. Moreover, according to trigonometric formula, the x value and orientation of the AE source could be determined.

Additionally, the developed AE sensor has much better SNR compared to the commercial sensor (Fig. 7). The fabricated and commercial sensors were installed on an aluminum plate with equal distances away from the AE source. Both detected signals without amplification had similar trends. However, the signal of developed sensor was two orders of magnitude greater than that of commercial sensor, causing a good signal-to-noise ratio performance.

Finally, to show the potential of the fabricated AE sensor for biomedical application, an experiment for inspecting the health and injured elbows was conducted. The healthy pig elbows were intentionally performed excess excise through the elbow-connected upper and lower limbs executing a repeated large-scaled back-and-forth motion. This caused the obviously elbow injuries as shown in the X-ray images (Fig. 8). Results of AE signals indicated the trend of characteristic frequency drop and the consistency for repeated testing.

Conclusion

We successfully demonstrated the novel fabricated device possessing better AE detection performance than the commercial sensor. Also, the potential for biomedical application in monitoring the joint health condition is verified.

Acknowledgments

The authors would like to thank the Ministry of Science and Technology in Taiwan for financial supports under contract no. 104-2221-E-194-006-MY2 and 104-2221-E-194-005-MY3.

References

[1] A. K. Rao, "Acoustic Emission and Signal Analysis," Defense science journal, Vol. 40, No. 1, p. 55 (1990).

[2] L.-K. Shark, H. Chena, J. Goodacreb, "Knee acoustic emission: a potential biomarker for quantitative assessment of joint ageing and degeneration," Medical Engineering & Physics, Vol. 33, p. 534 (2011).

[3] J. Liang, J., H. Zhang, H., D. Zhang, H. Zhang, W. Pang, "Lamb Wave AlN Micromechanical Filters Integrated With On-chip Capacitors for RF Front-End Architectures," IEEE Journal of the Electron Devices Society, Vol. 3, No. 4, p. 361-364 (2015).

[4] Gh.H. Khorrami, A. K. Zak, A. Kompany, and R. Yousifi, "Optical and structural properties of X-doped (X = Mn, Mg, and Zn) PZT nanoparticles by Kramers–Kronig and size strain plot methods," Ceramics International, Vol. 38, p. 5683 (2012).

[5] G. H. Feng, W. M. Chen, "Piezoelectric-film-based acoustic emission sensor array with thermoactuator for monitoring knee joint conditions," Sensors and Actuators A: Physical, Vol. 246, p. 180-191 (2016).

978-1-5090-4661-4/17 $31.00 © 2017 IEEE

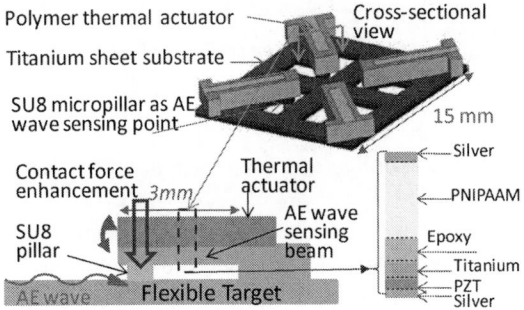

Fig. 1: Conceptual diagram and working principle of the developed card-type AE sensor.

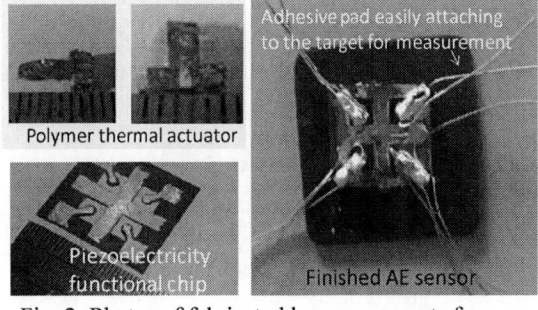

Fig. 2: Photos of fabricated key components for integrated AE sensor.

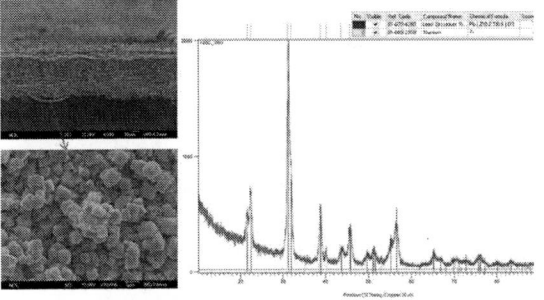

Fig. 3: SEM and XRD results confirm the better quality of PZT film using "double hydrothermal growth".

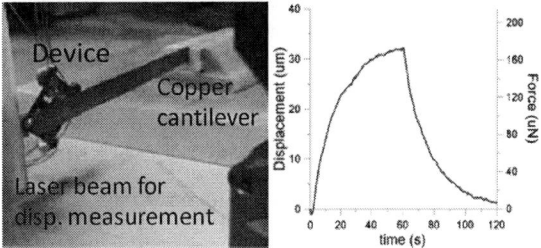

Fig. 4: Experimental setup of acquiring the force applied by the thermal actuator. The driving condition is set as 1A (1.5 V dc) (Left). Results of displacement and applied force (Right).

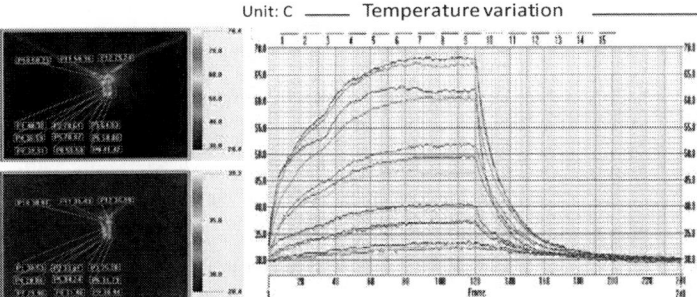

Fig. 5: Result of temperature distribution images of thermal actuator driven by 1A (1.5V dc) at 60 s (1 frame=0.5 s). The dynamic response of temperature variation at selected points.

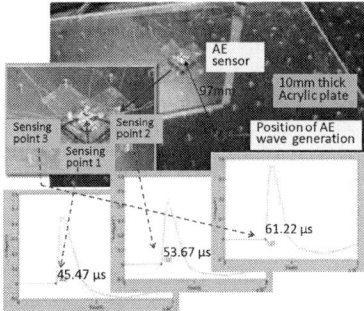

Fig. 6: Experiment confirm detecting AE source by 3 sensing points of a single sensor.

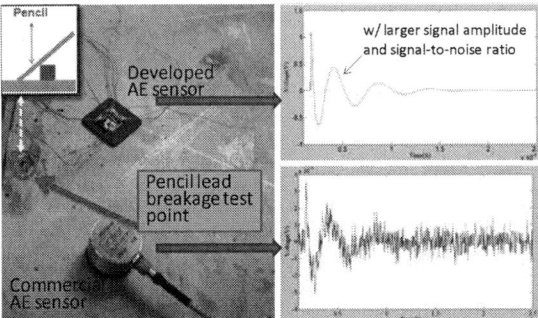

Fig.7: AE signal comparison of fabricated and commercial sensors.

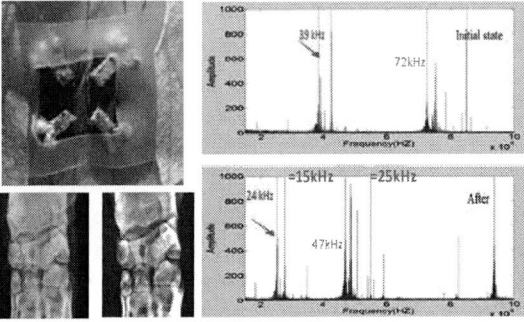

Fig. 8: Demonstrating the fabricated AE sensor for biomedical application.

978-1-5090-4661-4/17 $31.00 © 2017 IEEE

Electroactive Polymer Actuated Gripper Enhanced with Iron Oxide Nanoparticles and Water Supply Mechanism for Millimeter-Sized Fish Roe Manipulation

Guo-Hua Feng [1,2] and Shih-Chieh Yen [1]

[1] Department of Mechanical Engineering, National Chung Cheng University, Taiwan, imeghf@ccu.edu.tw

[2] Advanced Institute of Manufacturing with High-Tech Innovations, National Chung Cheng University, Taiwan

Abstract

This paper presents an electroactive polymer (EAP) actuator controlled scissor-type gripper (Fig. 1). The EAP actuator is made of ion polymer metal composite (IPMC). By doping Fe_2O_3 nanoparticles into IPMC, the actuator exhibits larger force output and 2-fold increase of maximum deformation compared to the pure IPMC actuator. Using water retaining poly(N-isopropylacrylamide) (PNIPAAM) made water supply mechanism results in a 10-fold increase of actuation capability. Using the gripper to manipulate the millimeter-sized fish roe is verified. (Keywords: Actuator, electroactive, nanoparticle)

Introduction

EAP actuators exhibit outstanding potentials to handle micro-objects due to their greater flexibility compared to common metal or glass devices [1, 2]. Especially, the EAP devices made of IPMC materials exhibit large dynamic deformation under low electric fields have been broadly studies as soft robotic actuators [3, 4]. We employ the lever principle to convert the actuator's displacement to a larger force output at the front tips of gripper [5]. In addition, a crucial approach to enhance the mechanical property of EAP actuator by doping nanoparticles into IPMC is presented. The constructed water supply mechanism provides proper amount of water to the EAP actuator, which significantly increases the lifetime of actuator working in air.

Device Fabrication

We first fabricated the open ring sockets for hinge joints (Fig. 1). The photoresist SU-8 made sockets were fabricated with photolithography. The IPMC actuator was implemented utilizing a PDMS micromold. The fabricated sockets were aligned and placed in the edge portion of the micromold grooves. Fe_2O_3 nanoparticles were added to liquid Nafion with a weight ratio 1:1000. After ultrasonic mixing, the doped Nafion was applied to fill grooves to cover the protrusions of the socket devices. After solidification, the shaped Nafion devices were trimmed and applied photoresist to define the inactive region of the actuator for subsequent electrode plating. Chemical processing was executed to plate platinum electrodes [6]. Ion exchange process was followed to complete the fabrication of IPMC actuator.

The two links of scissor-like gripper were micromachined with SU-8. The complementary links were connected through a pin-hole joint to form the gripper, which was then fitted to the socket of the IPMC actuator. A water retaining cap made of PNIPAAM was integrated with a Tygon tube as the water supply mechanism for the actuator. The PNIPAAM cap was fabricated by mixing PNIPA powder with cross-linking agent (EDGMA), alcohol and DI water to form the PNIPAAM gel. Subsequently, proper amount of APS and TEMED chemical solutions were added to solidify the PNIPAAM gel as designed shape [7]. Fabrication results are shown in Fig. 2.

Experimental Results

To drive the EAP actuator with electrical signals, a novel home-made gadget modified from a probe retractable hook was used. Two laser displacement meters were employed for motion characterization by simultaneously shooting two laser beams to the individual front ends of gripper. The water supply mechanism contacted the center portion of IPMC actuator during operation (Fig. 3). This mechanism significantly reduced the rising temperature of actuator during operation, which effectively increases the lifetime of actuator (Fig. 4). Figure 5 shows the device using the water supply mechanism could last 300 s continuous operation without displacement decrease.

In addition, the micro tensile/compressive test machine was utilized to find the force-displacement relation of the IPMC actuator with different compositions. Experimental results show the Fe_2O_3 doped actuator possesses better stiffness and larger deformation compared to that without doping (Fig. 6). Figure 7 indicates the gripper of doped actuator using water supply greatly improved the working lifetime and force output compared to the gripper of pure IPMC.

Moreover, to show the potential application of the developed device, the manipulation of millimeter-sized soft bio-particles was performed. The gripper was driven with an 8V dc to apply different force levels to hold, squeeze, and crush a single fish egg with a diameter of approximately 1

978-1-5090-4661-4/17 $31.00 © 2017 IEEE

mm (Fig. 8).

Conclusion

A novel EAP actuated gripper was successfully fabricated and tested. Improved force output and superior operation lifetime were achieved compared to conventional IPMC device. Useful manipulation of biological particles by the developed gripper with electrical control is demonstrated.

Acknowledgments

The authors would like to thank the Ministry of Science and Technology in Taiwan for financial supports under contract no. 104-2221-E-194-006-MY2 and 104-2221-E-194-005-MY3.

References

[1] M. Shahinpoor, "Ionic Polymer Metal Composites (IMPCs): Smart Multi-Functional Materials and Artificial Muscles," Volume 2. Royal Society of Chemistry, ISBN: 978-1-78262-077-8 (2016).

[2] S. J. Mazlouman, A. Mahanfar, M. Soleimani, H. Chan, C. Menon, R. G. Vaughan, "Pattern Reconfiguration by Rotating Parasitic Structure Using Electro-Active Polymer (EAP) Actuator," IEEE Transactions on Antennas and Propagation, Vol. 62, No. 3, 1046-1055 (2014).

[3] Y. Bahramzadeh and M. Shahinpoor, "A Review of Ionic Polymeric Soft Actuators and Sensors", Int. Journal of Soft Robotics, Vol. 1, No. 1, pp. 38-52 (2013).

[4] M. Annabestani, M. Maymandi-Nejad, N. Naghavi, "Restraining IPMC Back Relaxation in Large Bending Displacements: Applying Non-Feedback Local Gaussian Disturbance by Patterned Electrodes," IEEE Transactions on Electron Devices, Vol. 63, No.4 , p. 1689-1695. (2016).

[5] G. H. Feng, S. C. Yen, "Arch-Shaped Ionic Polymer–Metal Composite Actuator Integratable With Micromachined Functional Tools for Micromanipulation," IEEE Sensors Journal, Vol. 16, No.19, p. 7109-7115 (2016).

[6] M. Shahinpoor, K. J. Kim, "Ionic polymer-metal composites: I. Fundamentals," Smart materials and structures, Vol. 10, No. 4, p. 819 (2001).

[7] G. H. Feng, W. M. Chen, "Micromachined lead zirconium titanate thin-film-cantilever-based acoustic emission sensor with poly (N-isopropylacrylamide) actuator for increasing contact pressure," Smart Materials and Structures, Vol. 25, No. 5, p. 055046 (2016).

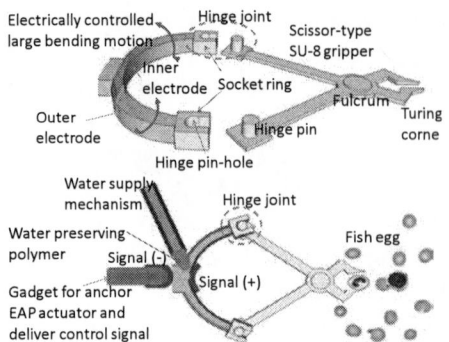

Fig. 1: Schematic diagram and working principle of developed EAP actuated gripper.

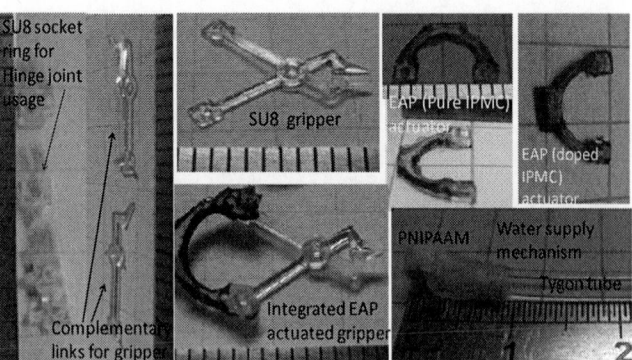

Fig. 2: Photographs of fabrication results.

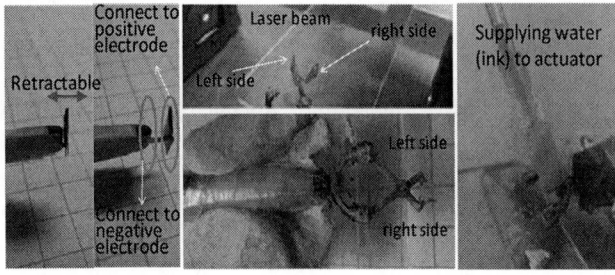

Fig. 3: Experimental setup for driving the developed gripper and measure the gripper tips displacement by laser displacement meters.

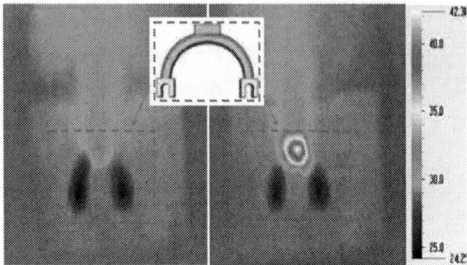

Fig. 4: Thermal images for (left) with (right) without water supply mechanism while the IPMC actuator is driven by 8V dc for 10 sec.

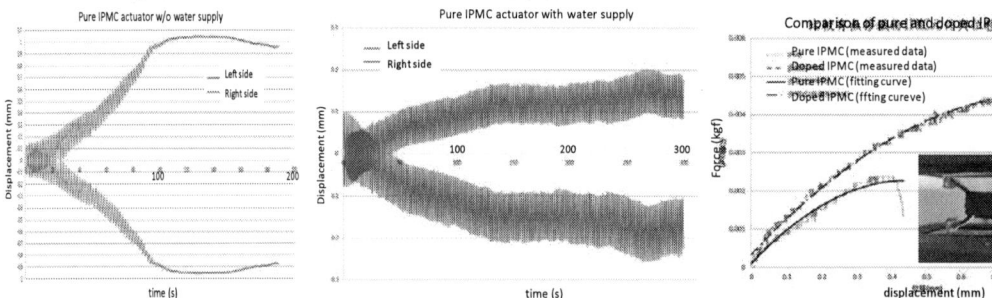

Fig. 5: Comparison of measured displacements for the Pure-IPMC based grippes driven with 0.5Hz, 50% duty cycle square waves of 8V (left) w/o (right) with water supply mechanism usage.

Fig. 6: Relation between force output and displacement for pure and doped U-shaped IPMC actuator.

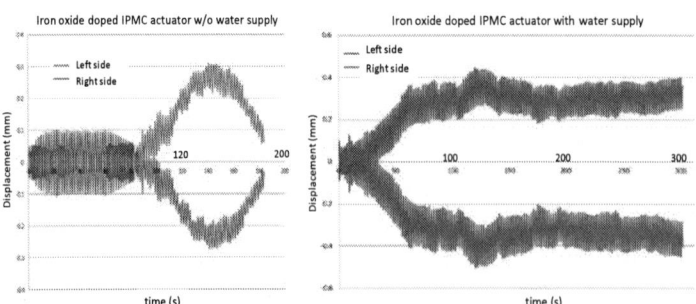

Fig. 7: Comparison of measured displacements for the gripper of iron oxide doped IPMC actuator using the same driving signals as Fig. 5.

Fig. 8: Images of micromanipulation acquired from the video file.

978-1-5090-4661-4/17 $31.00 © 2017 IEEE

P-21

Low Temperature Hermetic Sealing by Aluminum Thermocompression Bonding using Tin Intermediate Layer

Shiro Satoh[1], Hideyuki Fukushi[1], Masayoshi Esashi[1] and Shuji Tanaka[1,2]

[1]Micro System Integration center, Tohoku University, Sendai, JAPAN, shiro@mems.mech.tohoku.ac.jp

[2]Graduate School of Engineering, Tohoku University, Sendai, JAPAN

Abstract

Aluminum with Sn intermediate layer shows very large deformation even below 400°C. Using this new layer structure as sealing metal, high yield hermetic package of MEMS was demonstrated at only 370°C without any treatment of surface oxide removal. During bonding, the bonding metal is significantly pressed (the reduction rate of thickness ~90%), which guarantees hermeticity at high yield. Based on SEM, EDX and TEM analysis, the role of tin for hermetic sealing and the mechanism of softening of this layer structure were discussed.

Introduction

The chip size of MEMS is continuously shrinking. Such a dramatic size reduction has been achieved by advanced wafer-level packaging and MEMS-ASIC integration, where metal-based bonding is a key technology due to the much smaller sealing width and also has to be achieved in allowable temperature of back end process(<400°C). Al is common metal in CMOS process and not expensive. Therefore, low temperature (<400°C) Al-Al hermetic wafer bonding is strongly required in industry. Al-based wafer bonding has been reported [1][2]. However bonding was achieved over 450 °C because of strong surface oxide of Al. Satoh *et al.,*[3] reported high-yield hermetic Al-Al bonding at 370~380°C using thin Sn layer as an antioxidation covering layer on Al. They also observed very thin bonding interface metal in hermetic samples. Its thickness before bonding (4.4μm) reached 0.5~0.6μm after bonding. Therefore, it is necessary to clarify the roles of Sn and of the thinning interface metal to achieve hermetic sealing.

Experimental Procedure and results

We investigated hermetic sealing using Al exposed in air before bonding with and without Sn. Figure 1 shows the structure of samples. Three kinds of bonding metal layer were prepared, 1) Al, 2) Al with intermediate Sn layer shown in Fig. 1(c), 3) Al with 1.8at%Mg. In addition, we also investigated sealed interface thickness of similar samples reported by Satoh *et al.,* shown in Fig. 1(d) [3]. One substrate has 9~16 sealing frames of 30μm width and 20μm height, and the other 9~16 diaphragms of 800μm diameter and 8~10μm thickness to confirm the hermeticity of sealed cavity. These substrates were

bonded in vacuum at different pressures (40~82MPa) and temperatures (360~395°C). The reasons for Sn and Mg addition are follows. Sn has a low-melting-point at 232 °C, 97.6at%Sn-Al eutectic at 228°C, non-intermetallic compound and non-solid solution with Al at room temperature (Fig. 2) [4], and the vapor pressure is as low as 10^{-12}Pa [5] even at 400°C. Mg is reported to dissolves Al-oxide over 450°C [6].

The diaphragm deformation was evaluated by white light interferometer and probe type step profiler. Figure 3 shows the results of Al-Al bonding. Every shape of diaphragm was convex, so, hermetic sealing was not achieved. For Al layer with Mg, hermetic sealing was not also achieved. Figure 4

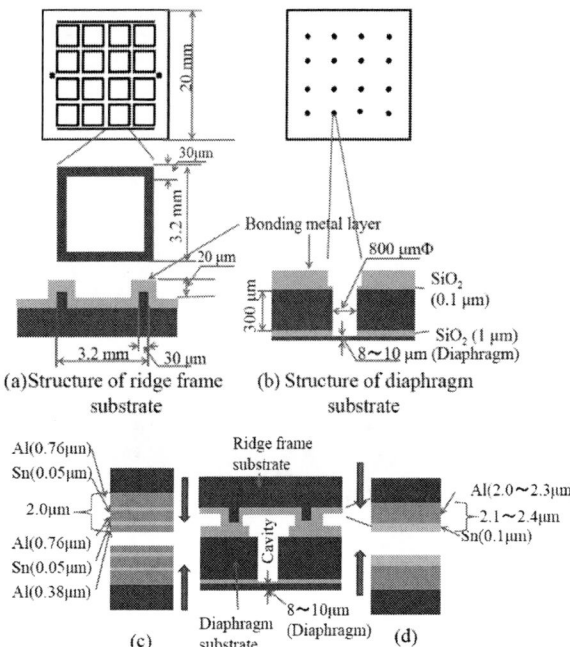

Fig. 1: Structure of bonding sample and metal layer.

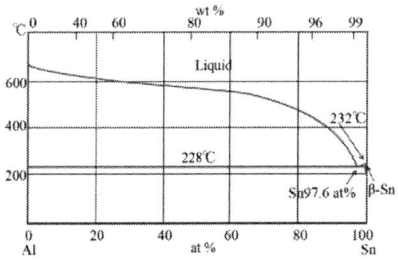

Fig. 2 : Al-Sn phase diaphragm.

978-1-5090-4661-4/17 $31.00 © 2017 IEEE   219

shows the results of Al-Al bonding with inserting Sn intermediate layer. 15 out of 16 diaphragms maintained same concave deformation over 1000 h demonstrating the hermeticity of cavities. Stable Al-Al hermetic sealing was achieved by Al with intermediate Sn layer.

We investigated the thickness change of interface metal before and after bonding by SEM and EDX analysis. We represent the thickness change as "Reduction rate", which is defined as follows :

"Reduction rate (%) = [(Total thickness of bonding layer before bonding – thickness after bonding) / Total thickness of bonding layer before bonding] ×100". Figure 5 shows the cross sections of hermetic sealed interface. Reduction rate of interface metal reached 87~89% for hermetic samples, whereas 63~69% for non-hermetic samples as shown in Fig. 6. We also observed the squeezed Al out of bonding interface at the both sides of ridge

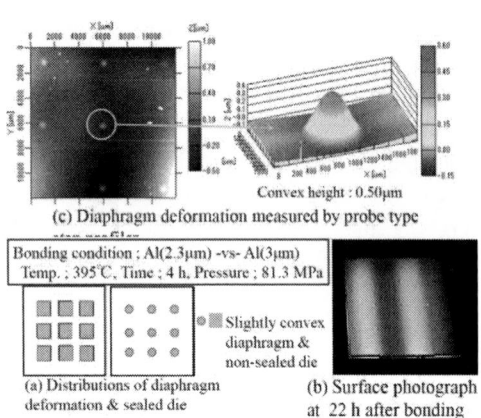

(c) Diaphragm deformation measured by probe type

Bonding condition ; Al(2.3μm) -vs- Al(3μm)
Temp. ; 395℃, Time ; 4 h, Pressure ; 81.3 MPa

● Slightly convex diaphragm & non-sealed die

(a) Distributions of diaphragm deformation & sealed die

(b) Surface photograph at 22 h after bonding

Fig. 3: Diaphragm deformation of Al to Al bonding.

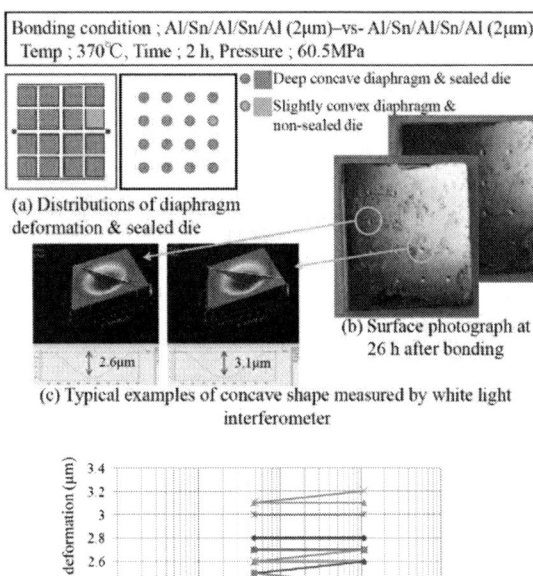

Bonding condition ; Al/Sn/Al/Sn/Al (2μm)–vs- Al/Sn/Al/Sn/Al (2μm)
Temp ; 370℃, Time ; 2 h, Pressure ; 60.5MPa

● Deep concave diaphragm & sealed die
● Slightly convex diaphragm & non-sealed die

(a) Distributions of diaphragm deformation & sealed die

(b) Surface photograph at 26 h after bonding

2.6μm   3.1μm

(c) Typical examples of concave shape measured by white light interferometer

(d) Time dependence of diaphragm deformation

Fig. 4 : Diaphragm deformation of Al/Sn/Al/Sn/Al to

Al/Sn/Al/Sn/Al bonding

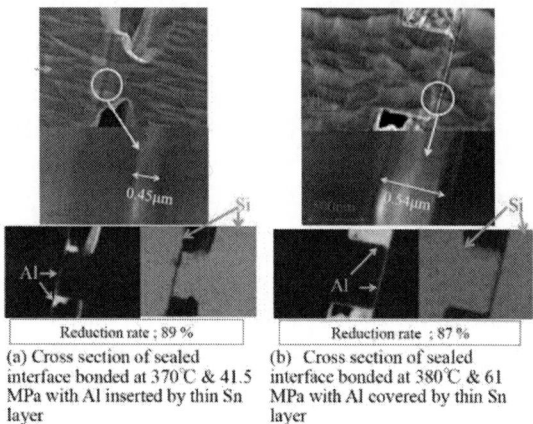

Reduction rate ; 89 %   Reduction rate ; 87 %

(a) Cross section of sealed interface bonded at 370℃ & 41.5 MPa with Al inserted by thin Sn layer

(b) Cross section of sealed interface bonded at 380℃ & 61 MPa with Al covered by thin Sn layer

Fig. 5: SEM and EDX analysis of sealed interface.

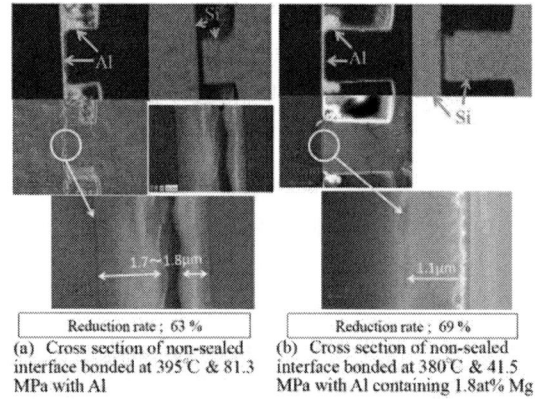

Reduction rate ; 63 %   Reduction rate ; 69 %

(a) Cross section of non-sealed interface bonded at 395℃ & 81.3 MPa with Al

(b) Cross section of non-sealed interface bonded at 380℃ & 41.5 MPa with Al containing 1.8at% Mg

Fig.6: SEM and EDX analysis of non-sealed interface.

Bonding Temp.(℃) Press.(MPa)	Bonding layer structure	Thickness of each layer (μm)	Total thick. of layer(μm) (h1)	Thick. after bond. (μm) (h2)	Reduction rate (%) (h1-h2) / h1×100	Sealed or non-sealed
390 78	Sn/Al/Sub	2.1/0.1	4.4	0.5~0.7	84~89	Sealed
380 61	Sn/Al/Sub	2.0/0.1	4.2	0.4~0.54	87~90	Sealed
370 41.5	Al/Sn/Al /Sn/Al	0.76/0.05/0. 76/0.05/0.38	4.0	0.45~0.56	86~89	Sealed
395 81.3	Al	2.3	4.6	1.5~1.8	67~63	Non-sealed
380 41.5	Al(1.8at% Mg)	1.75	3.5	1.1~1.2	66~69	Non-sealed

Table 1: Bonding condition and reduction rate.

978-1-5090-4661-4/17 $31.00 © 2017 IEEE

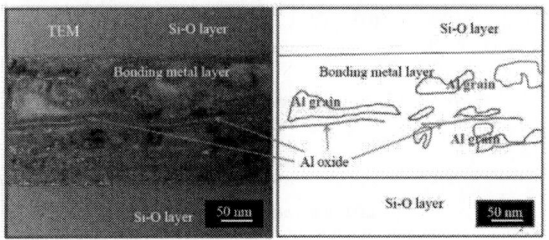

Reduction rate : >90%

Fig. 7: TEM analysis of bonding interface.

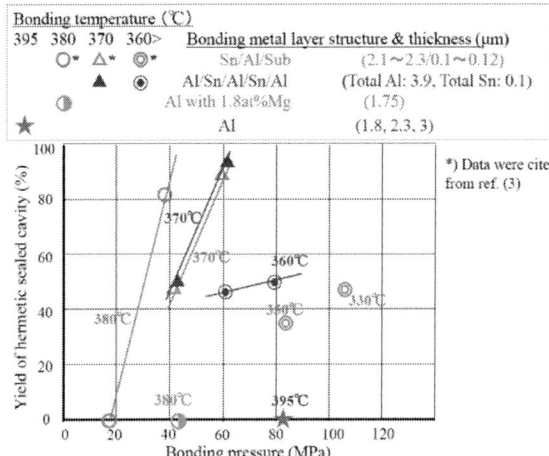

Bonding temperature (℃)				Bonding metal layer structure & thickness (μm)	
395	380	370	360>		
○*	△*	◎*		Sn/Al/Sub	(2.1~2.3/0.1~0.12)
	▲	◉		Al/Sn/Al/Sn/Al	(Total Al: 3.9, Total Sn: 0.1)
		◑		Al with 1.8at%Mg	(1.75)
★				Al	(1.8, 2.3, 3)

*) Data were cited from ref. (3)

Fig. 8 : Yield of hermetic sealed cavity vs. bonding pressure.

structure. Table 1 summarizes the bonding results and the reduction rates, and also shows the bonding conditions and layer structures. Al with covering Sn layer is represented as Sn/Al/Sub. We concluded the high reduction rate is needed to achieve the stable hermetic sealing. Figure 7 shows TEM image of sealed interface. We observed broken Al oxide and

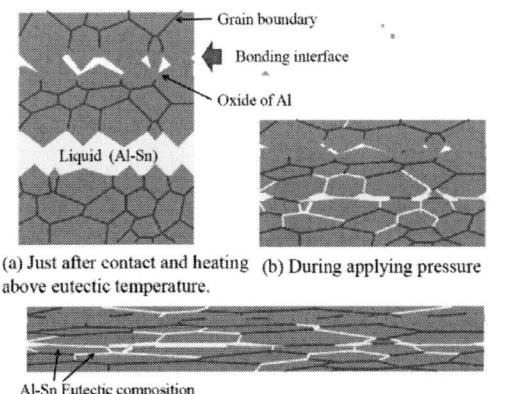

(a) Just after contact and heating above eutectic temperature.
(b) During applying pressure

(c) After compression

Fig. 9 : Transition model of bonding interface

small Al grains in the bonding metal layer. With compression, most Al grains are tore off and become small. Relatively large grains are only represented. Figure 8 summarizes the yield of hermetic sealing vs. bonding pressure. The results for Al with intermediate Sn layer coincide with those with covering Sn layer [3] shown as Sn/Al/Sub.

Discussion and Conclusion

Sn should exist among Al grains as liquid Sn-Al eutectic at bonding temperature, judging from Al-Sn phase diagram (Fig. 2). Figure 9 shows the transition model of bonding interface. At bonding temperature, bonding metal consists of solid Al and liquid with composition of near liquidus line. With reducing temperature, liquid composition changes along with liquidus line to eutectic point and liquid simultaneously diffuses into Al grains. This suggests grain slip of Al, and then large thinning of interface metal easily occur under applying pressure.

We conclude the addition of Sn creates a softening effect to Al bonding layer. Because of large deformation, surface oxide of Al is broken and direct contact and inter diffusion of Al of both substrates should appear. Therefore, Al-Al stable and high yield hermetic sealing was achieved with oxidized Al bonding surface before bonding.

Acknowledgments

This work was supported by "Creation of innovation centers for advanced interdisciplinary research areas program" from Japanese Ministry of education, Culture, sports, Science.

References

[1] C. H. Yun et al., *Proc. IEEE MEMS 200*8, pp.13-17.

[2] Nz. Malik et al., *Sens. Actuators A*, 211 (2014) pp. 115-120.

[3] S. Satoh et al., Proc. IEEE MEMS 2016, Shanghai, China (2016) pp.581-584.

[4] T. B. Massalsk, "Binary Phase Diaphragm, II Ed.", McAlister A.J. (1990)

[5] A. N. Nesmeyanov, Vapor Pressure of the Chemical Elements, (American Elesvier Pub. Co., Ltd, 1963), pp. 269-273

[6] H. Ikezawa et al., J. Japan. Inst. Metals (2005) pp. 739-742

978-1-5090-4661-4/17 $31.00 © 2017 IEEE

Low-carrier density sputtered-MoS₂ film by H₂S annealing for normally-off accumulation-mode FET

Jun'ichi Shimizu[1], Takumi Ohashi[1], Kentaro Matsuura[1], Iriya Muneta[1],

Kuniyuki Kakushima[1], Kazuo Tsutsui[1], Nobuyuki Ikarashi[2] and Hitoshi Wakabayashi[1]

[1]Tokyo Institute of Technology, Yokohama, Japan, shimizu.j.aa@m.titech.ac.jp,

[2]Nagoya University, Nagoya, Japan.

Abstract

We investigate low-temperature formation process of sputtered-MoS₂ film. The MoS₂ film was formed by radio frequency (RF) sputtering. Then the sputtered-MoS₂ was annealed in H₂S at from 200 to 400°C. We find that the hydrogen sulfur (H₂S) annealing compensate for sulfur defects at low temperature significantly, resulting in a lower carrier density of $2 \cdot 10^{16}$ cm^{-3}.

Introduction

Accumulation-mode MoS₂ FET, whose channel material is formed by RF sputtering, can be considered as a promising candidate for upper portion devices in 3D-IC, because of its low-temperature process [1-3]. However, carrier density of as-sputtered MoS₂ is not low as expected because of its sulfur defects [3] which is widely discussed as n-type dopants [4]. Thus, post-annealed process is needed to reduce carrier density, in order to realized normally-off accumulation-mode MoS₂ FET. There are some candidates for sulfur compensation process, such as forming gas (F.G.: 3% H₂ in N₂), sulfur powder, and di-tertiary-butyl disulfide annealing [3,5-6]. Furthermore, H₂S and sulfur powder are commonly used to synthesized MoS₂ in chemical vapor deposition. In this paper, we introduce H₂S annealing for sputtered-MoS₂, aiming for sulfur compensation and hydrogen termination to reduce carrier density.

Experimental methods

Figure 1 shows a process flow and a schematic image of post-annealing process. The MoS₂ film of 5-nm thick was deposited on an SiO₂/Si substrate by the RF sputtering. Then MoS₂ film was annealed in 1% H₂S in Ar or F.G.. Raman spectroscopy, x-ray photoemission spectroscopy (XPS), scanning transmission electron microscopy (STEM) and Hall-effect measurement were used to evaluate physical and electrical properties.

Results and discussion

A typical cross-sectional high-angle annular dark field (HAADF)-STEM image of the H₂S annealed MoS₂ film (300 Pa, 400°C) is shown in Figure 2. A layered structure of the film can be seen at atomic resolution. Figure 3 (a) shows Raman spectra of MoS₂ films which are annealed in H₂S or F.G. at 400°C in various pressures and (b) shows Raman of MoS₂ which are annealed in H₂S in 1000 Pa at various temperatures. E^1_{2g}-mode peak position become closer to that of bulk one, as MoS₂ is annealed in H₂S. This indicates a reduction of in-plane-direction strain because of reduction of sulfur defects. Figure 4 shows the value of full width at half maximum (FWHM) in E^1_{2g}- and A_{1g}-mode peaks of Raman spectra. The values of FWHM decrease significantly by H₂S annealing with increases in annealing temperature and decreases in annealing pressure. Figures 5 (a) and (b) are XPS spectra of MoS₂ films which are non-annealed and annealed in H₂S in 1000 Pa at 400°C, respectively. Figure 6 shows a composition ratio of MoS₂ films, which extracted from XPS measurements. Residual molybdenum and sulfur were found in as-sputtered-MoS₂ film. The residual molybdenum was sulfurized completely by H₂S annealing at 400°C. However, molybdenum oxide (MoO₃) was observed in each samples constantly. This indicates that MoS₂ films was oxidized after post-annealing process by O₂ and H₂O in nature. Eventually, the carrier density of $2 \cdot 10^{16}$ cm^{-3} and Hall-effect mobility of 18 cm^2/V-s were achieved as shown in Figs. 7 and 8. The carrier density is considered as enough low to control threshold voltage of accumulation-mode MoS₂ FET.

Conclusion

The high-quality sputtered-MoS₂ film was obtained by the H₂S annealing at less than 400°C in low pressure. The carrier-density of $2 \cdot 10^{16}$ cm^{-3} was achieved, which is enough low to control threshold voltage of accumulation-mode MoS₂ FET.

Acknowledgments

This paper is partly supported by CREST / COI of JST and JSPS 26105014.

References

[1] P. Batude, et al., Symp. on VLSI Tech. 2015.

[2] T. Ohashi, et al., Jpn. J. Appl. Phys **54**, (2015).

[3] J. Shimizu, et al., SSDM 2016, PS13-04.

[4] C.-P. Lu, et al., Nano Lett **14**, 4628, (2014).

[5] K. Matsuura, et al, SISC 2016, 3.6, to be presented.

[6] S. Ishihara, et al, Jpn. J. Appl. Phys **55**, (2015).

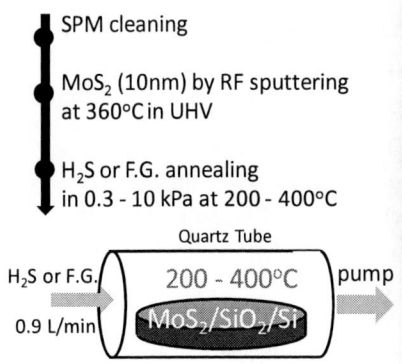

Fig. 1 Process flow and schematic image of our annealing process.

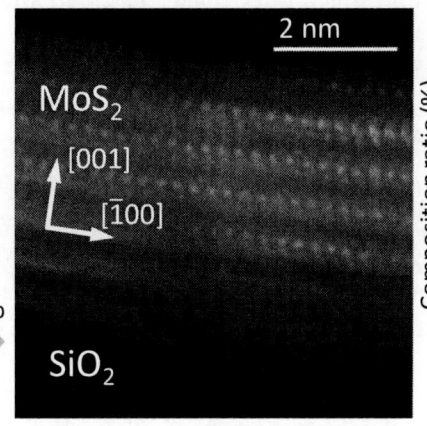

Fig. 2 Cross-sectional HAADF-STEM image of annealed MoS₂ film.

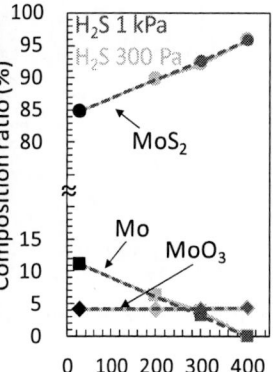

Fig. 6 Composition ratio of non-annealed and annealed MoS2 films.

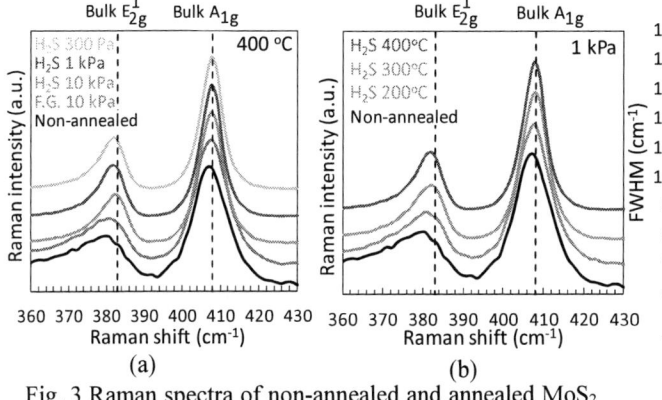

(a)   (b)

Fig. 3 Raman spectra of non-annealed and annealed MoS₂ films depending on (a) pressure and (b) temperature.

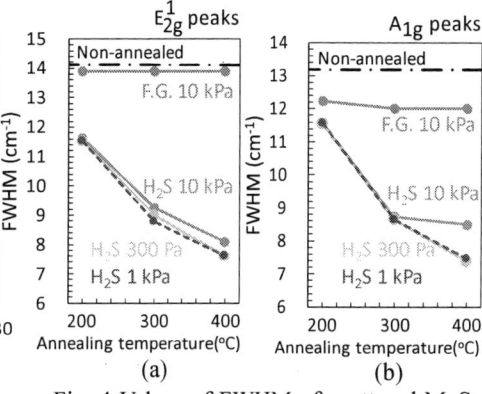

(a)   (b)

Fig. 4 Values of FWHM of sputtered-MoS₂ Raman spectra. (a) E_{2g}^1 and (b) A_{1g} modes.

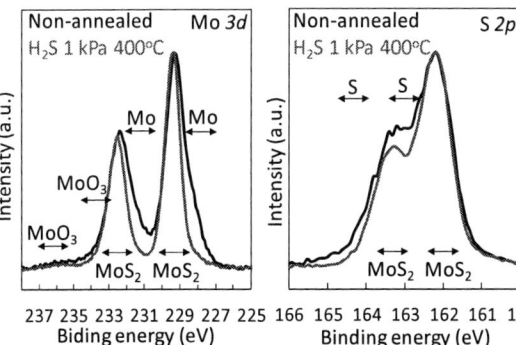

(a)   (b)

Fig. 5 XPS spectra of non-annealed and annealed MoS2 films. (a) molybdenum $3d$ and (b) sulfur $2p$.

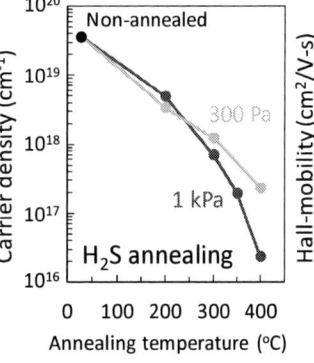

Fig. 7 Carrier density dependence on annealing temperature.

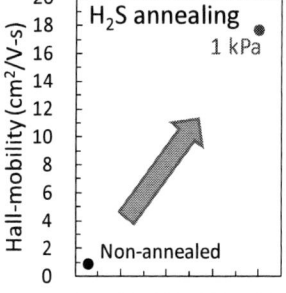

Fig. 8 Hall-effect mobility dependence on annealing temperature.

978-1-5090-4661-4/17 $31.00 © 2017 IEEE   223

Impact of Ferroelectric Domain Switching in Nonvolatile Charge-Trapping Memory

Chia-Chi Fan[1], Yu-Chien Chiu[1], Chien Liu[2], Guan-Lin Liou[3], Wen-Wei Lai[1], Yi-Ru Chen[1], Chun-Hu Cheng[3,*], and Chun-Yen Chang[1,4,#]

[1]Dept. of Electronics Eng., & [2]Dept. of Electro-physics, National Chiao-Tung University, Taiwan

[3]Dept. of Mechatronic Eng., National Taiwan Normal University, Taiwan

[4]Research Center for Applied Sciences, Academia Sinica, Taiwan

E-mail: *chcheng@ntnu.edu.tw, #cyc3562@gmail.com

Abstract

In this work, we proposal a ferroelectric domain to enhance program/erase/read efficiency of conventional charge-trapping nonvolatile memory. The ferroelectric-domain-dominated HfZrO/HfON memory shows the better subthreshold characteristics than control charge-trapping structure (HfO_2/HfON). Additionally, the memory speed with ferroelectric polarization (~800ns) is more than three orders of magnitude faster than that of control trapping type. (Keywords: Ferroelectric domain, HfZrO and HfON)

Introduction

The charge-trapping storage [1] with the advantages of CMOS-compatible process and 3D stackable architecture has been regarded as an effective solution for the development of storage-class memory. However, the charge-trapping nonvolatile memories (NVMs) face the technical issue such as high gate leakage and slow speed for continuously scaled memory structure with a growing need of high-density storage. Recently, there is significant progress in nonvolatile ferroelectric switching [2], [3]. Here, we propose a mixed storage mechanism dominated by ferroelectric domains to improve conventional charge-trapping memory.

Experiments

First, a stacked 30-nm-thick HfO_2 and 12-nm-thick HfON (HfO_2/HfON) were deposited by PVD process and annealed on Si substrates with a 3-nm tunneling SiO_2 as a control sample. The similar PVD process was also applied to the deposition of ferroelectric 30-nm-thick HfZrO (HZO) to form another HZO/HfON memory structure. Finally, the TaN gate was deposited and followed by source/drain implantation and activation.

Results and Discussion

Fig 1 and Fig. 2 show the I_d-V_g curves of control HfO_2/HfON and HZO/HfON NVMs, respectively. Under the same process and activation conditions, the off-state leakage of HfO_2/HfON NVM is significantly higher than that of HZO/HfON NVM. It is ascribed to the ZrO_2 doping into HfO_2 to reach a reduced gate leakage and improved subthreshold characteristics (Fig. 3). From previous studies, the HfZrO film had higher dielectric constant and less charge defects compared to HfO_2 dielectric [4]. As for memory switching properties, both of two structures can obtain the same memory window of ~10V under 16V DC sweep, as shown in Fig. 4. Even so, the HZO/HfON NVM with ferroelectric HZO domain switching still exist an intrinsic difference between these two NVM structures, as seen in the results of speed test (Fig. 5 & Fig. 6). The speed of 800ns for program/erase/read operations is significantly improved by more than three orders of magnitude in ferroelectric-domain-dominated HZO/HfON NVM. Also, the memory window measured at fast 800ns can be maintained after 10^6 program/erase cycles (Fig. 7), indicating the stabilization of ferroelectric domain polarization is less influenced by the location of electron trapping in HfON. Table 1 compares the memory characteristics of these two NVMs.

Conclusion

The ferroelectric-domain-dominated nonvolatile charge-trapping memory achieve a stable 10^6-cycled switching at a fast 800ns speed.

References

[1] D. C. Ahn, M. L. Seol, J. Hur, D. I. Moon, B. H. Lee, J. W. Han, J. Y. Park, S. B. Jeon, and Y. K. Choi, "Ultra-fast erase method of SONOS flash memory by instantaneous thermal excitation," IEEE Electron Device Lett., Vol. 37, no. 2, pp. 190-192, 2016.

[2] C. H. Cheng, A. Chin, "Low-leakage-current DRAM-Like memory using a one-transistor ferroelectric MOSFET with a Hf-Based gate dielectric," IEEE Electron Device Lett., Vol. 35, pp. 138-140, 2014.

[3] Y. C. Chiu, C. H. Cheng, C. Y. Chang, M. H. Lee, H. H. Hsu and S. S. Yen "Low power 1T DRAM/NVM versatile memory featuring steep sub-60-mV/decade operation, fast 20-ns speed, and robust 85C-extrapolated 10^{16} endurance," in Symp. on VLSI Tech., 2015, pp. 184-185.

[4] S. Heo, D. Tahir, J. G. Chung, J. C. Lee, K. Kim, J. Lee, H. Lee, G. S. Park, S. K. Oh, H. J. Kang, P. Choi, and B. D. Choi, "Band alignment of atomic layer deposited $(HfZrO_4)_{1-x}(SiO_2)_x$ gate dielectrics on Si(100)," App. Phys. Lett., Vol. 107, pp. 182101, 2015.

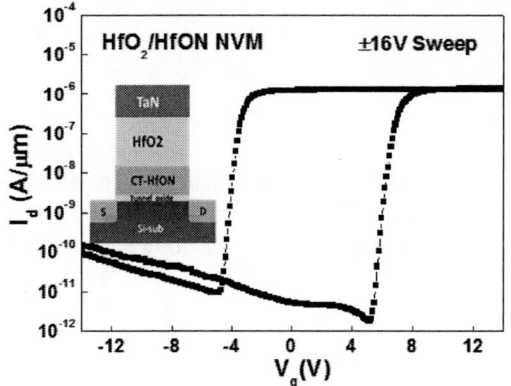

Fig. 1. I_d-V_g characteristic of TaN/HfO$_2$/HfON NVM device.

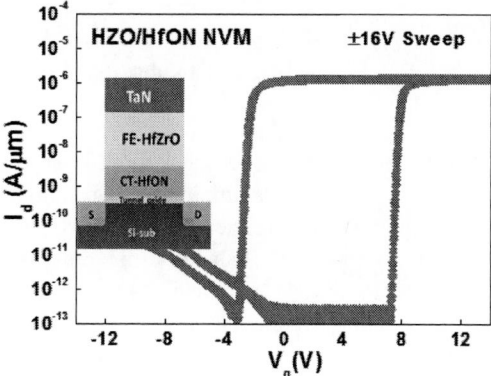

Fig. 2. I_d-V_g characteristic of TaN/HZO/HfON NVM device.

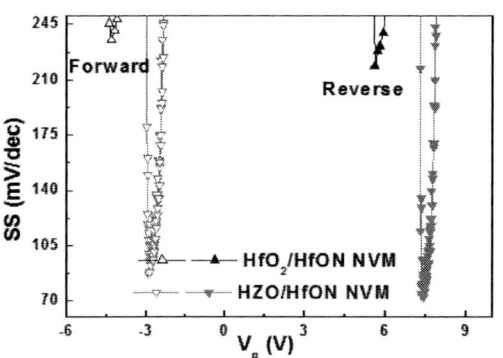

Fig. 3. SS characteristics of HfO$_2$/HfON and HZO/HfON NVMs under ±16 V DC sweep.

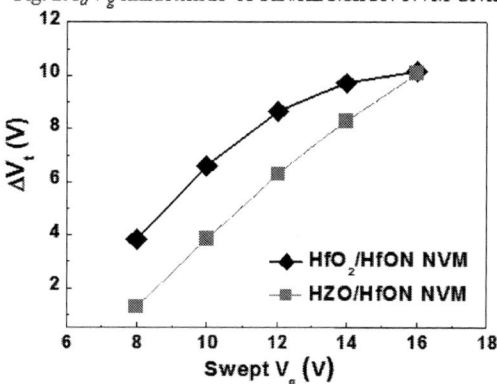

Fig. 4. Relation between V_t shift and V_g sweep (±8V~±16V) of HfO$_2$/HfON NVM and HZO/HfON NVM.

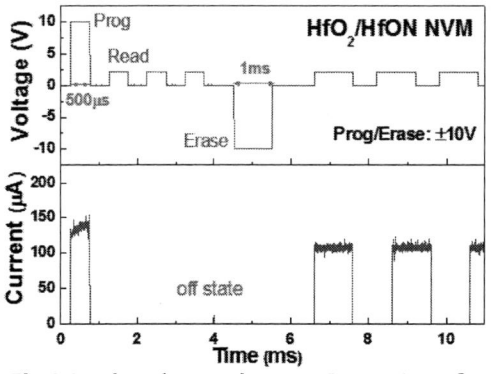

Fig. 5. Impulse voltages and response I_d current waveforms of HfO$_2$/HfON NVM measured under ±10V P/E and 1ms.

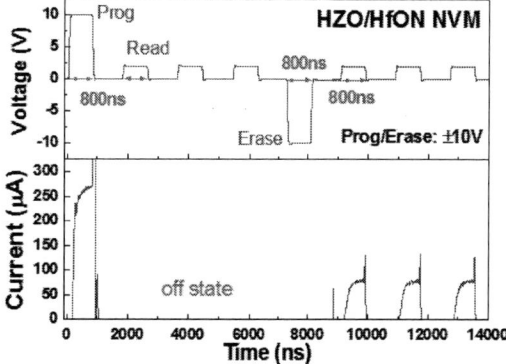

Fig. 6. Impulse voltages and response I_D current waveforms of HZO/HfON NVM measured under ±10V P/E and 800ns.

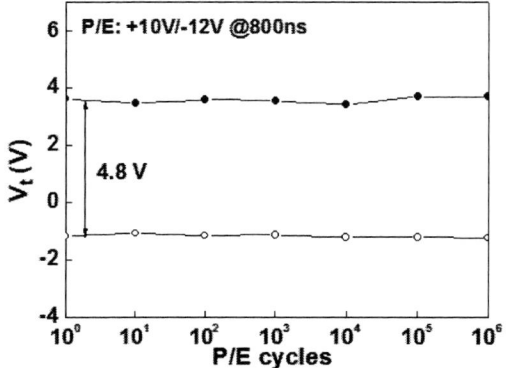

Fig. 7. Endurance characteristics of HZO/HfON NVM. The large memory window allows multi-level operation.

	HfO$_2$/HfON NVM	HZO/HfON NVM
Storage Mechanism	Conventional Charge-Trapping	Ferroelectric-Dominated Charge-Trapping
P/E Voltages	±16V	±16V
SS_{min}	182 mV/dec	74 mV/dec
Max. ΔV_t	10.2 V	10.1 V
Speed	1ms	800ns

Table 1 Characteristics comparison for HfO$_2$/HfON NVM and HZO/HfON NVM.

978-1-5090-4661-4/17 $31.00 © 2017 IEEE

An In-Line MOSFET Process With Photomask Fabrication Process In A Minimal Fab

Norio Umeyama[1,2], Sommawan Khumpuang[1,2], and Shiro Hara[1,2]

[1]AIST, Tsukuba, Ibaraki, Japan, n-umeyama@aist.go.jp, [2]MINIMAL

Abstract

This paper describes a cleanroom-less device fabrication process of a minimal fab including an in-line photomask fabrication process, where half-inch wafers and half-inch photomasks are transferred in airtight containers and loaded into fabrication tools to prevent invasion of air-borne particles, gases and UV light from outside. Photomasks are fabricated using a maskless exposure system in the minimal fab. We have fabricated two types of MOSFET using the photomasks or maskless exposure system, and found that the two types of MOSFET have the almost same electric performance with the density of interface states (D_{it}) of the order of 10^{10} states/cm^2.

(Keywords: Photomask, MOSFET, Lithography, Minimal Fab)

Introduction

When fabricating a device in a conventional fab, the maintenance cost of the clean room and the photomask cost are very expensive. In Minimal Fab, their costs are ignored, since local-clean technology using airtight wafer container and airtight wafer loading system to remove clean room was developed. Also, a maskless exposure system was developed [1]. One of disadvantages in the maskless exposure system is a longer exposure time of around 10 min for a half-inch wafer. Therefore, we have developed a photomask aligner with an exposure time of a few seconds. The diameter of the photomask is the same as that of the half-inch wafer. Thus, we can use the wafer container and wafer load loading system also for the photomask. The photomask itself can be fabricated in the maskless exposure system in the minimal fab. In this paper, we show fabrication times, costs, and electric performances of two types of MOSFET formed by a set of photomasks or formed by the maskless exposure system.

Preparation of an In-Line Fab Process

The wafer and the photomask have the same size of φ 12.5 mm as shown in Fig. 1. This schematic in-line process of fabricating photomasks and MOSFET is indicated in Fig. 2. We can select the photomask process or the maskless exposure process for MOSFET fabrication. In the maskless exposure system, UV LED of 365 nm is used. The resolution on the wafer is 0.5 μm.

Experimental Methods

All of the processes were carried out with tools of the minimal fab. To fabricate the masks, aluminum was sputtered on the half-inch glass plate. The sputtered aluminum layers on the glass plates were patterned to form the masks of MOSFET patterns by the minimal maskless lithography system. The formed four masks are shown in Fig. 3.

Results and Discussion

In Fig. 4, optical images of the MOSFET patterns fabricated by the mask aligner process using the photomasks (right hand side) and the maskless exposure process (left hand side) are shown. There is no distinguishing difference between them. The resolution of line and space on the wafer in the mask aligner process was ~1 μm or lower. I_d-V_d characteristics obtained MOSFETs for maskless exposure and mask aligner processes are shown in Fig. 5. Typical transistor characteristics were obtained. The C-V characteristics are also shown in Fig. 6. These results reveal that the density of interface states (D_{it}) is 4.3×10^{10} states/cm^2 and 2.3×10^{10} states/cm^2 for the maskless exposure process and the mask aligner process, respectively. This indicates that the both processes have low contamination levels. Moreover, we have measured the production times, and estimated fabrication costs for the two processes (Table 1). The exposure time of the mask aligner was 2 sec and that of the maskless exposure system was 240 sec. We have found that the cost of the mask aligner process is lower over the wafer number of 52.

Conclusion

We have produced photomasks using the minimal fab process. MOSFET was formed using these photomasks, indicating the contamination level was quite good, and the cost performance is better than the maskless process for the production over 52 wafers.

Acknowledgments

The authors gratefully acknowledge the research support from Ministry of Economy, Trade and Industry (METI) under the project 'Innovative manufacturing process technology development (Minimal Fab)' during the fiscal year of 2012-2014.

References

[1] Sommawan Khumpuang and Shiro Hara, "A MOSFET Fabrication Using a Maskless Lithography System in Clean-Localized Environment of Minimal Fab," IEEE Transactions on Semiconductor Manufacturing, 2015, Vol. 28, No. 3, 393-398.

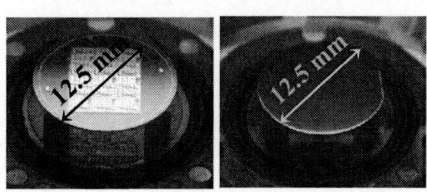

Fig.1: Photos of a photomask (left) and a wafer (right) of a half-inch size.

Producing Photomasks

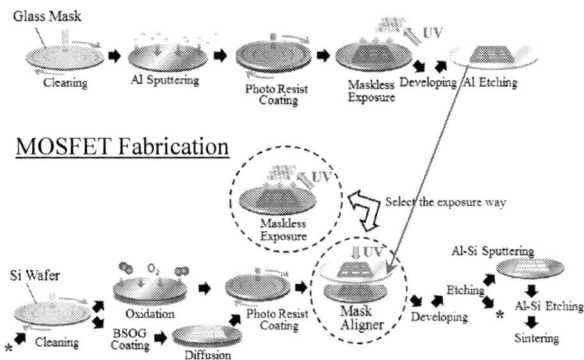

MOSFET Fabrication

Fig.2: Schematic image of an in-line process from producing photomasks to MOSFET fabrication.

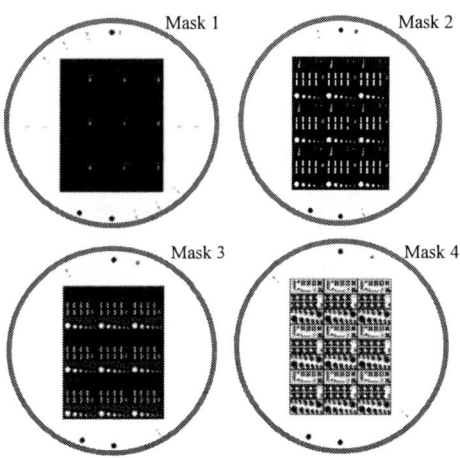

Fig.3: Patterns of the photomasks for the MOSFET fabrication.

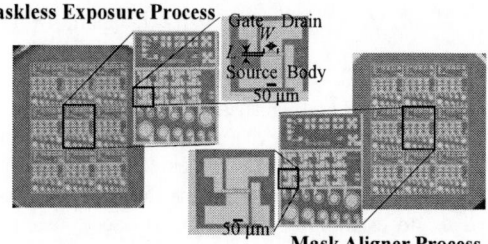

Fig.4: Photos of MOSFETs fabricated by the minimal maskless exposure process (left), and fabricated by the minimal mask aligner process (right).

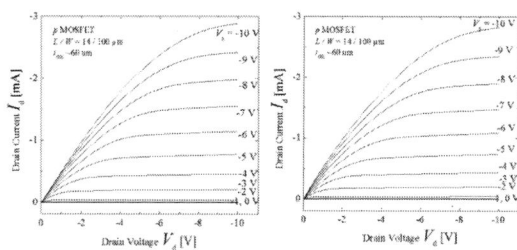

Fig.5: Plots of I_d-V_d characteristics of obtained MOSFETs in the assembly center by 'Maskless Exposure' (left) and 'Mask Aligner' (right) as lithography.

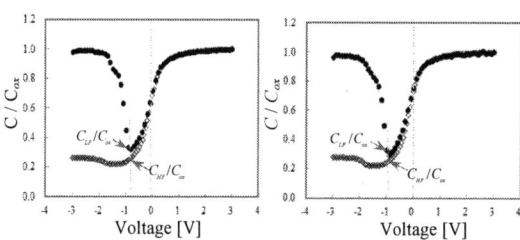

Fig.6: Plots of C-V curves of obtained MOSs in the assembly center by 'Maskless Exposure' (left) and 'Mask Aligner' (right) as lithography.

	Maskless Exposure	Mask Aligner
Exposure Time (sec)	240	2
Calculated time at MOSFET fabrication		~20 minutes short
(Gain amount when calculated by personnel cost of 5000 JPY / hour)		(~1300 JPY)
One mask production trial cost at Minimal Fab		~18000 JPY
Total mask production trial cost at Minimal Fab		~68500 JPY
Manufactured number that benefits by using a mask aligner machine		~52 (≒ 68500 JPY / 1300 JPY)

Table 1: List of trial cost calculations for the maskless exposure and the mask aligner processes.

978-1-5090-4661-4/17 $31.00 © 2017 IEEE

Development of a Half-Inch Wafer for Minimal Fab Process

Norio Umeyama[1,2], Atsushi Yamazaki[3], Takaaki Sakai[3], Sommawan Khumpuang[1,2], and Shiro Hara[1,2]

[1]AIST, Tsukuba, Ibaraki, Japan, n-umeyama@aist.go.jp, [2]MINIMAL [3]Fujikoshi Machinery

Abstract

Specifications and fabrication process suitable for a small wafer with the diameter of half-inch, which is used for a minimal fab, is presented. We beveled wafer edge by rapping and polishing in order to clean the edge and to suppress the strong surface tension at the edge. To show the crystallographic orientation of the wafer, we introduced laser marking process. By the processes, we have formed silicon wafer that is suitable for fabricating electronic devices. Formed MOS capacitor on the wafer revealed that the density of interface states (D_{it}) was 7.8×10^{10} states/cm^2 as the average of 37 wafers.

(Keywords: wafer, cutting, minimal fab)

Introduction

Until now, the semiconductor industry has been increasing the wafer size. This caused the investment inflation and incompatibility for low volume device markets. On the other hand, 'minimal fab' using a half-inch wafer suppress the investment and suitable for the low volume markets. However, in a small wafer there are significant problems to be solved. One of the most important problem is that the surface tension is strong. Also, the quality of flat surface area to fabricate devices is easy to be affected by the quality of the wafer edge. The purpose of this paper is to report our approach we developed for the half-inch wafer.

How to Determine Wafer Specifications

A. Thickness

The trend between the diameter and the thickness of the silicon wafer is shown in Fig. 1. According to the trend, the thickness for a half-inch wafer seems to be about 50 µm. For this kind of thin wafer, it is difficult to perform double-sided polishing because the spacer to separate and hold wafers during the polishing is too thin to polish. For double-sided polishing the wafer thickness should be above 200 µm. Therefore, the thickness of our half-inch wafer was defined to be 250 µm.

B. Edge shape

In a spin coating process of photoresist, resist layer is thinned and becomes homogeneous by spilling the residual resist solution over the wafer. Here, the centrifugal force blocks the resist spill. For a smaller wafer spin coating becomes difficult because of a stronger centrifugal force [1]. Then, we optimize the shape of wafer edge by beveling the edge to suppress the centrifugal force. Although an ideal edge shape would be a round shape, it is difficult to form by a rapping process. We found that the optimized edge shape was a bullet shape as shown in Fig. 2F. In the bullet shape, the height of resist edge bead was around 1.5 µm, which can be removed by an edge rinsing technique [2].

C. Crystal orientation display methods

Since the half-inch wafer is very small, it is difficult to fabricate an oriental flat or a notch and to detect them in a device process tool. Instead, we drew a laser mark on the wafer surface. The direction of the mark is [100], which is different from cleavable [110] direction by 45 degrees. The image after photoresist coating is shown in Fig. 3. It is found that the coated resist is homogeneous even near the mark groove.

Experimental Methods

A basic flow diagram of the half-inch silicon wafer fabrication is shown in Fig. 4. Cutting after grinding double-sided, then we polished the edge face and let the edge surface smoother and relieved internal stress by alkali etching. After that, wet cleaning and vapor IPA drying were performed. Lastly, all wafers were tested by particle inspection. We fabricated a MOSFET for silicon wafers produced using this process flow and measured its electrical characteristics.

Results and Discussion

The strong surface tension becomes an obstruction in rinsing and drying processes in wet cleaning of a wafer. Even for the well-beveled and polished wafer as described above, rinsing and drying is difficult when the wafer is touched to wafer fingers of its holder. We have developed fine fingers whose contact areas to the wafer are small. After the improvement of the finger structure, rinsing and drying became perfect and the water marks before the improvement was disappeared as shown in Fig. 5. Using this wafer, we fabricated MOS capacitors. Measured C-V characteristics is shown in Fig. 6, the density of interface states (D_{it}) was 7.8×10^{10} states/cm^2 as an average of 37 wafers. The quality of the wafers turns out to be in the device fabrication level.

Conclusion

Specification and fabrication processes of a half-inch wafer suitable for the minimal fab were developed. We determined thickness, edge shape, and crystal orientation displaying methods. By beveling and polishing the wafer edge, we have formed well-cleaned wafers which have homogeneous electric and electronic device performances.

Acknowledgments

The authors acknowledge the research support from Ministry of Economy, Trade and Industry under the project during the fiscal year of 2012-2014.

References

[1] Harry. J. Levinson, "Principles of Lithography", SPIE Press Monograph, 2005, Chapter 3, p. 62,

[2] N. Umeyama, et al., Clean Technology, 2014, vol. 24, No. 12, 21-24,

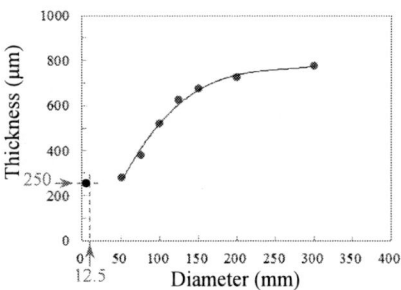

Fig.1: Plots of diameter vs. thickness of silicon wafer.

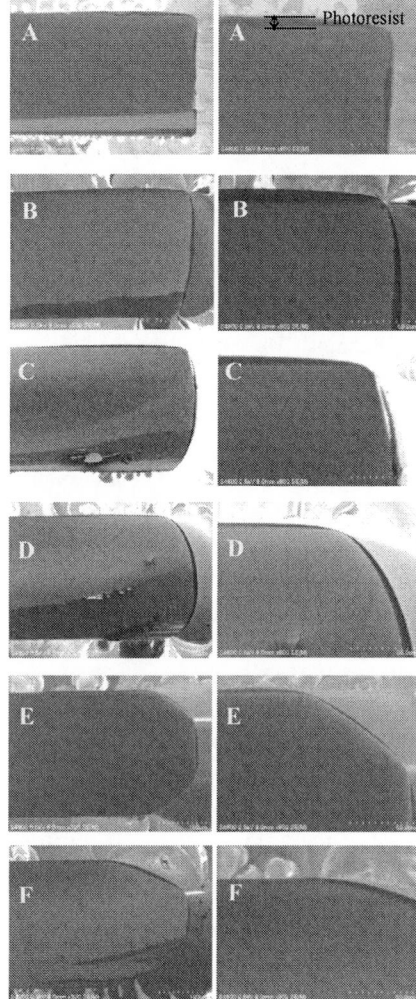

Fig.2: The wafers with various edge shapes were formed. After photoresist was spin-coated on the wafers, they were cleaved and the edge beats of the resist were observed by SEM. In the image of F with the bullet edge shape, the height of edge beat was lowest.

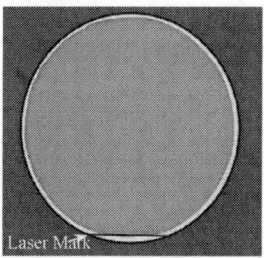

Fig.3: Photo of photoresist coated on a half-inch wafer. It is observed that coating is homogeneous up to the resist stripping boundary and the location of the laser mark.

Fig.4: The basic flow diagram of the half-inch silicon wafer fabrication.

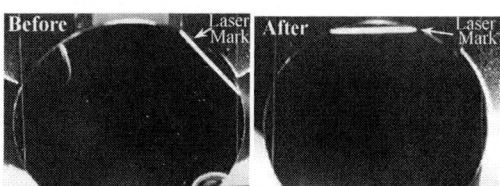

Fig.5: Photos of the wafer surface before and after improvement for cleaning jig. There are dot shaped watermarks in the photo of 'Before'.

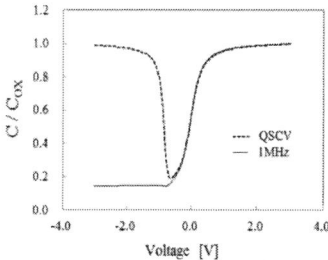

Fig. 6: The typical C-V characteristics for MOS capacitors.

978-1-5090-4661-4/17 $31.00 © 2017 IEEE

P-26

Helium ion microscopy (HIM) for imaging
fine line features patterned organic film with less damage

Shinichi Ogawa[1], Tomoya Ohashi[2], Shigeki Oyama[2], Yuki Usui[2]

[1]National Institute of Advanced Industrial Science and Technology, Japan, ogawa.shinichi@aist.go.jp

[2]NISSAN CHEMICAL INDUSTRIES, LTD., Japan

Abstract

Helium ion microscopy (HIM) was applied to image an organic film filled into narrow trenches. The film was characterized to examine structural changes after the HIM helium ion irradiation comparing with a SEM electron beam irradiation. In the HIM case, the change was seen in a deep region of the film, while it occurred at the surface in the SEM case. This depends on penetration properties of helium ions and electrons to the material, and surface imaging of the film looks more realistic by the HIM than the SEM.

(Keywords: Helium ion microscopy, Imaging, Organic film, Structural change, Damage, SEM)

Introduction

Imaging of fine line features patterned organic materials is required during ULSI processing. Secondary electron microscopy (SEM) imaging of organic materials often results in changes to the material line width, edge roughness, and shape. This makes SEM interpretation of the real shape difficult because of structural changes, or damages, during an electron beam irradiation at SEM observation.

In this paper, we investigated structural change of the organic film after the helium ions and electron beams irradiations and showed the ability of a helium ion microscopy (HIM) to provide an organic film surface images that contain pattern information not available with a conventional SEM.

Experiments

Acrylic organic films of 100 nm thick were spin coated on Si substrates, and then helium ion or electron beam was irradiated at a dose of 1.44×10^{14} - 9.96×10^{15} at 30 kV and $0.30 - 1.42 \times 10^{17}$/cm^2 at 0.7 kV, respectively. Morphological and structural changes after the beam irradiations were characterized by AFM, TOF-SIMS for surface region, IR and Raman for inside the film. Cross sections of the film filled into narrow trenches were observed by SEM and HIM at optimum conditions.

Results and Discussions

The AFM showed decrease in the film thickness with the increase of the dose as shown in Fig.1, and it indicated that the electrons might damage heavily than the helium ions at each optimum observation condition. Although the helium beam energy is higher, the combination of the lower helium ion current and the longer range of the ions means that the power density to the organic materials is nearly a factor of 10^3 lower with the 30 keV helium ion beam than it is with the 0.7 keV electron beam [1]. FTIR result showed larger signal intensity decrease with less peak broadening and OH system remained in a case of the electrons, while OH system was much destroyed in a case of the helium ions as shown in Fig.2. Raman result showed more amorphous carbon in the helium ions irradiated film (Fig.3), which was probably formed by destruction of CH system. Those results mean that helium ions irradiations brought about less surface morphological transformation while it resulted in larger chemical damage in a deeper region of the film than the electron irradiations. This phenomenon is probably because heavier helium ions with higher energy came into deeper than electrons to the organic film cutting chains of the organic material with amorphous. TOF-SIMS showed the similar results for larger decrease of signal intensities of CH and CNO (cross-linker) systems by the electron irradiation (Fig.4 and Fig.5). Those results mean that there were a lot of trade-off between irradiations of helium ions and electrons. Cross section images of filling of the organic materials into trenches by the SEM and HIM were shown in Fig.6. Helium ions damaged the organic materials heavier in depth direction than electrons, while it kept original surface morphology with less transformation or shrink, so imaging of the filled organic materials into trenches by the HIM presumably shows more realistic, such as voiding during filling process itself, than the SEM imaging, and SEM imaging might occurred the voiding during the observation.

Conclusion

Imaging of the organic materials filled in narrow trenches by the HIM presumably shows more realistic than the SEM imaging with less damage such as voiding during the observation.

Acknowledgments

The authors would like to thank T. Kishioka from NISSAN CHEMICAL INDUSTRIES, LTD. for his fruitful discussions, and T. Iijima and S. Migita from AIST SCR Facilities for the usage of the HIM.

978-1-5090-4661-4/17 $31.00 © 2017 IEEE

References

[1] S. Ogawa, W. Thompson, I. Stern, L. Scioponi, J. Notte, L. Farkas, L. Barris, "Helium ion secondary electron mode microscopy for interconnect material imaging", Jpn. Appl. Phys. 49, 04DB12 (2010)

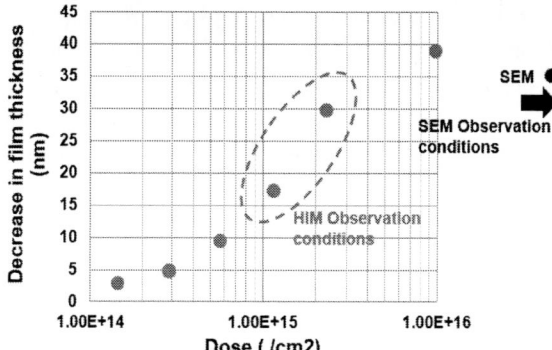

Fig. 1. Dependency of the decrease in the film thickness on the increase of the helium ions dose.

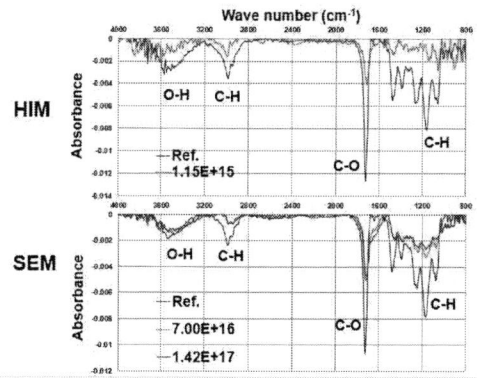

Fig.2 FTIR spectra from the helium ions (HIM) and electrons (SEM) irradiated acrylic organic films.

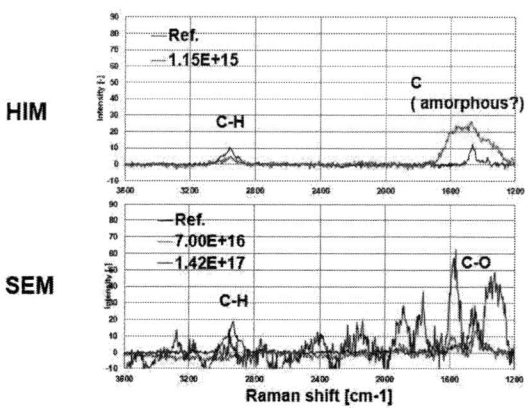

. Fig.3 Raman spectra from the helium ions (HIM) and electrons (SEM) irradiated samples.

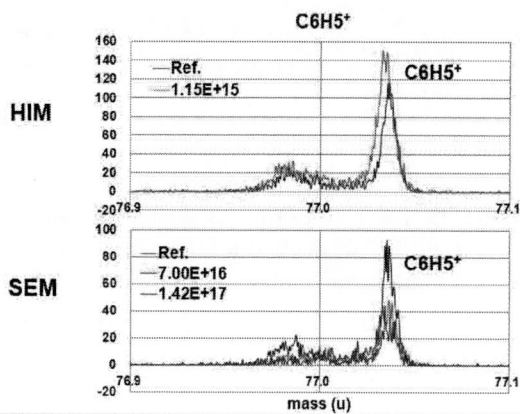

Fig. 4 TOF-SIMS spectra from the surface of helium ions (HIM) and electron (SEM) irradiated samples

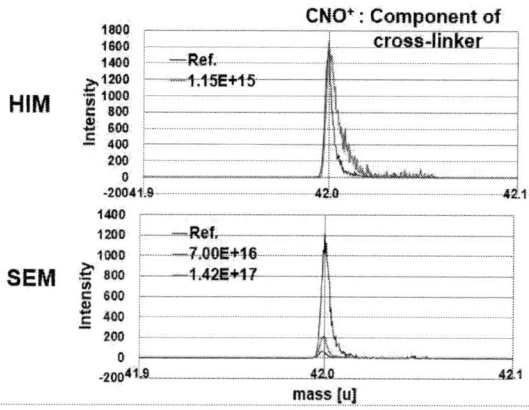

Fig.5 TOF-SIMS spectra from the surface of helium ions (HIM) and electron (SEM) irradiated samples

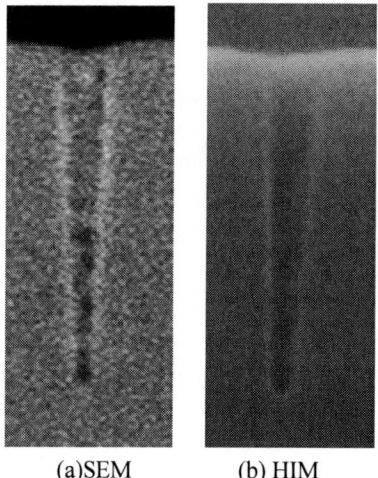

(a)SEM   (b) HIM

Fig.6 Cross section images of filling of the organic materials into trenches by the (a) SEM and (b) HIM. Voids in (a) might be generated during the SEM observation.

Supercritical fluid deposition of conformal oxide films: 3-dimentionally-stacked $RuO_2/TiO_2/RuO_2$ structures for MIM capacitors

Yu Zhao, Yusuke Shimoyama, Takeshi Momose, Yukihiro Shimogaki

The University of Tokyo, Tokyo, Japan, zhao@dpe.mm.t.u-tokyo.ac.jp

Abstract

Supercritical fluid deposition (SCFD), which utilizes the oxidation/reduction of organic compounds in supercritical CO_2 ($scCO_2$), is a promising thin film process for the fabrication of 3-dimenotional (3D) structure due to its conformal deposition capability on high aspect ratio features with moderate growth rate. In this work, SCFD was used to fabricate a conformally-stacked $RuO_2/TiO_2/RuO_2$ structure as an MIM capacitor. (Key words: supercritical fluid, thin film and MIM capacitor)

Introduction

In the past decade, as the integration density of semiconductor devices constantly increases, introduction of the 3D configuration to minimize the unit size has become a main trend of the ultra-large scale integration (ULSI) development [1]. For example, dynamic random access memory (DRAM) has adopted 3D capacitors with 30~50 aspect ratio (AR), where higher AR will be used in future. On the other hand, SCFD, a novel deposition technique involving the oxidation/reduction of organic compounds in $scCO_2$, has promising prospects of thin film fabrication on high AR structures in ULSI and micro-electro-mechanical systems (MEMS) [2]. Compared with traditional chemical vapor deposition (CVD) and atomic layer deposition (ALD), SCFD enables highly conformal film deposition on high AR features without compromising the growth rate due to the unique characteristics of the $scCO_2$ as reaction medium that combines liquid-like properties such as high fluid density and gaseous properties such as high diffusivity, as shown in Fig. 1-3. In this work, the $RuO_2/TiO_2/RuO_2$ MIM structure was fabricated on the Si trenches using SCFD.

Experiment

$Ti(O\text{-}iPr)_2(tmhd)_2$ and $Ru(tmhd)_3$ were used as the precursor, while Si wafer with 10AR trench was used as substrate. Fig. 4 shows the schematic diagram of the TiO_2-SCFD equipment [3]. A cold-wall flow-type reactor with face-down configuration was adopted. Before deposition, all the apparatus including the reactor was pressurized up to 10 MPa at 60°C to secure the supercritical state (critical point: 7.38 MPa, 31.1°C). The substrate placed face down in the reactor was heated up to 250°C. During the deposition, $scCO_2$ dissolving precursor was injected into the reactor with the other stream of $scCO_2$, where concentration of the precursor was diluted to the desired value. Conformal TiO_2 was then thermally deposited onto the heated face-down substrate. On another hand, RuO_2 was deposited in a closed-type reactor, where precursor was put in a sealed reactor and the deposition was enabled in $scCO_2$ of 250°C and 10 MPa, as shown in Fig. 5 [4]. The conformality of the sample was confirmed by the cross-sectional images of SEM.

Results and discussions

TiO_2 and RuO_2 were first deposited on planar substrate individually, as shown in Fig. 6-9. Since both of TiO_2 and RuO_2 deposition were enabled at 250°C, the same temperature (250°C) was used for multi-layer process to avoid the possible detachment caused by different expansion coefficients of TiO_2 and RuO_2. The cross section of the planar $RuO_2/TiO_2/RuO_2$ stacked structure is shown in Fig. 10. The multi-layer structure is clear, and it seems that RuO_2 and TiO_2 layers have a good adherence. Finally, conformal RuO_2, TiO_2 and 3D capacitor structure was successfully formed, as shown in Fig. 11-14. The process is supposed to be applied as a 3D MIM capacitor, where the 3D configuration is expected to promote the capacitance density and minimize the unit size.

Acknowledgments

This work was supported by JSPS KAKENHI Grant Number 15J09484.

References

[1] http://www.itrs2.net/

[2] Alvin H. Romang and James J. Watkins, "Supercritical fluids for the fabrication of semiconductor devices: emerging or missed opportunities?", Chemical Reviews, Vol. 110, 459-478, 2010, doi: 10.1021/cr900255w.

[3] Yu Zhao, Kyubong Jung, Takeshi Momose, and Yukihiro Shimogaki, "Smooth and conformal TiO_2 thin-film formation using supercritical fluid deposition", ECS Journal of Solid State Science and Technology, Vol. 2, N191-N195, 2013, doi: 10.1149/2.003311jss.

[4] Kyubong Jung, Takeshi Momose, and Yukihiro Shimogaki, "Strontium ruthenium oxide deposition in supercritical carbon dioxide using a closed reactor system", The Journal of Supercritical Fluids, Vol. 79, 244-250, 2013, doi: 10.1016/j.supflu.2013.03.008

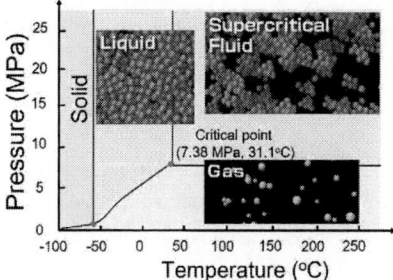

Fig. 1. Phase diagram of CO_2.

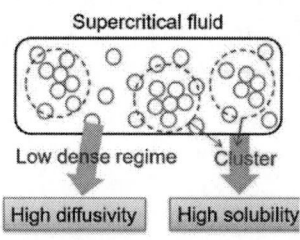

Fig. 2. The gas- and liquid-like supercritical CO_2.

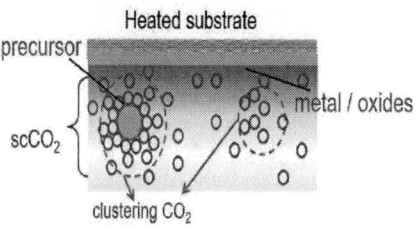

Fig. 3. Thermal decomposition of precursor in $scCO_2$ for film deposition.

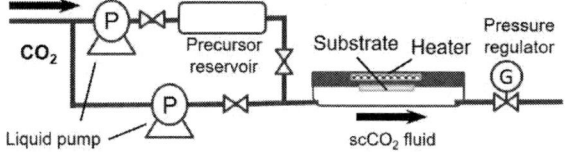

Fig. 4. Schematic diagram of flow-type TiO_2-SCFD.

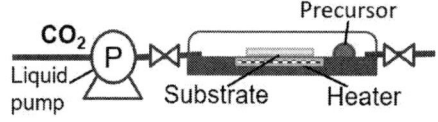

Fig. 5. Schematic diagram of closed-type RuO_2-SCFD.

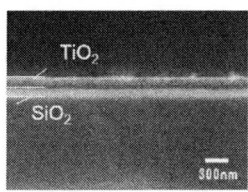

Fig. 6. TiO_2 film on SiO_2 by SCFD.

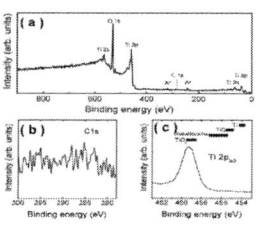

Fig. 7. Stoichiometric TiO_2 by XPS.

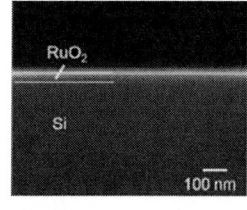

Fig. 8. RuO_2 film on Si by SCFD.

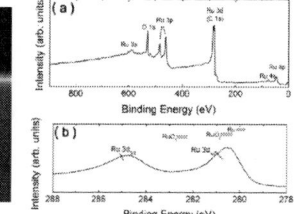

Fig. 9. Stoichiometric RuO_2 by XPS.

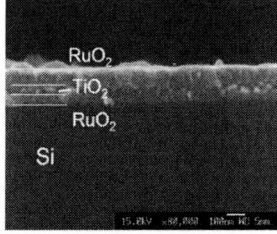

Fig. 10. $RuO_2/TiO_2/RuO_2$ multi-layers on Si substrate.

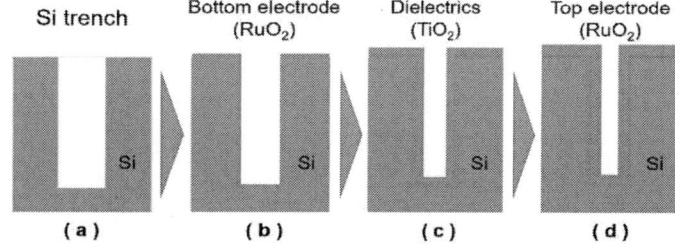

Fig. 11. Deposition procedure of fabricating conformal $RuO_2/TiO_2/RuO_2$ stacks on high aspect ratio Si trenches.

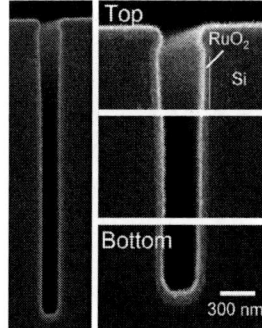

Fig. 12. Conformal SCFD-TiO_2.

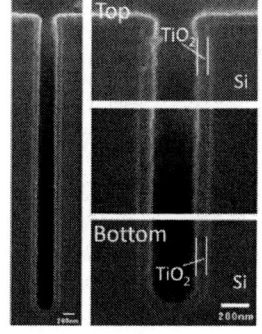

Fig. 13. Conformal SCFD-RuO_2.

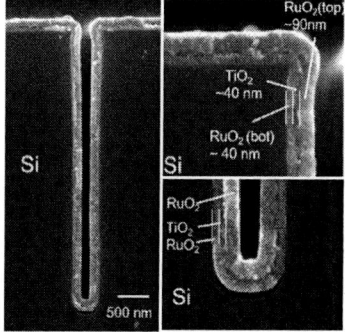

Fig. 14. Conformal $RuO_2/TiO_2/RuO_2$ structure on 10 AR trench.

978-1-5090-4661-4/17 $31.00 © 2017 IEEE

Crystallinity Improvement using Migration-Enhancement Methods for Sputtered-MoS$_2$ Films

Shin Hirano, Jun'ichi Shimizu, Kentaro Matsuura, Takumi Ohashi, Iriya Muneta,
Kuniyuki Kakushima, Kazuo Tsutsui and Hitoshi Wakabayashi

Tokyo Institute of Technology, Yokohama, Japan, hirano.s.ad@m.titech.ac.jp

Abstract

We investigate crystallinity of sputtered MoS$_2$ films formed in various sputtering conditions to enhance the migration. We found that high substrate temperature, high radio frequency (RF) power and long throw were effective for crystallinity improvement of sputtered MoS$_2$ films and in these conditions higher Hall-effect mobility of 12 cm^2/V-s and lower carrier density of 10^{18} cm^{-3} were achieved.

(Keywords: Transition-metal di-chalcogenide, Molybdenum di-sulfide, Sputtering, Migration)

Introduction

Advanced transistors for logic LSI, NAND flash, image sensor and others have been achieved by three-dimensional structure, to enhance a number of functionalities per unit area, even if a process to form fine pattern has been faced at a red brick wall [1]. Furthermore, a recent 3D transistor, FinFET, has been performed with a thin body thickness less than 10 nm. To suppress a variability of performance such as threshold voltages, an atomic-scale control of structures is required. Therefore, transition-metal di-chalcogenide (TMDC) such as molybdenum disulfide has been considered as an attractive channel material, having unique characteristics such as layer structures, transparency, flexibility, high mobility ($\sim$ 700 cm^2/V.s) even in atomically thin region, and has energy band gap (E_g = 1.9 eV) [2,3]. The MoS$_2$ films in application have been synthesized by the exfoliation method or CVD. However, the exfoliation is not suitable to synthesize on the 3D structure. Moreover, it is difficult to fabricate large and uniform MoS$_2$ films by using CVD because of its triangular structure [4]. In this paper therefore, we synthesize MoS$_2$ by sputtering process controlling a nucleation and growth using various sputtering conditions.

Experimental method

Figure 1 shows a process flow and a schematic image of our sputtering process. The MoS$_2$ films of 10-nm thick are deposited on a SiO$_2$/Si substrate by the RF sputtering. The sputtering conditions are substrate temperature set from 200 to 600°C, RF power of 20 to 50W, distance between the MoS$_2$ target and substrate from 150 to 200 mm, base pressure of 10^{-6} Pa, argon (Ar) partial pressure of 0.55 Pa, Ar flow of 7.0 sccm and the MoS$_2$ target is 99.79% pure with an effective diameter of 80 mm. The Raman spectroscopy and Hall-effect measurement are used to evaluate physical and electrical properties.

Results and discussion

Figure 2 (a) shows the Raman spectra of the MoS$_2$ films sputtered with various substrate temperatures from 200 to 600°C. Figure 2 (b) shows the Raman intensity of peak of A$_{1g}$ mode. The Raman intensity increases with an increase in substrate temperature from 200 to 400°C, and decreases from 500 to 600°C, maybe caused by sulfur defects in MoS$_2$. The high-quality sputtered-MoS$_2$ films were obtained at the temperature of 400°C.

Figures 3 (a) and (b) show the Raman spectra of the MoS$_2$ films sputtered with various RF powers of 20-50 W and the Raman peak intensity of A$_{1g}$ mode, respectively. The Raman intensity increases with an increase in RF power. Table 1 shows the Hall-effect results of the MoS$_2$ films sputtered with 20 and 50 W of RF power. The mobility of the MoS$_2$ film sputtered at 50 W is significantly higher than that at 20 W, and in contrast the carrier density at 50 W is considerably lower than 20 W.

Figure 4 (a) shows the Raman spectra of the MoS$_2$ films sputtered with a distance of 150 and 200 mm between the MoS$_2$ target and substrate [5]. Figure 4 (b) shows values of full width at half maximum (FWHM) of E$^1_{2g}$ and A$_{1g}$ modes of the Raman spectra. Although there is only slight difference in the Raman peak intensity, the values of FWHM decrease with a decrease in the distance between the target and substrate from 200 to 150 mm. We can speculate that these results are caused by the enhancement of the migration with a higher average energy of sputter particles without any increase in the process temperature, as shown in Fig. 5.

Conclusion

The high-quality sputtered-MoS$_2$ films were obtained by migration-enhanced sputtering such as high substrate temperature, high RF powers and long-throw methods. They are helpful to achieve the 2D-TMDC FET on 3D structure.

Acknowledgments

This work was partly supported by JSPS KAKENHI of 16K14247/26105014 and COI of JST.

References

[1] Yoshiaki Fukuzumi, et al., IEDM (2007), 449-452

[2] H. Wang, et al., IEDM, 4.6 (2012) 88-91.

[3] S. Das, et al., Nano Lett. 13, 100 (2013).

[4] T. Ohashi, et al., Jpn. J. Appl. Phys 54, (2015).

[5] Nobuhiro Motegi, et al., Journal of Vacuum Science & Technology B 13, 1906 (1995).

Fig. 1: Process flow and schematic image of our sputtering process.

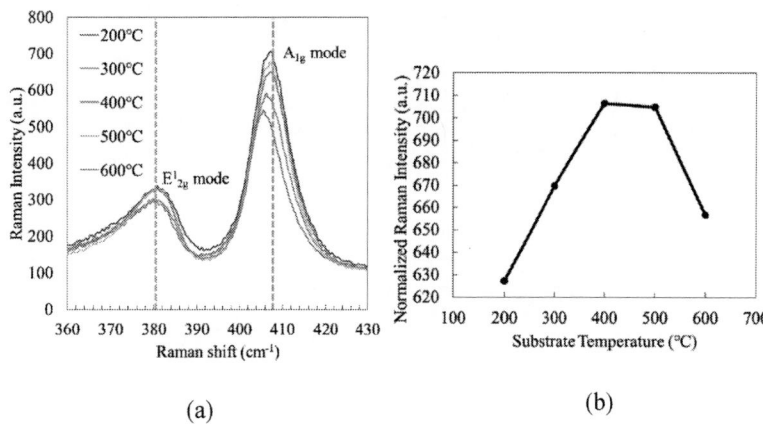

(a)   (b)

Fig. 2: Raman scattering results of the MoS_2 films depending on substrate temperature with RF power of 50 W and distance of 150 mm between the MoS_2 target and substrate. (a) spectra with bulk data by dashed lines and (b) peak intensity of A_{1g} mode.

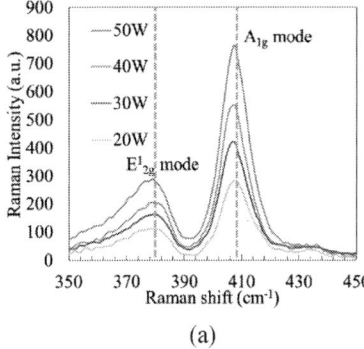

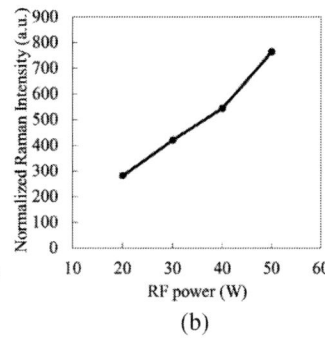

(a)   (b)

Fig. 3: Raman scattering results of the MoS_2 films depending on RF power with substrate temperature of 400°C and distance of 150 mm between the MoS_2 target and substrate. (a) spectra with bulk data by dashed lines and (b) peak intensity of A_{1g} mode.

Table. 1: Hall-effect measurement results of MoS_2 films depending on RF power.

RF power [W]	Hall mobility [cm²/V-s]	Carrier density [cm⁻³]
20	0.61	2.5×10^{20}
50	12	7.5×10^{18}

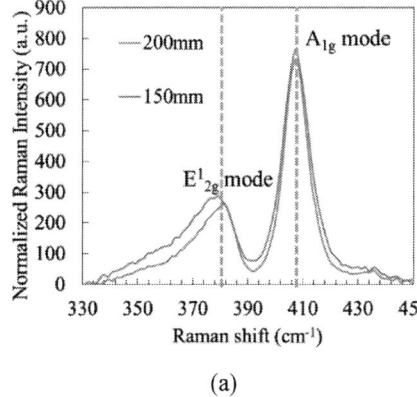

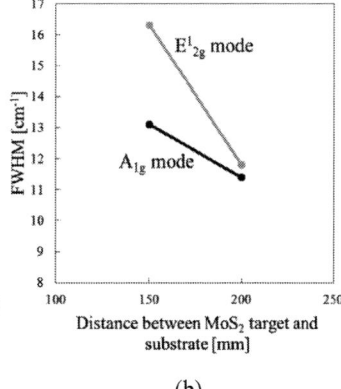

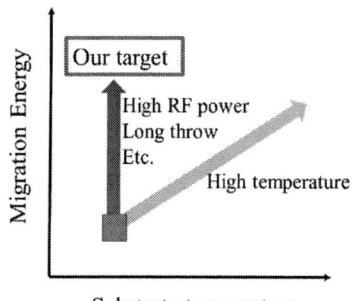

(a)   (b)

Fig. 4: Raman scattering results of the MoS_2 films depending on distance between the MoS_2 target and substrate with substrate temperature of 400°C and RF power of 50 W. (a) spectra with bulk data by dashed lines and (b) values of FWHM of E^1_{2g} and A_{1g} modes.

Fig. 5: Schematic figure of our experimental concept. High RF power and long throw methods are preferable rather than high temperature process to maintain the migration energy high and the process temperature low, simultaneously.

978-1-5090-4661-4/17 $31.00 © 2017 IEEE   235

Gap in pagination due to unavailable paper.

Pages 236-237

This page intentionally left blank.

Photoresist development for wafer-level packaging process

Makiko Irie, Toshiaki Tachi, Atushi Sawano

Tokyo Ohka Kogyo Co., LTD

Abstract

Semiconductor assembly process has been improved by wafer level packaging (WLP) introduction in high volume manufacturing. Cu-Redistribution layer (RDL) miniaturization is one of key process for small, thin and light chip manufacturing. Development of high resolution and transmittance control of photoresist is required to this technology realization on topology substrate.

Photoresist development with high resolution and transmittance is required to this technology realization on the topology substrate.

Chemical amplified (CA) type photoresist indicated stable sensitivity at various thickness because of high transparency at 365nm wavelength.

Below 2um L&S resolution was achieved by the optimized formulation in this research.

(Keywords: photolithography, photoresist, NQD, CA, ultra-fine-pitch)

Introduction

Scaling down of semiconductor device has been required to thinner, smaller and lighter for the assembled chip in the recent cell phone or information terminal tool market trend.

Latter 1990s, packaging process was innovated from chip scaled packaging (CSP) to wafer level packaging (WLP) and that is improved total cost, performance and productivity as a recent example of fan-out wafer level packaging (FOWLP) technology.

Cu-Redistribution layer(RDL) patterning is one of key processes and has been faced at miniaturization requirement with high pattern accuracy.[1] Fine patterned photoresist is only able to realize the target structure and several types of photoresist have been discuss to achieve the technology target.

Experimental

Two types of photoresist were compared each other in order to achieve below 2 um L&S resolution in this research.

Positive-tone type photoresist was selected in the view of high resolution and wide process latitude in photolithography process.

Photoresist based on naphthoquinone diazide (NQD) is good candidate due to excellent process stability control in the environment. On the other hand, chemically amplified(CA) type which controls development contrast by decomposing protecting group with catalytic activated photo acid generator(PAG) has advantages as higher resolution and as good profile control under high aspect ratio condition due to good transparency at irradiation sources, 365 nm wavelength.

Stability of photo sensitivity against irradiation energy and transparency of photoresist are very important parameters at manufacturing process that is related to profile control at the topology substrate. Photo sensitivity was measured as "Eth" by total film developed exposure energy. (Fig.1)

Eth curve of CA type showed more gradual slope than NQD type at various film thickness which indicated insensitive for exposure deviation through film thickness range.

CA type absorbance at 365 nm wavelength is quite lower than NQD type. (Fig.2)

Photo insensitive dependency at various thickness and high transparency of CA type are expected to achieve high resolution and no topology dependency at manufacturing process.

Formulated photoresist based on CA type was evaluated in details and it was confirmed to be achieved to our target performance. Resist pattern profile at 2 um L&S was introduced as an example. (Fig.3)

Conclusion

Advanced WLP technology brought us to continue developing high resolution RDL positive photoresist with high transparency, especially CA type indicated good potential to apply further next generation process.

References

[1] N. Motohashi, "System in Wafer-Level Package Technology with RDL-first Process," in *Proc. IEEE*
Electronic Components and Technol. Conf. (ECTC)

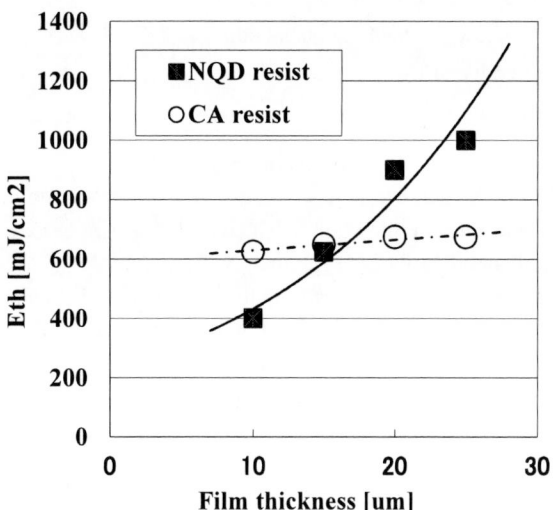

Fig1   Photo sensitivity dependency for
              photoresist thickness

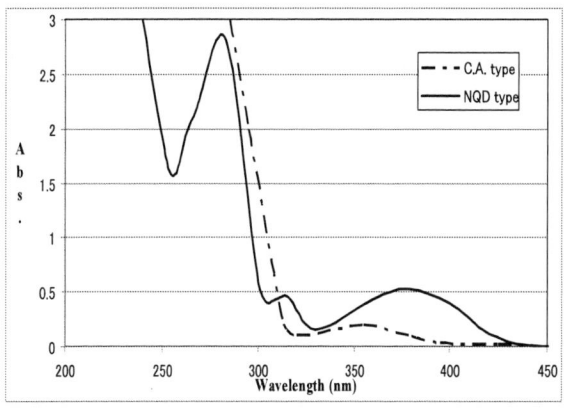

Fig2   Photoresist absorbance at each
              wavelength

Fig.3   SEM image of L&S pattern
             2umL&S
             8um film thickness

P-31

Second-harmonic susceptibility enhancement in Gallium nitride nanopillars

Kangwei Wang[1], Haoliang Qian[1], Zhaowei Liu[1,2], Paul K. L. Yu[1,2]

[1]Dept. ECE, University of California, San Diego, kaw032@ucsd.edu

[2]Materials Science and Engineering, University of California, San Diego

Abstract

Second-harmonic generation (SHG) from single GaN nanopoillar is reported. A model for the SHG processes in the GaN nanopillar as a function of diameter is presented; the analysis showed quantitatively that the SHG is dominated by its surface area. The effective second order nonlinear optical susceptibility increases as the diameter of the GaN nanopillar decreases, reaching a value of 136 pm/V at 150nm diameter, making them attractive for modulator applications.

(Keywords: GaN nanopillar, electro-optic materials, photonic integrated circuits, and optoelectronic modulator)

Introduction

GaN/AlGaN electronic and opto-electronic materials have been considered for future generations of silicon photonic integrated circuits, due largely to their desirable and resilient properties for hetero-epitaxy, for example, on Si. For opto-electronic modulation purpose, materials with large second-order nonlinear optical response is in great demand, as they can be exploited for strong linear electro-optic effect for applications which requires high linearity electro-optic modulation. In our prior study, we observed enhancement of linear EO coefficient in InP nanowires [1]. However, due to the bottom-up nanowires growth employed, the nanowire length showed high variation, which is not desirable for manufacture. Recent publications suggest that the second harmonic generation (SHG) could be enhanced by using nanowires/ nanopillars instead of bulk compound semi- conductor materials [2]. GaN belongs to wurtzite crystal and has a different crystal symmetry from that of InP (zinc-blende), their SHG properties can be different.

In this work, we investigate second harmonic response in single GaN nanopillar which is fabricated via a top-down approach, ensuring the nanopillar is vertical to the substrate, and also facilitates the fabrication of the electrooptic modulator. GaN is a high bandgap material and has small absorption in the visible and infrared wavelengths. Furthermore, it has wide application in high speed, high power electronics, and light emitters. In our measurement, the effective second harmonic susceptibility of GaN nanopillar at 900nm fundamental excitation is found to be ~7 times higher than GaN bulk materials, and our analysis shows that the enhancement of SHG is most likely comes from the nanowire surface area.

Experiments

A GaN layered structure was grown by metal-organic vapor phase epitaxy on a 2-inch diameter single side polished c-plane sapphire wafer. Single GaN nanopillars were patterned by using E-beam lithography followed by dry etching using a reactive ion etching apparatus. The nanopillars shown in Fig.1 have the same height of 432nm as determined from the RIE etching time. We performed the reflective SHG measurement of the samples at room temperature in a back scattering geometry (see Fig. 2) with mode-locked Ti: Sapphire laser operating at 900nm wavelength.

Results and Discussion

Table 1 listed the intensity of the SHG signals from single GaN nanopillar with an average diameter ranged from 150-400nm. The second order non-linearity is obtained from the second order susceptibility via

$$P^{(2)}(2\omega) = \epsilon_0 \chi^{(2)} E(\omega)^2 \qquad (1)$$

where $P^{(2)}$ is the amplitude of the component of nonlinear polarization oscillation at frequency 2ω, $\chi^{(2)}$ is the second order susceptibility. For reference, the GaN film grown on c-sapphire substrate has a $\chi^{(2)}$ of ~20pm/V [3]. Its nonlinear effect originates from the spontaneous polarization and piezo- electricity [3]. In the case of GaN nanopillar, the dangling bonds at the surface lead to an additional contribution to the second order nonlinearity. Consequently, the corresponding intensity of the second harmonic signals, $I(2\omega)$, exhibits both a volume-dependent component and a surface area-dependent component. By varying the diameter they can be separated via

$$I(2\omega) \propto A_{sur} \cdot S^2 + B_{bulk} \cdot V^2 \qquad (2)$$

As can be seen from Fig. 3, the narrower the nanopillar is, the higher is the surface-to-volume ratio, and therefore, the larger is the contribution of the surface term to the second order susceptibility.

Conclusion

Our results show the surface polarization contribution to the second order nonlinearity becomes dominant as GaN nanopillar diameter decreases and the GaN nanopillar can be further exploited for electro-optic modulation applications.

Acknowledgments

The authors like to acknowledge the funding support

978-1-5090-4661-4/17 $31.00 © 2017 IEEE

of the Multidisciplinary University Research Initiative (MURI) (N00014-13-1-0678).

References

[1] C. J. Novotny, C. T. DeRose, R. A. Norwood, and P. K. L. Yu, "Linear electrooptic coefficient of InP nanowires," Nano Lett., Vol. 8, 1020, March 2008, doi: 10.1021/nl072688k

[2] W. Liu, K. Wang, Z. Liu, G. Shen, and P. Lu, "Laterally emitted surface second harmonic generation in a single ZnTe nanowire," Nano Lett., Vol. 13, 4224, Aug 2013, doi: 10.1021/nl401921s.

[3] M. Abe, H. Sato, J. Suda, M. Yoshimura, Y. Kitaoka, Y. Mori, I. Shoji, T. Kondo, and H. Gan, "Accurate Measurement of Nonlinear Optical Coefficients of Gallium Nitride," J. Opt. Soc. Am. B, Vol. 27, 1320, October 2009, doi: 10.1364/JOSAB.27.002026

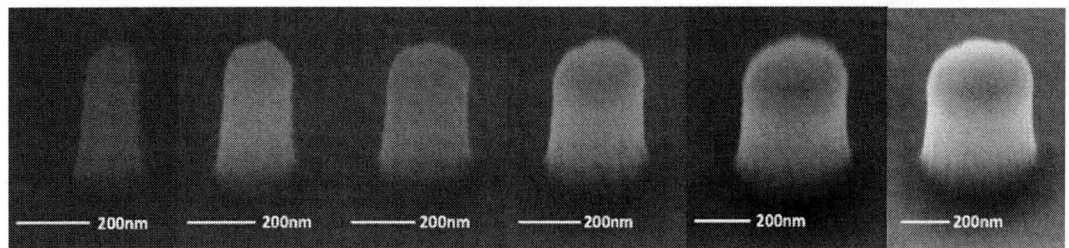

Fig. 1. 45° titled SEM images of single nanopillar. (diameter (from left to right):150, 200, 250, 300, 350, and 400 nm)

Nanopillar diameter (nm)		150	200	250	300	350	400
Input Parameters	Input Power (mW)	127.6	127.6	127.6	127.6	127.6	127.6
	Transmission (%)	85.2	86.1	85.9	86.6	87.1	85.2
SHG signal	SHG (counts)	80	95	85	245	260	400
	Reflectivity (%)	17.4	14.8	15.2	17.3	17.2	17.4
	$\chi_{eff}^{(2)}$ (pm/V)	136.0	90.2	53.7	58.6	44.2	42.7

Table 1. SHG measurement results for single GaN nanopillar with diameter in the range of 150-400nm.

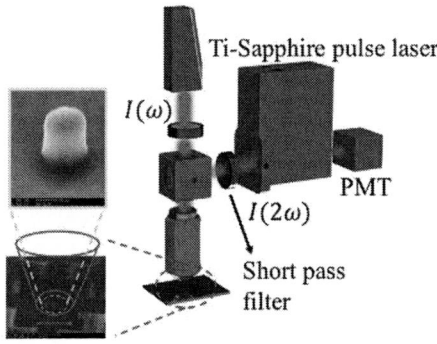

Fig. 2. Setup for reflective second harmonic generation measurement.

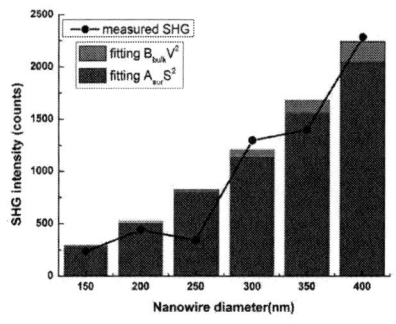

Fig. 3. The measured SHG and fitted results (blue: surface contribution; red: bulk) based on Eq.2.

Defect Formation in SiO$_2$ Formed by Thermal Oxidation of SiC

Kenta Chokawa[1], Masaaki Araidai[1,2], Kenji Shiraishi[1,2]

[1]Graduate School of Engineering, Nagoya University, chokawa@fluid.cse.nagoya-u.ac.jp

[2] Institute of Materials and Systems for Sustainability, Nagoya University

Abstract

In general, an insulating film of SiC MOSFET is SiO$_2$ formed by thermal oxidation of SiC. This SiO$_2$ film has many defect structures which induce much larger threshold voltage shift. In this paper, we investigated the defect formation of an oxygen vacancy in SiO$_2$ by SiC thermal oxidation, using the first-principles calculations, and we found that the oxygen vacancy defect can be generated in the amorphous SiO$_2$ at high temperature.

(Keywords: SiC, SiO$_2$, Defect, Thermal Oxidation)

Introduction

Silicon Carbide (SiC) is one of the candidates for next generation power device material because of its superior characteristics. Among a lot of poly-types of SiC, 4H-SiC has the largest band-gap and the larger wafer with low defect density is available. These characteristics meet the demands for the high performance MOSFET. To realize this device, selection of the insulator is very important.

In general, the insulating film in SiC MOSFET is silica (SiO$_2$). The SiO$_2$ film is formed by thermal oxidation of SiC or SiO$_2$ deposition. However, it has been reported that SiO$_2$ formed by SiC thermal oxidation has many defects near the 4H-SiC/SiO$_2$ interface [1]. These defects cause defect levels near the band-gap edge of SiC, resulting in threshold voltage shift. In this study, we investigated the defect formation of an oxygen vacancy in SiO$_2$ formed by SiC thermal oxidation, using the first-principles calculations.

Method

All calculations were performed by VASP code [2]. We used the Perdew-Burke-Ernzerhof (PBE) exchange-correlation functional and the PAW potentials. Structural optimizations were performed until all the atomic forces were less than 0.001 eV/Å. We defined the defect formation energy as

$$E_{form} = E_{vacancy} + \mu_{CO2} - (E_{SiO2} + \mu_{CO}).$$

$E_{vacancy}$ and E_{SiO2} are the total energies with and without vacancy, respectively. μ_{CO2} and μ_{CO} are the molar Gibbs free energies for CO$_2$ and CO [3]. The defect formation is favorable if E_{form} is negative. Owing to μ_{CO2} and μ_{CO}, we can discuss about the finite-temperature effects.

Result

In SiC oxidation process, it is predicted that CO molecules are emitted from the oxidation front. Consequently, we assumed that a CO$_2$ molecule is generated by a reaction between the CO molecule and an O atom in SiO$_2$, resulting in an oxygen vacancy defect [4].

In order to verify our assumption (Fig.1), we modeled one α-quartz with an O vacancy as shown in Fig.2(a) and two amorphous SiO$_2$ (a-SiO$_2$) with an O vacancy as shown in Fig.2 (b) and (c). The formation energies of Fig.1(a), (b) and (c) at zero temperature are E_{form}(a) = 1.86eV, E_{form}(b) = 0.47eV and E_{form}(c) = 1.49eV, respectively. The energy difference comes from the bond length of resulting Si-Si bonding. We found from the formation energies that CO$_2$ molecules does not arise at zero temperature. However, at finite temperatures, the sign of E_{form} (pt1) changed around the 1400K (Fig.3 (a)), which means that CO$_2$ molecules are generated over 1400K. We found from the detailed analyses that the energy gain originates from the rotational and vibrational modes of CO$_2$ molecules at high temperature (Fig.3 (b)). This result means that an emitted CO molecule extract an O atom in a-SiO$_2$ at high temperature, resulting in an O vacancy defect.

Conclusion

In Si oxidation, Si atoms or SiO molecules emitted from the oxidation front do not affect the resulting SiO$_2$, which means that the quality of the SiO$_2$ film is very high [5,6]. On the other hands, in SiC oxidation, emitted CO molecules degrade the quality of SiO$_2$. This implies that the number of oxygen vacancies and related defects in SiO$_2$ by SiC oxidation is much larger than Si oxidation. If the SiO$_2$ by thermal oxidation is used as an insulating film in SiC MOSFET, special treatments to decrease the defects are essential.

References

[1] H. Yano, et al., IEEE Trans. Electron Devices **62**, 324 (2015).

[2] G. Kresse and J. Hafner, Phys. Rev. B**47**, 558 (1993).

[3] Y. Kangawa, et al., Surf. Sci. **493**, 178 (2001).

[4] A. Oshiyama, J. Appl. Phys. **37**, 232 (1998).

[5] H. Kageshima and K. Shiraishi, et al., Phys. Rev. Lett. **81**, 5936 (1998).

[6] Z. Ming, et al., Appl. Phys. Lett. **88**, 153516 (2006).

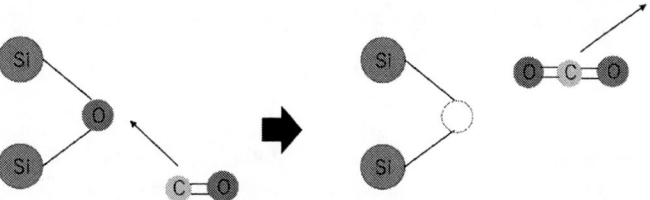

Fig. 1 Schematic illustration of our assumption for the defect formation of the oxygen vacancy in SiO$_2$.

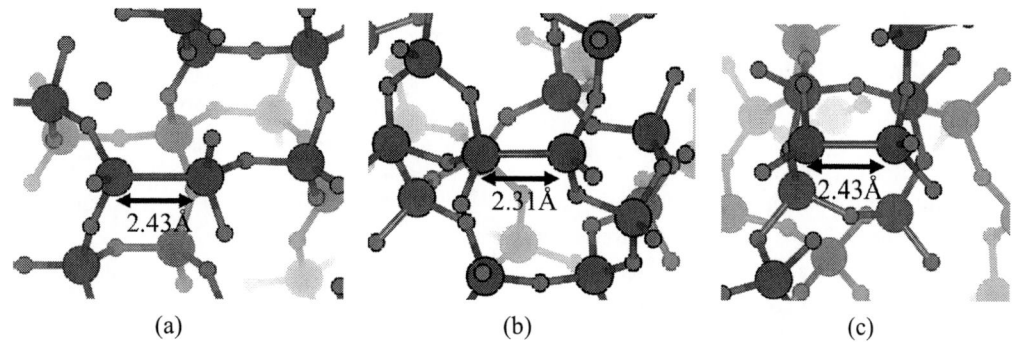

(a)   (b)   (c)

Fig. 2 Calculation models of SiO$_2$ with an oxygen vacancy defect. (a) α-quartz structure and (b) amorphous SiO$_2$ structure pattern1 and (c) pattern2. Red small spheres and blue large spheres represents oxygen atoms and silicon atom, respectively.

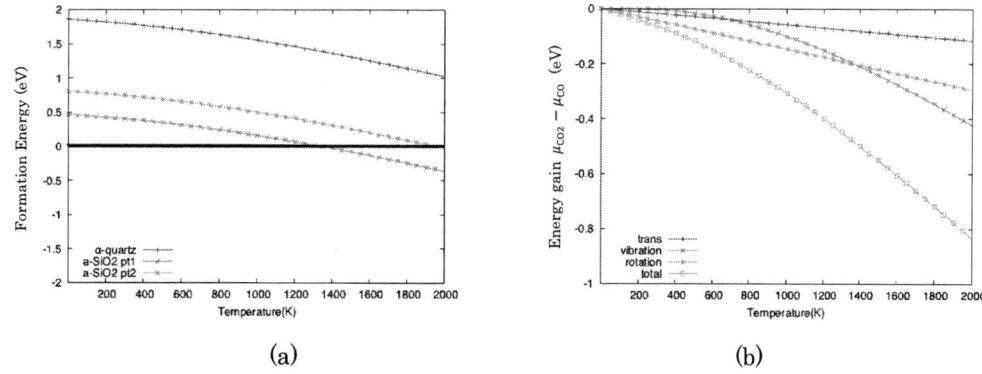

(a)   (b)

Fig. 3 (a) Formation energy of the oxygen vacancy defect. (b)The energy gain of CO$_2$ molecule, which is defined as the difference of molar Gibbs free energy, $\mu_{CO2} - \mu_{CO}$. The partial pressure of CO and CO$_2$ are set to P=0.10 atm.

Current Enhanced Solid Phase Precipitation (CE-SPP) for Direct Deposition of Multilayer Graphene on SiO$_2$ from a Cu Capped Co-C Layer

Hiroyasu Ichikawa[1] and Kazuyoshi Ueno[1,2]

[1] Graduate School of Engineering and Science, Shibaura Institute of Technology (SIT), Tokyo, Japan

[2] Research Centre for Green Innovation, SIT, Tokyo, Japan, ueno@shibaura-it.ac.jp

Abstract

To improve the crystallinity of multilayer graphene (MLG) directly deposited on SiO$_2$ for interconnect applications, a new solid phase precipitation (SPP) process involving current stress is investigated. It is found that the MLG crystallinity precipitated from a Cu capped Co-C layer can be improved by the vertical current to the Cu/Co-C but not by the horizontal current. The current enhanced SPP (CE-SPP) is expected as a mean to improve the MLG crystallinity directly deposited on SiO$_2$.

(Keywords: multilayer graphene, interconnect)

Introduction

Nano-carbons such as multilayer graphene (MLG) are expected as the candidate for advanced interconnect materials, since they have a longer mean free path and higher electro-migration (EM) resistance than Cu [1]. One of the issues for the MLG interconnects is the deposition process for MLG. Solid phase precipitation (SPP) by annealing Co/C has an advantage in production, since MLG can be deposited directly on a dielectric surface such as SiO$_2$ without using any transfer process [2]. However, high temperature and a thick Co/C layer which leads to a thick MLG were required to deposit a uniform MLG with good crystallinity [2]. We have developed an improved SPP process for thin and uniform MLG precipitation involving carbon-doped cobalt with Cu capping layer (Cu/Co-C), in which the Cu capping layer avoids the Co agglomeration [3]. However, improvement of the MLG crystallinity has been an issue. Here, current stress is introduced during annealing the Cu/Co-C to improve the MLG crystallinity for the first time, based on our previous study on the current enhanced CVD (CE-CVD) for MLG [4, 5]. Vertical or horizontal current to the Cu/Co-C was applied during annealing, and the crystallinity of the precipitated MLG is compared with those without current.

Experiment

Fig.1 illustrates schematic diagram of the sample structure. As the substrates, SiO$_2$ / Si wafers were cut into 2 $\times$ 1.5 cm^2. Cu (50 nm) / Co-C (100 nm) was deposited on the substrate by magnetron sputtering. The C concentration in the Co was 20 atomic %. The Cu electrodes for supplying current to the sample are connected as an extra feature in a conventional annealing apparatus as shown Fig.2. MLG was grown by annealing with or without current at the set-up temperature (Table.1) in an Ar atmosphere. The current stress was applied horizontally or vertically. In the case of horizontal current, the applied current was between 4 and 10 A, and the vertical current was 10A. The temperatures without current were set to be the same temperature of the samples with current including the Joule heating. After removing the metal layer, MLG crystallinity was determined by Raman spectroscopy using an exciting LASER of 532 nm. Surface morphology was observed by scanning electron microscope (SEM).

Results and discussion

Fig. 3 shows the Raman spectra of the MLG with and without current stress. The G-peak at 1580 cm^{-1} and the D-peak at 1350 cm^{-1} correspond to graphitic and defective structures of sp^2 carbons, respectively. Since the G peak intensities are higher than those of the 2D-peaks, the deposited films are considered to be multilayers. The peak intensity ratios (G/D ratios) can be used as the index of MLG crystallinity, and they are almost the same between with and without current for the horizontal current. On the other hand, the G peak intensity was enhanced by the vertical current as shown in Fig.3.

Fig. 4 shows the comparison of the G/D ratios at the same temperature including joule-heating by the current, showing that the G/D ratio is very high with the vertical current comparing to those with horizontal current or without current. In the case of thermal CVD on a Co catalyst, horizontal current enhanced the CVD growth of MLG [4, 5]. However, it is interesting that the horizontal current had almost no effects in the SPP. The reason for the difference may be due to the different metal structure in the SPP process. Since current is considered to flow preferentially in the Cu layer owing to the lower resistance, the current density in the Co-C layer may become low. On the other hand, C diffusion should be enhanced by the vertical current since current density in the Co-C layer is not affected by the Cu capping layer. Further investigation on the mechanism will be a future study.

The SEM image of MLG on SiO_2 for the vertical current is shown in Fig.5, and Raman maps of MLG in the range of 5 μm × 5 μm are shown in Fig.6. In the Raman mapping, it is possible to observe G-peak in all of the area. Similar uniform mapping for the 2D-peak was obtained. Therefore, The Raman mapping of Fig.6 indicate that MLG was grown uniformly and covered the SiO_2 along with SEM observation.

Conclusion

We investigated the effect of current stress for improvement of MLG crystallinity by solid-phase precipitation. The horizontal current had no almost effect, however vertical current enhanced the crystallinity of MLG. With further optimization of processes, current stress might be applied to the fabrication of MLG with improved crystallinity and uniformity at a lower temperature without the transfer process.

Acknowledgement

This research was supported by JSPS KAKENHI Grant Number JP26420319.

References

[1] H. Li, C. Xu, N. Srivastava and K. Banerjee, IEEE Trans. Electron Dev. **56** (2009) 1799.

[2] M. Sato, M. Takahashi, H. Nakano, Y. Takakuwa, M. Nihei, S. Sato and N. Yokoyama, Jpn. J. Appl. Phys. **53** (2014) 04EB05.

[3] S. Sano et al., IITC-MAM (Grenoble, 2015).

[4] L. A. Razak, D. Tobino and K. Ueno, Microelectronic Eng. **120** (2014) 200.

[5] K. Ueno, H. Ichikawa, and T. Uchida, Jpn. J. Appl. Phys. **55** (2016) 04EC13.

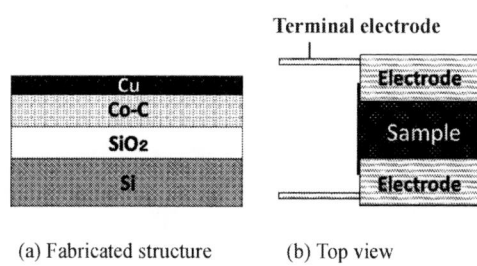

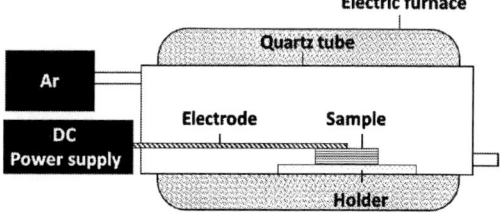

Fig.2 : Schematic diagram of annealing system.

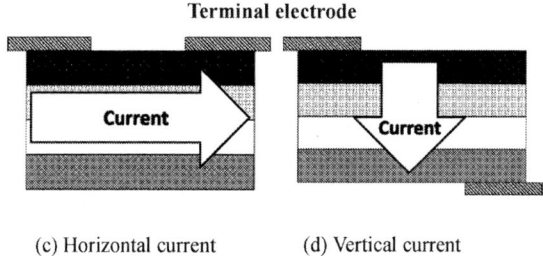

(c) Horizontal current   (d) Vertical current

Fig.1 : Schematic cross-sections of (a) fabricated structure, (b) top view of sample, (c) horizontal current flow, and (d) vertical current flow.

Table.1 : Summary of experimental conditions. The temperature without current were set to be the same of those with current including the Joule heating.

Without Current	With Current (Horizontal)	With Current (Vertical)
500 ℃		
620 ℃	500 ℃ + 4 A	
650 ℃	500 ℃ + 6 A	
710 ℃	500 ℃ + 10 A	500 ℃ + 10 A

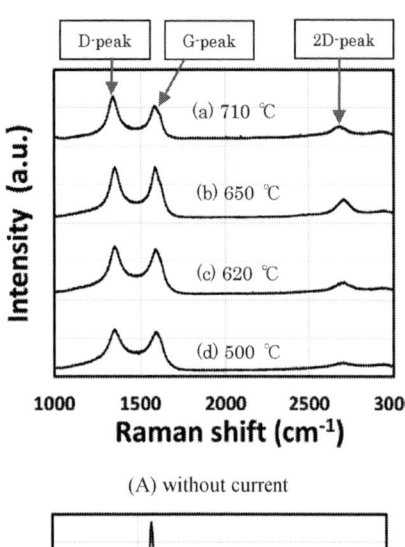

(A) without current

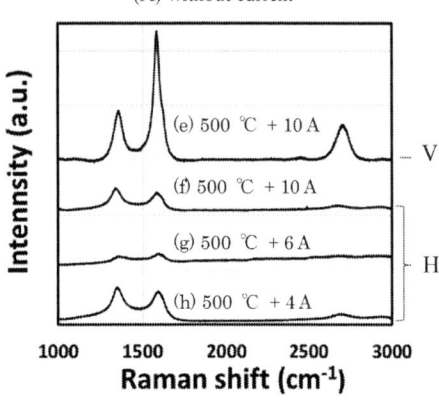

(B) with current

Fig.3 : Raman spectra of the MLG films for (a) , (b) (c), (d) without current, (e) with vertical current, and (f), (g), (h) with horizontal current.

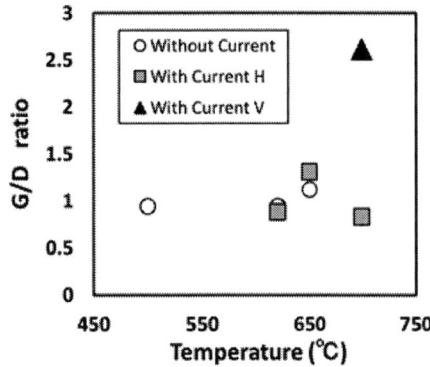

Fig.4 : Intensity ratios of G and D peaks for without current, with horizontal current (H), and with vertical current (V).

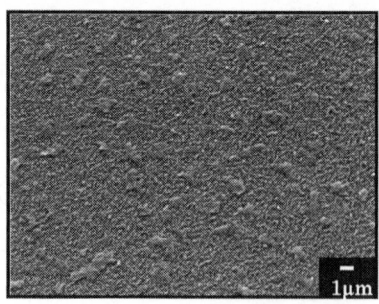

Fig.5 : SEM image of MLG for Cu (50 nm) / Co-C (100 nm) applied vertical current after removing metal.

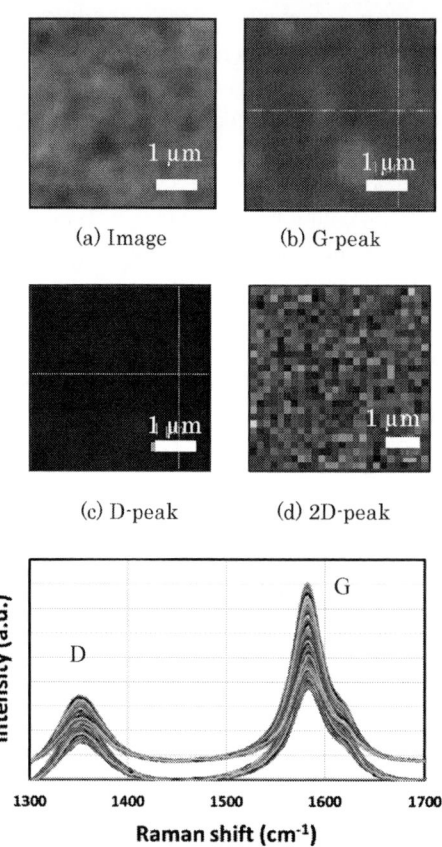

(a) Image   (b) G-peak

(c) D-peak   (d) 2D-peak

Fig.6 : Raman mapping of MLG for the applied vertical current after removing the metal. (a) Optical microscope image, (b) distribution mapping of G-peak, and (c) D-peak. The brighter areas correspond to higher intensities of the G- and D-peaks in (d) the Raman spectra.

Enhanced Photoresponse of InGaZnO TFT to Ultraviolet Illumination by Using a High-*k* Dielectric

Libin Liu[1], Renrong Liang[1], Jing Wang[1], and Jun Xu[1],

[1]Tsinghua National Laboratory for Information Science and Technology, Institute of Microelectronics,
Tsinghua University, Beijing 100084, China, liangrr@tsinghua.edu.cn

Abstract

High performance InGaZnO TFT with high-*k* Al_2O_3 gate dielectric were fabricated and its photoresponse to ultraviolet illumination was compared to a control TFT with SiO_2 dielectric. Due to the incorporation of Al_2O_3, its saturation mobility is increased for 7.4 times, and its photoresponsivity (~400A/W) is boosted for 10^3 times along with a significantly improvement in linearity and contrast ratio (~10^7).
(Keywords: InGaZnO, thin film transistor, high-*k*, ultraviolet photoresponse)

Introduction

Indium gallium zinc oxide (InGaZnO) has been found to be a promising channel material in thin film transistors (TFTs) due to low process temperature, good uniformity, high mobility [1]. Recently, InGaZnO-based photodetectors have drawn substantial attentions for ultraviolet (UV) detection due to its large bandgap (~3.4 eV) [2][3]. However, few efforts have been done about the phototransistors with high-*k* gate dielectric. In this work, the photorespose of the InGaZnO TFTs with Al_2O_3 dielectric was investigated.

Experiment

Fig. 1 shows the schematic diagram and process flow of the InGaZnO TFTs. Heavily doped P-type silicon wafer was used as substrate and the back gate of the TFTs. For Sample A, a 100-nm thick thermal oxide was grown as the gate dielectric, while for Sample B, a 30-nm thick Al_2O_3 was deposited instead using an atomic layer deposition (ALD) system. Then, a 30-nm thick amorphous InGaZnO (In:Ga:Zn=1:1:1 mol%) film was sputtered. The active region was defined using lithography followed by HCl solution etching. Afterwards, 200-nm thick Mo was deposited and patterned as the source and drain electrodes by lift-off. Finally, the TFTs were annealed in N_2 ambient at 300℃ for 1 h to complete the device fabrication. An UV laser of λ=365 nm wavelength was used as the illumination source. All measurement was carried out at room temperature.

Results and Discussion

The transfer characteristics (I_{DS}-V_{GS}) are shown in Fig. 2 and the key parameters are listed in Table 1. The two samples both exhibit excellent switching characteristics with an On/Off ratio higher than 10^8.

Due to the incorporation of Al_2O_3 dielectric and the reduced EOT, the Sample B shows a high saturation mobility of 13.37 $cm^2V^{-1}s^{-1}$, which is 7.4 times higher than that of Sample A. Moreover, Sample B shows a lower threshold voltage (V_{TH}) and subthreshold swing (SS). The I_{DS}-V_{GS} curves of Sample A under UV illumination with different intensities are shown in Fig. 3. It is shown that its off-state current (I_{off}) was sensitive to the UV light and exhibits a high contrast ratio of 10^5 at V_{GS}<0, which is defined as the ratio of photocurrent to dark current. However, Sample A exhibited a bad linearity and low photoresponsivity (S), which is 0.35 A/W at V_{GS}=0 V. For comparison, the I_{DS}-V_{GS} curves of Sample B is shown in Fig. 4. A contrast ratio of 10^7 is clearly seen and the light induced photocurrent is uniform over the full range in the off-state. The photocurrent as the function of UV intensities is plotted in Fig. 5 at various V_{GS}. It is obviously shown that the photocurrent increases as the increase of intensity with an excellent linearity. The slope of the line is extracted as the S value, which is increased as the increase of V_{GS}. The S as a function of V_{GS} at V_{DS}=5 V and 0.1 V is presented in Fig. 6. It is shown that the S increases as the increase of V_{DS} and V_{GS} in the off-state regime. Most importantly, the S value of Sample B is up to around 400 A/W at V_{GS}=0 V and V_{DS}=5 V, which is 10^3 times higher than that of Sample A at the same situation. For Sample B, the improvement in linearity, S and contrast ratio may originate from the increased electric field in the InGaZnO film. The dynamic response of Sample B to a periodical UV pulse is also shown in Fig. 7 and a fast real time response is obtained.

Conclusion

The incorporation of high-*k* gate dielectric can not only increase the electric performance of InGaZnO TFTs, but also significantly improve the contrast ratio and photoresponsivity, making such devices prospective candidates for UV light sensors.

References

[1] K. Nomura, et al., *Nature*, vol. 432, pp.488, 2004.

[2] D. L. Jiang, et al., *Applied Physics Letters*, vol. 106, pp. 171103, 2015.

[3] C. W. Lin, et al., *Thin Solid Films*, vol. 618, pp. 73, 2016.

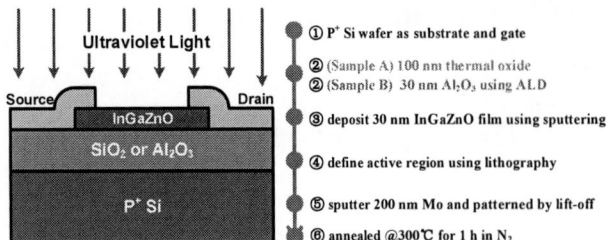

① P⁺ Si wafer as substrate and gate
② (Sample A) 100 nm thermal oxide
② (Sample B) 30 nm Al₂O₃ using ALD
③ deposit 30 nm InGaZnO film using sputtering
④ define active region using lithography
⑤ sputter 200 nm Mo and patterned by lift-off
⑥ annealed @300℃ for 1 h in N₂

Fig. 1: Schematic diagram and process flow of the fabricated InGaZnO TFTs with different gate dielectrics.

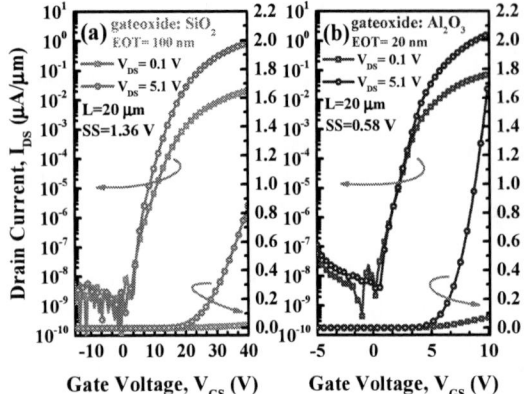

Fig. 2: I_{DS}-V_{GS} curves of (a) Sample A and (b) Sample B.

Table 1: Comparison of Sample A and B.

Parameters (unit)	Sample A	Sample B
Gate Length, L (μm)	20	20
Gate Width, W (μm)	80	100
Gate Dielectric	SiO₂	Al₂O₃
Thickness of gate oxide, t_{ox} (nm)	100	30
Equivalent oxide thickness, EOT (nm)	100	20
Mobility, μ_{sat} (cm²V⁻¹s⁻¹)	1.80	13.37
Threshold voltage, V_{TH} (V) *$I_{DS}(V_{TH})$=100 W/L nA	16.60	4.77
Subthreshold swing, SS (V/decade)	1.36	0.58
On/Off ratio @V_{DS}=5.1 V, I_{on}/I_{off}	$10^{9.7}$	$10^{8.6}$

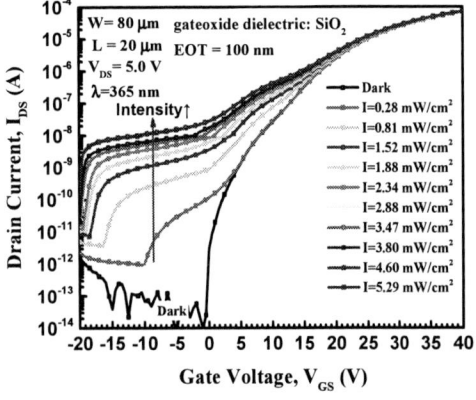

Fig. 3: I_{DS}-V_{GS} curves of the InGaZnO TFT with SiO₂ gate dielectric in the dark and under UV illumination with different intensities.

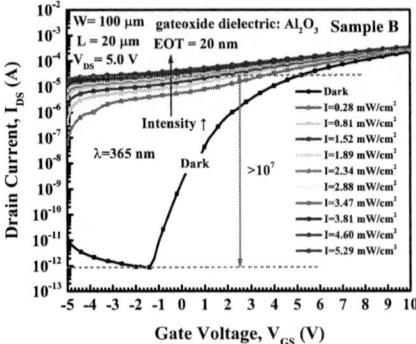

Fig. 4: I_{DS}-V_{GS} curves of the InGaZnO TFT with Al₂O₃ gate dielectric in the dark and under UV illumination with different intensities.

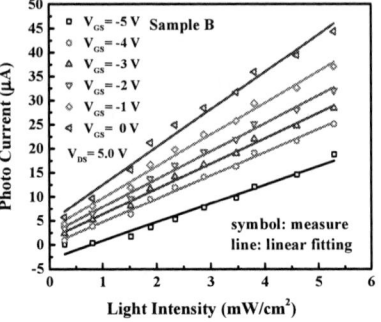

Fig. 5: Photocurrent as a function of light intensity at different V_{GS}. The lines stand for the linear fitting results, and the photoresponsivity is defined as the slope of the lines.

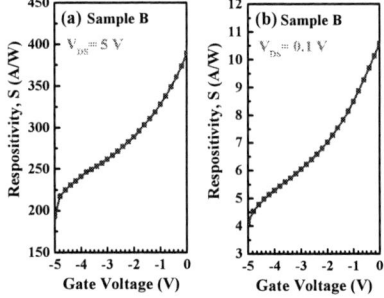

Fig. 6: The extracted photoresponsivity vs. V_{GS} at (a) V_{DS}= 5.0 V and (b) V_{DS}= 0.1 V.

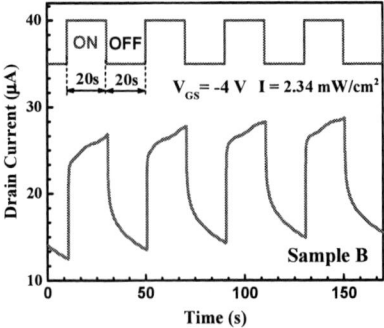

Fig. 7: The photoresponse of Sample B to periodical UV pulses.

978-1-5090-4661-4/17 $31.00 © 2017 IEEE

Development of *in-situ* Sb-Doped $Ge_{1-x}Sn_x$ Epitaxial Layers for Source/Drain Stressor of Strained Ge Transistors

Jihee Jeon[1], Akihiro Suzuki[1,2], Kouta Takahashi[1,2], Osamu Nakatsuka[1], and Shigeaki Zaima[1,3]

[1]Graduate School of Engineering, Nagoya University, Nagoya, Japan, jjeon@alice.xtal.nagoya-u.ac.jp

[2]Research Fellow of the Japan Society for the Promotion of Science, Japan,

[3]Institute of Materials and Systems for Sustainability, Nagoya University, Nagoya, Japan

Abstract

We have investigated the crystalline and electrical characteristics of heavily doped *n*-type $Ge_{1-x}Sn_x$ epitaxial layers with various Sb concentrations up to 10^{20} cm^{-3}. In this study, we focus the thermal stability of Sb doped $Ge_{0.94}Sn_{0.06}$ and Ge epitaxial layers and clarify the relationship between the crystalline and electrical characteristics. At the as-grown condition, the substitutional Sb concentration was achieved at higher than 2.6×10^{20} cm^{-3}. After the post-deposition annealing, Sb-doped $Ge_{1-x}Sn_x$ maintained their superior crystallinity up to 400 °C, while the sheet resistance increases with Sb segregation. We also found that Sb atoms doped in the $Ge_{0.94}Sn_{0.06}$ layer show a higher thermal robustness at 300 °C than those in the Ge layer. (Keywords: strain engineering, germanium tin, epitaxy, Sb doping, source/drain junction)

Introduction

$Ge_{1-x}Sn_x$ embedded source/drain (S/D) as stressor attracts attention for realizing uniaxial compressive strained Ge channel, which enhances the carrier mobility [1]. Aiming n^+-$Ge_{1-x}Sn_x$ S/D stressors, a Sn content more than 6% and heavy *n*-type doping are necessary to induce enough stress in Ge channel and to lower parasitic resistance [2]. However, heavy *n*-type doing technology for $Ge_{1-x}Sn_x$ is still in developing because of the low solubility of *n*-type dopants in Ge [3]. Previously, we reported the Sb heavy doping which has an electron concentration of 2.7×10^{19} cm^{-3} in $Ge_{1-x}Sn_x$ epitaxy higher than its maximum Sb solid solubility of 1.2×10^{19} cm^{-3} [3, 4]. Therefore, as the next step, we focus on the thermal stability of S/D junctions, which is also essential for practical application of $Ge_{1-x}Sn_x$ stressors. There is a possibility of the degradation of mobility and carrier concentration due to strain relaxation, Sn precipitation, and dopant segregation at post annealing process. Moreover, the influence of annealing on properties of heavily doped $Ge_{1-x}Sn_x$ layers has not yet been clarified in detail.

In this study, we investigated the thermal stability of Sb-doped $Ge_{1-x}Sn_x$ layers with various Sb concentrations from 10^{19} to 10^{20} cm^{-3}.

Sample preparation

P-type Ge(001) wafers were used as substrates. After chemical and thermal cleaning, an *in-situ* Sb-doped $Ge_{0.94}Sn_{0.06}$ or Ge epitaxial layer with various Sb concentrations was grown on Ge substrate using molecular beam epitaxy (MBE) method. Ge, Sn, and Sb were deposited using individual Knudsen cells (K-cells). The thickness of $Ge_{1-x}Sn_x$ layer was 100 nm. The growth temperature was 150 °C. The Sb concentration was controlled with the Sb K-cell temperature from 220 to 280 °C. Some samples were annealed at 300 and 400 °C for 1 min in a pure N_2 atmosphere.

Results and discussion

First, the depth profiles of the Sb chemical concentration in $Ge_{1-x}Sn_x$ epitaxial layers were measured by the secondary ion mass spectroscopy. The Sb chemical concentrations of $Ge_{1-x}Sn_x$ layers were estimated approximately to be 10^{18}, 10^{19}, and 10^{20} cm^{-3} for Sb K-cell temperatures of 220, 250, and 280 °C, respectively (*not shown*).

The sheet resistance of each sample was measured using micro four-point probe (M4PP) method. The Sb K-cell temperature dependence of the sheet resistance for as-grown $Ge_{0.94}Sn_{0.06}$/Ge and Ge/Ge samples are shown in **Fig. 1**. The decrease of the sheet resistance as the increase of the Sb K-cell temperature demonstrates convincingly that the effective increase in the electron concentration by *in-situ* Sb-doping, *i.e.* *n*-type doping. The *n*-type doping has been practically confirmed for $Ge_{0.94}Sn_{0.06}$ and Ge samples with Sb K-cell temperatures of 250 and 280 °C using the micro-Hall effect measurement [4]. The higher sheet resistance of the Sb-doped $Ge_{0.94}Sn_{0.06}$ layer than Ge can be deduced from the higher defect concentration of $Ge_{1-x}Sn_x$ than Ge [5].

Then, we examined the influence of the post-deposition annealing on the sheet resistance of $Ge_{1-x}Sn_x$/Ge samples as shown in **Fig. 2**. Both Sb-doped $Ge_{0.94}Sn_{0.06}$ and Ge layers grown with a Sb K-cell temperature of 280 °C show the increase of the sheet resistance with the annealing temperature especially for 400 °C, whereas both samples with a Sb K-cell temperature of 250 °C show a higher thermal robustness. This result indicates that the thermal instability of Sb atoms doped in the $Ge_{0.94}Sn_{0.06}$ heteroepitaxial layer is unignorable for a heavily doping as high as 10^{20} cm^{-3}, which was also observed in the heavily Sb-doped Ge homoepitaxial layer.

Figure 3 shows the typical Sb3d core-level spectrum measured by hard X-ray photoelectron spectroscopy (HAXPES). We can deconvolute 3 peaks related to chemical bonds of oxidized Sb, substitutional Sb (dopant), and isolated Sb (segregated or interstitial) as shown in **Fig. 3**. We estimated the chemical and substitutional concentrations of Sb from the ratios of the total area intensity of Sb and the area intensity of substitutional Sb, respectively, to the area intensity of Ge3d. The concentrations and the activation ratio of Sb in as-grown and annealed $Ge_{1-x}Sn_x$ layers prepared with a K-cell temperature of 280 °C are summarized in **Fig. 4**. In the as-grown sample, the activation ratios of Sb in $Ge_{0.94}Sn_{0.06}$ and Ge layers achieved to 74 and 76%, respectively. The carrier concentrations of both samples are higher than 2.6×10^{20} cm^{-3}, which exceed the solid solubility of Sb in Ge. In samples annealed at 300 °C, the Sb chemical concentration in Ge substantially decreased around 5.3×10^{19} cm^{-3}. In contrast, interestingly, the chemical and substitutional concentrations of Sb in the $Ge_{0.94}Sn_{0.06}$ layer slightly increased, maintaining an activation ratio of 73%. This result promises the higher thermal stability of substitutional Sb atoms doped in $Ge_{1-x}Sn_x$ epitaxial layer compared to Ge, while the sheet resistance slightly increased after annealing even at 300 °C. On the other hand, after annealing at 400 °C, the chemical concentration of Sb in Ge decreased consecutively. The activation ratio of Sb in $Ge_{0.94}Sn_{0.06}$ also decreased to 12%, even though the chemical concentration maintained at 3.2×10^{20} cm^{-3}.

The crystalline quality of Sb-doped $Ge_{1-x}Sn_x$ layers was also investigated with X-ray diffraction two-dimensional reciprocal space mapping (XRD-2DRSM). XRD-2DRSM results around the Ge$\overline{2}\overline{2}4$ Bragg reflection for $Ge_{0.94}Sn_{0.06}$ samples as-grown and annealed at 400 °C are shown in **Figs. 5(a) and 5(b)**, respectively. From the diffraction peak position of $Ge_{0.94}Sn_{0.06}$, we can confirm that the strain in the Sb-doped $Ge_{1-x}Sn_x$ layer is fully maintained even after the annealing at 400 °C. This result indicates that heavily Sb-doped $Ge_{1-x}Sn_x$ layer even with a high Sn content above the solid solubility of Sn and Sb in Ge promises a high crystallinity up to an annealing temperature of 400 °C.

Conclusions

We investigated the thermal stability of heavily doped $Ge_{1-x}Sn_x$ epitaxial layer prepared by *in-situ* Sb doping with a low-temperature MBE method on Ge(001) substrate. Although the sheet resistance of the $Ge_{1-x}Sn_x$ layer with a Sb doping higher than 10^{20} cm^{-3} slightly increased after annealing at 300 °C, substitutional Sb atoms were more thermally stable in $Ge_{0.94}Sn_{0.06}$ layer than those in Ge layer. Also, we found that the Sb-doped $Ge_{0.94}Sn_{0.06}$ epitaxial layer maintains its crystallinity up to the annealing temperature of 400 °C. Heavily Sb-doped $Ge_{1-x}Sn_x$ promises low-resistance S/D stressor for high-mobility uniaxial compressive strained Ge channel.

Acknowledgments

This work was partly supported by a Grant-in-Aid for Scientific Research (S) (No. 26220605) and (B) (No. 15H03565) from JSPS in Japan. The HAXPES measurement was performed at BL09XU in SPring-8 (No. 2016A1492).

References

[1] B. Vincent, Y. Shimura, S. Takeuchi, T. Nishimura, G. Eneman, A. Firrincieli, J. Demeulemeester, A. Vantomme, T. Clarysse, O. Nakatsuka, S. Zaima, J. Dekoster, M. Caymax, and R. Loo, "Characterization of GeSn materials for future Ge pMOSFETs source/drain stressors" Microelectronic Engineering, Vol. 88, p. 342, 2011, doi:10.1016/j.mee.2010.10.025.

[2] H. Miyoshi, T. Ueno, Y. Hirota, J. Yamanaka, K. Arimoto, K. Nakagawa, and T. Kaitsuka, "Low nickel germanide contact resistances by carrier activation enhancement techniques for germanium CMOS application", Japanese Journal of Applied Physics, Vol. 53, p. 04EA05, 2014, doi:10.7567/JJAP.53.04EA05.

[3] C. Claeys and E. Simoen, "Germanium-based technologies: from material to devices", Elsevier, ISBN: 978-0-08-044953-1 (2007).

[4] J. Jeon, T. Asano, Y. Shimura, W. Takeuchi, M. Kurosawa, M. Sakashita, O. Nakatsuka, and S. Zaima, "Effect of in situ Sb doping on crystalline and electrical characteristics of n-type $Ge_{1-x}Sn_x$ epitaxial layer" Japanese Journal of Applied Physics, Vol. 55, p. 04EB13, 2016, doi:10.7567/JJAP.55.04EB1

[5] T. Asano, N. Taoka, K. Hozaki, W. Takeuchi, M. Sakashita, O. Nakatsuka, and S. Zaima, "Impact of hydrogen surfactant on crystallinity of $Ge_{1-x}Sn_x$ epitaxial layers", Japanese Journal of Applied Physics, Vol. 54, 04DH15, 2015.

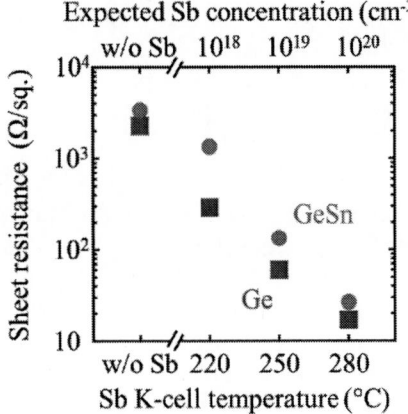

Fig. 1: The Sb K-cell temperature dependence of the sheet resistance of Sb doped $Ge_{0.94}Sn_{0.06}$/Ge and Ge/Ge samples.

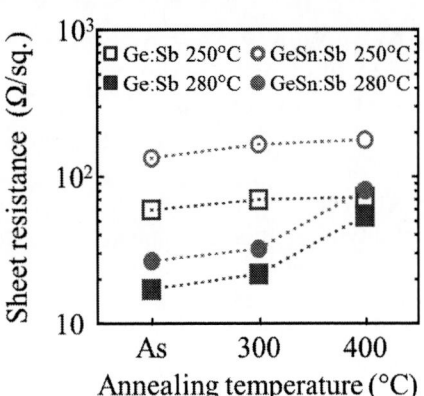

Fig. 2: The sheet resistance of Sb doped $Ge_{0.94}Sn_{0.06}$/Ge and Ge/Ge samples for as-grown and annealing cases.

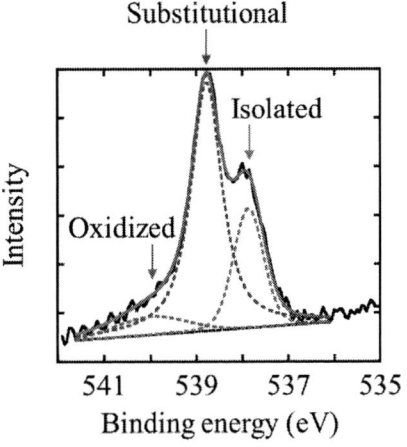

Fig. 3: The typical Sb3d core-level spectrum measured using HAXPES.

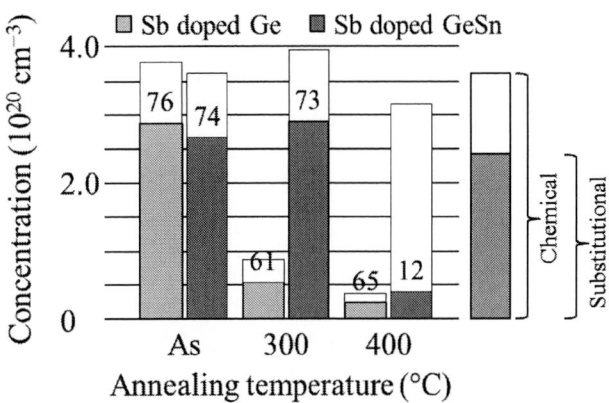

Fig. 4: The chemical and substitutional concentrations of Sb in as-grown and annealed $Ge_{0.94}Sn_{0.06}$ and Ge layers prepared with a K-cell temperature of 280 °C. Number on bar is the activation ratio (%) of Sb for the total concentration.

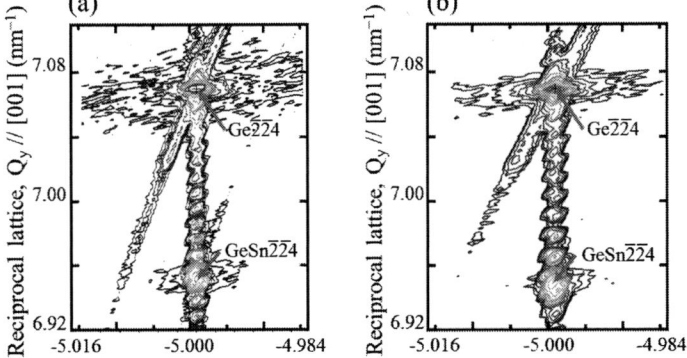

Fig. 5: XRD-2DRSM results around the $Ge\overline{2}\overline{2}4$ Bragg reflection of $Ge_{0.94}Sn_{0.06}$ layers (a) as-grown and (b) annealed at 400 °C. The Sb K-cell temperature was 280 °C.

978-1-5090-4661-4/17 $31.00 © 2017 IEEE

Novel In-situ Passivation of $MoCl_5$ Doped Multilayer Graphene with MoO_x for Low-Resistance Interconnects

K. Kawamoto[1], Y. Saito[1], M. Kenmoku[1], and K. Ueno[1, 2]*

[1] Shibaura Institute of Technology (SIT), 3-7-5 Toyosu, Koto, Tokyo, Japan

[2] Research Center for Green Innovation, SIT, 3-7-5 Toyosu, Koto, Tokyo, Japan,

ueno@shibaura-it.ac.jp

Abstract

To improve the stability of multilayer graphene (MLG) doped with molybdenum pentachloride ($MoCl_5$) for low-resistance interconnects, we have newly developed an in-situ passivation process with molybdenum oxides. The improved air stability of dopants was confirmed with Raman spectroscopy by the direct MoO_x passivation at room temperature.

(Keywords: graphene, doping, interconnect)

Introduction

MLG is expected as a low-resistance material for narrow interconnects below 20 nm due to its potential of lower resistivity than copper (Cu) in narrow lines [1]. However, carrier doping is necessary to obtain such low resistivity [2]. $MoCl_5$ intercalation doping to MLG is reported as a stable and effective doping method [3]. In addition, partial oxidation of MLG by O_2 exposure is important for improving air stability [4]. Considering the carrier concentration, however, O_2 exposure of $MoCl_5$ doped MLG potentially reduces the carrier concentration of MLG edges. Therefore, we investigated changes in the carrier concentration of the $MoCl_5$ doped MLG by the O_2 exposure and in-situ passivation process for the $MoCl_5$ doped MLG was investigated using a MoO_3 layer which was reported to have a charge doping effect to graphene [5]. We report difference in the $MoCl_5$ doped MLG between with and without O_2 exposure prior to the in-situ passivation with MoO_x, and it is found that higher doping is obtained by the in-situ passivation without O_2 exposure.

Experimental

Fig. 1 shows the experimental apparatus used in this study. The $MoCl_5$ intercalation doping and the successive MoO_x passivation were carried out in the same glass tube without air-exposure. Fig. 2 shows the schematic cross-section of the fabricated samples and the processing flow. The MLG flakes were transferred on the SiO_2/Si substrate from a HOPG block using a Scotch tape. After setting the substrate in the glass tube, it was evacuated to about 0.6 Pa, and the $MoCl_5$ intercalation was carried out by a two-zone method at 300℃ for 1 or 24 hours. To enhance the $MoCl_5$ intercalation, we developed a "High concentration of $MoCl_5$" condition using the capsule. We call this condition as "high doping condition" in this paper. When "low doping condition", $MoCl_5$ intercalation was carried out for 24 hours and "high doping condition" was carried out for 1 hour in a glass capsule which keep the high vapor pressure of $MoCl_5$ during the intercalation. After evacuating again, O_2 gas was introduced at room temperature for 5 min or 24 hours for some of the samples as in the previous study [4] to see the effect of O_2 exposure before the MoO_x passivation. Moreover, O_2 gas was introduced at 300℃ for a condition of enhanced oxidation. Then, the MoO_x layer was deposited by sublimation of MoO_3 powder at 400℃. The substrate was kept at the cool zone in the glass tube outside of the furnace during the MoO_x deposition. Raman spectroscopy measurements were carried out to determine the structure and the stability of doped MLG. The thickness of MoO_x passivation layer was analyzed by TEM cross-section of the MLG with in-situ passivation. The composition of MoO_x was determined by X-ray photoelectron spectroscopy (XPS) measurements.

Results and discusson

Fig. 3 shows the Raman spectra comparison of low-doped MLG between 5 min and 24 hours exposure to O_2 oxidatioon at room temperature. The GIC peak intensity which features the doped MLG decreased with the O_2 exposure than without the O_2 exposure, and longer the oxidation for 24 hours decreased more than the oxidation for 5 min. The results of the O_2 oxidation at different temperature are shown in Fig. 4. The GIC peak intensity was found to be decreased with O_2 exposure at 300 ℃ in comparison to that of annealing in vacuum without O_2 exposure. From Fig. 3 and 4, the GIC peak intensity decreased with longer oxidation time and decreased for the 300℃ oxidation in comparison to the room temperature oxidation. Changes of GIC/G peak ratio and the shift of G peak position which compared with the after

978-1-5090-4661-4/17 $31.00 © 2017 IEEE

doping without O_2 exposure are shown in Fig. 5. The GIC/G peak ratio and the G peak position which obtained by curved-fitted to Raman spectra are the average values of 10 samples under each condition. The GIC/G peak ratios decreased with the O_2 exposure time, and the G peak position shifts from the pristine graphene ($\sim$1580 cm^{-1}) also decreased. The results indicate that the O_2 exposure leads to decrease in doping effect since the shift of G peak position represents the charge transfer doping from dopants.

Fig. 6 shows the optical microscope image of Ni surface. The Ni surface is corroded by $MoCl_5$ exposure without MoO_x layer. From the results, we confirmed the barrier effect of MoO_x layer against $MoCl_5$ diffusion. The XPS spectra of the Mo 3d is shown in Fig. 7. Mo 3d region of MoO_3 is reported that the binding energies of Mo $3d_{5/2}$ and Mo $3d_{3/2}$ are 232.5 and 235.6 eV. MoO_3 is transformed to MoO_2 by annealing with hydrogen at 400℃ [6]. The MoO_x passivation layer has a two layer of MoO_3 and MoO_2.

Fig. 8 shows the comparison of the Raman spectra of the low-doped MLG between with and without the O_2 exposure. Each Raman maps is shown in Fig. 11. In the case of "low doping condition", GIC/G peak ratio of the MoO_x passivation without O_2 exposure is higher than that with O_2 exposure. It indicates that the MoO_x passivation without the O_2 exposure is preferrable to the O_2 oxidation. The thickness of the MoO_x passivation layer was 70 nm and MLG edges were covered with the passivation layer as shown in Fig. 9. Fig. 10 shows the comparison of the Raman spectra of the high-doped MLG only with MoO_x deposition. The highest GIC/G peak ratio was obtained by the MoO_x passivation without the O_2 exposure in the case of "high doping condition". In this case, the distribution of the $MoCl_5$ doped MLG is considered that the $MoCl_5$ intercalated to one side of a graphene layer (Stage 2) since the GIC peak is around 1600 cm^{-1} [4].

From the results, the direct MoO_x passivation without O_2 exposure is considered promising in improving the stability $MoCl_5$ doped MLG for low-resistance MLG interconnect.

Conclusion

We have developed an in-situ passivation process for $MoCl_5$ doped MLG with MoO_x layer. It is suggested that MoO_x passivation without O_2 exposure is considered to improve the stability of $MoCl_5$ doped MLG. This process can prevent for $MoCl_5$ desorption from narrow MLG lines

and leads to low-resistance interconnects.

Acknowledgement

This work was partially supported by "Ultra-Low Voltage Device Project" funded and supported by METI and NEDO. This presentation is supported by the Research Center for Green Innovation of SIT. The authors thank H. Miyazaki, R. Matsumoto, T. Sakai, A. Kajita, A. Isobayashi, and Y. Awano for discussion.

References

1. A. Naeemi and J. D. Meindle, IEEE Electron Devices Lett. **28**, 428 (2007).
2. C. Xu, H. Li, and K. Banerjee, IEEE Trans. Electron Devices **56**, 1567 (2009).
3. H. Miyazaki, et. al, Extended Abst. 2015 Solid State Devices and Materials (Sapporo, 2015) p. 766.
4. H. Miyazaki, et. al, Extended Abst. 2016 Solid State Devices and Materials (Tsukuba, 2016) p. 493.
5. L. D'Arsie, et. al, Appl. Phys. Lett. **105**, 103103 (2014).
6. J. G. Choi and L. T. Thompson, Appl. Surf. Sci. **93**, 143 (1996).

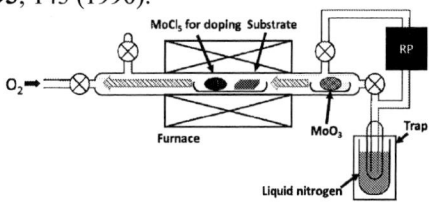

Fig. 1. Experimental apparatus for in-situ passivation. After the $MoCl_5$ doping, the substrate and the MoO_3 was moved to the left for the MoO_3 deposition.

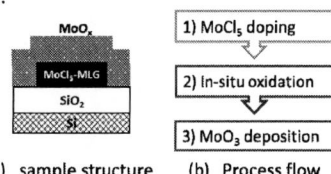

Fig. 2. Sample structure and the process flow.

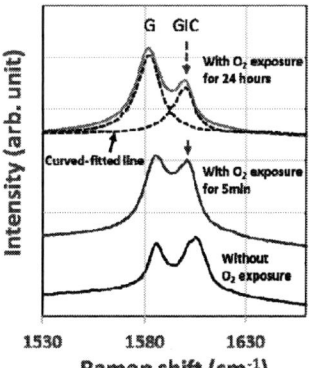

Fig. 3. Raman spectra for the low-doped MLG with O_2 exposure at room temperature.

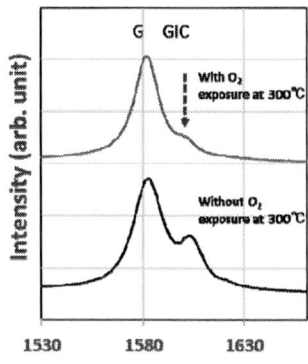

Fig. 4. Raman spectra for the low-doped MLG with or without O_2 exposure at 300℃.

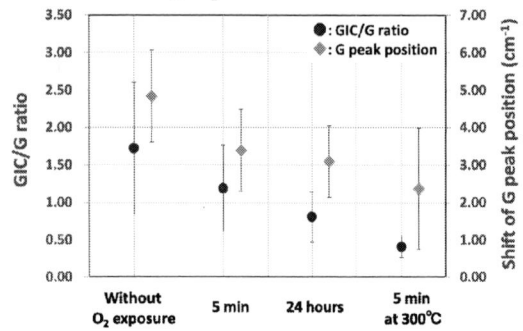

Fig. 5. GIC/G ratio and the shift of G peak position for O_2 exposure between 5 min and 24 hours at room temperature or 300℃.

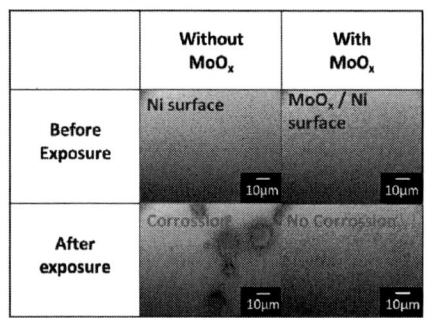

Fig. 6. Diffusion barrier effect of MoO_x against $MoCl_5$-exposure (Ni corrosion tests)

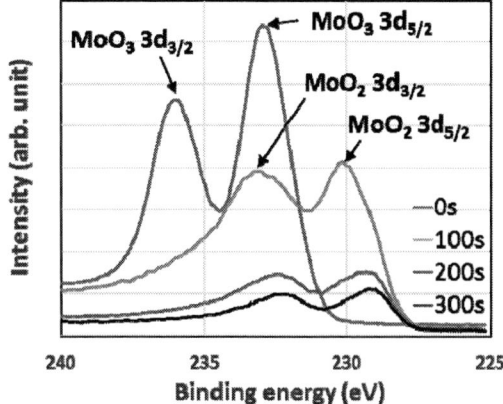

Fig. 7. XPS spectra of the Mo 3d region for the MoO_x passivation layer for different etching time.

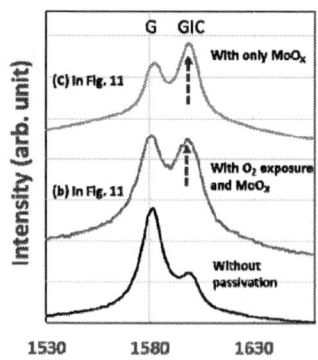

Fig. 8. Raman spectra for the low-doped MLG with or without O_2 oxidation.

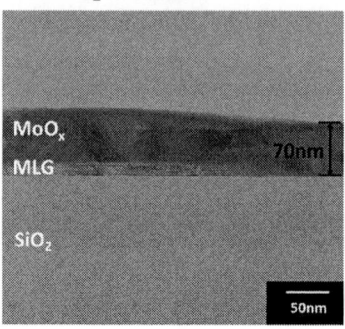

Fig. 9. TEM cross-section of the MLG with in-situ passivation.

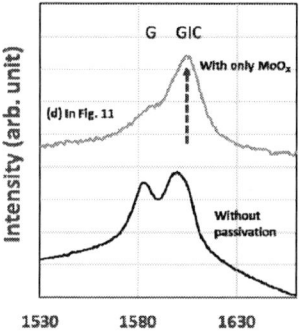

Fig. 10. Raman spectra for the high-doped MLG without the O_2 oxidation.

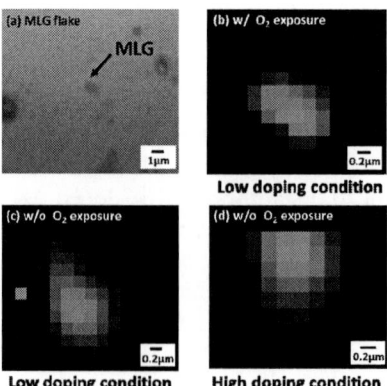

Fig. 11. Raman maps for the high- or low-doped MLG with or without the O_2 oxidation.

P-37

Thickness-Independent Behavior of Coercive Field in HfO_2-based Ferroelectrics

Shinji Migita[1], Hiroyuki Ota[1], Hiroyuki Yamada[1], Akihito Sawa[1], and Akira Toriumi[2]

[1]National Institute of Advanced Industrial Science and Technology, Tsukuba, s-migita@aist.go.jp

[2]The University of Tokyo

Abstract

Electrical properties of Hf-Zr-O films with several thicknesses are studied. Decent ferroelectric performances are obtained in a wide range of film thickness between 6.8 and 41 nm. A drawback in Hf-Zr-O system is that the operation electric filed for the saturated ferroelectric polarization is close to the breakdown electric field. An investigation of coercive field vs. thickness reveals that HfO_2-based ferroelectrics are distinct from conventional ones.
(Keywords: Ferroelectric, HfO_2, Hf-Zr-O, MIM capacitor, and coercive field)

Introduction

HfO_2-based ferroelectrics [1-5] have been attracting much attention from viewpoints of emergence of ferroelectricity in sub-10 nm-thickness region and compatibility with LSI process. The ferroelectricity appears only in the metastable orthorhombic crystal phase, and usually other crystal grains such as monoclinic and tetragonal phases coexist in the films. Identification of physical properties and exploration of application fields of HfO_2-based ferroelectrics are now underway.

In this work, we examined the electrical properties of Hf-Zr-O ferroelectric films with several thicknesses. It is confirmed that due to the large coercive field of HfO_2-based ferroelectrics, its polarization is liable to breakdown of the film. Furthermore, it is cleared that the correlation between the coercive field and thickness in HfO_2-based ferroelectrics is much different from the trend of conventional ferroelectric oxides.

Experiment

A multi-chamber vacuum system that consists of *RF* sputtering chamber, *DC* sputtering chamber and RTA chamber is used for the fabrication of MIM capacitors. The bottom electrode of MIM capacitor is 10-nm-thick TaN film deposited by *DC* sputtering on heavily doped n-type Si substrates. Hf-Zr-O films were deposited by *RF* sputtering using the HfO_2 and ZrO_2 targets. The chemical composition was adjusted to Hf:Zr=50:50 by controlling the sputtering condition. Following to the deposition of 10-nm-thick TaN top electrode, the samples were annealed at 600 °C for 1 min in a vacuum condition (<1 Pa). This is the so-called Capped-anneal process. Afterwards 100-nm-thick Al film is deposited by another tool, then top electrodes were patterned by photolithography and dry etching. Electrode size is 100 μm x 100 μm. Finally aluminum films were deposited on the backside of wafers.

Results and Discussion

Polarization behaviors, measured by TF Analyzer 2000E system, are shown in **Fig. 1**. Decent ferroelectric behaviors are obtained in Hf-Zr-O films with thicknesses between 6.8 nm and 41 nm. The remnant polarizations (P_r) and the coercive fields (E_C) are changing with film thickness. I-V characteristics of these samples are summarized in **Fig. 2**. Although accidental breakdown sometimes occurs in thinner films, which might be caused by droplets and particles on the wafer, the intrinsic breakdown field was evaluated through the measurement of many samples. The electrical fields of breakdown (E_{bd}), maximum field for ferroelectric operation (E_{Max}), and coercive field (E_C) against the film thicknesses are plotted in **Fig. 3**. It clearly shows that operation of ferroelectric switching in Hf-Zr-O films is close to the breakdown. Thus the large coercive field in HfO_2-based ferroelectrics is a key parameter for the device application. Finally, the correlation between coercive field and the film thickness are compared in **Fig. 4** for several ferroelectric oxides. In addition to our work, Hf-Zr-O films prepared by ALD [6] and Y-doped HfO_2 films prepared by chemical solution deposition [7] also show that E_c is independent of the film thickness. This is much different from conventional ferroelectrics such as perovskite-type oxides, where E_c increases with scaling the film thickness in a manner of $E_c \sim d^{-2/3}$ [8]. We believe that this difference comes from the excellent thermal stability of HfO_2-based oxides and the unique motion of ferroelectric domain in the films.

Conclusion

The mechanism of ferroelectric switching in HfO_2-based oxides might be distinct from conventional ones. Understanding of this behavior is mandatory for the application of this material in reliable devices.

Acknowledgments

This work was supported by JST-CREST.

References

[1] T. S. Böscke, J. Müller, *et al.* , *Appl. Phys. Lett.* **99**, 102903 (2011).

[2] J. Müller, T. S. Böscke, *et al.*, *Appl. Phys. Lett.* **99**, 112901 (2011).

[3] J. Müller, T. S.Böscke, *et al.*, *Nano Lett.* **12**, 4318−4323 (2012).

[4] J. Mueller, *et al.*, *ECS J. of Solid State Science and Technol.*, **4** (5) N30-N35 (2015).

[5] M. H. Park, U. Schroeder, and C. S. Hwang *et al.*, *Adv. Mater.* **27**, 1811−1831, (2015).

[6] M. H. Park, C. S. Hwang et al., *Appl. Phys. Lett.* **102**, 242905 (2013).

[7] S. Starschich, R. Waser, and U. Böttger *et al.*, *Appl. Phys. Lett.* **104**, 202903 (2014).

[8] M. Dawber, P. Chandra, P.B. Littlewood and J.F. Scott, J. Phys. Cond. Matt. **15**, L393 (2003).

978-1-5090-4661-4/17 $31.00 © 2017 IEEE   255

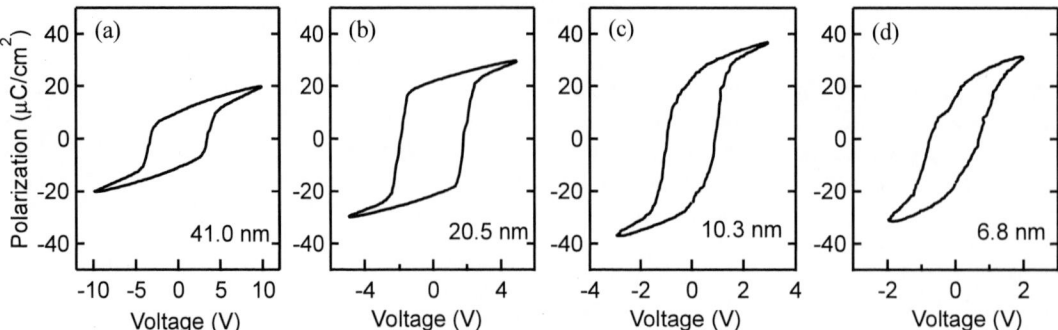

Fig. 1: Polarization-voltage characteristics of TaN/Hf-Zr-O/TaN MIM capacitors. Thicknesses of Hf-Zr-O films are (a) 41.0 nm, (b) 20.5 nm, (c) 10.3 nm, and (d) 6.9 nm. The remnant polarization (P_r) and the coercive field (E_c) are evaluted to be (a) 10.5 μC/cm² and 0.826 MV/cm, (b) 21.65 μC/cm² and 0.914 MV/cm, (c) 23.57 μC/cm² and 0.909 MV/cm, and (d) 16.35 μC/cm² and 1.045 MV/cm, respectively.

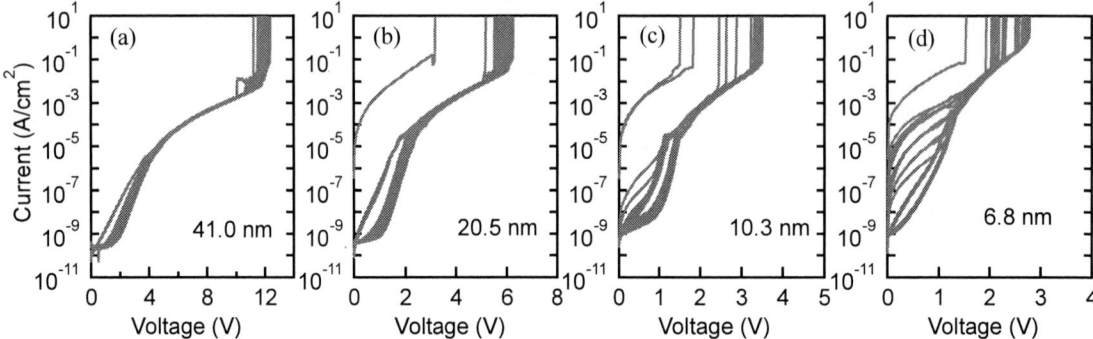

Fig. 2: Current-voltage characteristics of TaN/Hf-Zr-O/TaN MIM capacitors. Thicknesses of Hf-Zr-O films are (a) 41.0 nm, (b) 20.5 nm, (c) 10.3 nm, and (d) 6.9 nm. Results of 32 samples are plotted. The breakdown voltages are defined at voltages where the currents drastically increased to more than 1 A/cm². Accidental breakdown behaviors at low voltages in some thinner film samples are ignored, and the intrinsic breakdown voltgaes are defined at the voltages where the frequencies are the maximum. Thus the breakdown fields (E_{bd}) are evaluated to be (a) 2.85 MV/cm, (b) 2.85 MV/cm, (c) 3.35 MV/cm, and (d) 3.90 MV/cm, respectively.

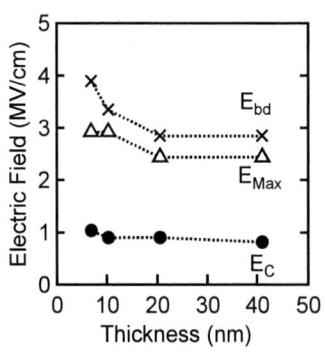

Fig. 3: Breakdown fields (E_{bd}), maximum applied field for P-V measurements (E_{Max}), and coercive fields (E_c) plotted against the Hf-Zr-O thicknesses.

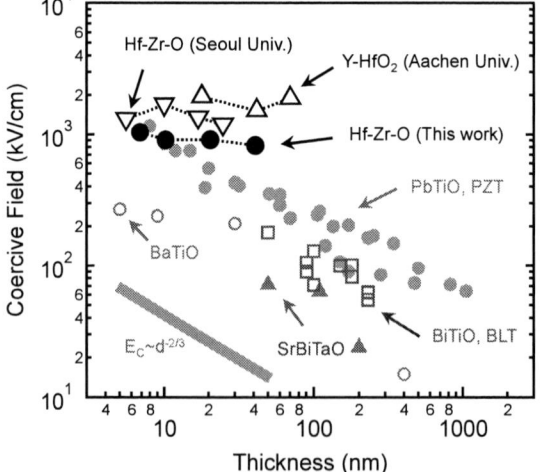

Fig. 4: Correlation between coercive field, E_c, and film thickness, d, for various ferroelectric oxides.

978-1-5090-4661-4/17 $31.00 © 2017 IEEE

P-38

Crystallinity study of Si single crystal stripe on bended glass substrate fabricated by micro-chevron laser beam scanning method

Wenchang Yeh and Seigo Moriyama

Shimane University, Interdisciplinary Graduate School of Science and Engineering,

1060 Nishikawatsu-cho, Matsue-shi, Shimane 690-8504, Japan

Phone/Fax: +81-852-32-8504   E-mail:yeh@riko.shimane-u.ac.jp

Abstract

Si single crystal stripe on bended glass substrate was fabricated by micro-chevron laser beam scanning method. The single crystal stripe had a dimension of about 6 μm in width and several hundreds of microns in length. The crystal quality was evaluated by EBSD, to reveal that the crystal orientation rotated about the transverse direction in forward direction, and only Σ3 and Σ9 coincidence site lattice (CSL) twins transvers the single crystal stripe.

(Keywords: TFT, micro chevron laser beam, laser annealing)

Introduction

Figure 1 shows evolution of displays in relation to display size and pixel density. For a mobile information terminal, display size is limited within 6 inches for portability, and the evolution trend is toward increment of pixel density for a larger amount of information. The mobile display was used not only as direct-view-type display but also was used as image source for head mounted display (HMD). Because of the later application, ultimately 3000dpi is demanded to satisfy resolution limit of retina cells of human. For such a fine definition displays, dimension of sub pixel will be several microns and dimension of thin-film transistors (TFTs) correspondingly will reduce to below 1 micron. In such a display, there are two reasons for single crystalline Si as a promising material for TFTs. First, high mobility TFTs were demanded for peripheral circuit to drive huge amount of pixels at a sufficiently high refreshing rate. Second, single crystalline was demanded for ensuring uniformity of pixel driving current among pixels especially in sub-micron TFTs. Attempts to grow a single grain in Si film on amorphous substrate have been performed since the early 1980s using Ar lasers on 400 nm-thick Si films, in which either donut-shaped[1] or twin Gaussian [2] cw laser beams were proposed to scanning Si films. On the other hand, we have proposed micro-chevron-shaped laser beam scanning (μCLBS) method for forming single crystal Si stripe [3], in which the μCLB was formed by passing an output of a 405nm-wavelength multimode laser

diode through a novel one-side Dove prism (OSDP), as was shown in Fig.2. In this study, μCLBS was applied to Si film on bended glass substrate, and the crystallinity of single crystalline stripe was investigated.

Experiments

Figure 3 shows μCLBS system in this study. 0.14-0.17mm borosilicate flexible glass was used as the substrate. 300nm-thick SiO_2 and 100nm-thick Si was deposited successively by sputtering method at 400°C. The substrate was closely attached to a rotating roll with a radius of 250mm by a vacuum chuck. μCLB was focused on the Si film while the substrate move at a circumferential speed of 0.013 m/s. The sample after laser annealing was evaluated by optical microscopy (OM), scanning electron microscope (SEM) and electron backscattering diffraction (EBSD).

Experimental results

Figure 4 shows an EBSD image of single crystalline Si stripe, in which the color shows crystal orientation along traveling direction of grain growth. Crystal orientation rotated about transverse axes toward the growth direction. The rotation rates was about 0.47°/μm. This result suggests different level of strain between Si film top surface and bottom surface. No grain boundaries (GBs) traverse single crystal stripe, and only Σ3 and Σ9 coincidence site lattice (CSL) twins observed traversing the single crystal stripe. Since Σ3 and Σ9 CSL twins are electrically inactive, TFTs with high mobility and high uniformity is expected by using μCLBS to form single crystalline stripe.

Conclusion

Si single crystal stripe on bended glass substrate was fabricated by μCLBS. The single crystal stripe had a dimension of about 6 μm in width and several hundreds of microns in length, and there are only Σ3 and Σ9 CSL twins transverse the single crystal stripe. This method provide a promising method for fabricating single crystalline Si TFTs for next generation active matrix flat panel displays.

Acknowledgments

The authors gratefully acknowledge Shimane prefecture for financial support on μCLBS system.

978-1-5090-4661-4/17 $31.00 © 2017 IEEE

References

[1] S. Kawamura, J. Sakurai, M. Nakano, and M. Takagi, Appl. Phys. Lett. 40, 394 (1982).

[2] N. Sasaki, R. Mukai, T. Izawa, M. Nakano, and M. Takagi, Appl. Phys. Lett. 45, 1098 (1984).

[3] Wenchang Yeh et. al., APEX 9 (2016) 025503.

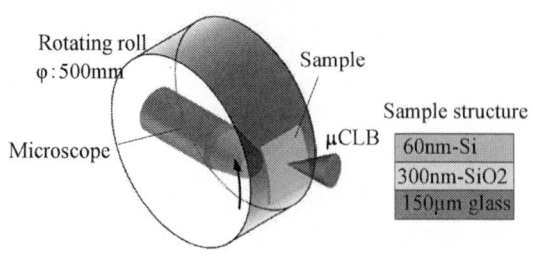

Fig. 3 Schematic view of μCLBS system and sample structure

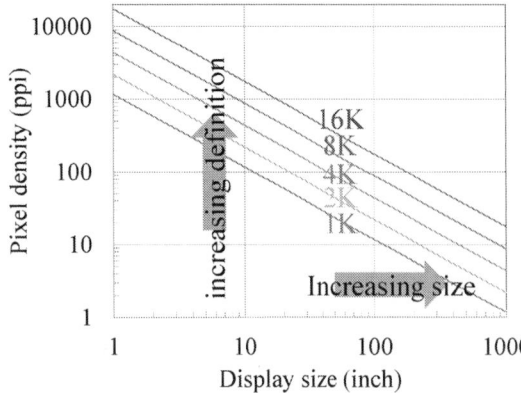

Fig.1 Trends of display in relation to size and pixel density

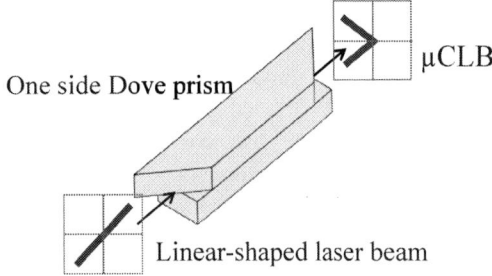

Fig. 2 A schematic view of forming a chevron-shaped laser beam.

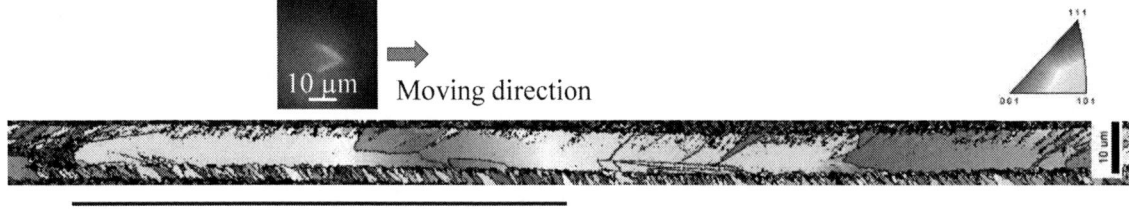

Fig.4 photo image of μCLB and EBSD image of formed Si single crystal stripe

P-39

Physics Based System Simulation for Robot Electro-Mechanical Control Design

T. K. Maiti[1], L. Chen[1], M. Miura-Mattausch[1], S. K. Koul[2], and H. J. Mattausch[1]

[1]HiSIM Research Center, Hiroshima University, Higashi-Hiroshima, 7398530, Japan, tkm@hiroshima-u.ac.jp

[2]Centre for Applied Research in Electronics, Indian Institute of Technology-Delhi, New Delhi, 110016, India

Abstract

The aim of our investigation is to develop complete electro-mechanical system simulation. The present developed prototype system includes pressure sensors, amplifiers, a controller, servo motors, and a robot-body. Sensor signals such as transduced voltage, servo motor actuation signals such as shaft angle, velocity, and acceleration are modeled in an analytical way. The entire system structure is formed based on the equivalent circuit concept. Mechanical dynamics of a two-leg robot-body is verified with the developed simulation system.

Keywords: System-modeling, simulation, electro-mechanical, sensor and actuator

Introduction

The new generation of robotics is based on applications of sensing, information processing, control, manipulation, and IoT, which require the development of integrated electro-mechanical system design. Much robotics literatures cover the CAD based robotic mechanical-dynamics simulation and only scarcely the complete electro-mechanical simulation [1]. The reason is to design of electrical schematics and mechanical drawings of robots separately, as well as the lack of robotic system design software that can combine complex sensor, electrical, and mechanical systems [2]. In this paper, we present modeling of electro-mechanical dynamic system simulation of a walking robot by developing a physics based integrated system model. A piezoresistive pressure sensor model is developed to consider the effect of robot-foot pressure for controlling robot movement. Servo motor actuations, such as shaft angle, velocity, and acceleration are investigated by developing an electro-mechanical model. Robot-body movements with respect to foot pressure for smooth and non-smooth ground surfaces are investigated by formulating a robot dynamic model. The verified sensing based robot movement assists to balance the walking robot on both flat and texture surface.

Electro-Mechanical System Modeling

The robot system which we will develop is depicted in Fig. 1. The robot system employs two legs connected with (A) pressure sensors (Sense1 and Sense2) which transduce mechanical force to electrical signals (voltage), magnified using (B) amplifiers (Amp1 and Amp2). The outputs of the amplifiers is fed into (C) a controller which drives (D) servo motors (Motor1 and Motor2), resulting in (E) robot motion. Since the robot position and stability need to be controlled, a position and a velocity feedback loops are included in the system. A simple implementation of the system with the above close feedback loops and electro-mechanical controller is shown in Fig. 2(a). Detailed design of electro-mechanical controller system at component and device level model development is represented in Fig. 2(b).

The pressure sensor model is formulated in terms of electrical and mechanical disciplines in which the mechanical discipline relates the sensor structure deformation to foot forces. Here the sensors measure the contact force f_c between robot legs and environment by transducing z-direction (vertical to the ground) foot pressure into voltage V. The sensor model is implemented by developing the spring-mass-damping force-balance, mechanical stress σ and current I equations [see (1)-(3)] [3].

$$v_z = \frac{1}{m_s}\left(-b_s\frac{dz}{dt} - k_s z + f_c\right) \quad (1)$$

$$\sigma = \frac{96YHW}{5L^4}(z^2 - Lz) \quad (2)$$

$$I = \frac{V}{R_0(1+\pi\sigma)} \quad (3)$$

Here, pressure sensor has stiffness $k_s = YWH^3/4L^3$, damping b_s, and seismic mass m_s. Y is the Young's modulus of the sensor material, while L, W, and H are length, width, and height of the sensor, respectively. π and R_0 are piezoresistive coefficient and stress independent resistance of the sensor, respectively. Equation (1) solves z which then used in (2) to determine σ and finally solves I using (3). Schematic illustration of equivalent circuit element with energy conversion between electrical and mechanical components is presented in Fig. 3(a) [4]. Sensor structural deformations z with respect to right leg, R_{leg} and left leg, L_{leg} steps are shown in Fig. 3(b), and corresponding transduced voltage outputs (dotted lines) are shown in Fig. 3(c). The controller (C) is modeled with a simple gain equation $V = A_v \times (V_{in} - V_{ref})$. Here A_v is the amplification factor, while input V_{in} and reference V_{ref} are the two electrical input signals whose difference is amplified.

978-1-5090-4661-4/17 $31.00 © 2017 IEEE   259

Control signals (solid lines) for servo motor drive in response to sensor signals (dotted lines) for both legs are depicted in Fig. 3(c). The motor is modeled with electrical resistance R_m, inductance L_m, mechanical inertia J_m and rotational friction D_m [schematically depicted in Fig. 4(a)]. The back voltage, V_m which is generated by the motor, is K_m times of its angular velocity ω_m, and the torque τ_m is K_t times of current I_m. Equations (4) and (5) consider the electro-mechanical behavior of servo-motor.

$$V_m = I_m R_m + K_m \omega_m + L_m \frac{dI_m}{dt} \qquad (4)$$

$$\tau_m = K_t I_m - D_m \omega_m - J_m \frac{d\omega_m}{dt} \qquad (5)$$

Its mechanical effects are defined in terms of rotational discipline that relates shaft angle, velocity, and acceleration to torque. We show the simulated angle (rad), velocity (rad/s), and acceleration (rad/s^2) in response to control signals for both legs in Figs. 4(b), (c), and (d), respectively.

Walking Robot Simulation

The modules are assembled according to the block diagram of Fig. 1. The mechanical movement of the robot-body is modeled with the kinetic and rotational disciplines which relate translational and rotation position, velocity, and acceleration to force and torque, respectively. Here robot-body motion is controlled and driven along x-direction (horizontal to the ground) by the motor torque τ_m. The equivalent circuit network for the robot-body model is shown in Fig. 5(a) and the mechanical equations of motion for two legs, specified in (6) and (7), are independently solved.

$$m_{r1} \ddot{x}_1 + (b_{r1} + K_{t1} K_{m1} \omega_{m1} / R_{m1}) = I'_{m1} V_{m1} \qquad (6)$$

$$m_{r2} \ddot{x}_2 + (b_{r2} + K_{t2} K_{m2} \omega_{m2} / R_{m2}) = I'_{m2} V_{m2} \qquad (7)$$

Here m_r is the mass of a robot leg, b_r is the damping coefficient associated with its actuation and movement losses, and $I'_m = K_t / R_m$. The suffixes 1 and 2 represent the right leg, R_{leg} and left leg, L_{leg} respectively. Since the robot consists of two legs, it senses distributed pressure depending on leg-weight distribution, motion pattern, and ground surface.

The time-domain trajectories of robot-legs position for both legs with equal (for smooth surface) and unequal (for distributed weight and non-smooth surface) pressures are depicted in Figs. 5(b) and 5(d), respectively. Foot pressure in pascal (Pa) senses by the pressure sensor for non-smooth surface is depicted in Fig. 5(c). The pressure data is used to obtain the stable position of the robot-body. Fig. 5(d) show that the low pressure sensing leg is moved more step cycles than the high pressure sensing leg to balance the robot-body (R_{leg}: 2-cycles and L_{leg}: 3-cycles), i.e. the step cycle is modified in order to get stable body center. In this method, the predefined information about the ground surface is not required for the robot movement. Please note that with respect to shaft position [see Fig. 4(b)], robot-leg positions [see Fig. 5(b)] show the translational influence on their leg position due to the bidirectional effect of motor torque on robot, and robot-inertia load on motor.

Conclusion

A system level electro-mechanical model is developed to study the robot's electro-mechanical dynamics. Model development is based on system physics and analysis of the robot's equivalent circuits, as well as its mechanical movements. In particular, the developed model can be used to design and simulate robot control circuitry.

References

[1] B. Siciliano and O. Khatib, Springer Handbook of Robotics, Springer Berlin Heidelberg, 2008.

[2] J. Chen, M. Henrie, M. F. Mar, and M. Nizic, Mixed-Signal Methodology Guide, Cadence Design Systems Inc., 2012.

[3] A. Nathan and H. Baltes, Microtransducer CAD, Springer Vienna, ch. 9, 1999.

[4] T. K. Maiti, et al., "Modeling of electrostatically actuated fluid flow system for mixed-domain simulation," in Proc. Int. Conf. Sim. Semi. Process Dev. (SISPAD), pp. 190-193, Sept. 2015.

[5] https://www.hisim.hiroshima-u.ac.jp

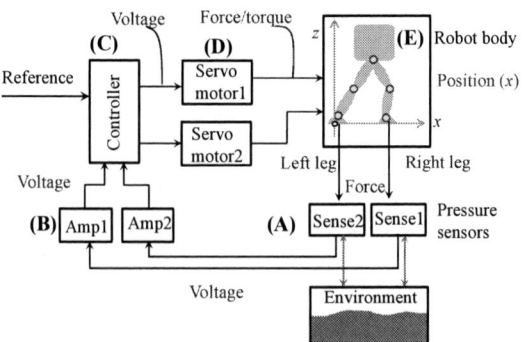

Fig. 1 Block diagram of the robot system, an assembly of modules, consisting of two pressure sensors, two amplifiers, a controller, two servo motors, and a robot-body.

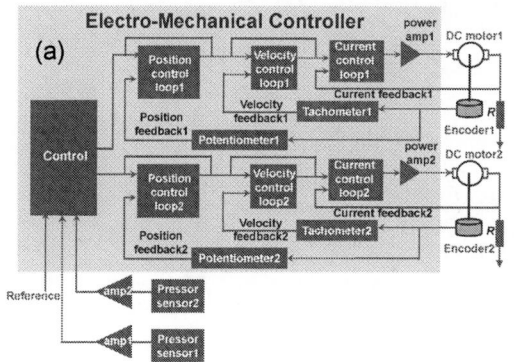

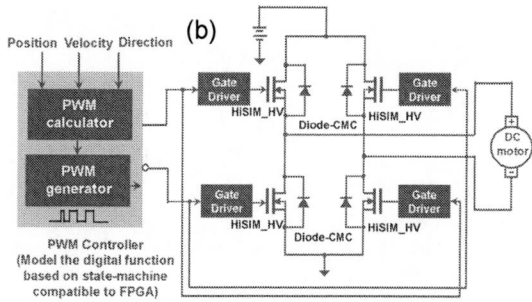

Fig. 2 (a) Implementation of electro-mechanical control system with position and velocity feedback loops and (b) detailed electro-mechanical control system at component and device level. H-bridge motor driver is design using HiSIM-HV and Diode-CMC models [5]. PWM controller is implemented using digital functional model.

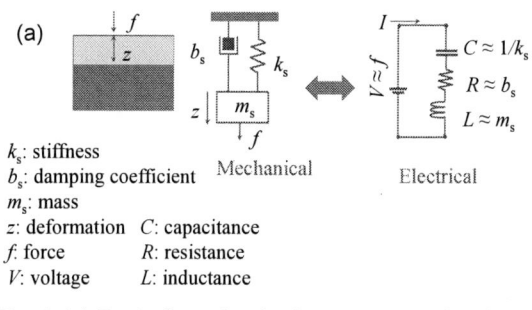

Fig. 3 (a) Equivalent circuit element conversion between electrical and mechanical components, (b) transient responses of sensor structural deformation (R_{deform} and L_{deform}) with respect to right and left leg steps (R_{leg} and L_{leg}), and (c) right and left leg sensor signals (R_{sense} and L_{sense}) (dotted lines) and corresponding right and left leg control signals ($R_{control}$ and $L_{control}$) (solid lines).

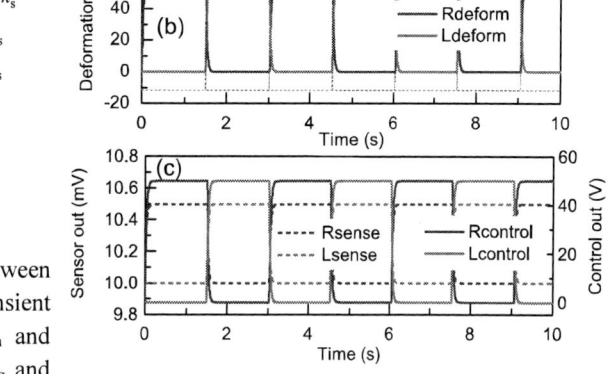

Fig. 4 (a) Schematic representation of a servo-motor model. Shaft angle, velocity, and acceleration versus time are shown in (b), (c) and (d), respectively.

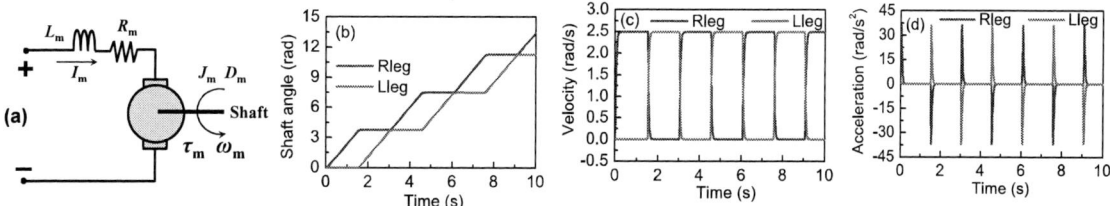

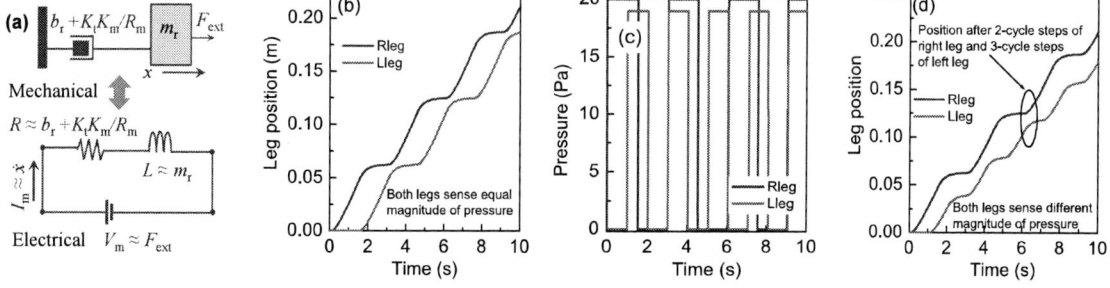

Fig. 5 (a) Equivalent circuit for a simple robot body, (b) right leg, R_{leg} and left leg, L_{leg} positions when both legs sense equal magnitude of pressure for robot movement on smooth surface, (c) unequal pressure due to non-smooth surface, sense by pressure sensors, and corresponding R_{leg} and L_{leg} positions is illustrated in (d).

978-1-5090-4661-4/17 $31.00 © 2017 IEEE   261

Gap in pagination due to unavailable papers.

Pages 262-270

IEEE
445 Hoes Lane
Piscataway, NJ 08854-4141

ISBN 978-1-5090-4661-4